AF723884

Diet Therapy and Nutritional Management of Phenylketonuria

Diet Therapy and Nutritional Management of Phenylketonuria

Editor

Anita MacDonald

MDPI • Basel • Beijing • Wuhan • Barcelona • Belgrade • Manchester • Tokyo • Cluj • Tianjin

Editor
Anita MacDonald
Dietetic Department,
Birmingham Children's
Hospital
Birmingham, UK

Editorial Office
MDPI
St. Alban-Anlage 66
4052 Basel, Switzerland

This is a reprint of articles from the Special Issue published online in the open access journal *Nutrients* (ISSN 2072-6643) (available at: https://www.mdpi.com/journal/nutrients/special_issues/Diet_Nutritional_Phenylketonuria).

For citation purposes, cite each article independently as indicated on the article page online and as indicated below:

LastName, A.A.; LastName, B.B.; LastName, C.C. Article Title. *Journal Name* **Year**, *Volume Number*, Page Range.

ISBN 978-3-0365-5319-1 (Hbk)
ISBN 978-3-0365-5320-7 (PDF)

© 2022 by the authors. Articles in this book are Open Access and distributed under the Creative Commons Attribution (CC BY) license, which allows users to download, copy and build upon published articles, as long as the author and publisher are properly credited, which ensures maximum dissemination and a wider impact of our publications.
The book as a whole is distributed by MDPI under the terms and conditions of the Creative Commons license CC BY-NC-ND.

Contents

About the Editor

Prof. Dr. Anita MacDonald

Dr Anita MacDonald OBE is a Consultant Dietitian in Inherited Metabolic Disorders at Birmingham Children's Hospital and an Honorary Professor in Dietetics at Plymouth University, UK. Although she semi-retired 6 years ago, she is even more involved in phenylketonuria, as a clinical dietitian, conducting research and doing voluntary work for the National Society for PKU (NSPKU). Her involvement in phenylketonuria has spanned all her working life (over 40 years). She is a member of the European PKU Guidelines group (which is aiming to standardise PKU care across Europe), is a member of ESPKU Scientific Advisory Committee, and is a member of the UK NSPKU Medical Advisory Committee.

Preface to "Diet Therapy and Nutritional Management of Phenylketonuria"

Phenylketonuria (PKU) is an established inherited amino acid disorder treated with a very traditional dietary therapy, but there is still more to learn and verify about its nutritional composition, application and overall effectiveness. Although in the 1950s, the first patient successfully treated with diet therapy patently established the need for a phenylalanine free/low phenylalanine protein substitute, in present times, it is still necessary to characterise the most effective source of artificial protein; defining its optimal amino acid profile; and identifying any nutrient modulation that will improve the functionality of protein substitutes. It is also important to understand the impact of a life-long synthetic diet on gut microbiota, metabolomics and inflammatory status.

In early-treated patients with PKU, it is unclear if co-morbidities such as overweight, obesity, hypertension and diabetes are higher than in the general population and if these are associated with increased cardiovascular risk. It is also uncertain if overweight and obesity in PKU is related to early dietary practices, the nutritional composition of protein substitutes and special low-protein foods, the impact of the dietary treatment on satiety, disordered eating patterns, non-adherence with the low phenylalanine diet and poor metabolic control, or if this is even a consequence of the disorder. In a generation of ageing patients, the impact of intermittent and suboptimal dietary adherence on nutritional status deserves systematic study.

Anita MacDonald
Editor

nutrients

MDPI

Article

Special Low Protein Foods in the UK: An Examination of Their Macronutrient Composition in Comparison to Regular Foods

Georgina Wood [1,*], Sharon Evans [2], Kiri Pointon-Bell [1], Júlio César Rocha [3,4] and Anita MacDonald [2]

1 Faculty of Health, Education & Life Sciences, Birmingham City University, City South Campus, Westbourne Road, Edgbaston, Birmingham B15 3TN, UK; Kiri.Pointon-Bell@bcu.ac.uk
2 Birmingham Women's and Children's NHS Foundation Trust, Steelhouse Lane, Birmingham B4 6NH, UK; sharon.morris6@nhs.net (S.E.); Anita.Macdonald@nhs.net (A.M.)
3 Nutrition & Metabolism, NOVA Medical School, Faculdade de Ciências Médicas, Universidade Nova de Lisboa, Campo Mártires da Pátria, 130, 1169-056 Lisbon, Portugal; rochajc@nms.unl.pt
4 Center for Health Technology and Services Research (CINTESIS), R. Dr. Plácido da Costa, s/n, 4200-450 Porto, Portugal
* Correspondence: georgiewood93@googlemail.com

Received: 29 May 2020; Accepted: 23 June 2020; Published: 25 June 2020

Abstract: Special low protein foods (SLPFs) are essential in a low phenylalanine diet for treating phenylketonuria (PKU). With little known about their nutritional composition, all SLPFs on UK prescription were studied (n = 146) and compared to equivalent protein-containing foods (n = 190). SLPF nutritional analysis was obtained from suppliers/manufacturers. Comparable information about regular protein-containing foods was obtained from online UK supermarkets. Similar foods were grouped together, with mean nutritional values calculated for each subgroup (n = 40) and percentage differences determined between SLPFs and regular food subgroups. All SLPF subgroups contained 43–100% less protein than regular foods. Sixty-three percent (n = 25/40) of SLPF subgroups contained less total fat with palm oil (25%, n = 36/146) and hydrogenated vegetable oil (23%, n = 33/146) key fat sources. Sixty-eight percent (n = 27/40) of SLPF subgroups contained more carbohydrate, with 72% (n = 105/146) containing added sugar. Key SLPF starch sources were maize/corn (72%; n = 105/146). Seventy-seven percent (n = 113/146) of SLPFs versus 18% (n = 34/190) of regular foods contained added fibre, predominantly hydrocolloids. Nine percent of SLPFs contained phenylalanine > 25 mg/100 g and sources of phenylalanine/protein in their ingredient lists. Stricter nutritional composition regulations for SLPFs are required, identifying maximum upper limits for macronutrients and phenylalanine, and fat and carbohydrate sources that are associated with healthy outcomes.

Keywords: phenylketonuria; special low protein foods; nutritional composition; UK; macronutrients

1. Introduction

In phenylketonuria (PKU), the only UK treatment option is a rigorous low phenylalanine diet that is essential to prevent neurotoxicity and irreversible brain damage [1]. Most patients with classical PKU tolerate < 10 g natural protein daily [2], with up to 80% of daily protein provided by minimal phenylalanine-containing protein substitutes which are derived from either L-amino acids or glycomacropeptide. Special low protein foods (SLPFs) are an integral part of dietary treatment. They contribute essential energy (up to 50% of intake), variety and bulk, helping to improve or maintain metabolic control and growth [3–5]. Given their importance in a low protein diet, their nutritional profile and food labelling should receive the same care and attention as regular foods.

The composition and labelling of SLPFs is regulated by European Commission (EC) legislation on "dietary foods for special medical purposes" [6]. It gives no guidance on the source, amount or even quality of the carbohydrate and fat added to SLPFs [6]. The EC and UK regulations require SLPFs to list the amounts of energy, carbohydrate (including sugars), fat, protein and salt per 100 g [6–9] but no upper nutrient limits are defined. As a consequence of protein removal, it is expected that lower protein foods will contain higher amounts of carbohydrate and possibly fat [4,10,11], but there is no research describing the nutritional composition of UK SLPFs.

Considering that SLPFs receive minimal regulation, and with limited research into their nutritional profile, it has been suggested that a detailed analysis of each country's SLPFs be conducted [4,10]. The present study aimed to analyse the nutritional composition of all SLPFs available by the Advisory Committee of Borderline Substances (ACBS) prescription system in the UK.

2. Materials and Methods

From January–May 2019, detailed nutritional composition data for all UK SLPFs available on ACBS prescription was collected from manufacturers and suppliers. Data was obtained from company websites or from information sheets provided directly from the companies. Nutritional data was obtained per 100 g/100 mL and per serving for cooked and dried weight of products for: energy, protein, phenylalanine, total carbohydrate, sugars, fibre, total fat, saturated fat and salt. If nutritional data was stated as less than a certain value, e.g., "<0.1" or "<0.5", 0.001 was deducted from these numbers and values of "0.099" or "0.499" were used. Product ingredients, sources of added fibre, starch, sugar, fat and phenylalanine were obtained. Information was stored on an excel spreadsheet. Products were divided into 10 groups in a similar way to Pena and colleagues [10], and included: bread products (bread, pizza bases), pasta/rice/noodles, flour/mixes, meat/meat replacers, breakfast products (cereals and bars), eggs/egg replacers, milk/milk replacers, snacks (biscuits, cakes, crisps, chocolate, rusks, hazelnut spread and crackers), desserts (rice pudding, flavoured desserts, yogurt, and jelly) and other snacks/meals (soups, potato cakes, cheese sauce and potato pots). These groups were then categorised into 40 subgroups of equivalent product types, e.g., burgers, sausages, cookies/biscuits, cake mixes. The mean and range values for every nutrient across subgroups of similar products were calculated.

The same information (except for sources of phenylalanine) was collected and calculated for at least 2 regular protein-containing comparable foods per subgroup, from major UK supermarkets with nutritional analysis data online (ASDA, Morrisons, Sainsburys, Tesco, Waitrose, Ocado and Marks & Spencer). Phenylalanine content was estimated by calculating that 1 g of protein contained 50 mg phenylalanine [12]. Taste, texture, recipe ingredients and food function were considered when choosing comparator foods. Where possible, only regular products that had nutritional analysis available in the same format as SLPFs were considered, e.g., dried format or after preparation. Percentage differences between SLPFs and regular foods for all mean nutritional values were then determined. Variations of ±0–10% were considered comparable.

3. Results

One hundred and fifty one SLPFs were identified on UK ACBS prescription. One SLPF was undergoing reformulation and regular comparators for four SLPFs were not available. Thus, 146 SLPFs were compared with 190 regular products. Appendix A displays all SLPF and regular product subgroups (n = 40) and the investigated variables.

3.1. Energy

Mean energy content (per 100 g) for all SLPFs (n = 146) was 292 kcal (range: 32–583 kcal) and for all regular foods (n = 190) was 298 kcal (range: 26–558 kcal). Energy content was comparable for 50% of the subgroups of products (n = 20/40). For SLPFs, mean energy values for low protein hazelnut spread, prepared sausage mixes, prepared burger mixes, egg white and egg replacers were 37–66%

lower than regular varieties. Low protein dessert pots, hot breakfast cereals, potato pots and fish substitutes contained 36–41% more energy than regular versions.

3.2. Protein and Phenylalanine

All SLPF subgroups contained between 43–100% less protein and 60–100% less phenylalanine than regular foods. Table 1 displays the mean and range for phenylalanine content and sources of phenylalanine for all SLPF subgroups. The main sources of phenylalanine found in SLPFs were milk (including milk protein) (32% of SLPFs; $n = 47/146$) and yeast (14% of SLPFs; $n = 21/146$). For 91% of SLPFs ($n = 133/146$), the phenylalanine content was either ≤25 mg per 100 g or no sources of phenylalanine/protein were identified in the product ingredient list (Table 1).

Table 1. Phenylalanine content and identified sources of natural protein for all special low protein food (SLPF) subgroups. Values displayed as mean (range).

SLPF Subgroup	Phenylalanine (mg) per 100 g of Product	Identified Sources of Natural Protein/Phenylalanine in Each SLPF Subgroup
Bread ($n = 13$)	15 (8–30)	Yeast ($n = 13$), fennel seeds ($n = 1$), anis seeds ($n = 1$)
Pizza base ($n = 2$)	13 (2–24)	Yeast ($n = 2$)
Pasta/rice/noodles ($n = 33$)	13 (8–25)	Rice flour ($n = 5$)
Pasta and sauces (prepared) ($n = 5$)	8 (3–14)	Milk ($n = 4$), yeast extract ($n = 1$), cheese powder ($n = 1$)
Risotto ($n = 1$)	6	Milk ($n = 1$)
xPots/pot noodles (prepared) ($n = 4$)	9 (6–15)	Peas (dried) ($n = 1$), milk ($n = 4$)
Bread mix ($n = 3$)	15 (4–20)	Yeast ($n = 1$)
Cake mix ($n = 4$)	14 (4–30)	Cocoa powder ($n = 1$), cocoa ($n = 1$)
Flour ($n = 4$)	5 (4–<10)	No sources identified
Pancake/waffle mix ($n = 1$)	22	No sources identified
Pizza mix ($n = 1$)	<31	No sources identified
Egg replacer (dried mix) ($n = 3$)	7 (<5–10)	No sources identified
Egg white replacer ($n = 1$)	Nil added	No sources identified
Milk (liquid) ($n = 4$)	6 (0–10)	Milk ($n = 4$), whey powder ($n = 2$)
Milk (powder) ($n = 1$)	20	Milk ($n = 1$), whey permeate ($n = 1$)
Burgers (prepared) ($n = 3$)	25 (16–31)	Milk ($n = 2$), yeast ($n = 1$)
Fish substitute (prepared) ($n = 1$)	38	Shrimps ($n = 1$), cod ($n = 1$), rice flour ($n = 1$), milk ($n = 1$)
Sausages (prepared) ($n = 3$)	33 (29–38)	Milk ($n = 3$), potato flake ($n = 3$)
Breakfast bar ($n = 4$)	17 (12–25)	Milk ($n = 4$), cocoa powder ($n = 1$)
Breakfast cereal (dried) ($n = 3$)	12 (6–22)	Cocoa powder ($n = 1$)
Fruit bar ($n = 1$)	16	Egg ($n = 1$)
Hot breakfast cereal (prepared with water) ($n = 4$)	4 (2–6)	Cocoa powder ($n = 1$), milk ($n = 4$)
Biscuits/cookies ($n = 9$)	10 (1–27)	Cocoa mass ($n = 1$), egg ($n = 1$), cocoa ($n = 2$)
Cake ($n = 3$)	6 (6–6)	No sources identified
Chocolate ($n = 2$)	12 (<10–14)	Milk ($n = 1$), cocoa powder ($n = 1$), carob flour ($n = 1$)
Crackers ($n = 3$)	12 (10–17)	No sources identified
Crisps ($n = 4$)	16 (8–22)	Wheat flour ($n = 2$), rice flour ($n = 1$), whey powder ($n = 2$), yeast extract powder ($n = 1$), cheese powder ($n = 1$), yeast powder ($n = 1$)
Crispbread crackers ($n = 1$)	6	Pea starch ($n = 1$)

Table 1. *Cont.*

SLPF Subgroup	Phenylalanine (mg) per 100 g of Product	Identified Sources of Natural Protein/Phenylalanine in Each SLPF Subgroup
French toast crackers (*n* = 1)	30	Baker's yeast (*n* = 1)
Hazelnut spread (*n* = 1)	19	Milk (*n* = 1), hazelnuts (*n* = 1), almonds (*n* = 1), cocoa paste (*n* = 1)
Rusks (*n* = 1)	4	Milk (*n* = 1)
Dessert pot (*n* = 2)	<4	No sources identified
Flavoured desserts (prepared) (*n* = 4)	5 (1–13)	Milk (*n* = 4), chocolate powder (*n* = 1)
Jelly (dried) (*n* = 2)	<2	No sources identified
Rice pudding (*n* = 4)	6 (5–8)	Milk (*n* = 4)
Yogurt (prepared) (*n* = 1)	2	No sources identified
Cheese sauce (prepared) (*n* = 1)	13	Milk (*n* = 1)
Potato cakes (prepared) (*n* = 1)	46	Potato flake (*n* = 1)
Potato pots/Smash (prepared) (*n* = 3)	25 (23–27)	Potato flake (*n* = 3), milk (*n* = 3)
Soup (prepared) (*n* = 4)	2 (1–2)	Milk (*n* = 4), peas (*n* = 2)

3.3. *Carbohydrate (Including Sugars)*

Overall, the carbohydrate content was higher in 68% (*n* = 27/40) of SLPF subgroups when compared to protein-containing foods, with the greatest differences for meat, fish and egg substitutes (281–9167%).

The percentage of foods containing added sugar is given in Figure 1. Only 35% (*n* = 14/40) of SLPF subgroups contained higher amounts of sugar with 45% (*n* = 18/40) containing less than regular foods. Fish substitute contained 1000% more sugar than regular fish, but the amount of sugar was small (sugar content in fish substitute 1.1 g/100 g). Low protein pizza bases, flour and breakfast cereals contained only 3–22% more total carbohydrate than regular foods, but 81–273% more sugar.

Over 70% (72%; *n* = 105/146) of SLPFs compared with 66% (*n* = 125/190) of regular foods contained an added sugar source (Figure 1), with low protein bread, milk and meat replacements commonly adding sugar where regular foods did not. Key sugar sources in both groups are given in Table 2.

Table 2. Key sources of added sugar identified from ingredient lists for SLPFs and regular protein-containing foods.

Key Sources of Added Sugar	% of SLPF (*n* = 146)	% of Regular Protein Containing Foods (*n* = 190)
Sugar	52% (*n* = 76/146)	58% (*n* = 111/190)
Glucose	29% (*n* = 43/146)	23% (*n* = 44/190)
Maltodextrin	23% (*n* = 33/146)	13% (*n* = 25/190)
Dextrose	15% (*n* = 22/146)	12% (*n* = 22/190)
Sucrose	3% (*n* = 5/146)	1% (*n* = 2/190)
Fructose	<1% (*n* = 1/146)	6% (*n* = 12/190)

Maize/corn and potato starch were the main types of starch used in SLPFs. Over 70% (*n* = 105/146) of SLPFs contained maize/corn starch whereas 56% (*n* = 82/146) included potato starch. Fifty-four percent (*n* = 79/146) of SLPFs contained both starches. Maize/corn starch was common in low protein pasta, rice and noodles (100%; *n* = 43/43) and snacks (80%; *n* = 20/25). In contrast, the most common starch sources identified in regular foods were wheat flour (*n* = 82/190); wheat semolina (*n* = 30/190) and rice or rice flour (*n* = 27/190). Maize/corn starch and potato starch were only listed in 13% (*n* = 24/190) of regular foods.

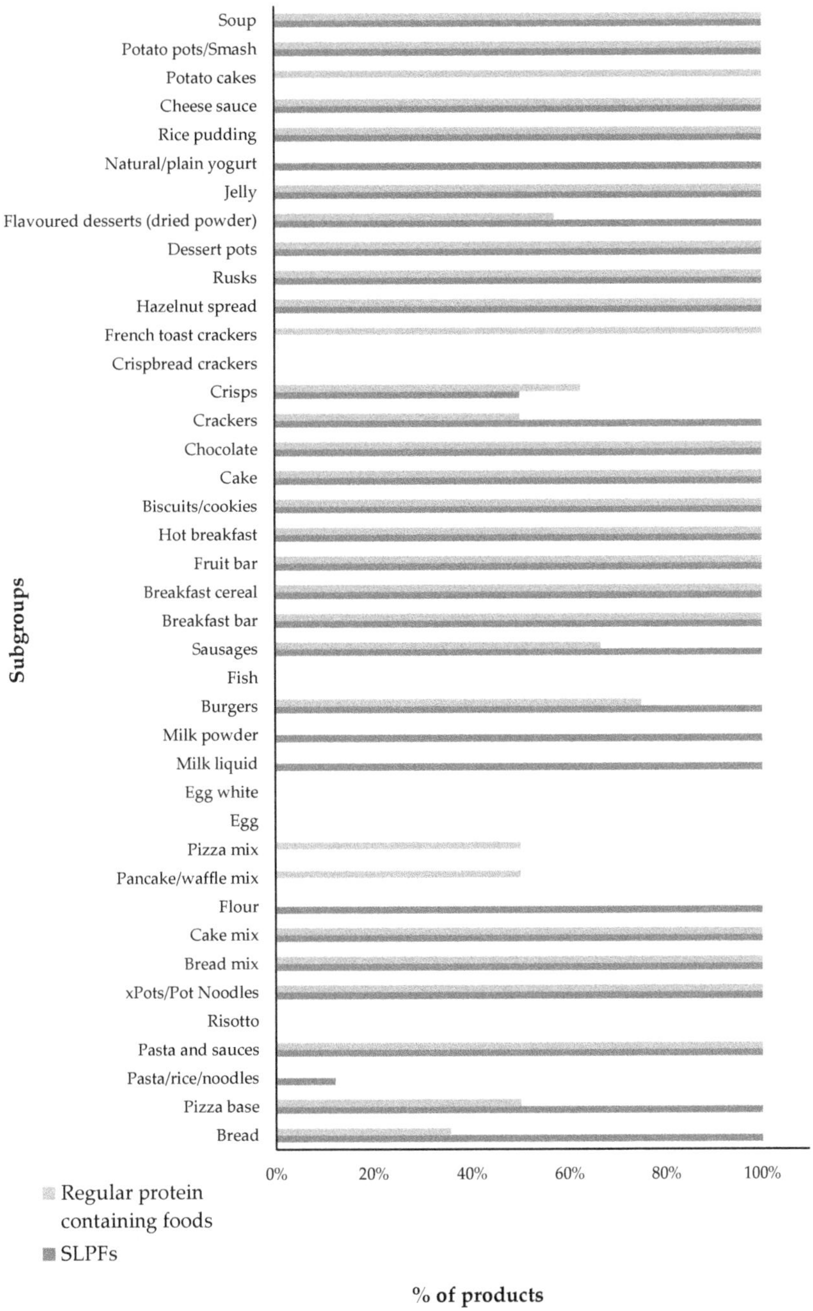

Figure 1. Percentage of regular and special low protein food (SLPF) products containing added sugar in their ingredient list by subgroup.

3.4. Total and Saturated Fat

Sixty three percent (n = 25/40) of SLPF subgroups contained less total fat (including egg substitutes, meat replacements, flour/mixes, flavoured desserts (dried powder), dried breakfast cereal, pasta, rice and noodles), whilst 28% (n = 11/40) contained 21–94% more total fat (including breads, pizza bases, breakfast bars, fruit bars, chocolate, pasta and sauces, risotto, dessert pots, rusks and liquid milk replacers) than regular foods. In 8% (n = 3/40) of the SLPF subgroups, total fat content was comparable to that found in regular foods. Calculation of percentage differences between SLPF egg whites and regular egg whites was not possible, due to SLPF egg whites reporting "nil added" for total fat content.

Thirty-five percent (n = 14/40) of SLPF subgroups contained more saturated fat (14–262%) than regular foods, including cakes, breakfast bars, pizza bases, fruit bars, bread and breakfast cereals. Conversely, 50% (n = 20/40) of SLPF subgroups contained less saturated fat (<−10%) than regular foods. SLPF pizza mixes, cake mixes, eggs and fish substitutes contained 85–100% less saturated fat.

Palm oil was the most common fat source found in 25% (n = 36/146) of SLPFs. Twenty-five (17%) of these SLPFs did not specify if palm oil was hydrogenated or non-hydrogenated but one food contained partially hydrogenated palm oil (<1%), one hydrogenated palm oil (<1%) and nine non-hydrogenated palm oil (6%) (Figure 2). Hydrogenated vegetable oil was another common fat source in SLPFs (23%, n = 33/146) (Figure 2). SLPFs with "hydrogenated vegetable oil" or "hydrogenated palm oil" were all produced by the same manufacturer and it was unclear if the sources were partially hydrogenated. The most prevalent fat sources in regular foods were milk (41%, n = 78/190) and palm oil (39%, n = 75/190), with no products listing hydrogenated oil sources (Figure 2). Palm oil was found in 80% (n = 20/25) of SLPF snacks compared with 58% (n = 23/40) of regular snacks.

In the SLPF subgroups containing less saturated fat (n = 20/40), hydrogenated vegetable oil was present in 35% (n = 7/20) (cheese sauce, soups, flavoured desserts, pasta and sauces, xPots and meat replacements).

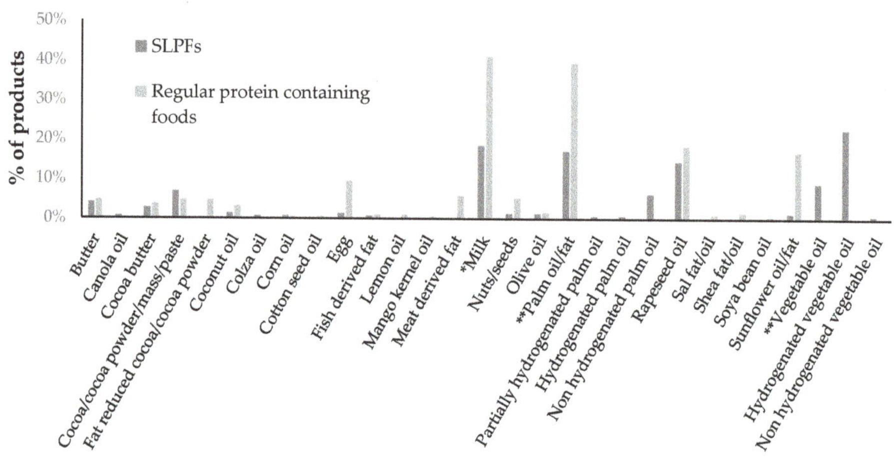

Figure 2. Percentage of SLPFs and regular protein containing foods containing different types of fat in their ingredient lists. * Not including milk protein (where products specified this as an ingredient) ** oil/fat, did not specify whether it was hydrogenated or non-hydrogenated.

3.5. Fibre

From the nutritional analysis, only 44% (n = 64/146) of SLPFs quantified a fibre amount compared with 82% (n = 156/190) of regular foods. When fibre content was listed, low protein milk (liquid) and egg substitutes contained more fibre than regular comparator foods which did not contain added fibre. Low protein French toast, chocolate, bread, pizza bases, cake mixes and fruit bars contained more fibre (16–189%) than regular foods. The largest differences were for egg white replacers, burger and fish substitutes (1645–5050%), with SLPFs containing higher amounts.

Some products contained natural fibre sources such as whole-wheat flour or apple flakes but only added fibre sources (e.g., barley/wheat/gluten-free wheat fibre, methylcellulose, pectin, guar gum etc.) were identified from the ingredient lists. Added fibre was found in 77% (n = 113/146) of SLPFs but only 18% (n = 34/190) of regular foods (Figure 3). The main fibre sources added to SLPFs were methylcellulose, guar gum, hydroxypropyl-methylcellulose, inulin and carob/locust bean gum. These were added to primarily improve texture and quality.

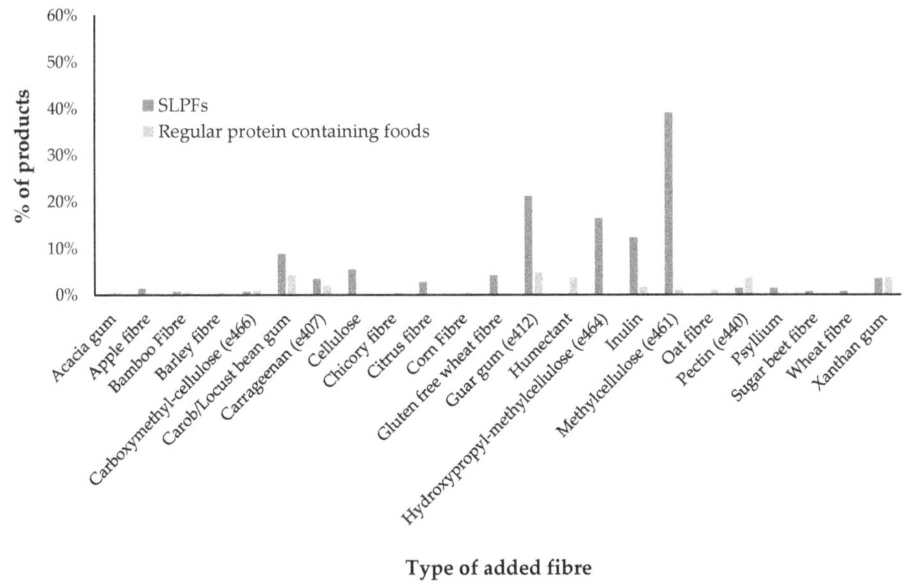

Figure 3. Percentage of regular and SLPF products containing added fibre in their ingredient lists by type of fibre.

3.6. Salt

Over 50% of SLPF subgroups contained 17–100% less salt than regular foods (n = 21/40), with low protein rice pudding, chocolate and jelly subgroups all containing 100% less. Salt content was higher in 33% of SLPF subgroups when compared to regular foods with higher amounts in low protein potato pots, xPots, hazelnut spread, crisps, cakes, hot breakfast cereal, fish substitute and pizza mix (100–1050%).

4. Discussion

This is the first study to investigate the nutritional composition of all SLPFs available on UK ACBS prescription, compared with regular protein-containing foods, examining macronutrients and their ingredient sources. The overall nutrient quality of SLPFs was variable with no consistent pattern. Some of the nutrients reported on food labelling were incomplete with 56% of foods not itemising fibre content. The energy content of 50% of SLPF subgroups was comparable to regular foods, with only 23% of SLPF subgroups containing a higher amount (>10%) than regular foods.

Sixty three percent of SLPF subgroups contain less total fat and 50% contain less saturated fat (<−10%) when compared to regular foods, including: milk powder, eggs, biscuits/cookies, crisps, crispbread crackers, flavoured desserts, yogurt, cheese sauce, soup, potato cakes, meat and certain flour/mixes subgroups. This appears advantageous. Some studies in PKU, have reported improved or similar biomarkers of cardiovascular disease when compared to healthy controls [13–17]. However, although 50% of SLPF subgroups contained less saturated fat than regular foods, some of the subgroups listed hydrogenated vegetable oil as a fat source and did not specify if this was "partially" or "fully" hydrogenated. Full hydrogenation of vegetable oil produces exclusively saturated fats, whereas partial hydrogenation of vegetable oil leads to a higher amount of trans fatty acids [18,19]. Consumption of trans fatty acids has been linked to the development of several health problems, including metabolic syndrome, coronary heart disease, obesity and diabetes [18–20]. Although dietary trans fatty acids may have a similar elevating effect on LDL-cholesterol to that of saturated fatty acids, the former will contribute to HDL-cholesterol reduction [21]. Low HDL-cholesterol has already been reported in PKU patients [14]. Therefore, some SLPFs that may appear "healthier" with a low saturated fat content may actually be higher in trans fats, but this information is not disclosed by the manufacturers. In contrast, 35% of SLPF subgroups contained more saturated fat than regular foods, particularly staple items such as breakfast cereal and breads, which is a concern. Common fat sources were palm oil and hydrogenated vegetable oil, both of which contain saturated fat [18,20,22,23]. The chain length of saturated fat is important, with longer-chain saturated fatty acids being more harmful, whilst short- and medium-chain fatty acids have potential benefits on metabolic risk, weight gain, obesity and gut microbiome [24]. In summary, more precise information on the type of fat added is required for SLPFs.

Over 70% of SLPFs on UK prescription contained added sugar but this percentage was only slightly higher than regular foods. When subgroups were examined more closely, it was apparent that certain SLPFs commonly added sugar when regular foods did not. Specifically, 100% of low-protein breads, pizza bases, flour, meats, crackers, flavoured desserts, yogurt, milks and some pastas contained added sugar. Maize/corn and potato starch were the most frequently used starch sources in SLPFs with most ingredient lists indicating that these starches were present in isolation. Isolated starches are more refined than regular flour and/or raw materials, and foods containing isolated starches may have a higher glycaemic index (GI) than those made from wheat flour [25,26]. In contrast, the addition of fat to a regular carbohydrate food is known to delay gastric emptying and lower GI [27]. The GI of SLPFs available on UK ACBS prescription has not been formally evaluated. This needs to be determined as it is uncertain how the isolated starches, added sugar and increased levels of fat found in some SLPFs impact on GI function.

In PKU, a high carbohydrate intake and the carbohydrate profile of SLPFs may contribute to higher levels of insulin resistance, as a relationship between the quality and amount of carbohydrate in SLPFs and peripheral insulin resistance has been reported [11,28]. An association between the overall glycaemic load and triglyceride glucose index in children with PKU has also been described [11]. In patients with increased abdominal obesity (waist circumference), which is a component of metabolic syndrome, increased triglycerides, lower HDL-cholesterol and increased HOMA-IR (homeostasis model assessment of insulin resistance) is documented [14]. Insulin resistance, a marker of metabolic syndrome, is linked to an increased risk of cardiovascular disease [29].

Gluten and other proteins in regular grains/cereals are important in maintaining structural integrity, texture and quality of regular foods [25]. However, with the majority of SLPFs based on maize/corn/potato starches, it is not surprising that 77% of SLPFs contained added fibre, predominantly in the form of hydrocolloids. Hydrocolloids are additives that improve the quality, formulation and texture of low protein and gluten-free products [25,26,30]. Their contribution as a source of dietary fibre has not been explored, despite the fibre content of hydrocolloids typically varying between 60–90% [31]. Generally, such additives are used in small amounts and are commonly not significant enough to make a fibre claim on a product [31]. However, in patients with PKU where approximately 50% of their energy intake may be from SLPFs [3] containing hydrocolloids, it is probable that these ingredients are significantly contributing to daily fibre intake, although this remains unreported. Therefore, regular consumption of SLPFs may also have an impact on gastrointestinal function and gut microbiome, with previous research reporting that 34% of patients with PKU suffer from digestive problems [2].

Over 30% of SLPF subgroups contained more salt than regular foods, with some containing 100–1050% extra. It is possible that their habitual consumption may contribute to nutritional co-morbidities such as hypertension [32–34], vascular stiffness [34,35], overweight/obesity [3,34,36–40] and an atherogenic lipoprotein profile [34].

For 91% of SLPFs, phenylalanine content was ≤25 mg/100 g of the product, or all product ingredients were "exchange-free", meaning these items can be eaten without measurement [41]. The remaining 9% of SLPFs contained phenylalanine >25 mg/100 g and included ingredients such as milk and potato flakes; and consequently, these foods must be restricted and given in controlled amounts in a low phenylalanine diet [41]. The few SLPFs containing >25 mg/100 g add complexity to a low phenylalanine diet as patients and caregivers may be unsure about their suitability.

Overall, there is limited research into the dietary patterns of patients with PKU, but evidence suggests that SLPFs contribute up to 47% of energy intake [11]. Many contemporary low phenylalanine protein substitutes have a low fat and carbohydrate content, meaning there is an increased reliance on SLPFs to provide these macronutrients [42,43]. With a "treatment for life" policy, it is essential that SLPFs have a nutritional profile that supports long term healthy eating patterns.

There are many recommendations required to improve standards in the nutritional composition and labelling of UK SLPFs. Transparency is necessary by SLPF manufacturers about the nutritional profile of their products. All ingredients should be clearly listed including sources of, at least, starch, sugar, fat and fibre and the amount of fibre added (per 100 g/100 mL) for all SLPFs. Nutritional analysis for both dried and prepared weights should be available. Packaging and website nutritional information should be accurate and consistent. To ensure that all SLPFs can be safely consumed without calculation and measurement, the phenylalanine content should be no more than 25 mg/100 g for all prescribed SLPFs; and no more phenylalanine than 5 mg/100 mL for milk replacements [44]. SLPF macronutrient composition regulations should be strengthened, ensuring similarity to regular protein-containing comparators. Upper limits should be set for carbohydrate and fat content. Fat sources should be predominantly poly- or mono-unsaturated rather than saturated or trans-fats; the addition of trans fatty acid sources should be clearly labelled. Fortunately, the EU Commission, 2019, has now adopted a regulation setting a maximum limit for trans-fats in industrially produced trans-fat of 2 g/100 g of fat [45]. Some isolated starches could be replaced by plants naturally low in phenylalanine such as cassava. In SLPFs, added sugar should be restricted if protein-containing comparators do not contain it. It is hypothesised that high sugar consumption may affect gut microbiota, disturbing the crosstalk between the gut and systemic metabolism, with a potentially harmful impact on metabolic health [46]. Reducing the salt content of some savoury products and replacing it with herbs and spices to improve or maintain the taste and flavour of SLPFs would be beneficial. A simple traffic light colour system has been proposed to categorise SLPFs based on their nutritional profile [10] and this may help patients reduce refined carbohydrate and salt intake and increase their consumption of healthier fats and complex carbohydrates.

In this evaluation of SLPFs, difficulties in accessing nutritional composition data has led to several limitations. Data was missing for some key nutrients such as fibre. Nutritional values were often reported as "<0.5" or "<0.1", and so the precise content was unclear. There were occasional discrepancies in nutritional information between SLPFs and regular foods. Some foods provided information for dried ingredients whilst others only for cooked/prepared products. The selection of protein-containing foods as comparators and how the products were grouped was subjective. Finally, this study only examined products accessible on UK prescription compared with protein-containing products available from UK supermarkets. Detailed nutritional composition analysis of SLPFs available on prescription compared with regular equivalent products in other countries is warranted to determine if findings are consistent.

5. Conclusions

In conclusion, this UK study shows that the nutritional content of SLPFs available on ACBS prescription differed to regular comparable foods but with no clear consistent pattern. Almost two thirds of SLPF subgroups contained less total fat but with palm oil and hydrogenated vegetable oil as key fat sources. Over two thirds of SLPF subgroups contained more carbohydrate commonly as isolated starches. More added fibre was identified in SLPFs but predominantly in the form of hydrocolloids. It is possible that habitual consumption of SLPFs higher in salt, sugars, isolated starches, or saturated fat may contribute to future nutritional comorbidities.

Stricter nutritional composition regulations, improvements in product labelling and access to full nutritional composition data will allow health professionals and patients to make informed decisions when prescribing and using SLPFs. Identifying upper limits for macronutrients, and improving fat and carbohydrate sources is essential in supporting patients with PKU in meeting their nutritional needs and improving health outcomes.

Author Contributions: Conceptualization, S.E. and A.M.; Data curation, G.W.; Formal analysis, G.W.; Investigation, G.W.; Methodology, G.W., S.E. and A.M.; Supervision, S.E., K.P.-B. and A.M.; Visualization, G.W., S.E., K.P.-B., J.C.R. and A.M.; Writing—original draft, G.W.; Writing—review and editing, G.E., S.E., K.P.-B., J.C.R. and A.M. All authors have read and agree to the published version of the manuscript.

Funding: This research received no external funding.

Acknowledgments: Thank you to the suppliers/manufacturers of special low protein foods for providing nutritional composition data for your products.

Conflicts of Interest: A.M. has been a member of the European Nutritionist Expert Panel Biomarin, Sapropterin Advisory Board Biomarin, the advisory board ELEMENT Danone-Nutricia, the advisory board for Arla and Applied Pharma Research, and received research funding and honoraria from Nutricia, Vitaflo International and Merck Serono. J.C.R. has been a member of the European Nutritionist Expert Panel (Biomarin), the Advisory Board for Applied Pharma Research and Nutricia, and received honoraria as a speaker from APR, Merck Serono, Biomarin, Nutricia, Vitaflo, Cambrooke, PIAM and Lifediet. S.E. receives research funding from Nutricia, and has received financial support and honoraria from Nutricia and Vitaflo to attend/speak at study days and conferences.

Appendix A

Table A1. Nutritional composition data for all low protein and regular subgroups analysed per 100 g of product. Values displayed as Mean (range).

Product		Energy (kcal) per 100 g	Protein (g) per 100 g	Phenylalanine (mg) per 100 g	Total Carbohydrate (g) per 100 g	Carbohydrate of Which Is Sugars (g) per 100 g per 100 g	Fibre (g) per 100 g	Total Fat (g) per 100 g	Saturated Fat (g) per 100 g	Salt (g) per 100 g
Breads/Pizza bases										
Bread	**SLPF** (n = 13)	244 (214–266)	0.6 (0.2–1.0)	15 (8–30)	47.1 (37.0–53.8)	3 (1.4–4.3)	9.2 (3.8–16.0)	4.3 (2.7–5.3)	1.2 (0.3–2.3)	0.5 (0.3–1.3)
	Regular (n = 14)	255 (221–285)	9.7 (8.4–11.9)	485 (420–595)	46.6 (31.4–58.8)	3.1 (2.2–4.1)	4 (1.6–12.4)	2.5 (0.6–7.5)	0.5 (0.2–1.4)	0.9 (0.7–1.1)
	% Difference	−4%	−94%	−97%	1%	−3%	130%	72%	140%	−44%
Pizza base	**SLPF** (n = 2)	290 (263–316)	0.9 (0.8–0.9)	13 (2–24)	55.7 (49.0–62.3)	4.9 (4.7–5.0)	7.8 (2.6–13)	5.4 (4.2–6.5)	1.7 (1.4–2.0)	0.5 (0.3–0.8)
	Regular (n = 2)	296 (288–304)	8.9 (8.8–9.0)	445 (440–450)	54.3 (51.6–57.0)	2.4 (2.4–2.4)	2.7 (2.6–2.8)	4.2 (2.1–6.3)	0.6 (0.3–0.9)	1.4 (1.4–1.4)
	% Difference	−2%	−90%	−97%	3%	104%	189%	29%	183%	−64%
Pasta/rice/noodles										
Pasta/rice/noodles	**SLPF** (n = 33)	356 (343–366)	0.3 (0.1–0.5)	13 (8–25)	85.8 (79.0–88.1)	0.6 (0.0–3.2)	3.1 (0.2–7.3)	0.9 (0.6–1.6)	0.5 (0.2–0.9)	0.1 (0.1–0.5)
	Regular (n = 23)	356 (336–380)	11.9 (7.2–14.0)	595 (360–700)	72.5 (68.6–78.1)	2.4 (0.0–4.8)	2.9 (1.0–4.3)	1.8 (0.7–3.3)	0.4 (0.2–0.9)	0.1 (0.0–0.8)
	% Difference	0%	−97%	−98%	18%	−75%	7%	−50%	25%	0%
Pasta and sauce	**SLPF** (n = 5)	123 (98–140)	0.5 (0.3–0.6)	8 (3–14)	25.9 (23.9–31.3)	1.9 (0.4–4.9)	No values	2 (0.1–1.1)	0.6 (0.0–1.1)	0.8 (0.6–1.0)
	Regular (n = 10)	104 (81–137)	3.5 (2.7–4.8)	175 (135–240)	18.1 (14.0–25.1)	2.5 (0.8–4.6)	1.2 (<0.5–2.1)	1.6 (0.6–4.1)	0.8 (0.1–2.4)	0.6 (0.3–0.8)
	% Difference	18%	−86%	−95%	43%	−24%	–	25%	−25%	33%
Risotto	**SLPF** (n = 1)	103	0.3	6	14	<0.2	0.5	5	1.3	0.7
	Regular (n = 2)	95 (93–97)	2.9 (2.3–3.4)	145 (115–170)	13.8 (13.6–14.0)	1.4 (1.0–1.7)	1.4 (1 value)	3 (2.3–3.6)	1.1 (1.1–1.1)	0.5 (0.4–0.5)
	% Difference	8%	−90%	−96%	1%	−86%	−64%	67%	18%	40%
xPots/pot noodles	**SLPF** (n = 4)	138 (136–140)	0.4 (0.3–0.6)	9 (6–15)	23 (22.6–23.5)	1.8 (1.6–2.1)	No values	4.8 (4.5–5.1)	0.9 (0.7–1.1)	1.9 (1.6–2.3)
	Regular (n = 8)	131 (83–145)	2.9 (2.3–3.5)	145 (115–175)	19.1 (17.4–21.6)	1.5 (0.9–2.3)	1.1 (0.8–1.3)	4.5 (0.3–5.7)	1.9 (<0.1–2.9)	0.5 (0.4–0.7)
	% Difference	5%	−86%	−94%	20%	20%	–	7%	−54%	280%

Table A1. *Cont.*

Product		Energy (kcal) per 100 g	Protein (g) per 100 g	Phenylalanine (mg) per 100 g	Total Carbohydrate (g) per 100 g	Carbohydrate of Which Is Sugars (g) per 100 g per 100 g	Fibre (g) per 100 g	Total Fat (g) per 100 g	Saturated Fat (g) per 100 g	Salt (g) per 100 g
Flour/mixes										
Bread mix	**SLPF** ($n = 3$)	347 (339–354)	0.4 (0.1–0.7)	15 (4–20)	83 (80.1–86.0)	2.8 (1.7–4.6)	2.5 (2.0–3.0)	0.6 (0.5–0.7)	0.3 (0.1–0.5)	0.4 (0.1–0.6)
	Regular ($n = 2$)	338 (334–341)	13.1 (13.0–13.1)	655 (650–655)	70.4 (67.6–73.1)	3.1 (1.8–4.4)	3.1 (2.8–3.4)	1.6 (1.2–2.0)	0.3 (0.3–0.3)	1.6 (1.3–1.9)
	% Difference	3%	−97%	−98%	18%	−10%	−19%	−63%	0%	−75%
Cake mix	**SLPF** ($n = 4$)	366 (365–367)	0.5 (0.2–0.9)	14 (4–30)	89 (84.6–92.1)	39.6 (35.7–47.9)	3.4 (1.8–7.1)	0.7 (0.1–1.3)	0.4 (0.0–0.9)	0.8 (0.6–0.8)
	Regular ($n = 4$)	381 (370–395)	5.8 (4.6–6.9)	290 (230–345)	77.4 (74.7–81.9)	43.5 (32.8–47.6)	2.1 (1.1–3.4)	5 (4.7–5.3)	2.7 (2.5–2.9)	1.5 (1.2–2.1)
	% Difference	−4%	−91%	−95%	15%	−9%	62%	−86%	−85%	−47%
Flour	**SLPF** ($n = 4$)	349 (339–361)	0.2 (0.1–0.3)	5 (4–<10)	85.3 (80.1–88.3)	5.6 (4.6–7.2)	2.6 (2.1–3.1)	0.3 (0.0–0.5)	0.1 (0.1–0.2)	0.6 (0.4–0.7)
	Regular ($n = 2$)	335 (330–340)	9.8 (9.6–9.9)	490 (480–495)	70 (67.9–72.0)	1.5 (1.3–1.7)	No values	1.1 (0.7–1.4)	0.2 (0.1–0.2)	0.9 (0.8–1.1)
	% Difference	4%	−98%	−99%	22%	273%	–	−73%	−50%	−33%
Pancake/waffle mix	**SLPF** ($n = 1$)	353	0.5	22	86.5	14	0.6	0.4	<0.1	0.2
	Regular ($n = 2$)	434 (335–532)	9.3 (8.5–10.0)	465 (425–500)	72.1 (70.2–74.0)	9.6 (4.2–14.9)	2.6 (2.6–2.6)	1.3 (1.1–1.4)	0.2 (0.0–0.3)	3.1 (2.3–3.8)
	% Difference	−19%	−95%	−95%	20%	46%	−77%	−69%	−51%	−94%
Pizza mix (dried powder)	**SLPF** ($n = 1$)	353	0.2	<31	86.9	<0.1	No value	<0.5	<0.1	1.4
	Regular ($n = 2$)	379 (372–386)	12.1 (11.3–13.0)	605 (565–650)	70 (69.4–70.6)	3.9 (3.9–3.9)	4.3 (4.0–4.6)	4.7 (4.4–4.9)	1.7 (1.5–1.9)	0.4 (0.4–0.4)
	% Difference	−7%	−98%	−95%	24%	−97%	–	−89%	−94%	250%
Eggs/replacers										
Egg	**SLPF** ($n = 3$) (prepared)	44 (32–68)	0 (0.0–0.0)	1 (1–1)	10.7 (7.5–16.8)	0 (0.0–0.0)	0.5 (0.3–0.7)	0 (0.0–0.0)	0 (0.0–0.0)	0.1 (0.0–0.1)
	Regular ($n = 2$)	131 (131–131)	12.6 (12.6–12.6)	630 (630–630)	0.2 (0.0–< 0.5)	0.2 (0.0–< 0.5)	0 Only 1 value	9 (9.0–9.0)	2.5 (2.5–2.5)	0.4 (0.4–0.4)
	% Difference	−66%	−100%	−100%	5250%	−100%	–	−100%	−100%	−75%
Egg white	**SLPF** ($n = 1$)	185	Nil added	Nil added	Nil added	Nil added	92.5	Nil added	Nil added	1
	Regular ($n = 2$)	354 (345–363)	83.3 (82.6–84.0)	4165 (4130–4200)	3.4 (<0.5–6.3)	0.3 (0.0–< 0.5)	5.3 Only 1 value	0.4 (0.2–< 0.5)	0.1 (<0.1–0.1)	2.6 (1.8–3.4)
	% Difference	−48%	–	–	–	–	1645%	–	–	−62%

Table A1. *Cont.*

Product		Energy (kcal) per 100 g	Protein (g) per 100 g	Phenylalanine (mg) per 100 g	Total Carbohydrate (g) per 100 g	Carbohydrate of Which Is Sugars (g) per 100 g per 100 g	Fibre (g) per 100 g	Total Fat (g) per 100 g	Saturated Fat (g) per 100 g	Salt (g) per 100 g
Milk/replacers										
Milk (liquid)	**SLPF** ($n = 4$)	62 (40–89)	0.2 (0.0–0.4)	6 (0–10)	7.6 (5.0–10.8)	4.5 (3.5–5.8)	0.5 (0.2–0.8)	3.3 (2.0–4.7)	1.7 (1.3–2.3)	0.1 (0.0–0.2)
	Regular ($n = 2$)	58 (50–65)	3.5 (3.4–3.6)	175 (170–180)	4.8 (4.7–4.8)	4.8 (4.7–4.8)	0 (0.0–0.0)	2.7 (1.8–3.6)	1.7 (1.1–2.3)	0.1 (0.1–0.1)
	% Difference	7%	−94%	−97%	58%	−6%	–	22%	0%	0%
Milk (powder)	**SLPF** ($n = 1$)	428	1.7	20	77.5	45.1	No value	12.3	6.2	0.7
	Regular ($n = 2$)	428 (353–503)	30.8 (25.7–35.9)	1540 (1285–1795)	43.5 (36.5–50.5)	43.2 (36.5–49.8)	0.5 (0.0–1.0)	14.4 (0.6–28.2)	9 (0.4–17.6)	1 (0.9–1.1)
	% Difference	0%	−94%	−99%	78%	4%	–	−15%	−31%	−30%
Meat/replacers										
Burgers	**SLPF** ($n = 3$)	155 (155–157)	0.7 (0.4–0.8)	25 (16–31)	27.9 (27.4–28.9)	2.2 (2.0–2.7)	7.4 Only 1 value	4.1 (2.7–4.9)	2.7 (1.5–3.2)	0.7 (0.5–0.8)
	Regular ($n = 4$)	249 (226–280)	22 (17.0–25.6)	1100 (850–1280)	5 (1.2–10.0)	1.2 (<0.5–3.3)	0.4 (0.0–0.7)	15.7 (13.0–17.4)	7.4 (6.2–7.9)	0.9 (0.7–1.1)
	% Difference	−38%	−97%	−98%	458%	83%	1750%	−74%	−64%	−22%
Fish	**SLPF** ($n = 1$)	138	1.1	38	27.8	1.1	10.3	0.2	0	2.3
	Regular ($n = 2$)	98 (98–98)	22.5 (21.8–23.1)	1125 (1090–1155)	0.3 (0.0–0.5)	0.1 (0.0–0.1)	0.2 (0.0–0.3)	0.8 (0.3–1.2)	0.3 (0.1–0.4)	0.2 (0.1–0.3)
	% Difference	41%	−95%	−97%	9167%	1000%	5050%	−75%	−100%	1050%
Sausages	**SLPF** ($n = 3$)	146 (140–150)	0.8 (0.6–0.9)	33 (29–38)	27.8 (27.1–28.4)	4.1 (1.8–6.3)	No values	4 (3.1–4.6)	2.6 (2.1–2.8)	0.8 (0.7–0.9)
	Regular ($n = 6$)	260 (200–309)	16.2 (14.0–21.2)	810 (700–1060)	7.3 (0.7–16.0)	3 (0.7–6.5)	1.9 (1.0–3.8)	18 (13.0–24.5)	7.9 (5.0–13.2)	1.4 (1.0–1.9)
	% Difference	−44%	−95%	−96%	281%	37%	–	−78%	−67%	−43%
Breakfast and cereal bars										
Breakfast bar	**SLPF** ($n = 4$)	472 (464–487)	0.3 (0.2–0.5)	17 (12–25)	67.2 (65.5–68.7)	30.5 (26.2–33.5)	No values	22.3 (20.8–24.4)	14.1 (13.0–15.4)	0.5 (0.5–0.6)
	Regular ($n = 8$)	413 (372–485)	7.5 (4.7–15.0)	375 (235–750)	62.7 (42.0–74.0)	24.6 (19.0–36.1)	5.9 (4.3–10.0)	13.9 (6.6–26.0)	4.1 (0.8–10.4)	0.4 (0.0–0.9)
	% Difference	14%	−96%	−95%	7%	24%	–	60%	244%	25%
Breakfast cereal (dried)	**SLPF** ($n = 3$)	380 (374–385)	0.4 (0.2–0.6)	12 (6–22)	92.5 (91.0–93.6)	35.6 (34.0–38.9)	1.6 (1.1–2.3)	0.9 (0.7–1.1)	0.8 (0.6–0.9)	0.2 (0.2–0.2)
	Regular ($n = 6$)	385 (378–398)	7.7 (6.0–9.4)	385 (300–470)	78.8 (72.0–84.0)	19.7 (8.0–35.0)	5 (2.5–8.9)	3.3 (0.9–4.6)	0.7 (0.2–0.9)	0.7 (0.2–1.1)
	% Difference	−1%	−95%	−97%	17%	81%	−68%	−73%	14%	−71%

Table A1. *Cont.*

Product		Energy (kcal) per 100 g	Protein (g) per 100 g	Phenylalanine (mg) per 100 g	Total Carbohydrate (g) per 100 g	Carbohydrate of Which Is Sugars (g) per 100 g per 100 g	Fibre (g) per 100 g	Total Fat (g) per 100 g	Saturated Fat (g) per 100 g	Salt (g) per 100 g
Fruit bar	**SLPF** (n = 1)	424	0.6	16	72	38	3.6	14	7	0.3
	Regular (n = 2)	358 (351–364)	4.1 (3.9–4.2)	205 (195–210)	69.7 (67.0–72.3)	34 (33.0–34.9)	3.1 (2.4–3.8)	7.2 (6.0–8.3)	2.8 (2.5–3.0)	0.5 (0.4–0.6)
	% Difference	18%	−85%	−92%	3%	12%	16%	94%	150%	−40%
Hot breakfast cereal (with water)	**SLPF** (n = 4)	137 (130–147)	0.1 (0.0–0.1)	4 (2–6)	31.5 (30.0–33.5)	8.3 (6.5–10.0)	No values	1.1 (1.0–1.4)	0.7 (0.6–1.0)	0.2 (0.1–0.2)
	Regular (n = 4)	97 (91–104)	3.8 (3.4–4.3)	190 (170–215)	16.6 (16.0–17.4)	5.6 (4.1–6.3)	1.6 (1.1–2.1)	1.3 (1.2–1.4)	0.3 (0.2–0.3)	0.1 (0.1–0.1)
	% Difference	41%	−97%	−98%	90%	48%	–	−15%	133%	100%
Snacks										
Biscuits/cookies	**SLPF** (n = 9)	488 (476–506)	0.5 (0.2–0.7)	10 (1–27)	75.1 (68.2–84.0)	17.5 (14.9–25.6)	0.7 (0.5–1.0)	20.5 (15.0–25.0)	8.8 (7.3–10.4)	0.3 (0.0–0.7)
	Regular (n = 14)	499 (475–531)	5.6 (4.1–7.1)	280 (205–355)	62.6 (48.0–72.6)	26.3 (16.2–38.1)	2.6 (1.3–5.5)	24.8 (19.0–32.1)	12.6 (2.8–19.0)	0.7 (0.5–1.1)
	% Difference	−2%	−91%	−96%	20%	−33%	−73%	−17%	−30%	−57%
Cake	**SLPF** (n = 3)	372 (372–372)	0.2 (0.2–0.2)	6 (6–6)	58 (58.0–58.0)	33.5 (33.5–33.5)	1.3 (1.3–1.3)	15.2 (15.2–15.2)	7.6 (7.6–7.6)	0.7 (0.7–0.7)
	Regular (n = 2)	422 (393–450)	4.3 (4.1–4.4)	215 (205–220)	55.2 (52.4–58.0)	27.4 (22.7–32.0)	No values	21.2 (19.4–23.0)	2.1 (1.8–2.4)	0.3 (0.1–0.4)
	% Difference	−12%	−95%	−97%	5%	22%	–	−28%	262%	133%
Chocolate	**SLPF** (n = 2)	566 (549–583)	0.3 (0.2–0.4)	12 (<10–14)	54.2 (47.0–61.4)	51.1 (43.0–59.1)	5.4 (0.9–9.8)	37.7 (33.4–42)	27.7 (25–30.4)	0 (0.0–0.0)
	Regular (n = 2)	540 (534–546)	7 (6.7–7.3)	350 (335–365)	56.5 (56.0–57.0)	55.7 (55.4–56)	2.1 Only 1 value	31.2 (30.0–32.4)	18.8 (18.0–19.6)	0.2 (0.2–0.3)
	% Difference	5%	−96%	−97%	−4%	−8%	157%	21%	47%	−100%
Crackers	**SLPF** (n = 3)	446 (444–450)	0.5 (0.4–0.5)	12 (10–17)	77.3 (77.0–78.0)	2.5 (1.5–3.0)	1 (0.9–1.1)	14.7 (14.6–15.0)	6.9 (6.9–7.0)	1.5 (1.3–1.6)
	Regular (n = 6)	426 (360–470)	8.7 (5.7–10.1)	435 (285–505)	72.6 (66.4–82.6)	5.1 (1.6–15.0)	3.5 (2.5–4.2)	11 (1.0–19.6)	3.6 (0.2–8.7)	1.8 (1.2–2.4)
	% Difference	5%	−94%	−97%	6%	−51%	−71%	34%	92%	−17%
Crisps	**SLPF** (n = 4)	437 (369–465)	0.3 (0.1–0.5)	16 (8–22)	77.8 (77.5–78.4)	<0.1 (<0.1–<0.1)	No values	16.2 (16.0–16.6)	2.3 (2.3–2.3)	3.2 (2.6–4.2)
	Regular (n = 8)	519 (499–536)	5.8 (3.6–6.7)	290 (180–335)	52.8 (51.0–57.1)	1.8 (0.2–3.6)	4 (3.1–4.3)	30.8 (28.7–33.0)	3.4 (2.5–8.9)	1.5 (1.0–2.5)
	% Difference	−16%	−95%	−94%	47%	−95%	–	−47%	−32%	113%

Table A1. *Cont.*

Product		Energy (kcal) per 100 g	Protein (g) per 100 g	Phenylalanine (mg) per 100 g	Total Carbohydrate (g) per 100 g	Carbohydrate of Which Is Sugars (g) per 100 g per 100 g	Fibre (g) per 100 g	Total Fat (g) per 100 g	Saturated Fat (g) per 100 g	Salt (g) per 100 g
Crispbread crackers *	**SLPF** ($n = 1$)	388	0.3	6	88	0.1	2.6	3.3	1.8	0.6
	Regular ($n = 2$)	441 (440–442)	9.9 (9.7–10.0)	495 (485–500)	67.5 (67.2–67.7)	1.5 (1.4–1.6)	3.5 (3.1–3.8)	13.9 (13.5–14.3)	6.4 (6.2–6.5)	1.3 (1.3–1.3)
	% Difference	−12%	−97%	−99%	30%	−93%	−26%	−76%	−72%	−54%
French toast crackers	**SLPF** ($n = 1$)	413	<1.0	30	76.3	5.2	7	10	5.6	0.1
	Regular ($n = 2$)	440 (440–440)	8 (7.8–8.2)	400 (390–410)	75 (74.5–75.5)	18.3 (18.0–18.6)	2.9 (2.8–2.9)	11.4 (11.0–11.7)	5.1 (5.0–5.1)	0.4 (0.4–0.4)
	% Difference	−6%	−99%	−93%	2%	−72%	141%	−13%	10%	−75%
Hazelnut spread	**SLPF** ($n = 1$)	347	0.5	19	42	7	0.5	19.6	10.6	0.2
	Regular ($n = 2$)	549 (539–558)	6.3 (6.3–6.3)	315 (315–315)	54.8 (52.0–57.5)	53.2 (50.0–56.3)	3.2 (3.2–3.2)	33.5 (30.9–36.0)	8.9 (7.2–10.6)	0.1 (0.1–0.1)
	% Difference	−37%	−92%	−94%	−23%	−87%	−84%	−41%	19%	100%
Rusks	**SLPF** ($n = 1$)	388	0.3	4	68.8	24.5	0.3	12.1	7.6	0.3
	Regular ($n = 2$)	410 (405–414)	7.8 (7.0–8.5)	390 (350–425)	75.2 (71.2–79.2)	26.5 (24.0–29.0)	4.6 (2.1–7.0)	7.6 (7.2–8)	3.6 (3.1–4.0)	0.2 (0.0–0.4)
	% Difference	−5%	−96%	−99%	−9%	−8%	−93%	59%	111%	50%
Desserts										
Dessert pot	**SLPF** ($n = 2$)	181 (181–181)	0 (0.0–0.0)	<4 (<4–<4)	27 (26.9–27.1)	12.1 (11.7–12.5)	No values	8.2 (8.1–8.2)	0.7 (0.6–0.8)	0.1 (0.1–0.1)
	Regular ($n = 4$)	133 (114–181)	3.7 (3.3–4.3)	185 (165–215)	17.6 (15.4–20.0)	14.8 (13.0–20.0)	0.2	5.3 (3.4–9.3)	3.4 (2.6–5.1)	0.1 (0.1–0.2)
	% Difference	36%	−100%	−98%	53%	−18%	–	55%	−79%	0%
Flavoured desserts (dried powder)	**SLPF** ($n = 4$)	406 (400–409)	0.4 (0.1–0.9)	20 (4–50)	91.5 (89.3–95.3)	48.3 (46.1–51.5)	No values	4.5 (2.2–5.4)	3.6 (2.1–4.3)	0.9 (0.1–3.1)
	Regular ($n = 7$)	424 (352–485)	2.2 (0.4–4.6)	110 (20–230)	77.9 (71.4–87.5)	35.8 (0.1–58.3)	0.1 (0.0–0.2)	11.4 (0.2–20.9)	9.4 (0.1–17.5)	1.2 (0.0–2.4)
	% Difference	−4%	−82%	−82%	17%	35%	–	−61%	−62%	−25%
Jelly (dried)	**SLPF** ($n = 2$)	356 (356–356)	0 (0.0–0.0)	<2 (<2–<2)	88 (88.0–88.0)	87 (87.0–87.0)	No values	0 (0.0–0.0)	0 (0.0–0.0)	0 (0.0–0.0)
	Regular ($n = 4$)	332 (296–375)	2.8 (0.0–5.5)	140 (0–275)	81 (68.5–94.5)	74.6 (57.4–94.5)	No values	0 (0.0–0.0)	0 (0.0–0.0)	0.3 (0.02–0.5)
	% Difference	7%	−100%	−99%	9%	17%	–	0%	0%	−100%
Rice pudding	**SLPF** ($n = 4$)	121 (119–122)	0.1 (0.1–0.2)	6 (5–8)	26.4 (26.3–26.6)	2.3 (1.3–2.8)	No values	1.6 (1.4–1.8)	1.5 (1.3–1.7)	0 (0.0–0.0)
	Regular ($n = 6$)	102 (96–107)	3 (2.7–3.4)	150 (135–170)	17.6 (15.8–18.8)	11.1 (8.9–13.3)	0.3 (0.0–0.5)	2.2 (1.9–2.6)	1.2 (1.0–1.5)	0.2 (0.1–0.2)
	% Difference	19%	−97%	−96%	50%	−79%	–	−27%	25%	−100%

Table A1. *Cont.*

Product		Energy (kcal) per 100 g	Protein (g) per 100 g	Phenylalanine (mg) per 100 g	Total Carbohydrate (g) per 100 g	Carbohydrate of Which Is Sugars (g) per 100 g per 100 g	Fibre (g) per 100 g	Total Fat (g) per 100 g	Saturated Fat (g) per 100 g	Salt (g) per 100 g
Yogurt **	**SLPF** ($n = 1$)	61	0.1	2	8	1	1	2.6	1	0.1
	Regular	75	4.4	220	4.5	4.5	No values	4.1	2.7	0.1
	($n = 2$)	(68–82)	(3.7–5.1)	(185–255)	(3.4–5.6)	(3.4–5.6)		(3.7–4.5)	(2.4–2.9)	(0.1–0.2)
	% Difference	−19%	−98%	−99%	78%	−78%	–	−37%	−63%	0%
Other snacks/meals										
Cheese sauce (prepared)	**SLPF** ($n = 1$)	86	0.8	13	18.4	0.9	No value	1	0.7	0.9
	Regular	76	1.4	70	9.5	1.5	0.5	3.4	2.4	0.9
	($n = 2$)	(65–86)	(1.2–1.6)	(60–80)	(9.2–9.8)	(1.4–1.5)	(<0.5–0.5)	(2.0–4.8)	(1.1–3.6)	(0.7–1.0)
	% Difference	13%	−43%	−81%	94%	−40%	–	−71%	−71%	0%
Potato cakes	**SLPF** ($n = 1$)	165	0.8	46	30.9	0.6	No value	3.7	0.5	0.6
	Regular	190	2.3	115	22.5	0.8	2	9.6	1.2	0.6
	($n = 2$)	(175–205)	(1.9–2.6)	(95–130)	(21.0–23.9)	(<0.5–1.2)	(1.8–2.1)	(8.6–10.5)	(1.1–1.2)	(0.5–0.7)
	% Difference	−13%	−65%	−60%	37%	−25%	–	−61%	−58%	0%
Potato pots/Smash	**SLPF** ($n = 3$)	112	0.5	25	22.8	0.6	No values	1.8	1	0.9
		(111–115)	(0.4–0.5)	(23–27)	(22.7–22.9)	(0.4–0.8)		(1.6–2.1)	(0.9–1.1)	(0.9–0.9)
	Regular	82	1.8	90	13.9	1	1.3	2	1.1	0.3
	($n = 4$)	(75–87)	(1.6–2.1)	(80–105)	(12.0–15.4)	(0.7–1.4)	(1.1–1.3)	(1.5–2.2)	(0.8–1.4)	(0.1–0.5)
	% Difference	37%	−72%	−72%	64%	−40%	–	−10%	−9%	200%
Soup	**SLPF** ($n = 4$)	37	0.2	2	7.5	1.5	No value	0.6	0.3	0.6
		(35–40)	(0.2–0.4)	(1–2)	(6.6–8.5)	(1.4–1.8)		(0.3–0.8)	(0.1–0.4)	(0.4–0.8)
	Regular	37	0.7	35	6.2	2	0.4	1	0.6	0.5
	($n = 8$)	(26–48)	(<0.5–1.1)	(<25–55)	(4.9–8.3)	(<0.5–4.5)	(0.1–0.5)	(0.4–1.9)	(0.1–1.4)	(0.5–0.6)
	% Difference	0%	−71%	−94%	21%	−25%	–	−40%	−50%	20%

* Compared to cream crackers; ** Compared to plain/natural yogurt.

References

1. Van Wegberg, A.M.J.; MacDonald, A.; Ahring, K.; Bélanger-Quintana, A.; Blau, N.; Bosch, A.M.; Burlina, A.; Campistol, J.; Feillet, F.; Giżewska, M.; et al. The complete European guidelines on phenylketonuria: Diagnosis and treatment. *Orphanet J. Rare Dis.* **2017**, *12*, 162. Available online: https://ojrd.biomedcentral.com/articles/10.1186/s13023-017-0685-2 (accessed on 20 May 2020). [CrossRef] [PubMed]
2. Ford, S.; O'Driscoll, M.; MacDonald, A. Living with phenylketonuria: Lessons from the PKU community. *Mol. Genet. Metab. Rep.* **2018**, *17*, 57–63. [CrossRef] [PubMed]
3. Rocha, J.C.; MacDonald, A.; Trefz, F. Is overweight an issue in phenylketonuria? *Mol. Genet. Metab.* **2013**, *110* (Suppl.), S18–S24. [CrossRef] [PubMed]
4. Rocha, J.C.; MacDonald, A. Dietary intervention in the management of phenylketonuria: Current perspectives. *Pediatr. Health Med. Ther.* **2016**, *7*, 155–163. [CrossRef]
5. Cochrane, B.; Schwahn, B.; Galloway, P.; Robinson, P.; Gerasimidis, K. A questionnaire survey on the usage of low protein staple foods by people with phenylketonuria in Scotland. *J. Hum. Nutr. Diet.* **2014**, *27*, 533–541. [CrossRef] [PubMed]
6. European Commission. Commission Directive 1999/21/EC on dietary foods for special medical purposes. *Off. J. Eur. Union* **1999**, *42*, 29–36.
7. Advisory Committee on Borderline Substances. Nutritional Legislation and Other Guidance with Which Products Must Comply. Available online: https://assets.publishing.service.gov.uk/government/uploads/system/uploads/attachment_data/file/792996/Appendix3.pdf (accessed on 16 February 2020).
8. Food Standards Agency. Food Information Regulations 2014: Summary Guidance for Food Business Operators and Enforcement Officers in Scotland, Wales and Northern Ireland. Available online: https://www.food.gov.uk/sites/default/files/media/document/fir$-$guidance2014.pdf (accessed on 16 February 2020).
9. Department of Health. The Medical Food (England) Regulations. Statutory Instrument No. 845. 2000. Available online: http://www.legislation.gov.uk/uksi/2000/845/made (accessed on 16 February 2020).
10. Pena, M.J.; Almeida, M.F.; Van Dam, E.; Ahring, K.; Bélanger-Quintana, A.; Dokoupil, K.; Gokmen-Ozel, H.; Lammardo, A.M.; MacDonald, A.; Robert, M.; et al. Special low protein foods for phenylketonuria: Availability in Europe and an examination of their nutritional profile. *Orphanet J. Rare Dis.* **2015**, *10*, 162. Available online: https://ojrd.biomedcentral.com/articles/10.1186/s13023-015-0378-7 (accessed on 20 May 2020). [CrossRef]
11. Moretti, F.; Pellegrini, N.; Salvatici, E.; Rovelli, V.; Banderali, G.; Radaelli, G.; Scazzina, F.; Giovannini, M.; Verduci, E. Dietary glycemic index, glycemic load and metabolic profile in children with phenylketonuria. *Nutr. Metab. Cardiovasc. Dis.* **2017**, *27*, 176–182. [CrossRef]
12. Dixon, M.; MacDonald, A.; White, F.; Stafford, J. Disorders of amino acid metabolism, organic acidaemias, and urea cycle disorders. In *Clinical Paediatric Dietetics*, 4th ed.; Shaw, V., Ed.; Wiley Blackwell: Chichester, UK, 2015; pp. 381–525.
13. Karam, P.E.; Majdalani, M.N.; Daher, R.T.; Barhoumi, A.; Yazbeck, N. Cardiovascular disease biomarkers in patients with inborn errors of protein metabolism: A pilot study. *J. Hum. Nutr. Diet.* **2015**, *28*, 344–349. [CrossRef]
14. Rocha, J.C.; van Spronsen, F.J.; Almeida, M.F.; Soares, G.; Quelhas, D.; Ramos, E.; Guimarães, J.T.; Borges, N. Dietary treatment in phenylketonuria does not lead to increased risk of obesity or metabolic syndrome. *Mol. Genet. Metab.* **2012**, *107*, 659–663. [CrossRef]
15. Evans, M.; Truby, H.; Boneh, A. The relationship between dietary intake, growth and body composition in phenylketonuria. *Mol. Genet. Metab.* **2017**, *122*, 36–42. [CrossRef] [PubMed]
16. Robertson, L.V.; Mcstravick, N.; Ripley, S.; Weetch, E.; Donald, S.; Adam, S.; Micciche, A.; Boocock, S.; MacDonald, A. Body mass index in adult patients with diet–treated phenylketonuria. *J. Hum. Nutr. Diet.* **2013**, *26* (Suppl. 1), 1–6. [CrossRef] [PubMed]
17. Verduci, E.; Banderali, G.; Moretti, F.; Lassandro, C.; Cefalo, G.; Radaelli, G.; Salvatici, E.; Giovannini, M. Diet in children with phenylketonuria and risk of cardiovascular disease: A narrative overview. *Nutr. Metab. Cardiovasc. Dis.* **2016**, *26*, 171–177. [CrossRef] [PubMed]
18. Dhaka, V.; Gulia, N.; Ahlawat, K.S.; Khatkar, B.S. Trans fats–sources, health risks and alternative approach—A review. *J. Food Sci. Technol.* **2011**, *48*, 534–541. [CrossRef] [PubMed]
19. Klonoff, D.C. Replacements for trans fats—Will there be an oil shortage? *J. Diabetes Sci. Technol.* **2007**, *1*, 415–422. [CrossRef] [PubMed]

20. Hosseinpour–Niazi, S.; Mirmiran, P.; Hosseini–Esfahani, F.; Azizi, F. Is the metabolic syndrome inversely associates with butter, non–hydrogenated– and hydrogenated–vegetable oils consumption: Tehran lipid and glucose study. *Diabetes Res. Clin. Pract.* **2016**, *112*, 20–29. [CrossRef]
21. Mach, F.; Baigent, C.; Catapano, A.L.; Koskinas, K.C.; Casula, M.; Badimon, L.; Chapman, M.J.; De Backer, G.G.; Delgado, V. 2019 ESC/EAS Guidelines for the management of dyslipidaemias: Lipid modification to reduce cardiovascular risk: The Task Force for the management of dyslipidaemias of the European Society of Cardiology (ESC) and European Atherosclerosis Society (EAS). *Eur. Heart J.* **2020**, *41*, 111–188. [CrossRef]
22. Boateng, L.; Ansong, R.; Owusu, W.B.; Steiner-Asiedu, M. Coconut oil and palm oil's role in nutrition, health and national development: A review. *Ghana Med. J.* **2016**, *50*, 189–196.
23. Montoya, C.; Cochard, B.; Flori, A.; Cros, D.; Lopes, R.; Cuellar, T.; Espeout, S.; Syaputra, I.; Villeneuve, P.; Pina, M.; et al. Genetic architecture of palm oil fatty acid composition in cultivated oil palm (Elaeis guineensis Jacq.) compared to its wild relative E. oleifera (H.B.K) Cortés. *PLoS ONE* **2014**, *9*, e95412. Available online: https://journals.plos.org/plosone/article?id=10.1371/journal.pone.0095412 (accessed on 21 May 2020). [CrossRef]
24. Mozaffarian, D. Dairy foods, obesity and metabolic health: The role of the food matrix compared with single nutrients. *Adv. Nutr.* **2019**, *10*, 917S–923S. [CrossRef]
25. Soltanizadeh, N.; Mirmoghtadaie, L. Strategies used in production of phenylalanine–free foods for PKU management. *Compr. Rev. Food Sci. Food Saf.* **2014**, *13*, 287–299. [CrossRef]
26. Schober, T.J. Manufacture of Gluten–Free Specialty Breads and Confectionery Products. In *Gluten—Free Food Science and Technology*; Gallagher, E., Ed.; Blackwell Publishing Ltd.: Oxford, UK, 2009; pp. 130–180.
27. Henry, C.; Lightowler, H.; Kendall, F.; Storey, M. The impact of the addition of toppings/fillings on the glycaemic response to commonly consumed carbohydrate foods. *Eur. J. Clin. Nutr.* **2006**, *60*, 763–769. [CrossRef] [PubMed]
28. Couce, M.L.; Sánchez-Pintos, P.; Vitoria, I.; De Castro, M.J.; Aldámiz-Echevarría, L.; Correcher, P.; Fernández-Marmiesse, A.; Roca, I.; Hermida, A.; Martínez-Olmos, M.; et al. Carbohydrate status in patients with phenylketonuria. *Orphanet J. Rare Dis.* **2018**, *13*, 103. Available online: https://ojrd.biomedcentral.com/articles/10.1186/s13023-018-0847-x (accessed on 21 May 2020). [CrossRef] [PubMed]
29. Ormazabal, V.; Nair, S.; Elfeky, O.; Aguayo, C.; Salomon, C.; Zuñiga, F. Association between insulin resistance and the development of cardiovascular disease. *Cardiovasc. Diabetol.* **2018**, *17*, 122. Available online: https://cardiab.biomedcentral.com/articles/10.1186/s12933-018-0762-4#citeas (accessed on 27 May 2020). [CrossRef] [PubMed]
30. El Khoury, D.; Balfour-Ducharme, S.; Joye, I.J. A review on the gluten–free diet: Technological and nutritional challenges. *Nutrients* **2018**, *10*, 1410. Available online: https://www.mdpi.com/2072-6643/10/10/1410 (accessed on 21 May 2020). [CrossRef]
31. Viebke, C.; Al-Assaf, S.; Phillips, G.O. Food hydrocolloids and health claims. *Bioact. Carbohydr. Diet. Fibre* **2014**, *4*, 101–114. [CrossRef]
32. Trefz, K.F.; Muntau, A.C.; Kohlscheen, K.M.; Altevers, J.; Jacob, C.; Braun, S.; Greiner, W.; Jha, A.; Jain, M.; Alvarez, I.; et al. Clinical burden of illness in patients with phenylketonuria (PKU) and associated comorbidities—A retrospective study of German health insurance claims data. *Orphanet J. Rare Dis.* **2019**, *14*, 181. Available online: https://ojrd.biomedcentral.com/articles/10.1186/s13023$-$019$-$1153$-$y (accessed on 21 May 2020). [CrossRef]
33. Burton, B.K.; Bradford Jones, K.; Cederbaum, S.; Rohr, F.; Waisbren, S.; Irwin, D.E.; Kim, G.; Lilienstein, J.; Alvarez, I.; Jurecki, E.; et al. Prevalence of comorbid conditions among adult patients diagnosed with phenylketonuria. *Mol. Genet. Metab.* **2018**, *125*, 228–234. [CrossRef]
34. Azabdaftari, A.; van der Giet, M.; Schuchardt, M.; Hennermann, J.B.; Plöckinger, U.; Querfeld, U. The cardiovascular phenotype of adult patients with phenylketonuria. *Orphanet J. Rare Dis.* **2019**, *14*, 213. Available online: https://ojrd.biomedcentral.com/articles/10.1186/s13023-019-1188-0 (accessed on 21 May 2020). [CrossRef]
35. Hermida-Ameijeiras, A.; Crujeiras, V.; Roca, I.; Calvo, C.; Leis, R.; Couce, M.L. Arterial stiffness assessment in patients with phenylketonuria. *Medicine* **2017**, *96*, e9322. Available online: https://journals.lww.com/md-journal/Fulltext/2017/12220/Arterial_stiffness_assessment_in_patients_with.94.aspx (accessed on 21 May 2020).

36. Rocha, J.C.; van Rijn, M.; van Dam, E.; Ahring, K.; Bélanger-Quintana, A.; Dokoupil, K.; Gokmen-Ozel, H.; Lammardo, A.M.; Robert, M.; Heidenborg, C.; et al. Weight Management in Phenylketonuria: What Should Be Monitored? *Ann. Nutr. Metab.* **2016**, *68*, 60–65. [CrossRef] [PubMed]
37. Albersen, M.; Bonthuis, M.; de Roos, N.M.; van den Hurk, D.A.M.; Carbasius Weber, E.; Hendriks, M.M.W.B.; de Sain-van der Velden, M.G.M.; de Koning, T.J.; Visser, G. Whole body composition analysis by the BodPod air–displacement plethysmography method in children with phenylketonuria shows a higher body fat percentage. *J. Inherit. Metab. Dis.* **2010**, *33* (Suppl. 3), S283–S288. [CrossRef] [PubMed]
38. Belanger-Quintana, A.; Martínez-Pardo, M. Physical development in patients with phenylketonuria on dietary treatment: A retrospective study. *Mol. Genet. Metab.* **2011**, *104*, 480–484. [CrossRef] [PubMed]
39. Burrage, L.C.; McConnell, J.; Haesler, R.; O'Riordan, M.A.; Sutton, V.R.; Kerr, D.S.; McCandless, S.E. High prevalence of overweight and obesity in females with phenylketonuria. *Mol. Genet. Metab.* **2012**, *107*, 43–48. [CrossRef] [PubMed]
40. Doulgeraki, A.; Skarpalezou, A.; Theodosiadou, A.; Monopolis, I.; Schulpis, K. Body composition profile of young patients with phenylketonuria and mild hyperphenylalaninemia. *Int. J. Endocrinol. Metab.* **2014**, *12*, e16061. Available online: https://sites.kowsarpub.com/ijem/articles/17675.html (accessed on 21 May 2020). [CrossRef]
41. Evans, S.; Ford, S.; Adam, S.; Adams, S.; Ash, J.; Ashmore, C.; Caine, G.; Carruthers, R.; Cawtherley, S.; Chahal, S.; et al. Development of national consensus statements on food labelling interpretation and protein allocation in a low phenylalanine diet for PKU. *Orphanet J. Rare Dis.* **2019**, *14*, 2. Available online: https://ojrd.biomedcentral.com/articles/10.1186/s13023$-$018$-$0950$-$z (accessed on 21 May 2020). [CrossRef]
42. MacDonald, A.; Lilburn, M.; Davies, P.; Evans, S.; Daly, A.; Hall, S.K.; Hendriksz, C.; Chakrapani, A.; Lee, P. 'Ready to drink' protein substitute is easier is for people with phenylketonuria. *J. Inherit. Metab. Dis.* **2006**, *29*, 526–531. [CrossRef]
43. MacDonald, A.; Lilburn, M.; Cochrane, B.; Davies, P.; Daly, A.; Asplin, D.; Hall, S.K.; Cousins, A.; Chakrapani, A.; Robinson, P.; et al. A new, low–volume protein substitute for teenagers and adults with phenylketonuria. *J. Inherit. Metab. Dis.* **2004**, *27*, 127–135. [CrossRef]
44. Evans, S.; Adam, S.; Adams, S.; Allen, H.; Ashmore, A.; Bailey, S.; Banks, J.; Churchill, H.; Cochrane, B.; Cook, J.; et al. Uniformity of food protein interpretation amongst dietitians for patients with phenylketonuria (PKU): 2020 UK National Consensus Statements. *Nutrients*, under Review.
45. Piepoli, M.F.; Abreu, A.; Albus, C.; Ambrosetti, M.; Brotons, C.; Catapano, A.L.; Corra, U.; Cosyns, B.; Deaton, C.; Graham, I.; et al. Update on cardiovascular prevention in clinical practice: A position paper of the European Association of Preventive Cardiology of the European Society of Cardiology. *Eur. J. Prev. Cardiol.* **2020**, *27*, 181–205. [CrossRef]
46. Zmora, N.; Suez, J.; Elinav, E. You are what you eat: Diet, health and the gut microbiota. *Nat. Rev.* **2019**, *16*, 35–56. [CrossRef]

© 2020 by the authors. Licensee MDPI, Basel, Switzerland. This article is an open access article distributed under the terms and conditions of the Creative Commons Attribution (CC BY) license (http://creativecommons.org/licenses/by/4.0/).

nutrients

MDPI

Article

Uniformity of Food Protein Interpretation Amongst Dietitians for Patients with Phenylketonuria (PKU): 2020 UK National Consensus Statements

Sharon Evans [1,*], Sarah Adam [2], Sandra Adams [3], Heather Allen [4], Catherine Ashmore [1], Sarah Bailey [5], Janette Banks [6], Harriet Churchill [7], Barbara Cochrane [2], Jennifer Cook [8], Clare Dale [9], Anne Daly [1], Marjorie Dixon [10], Carolyn Dunlop [11], Charlotte Ellerton [7], Anita Emm [12], Sarah Firman [8], Suzanne Ford [13], Moira French [14], Joanna Gribben [15], Anne Grimsley [16], Ide Herlihy [10], Melanie Hill [17], Shirley Judd [18], Karen Lang [19], Jo Males [20], Joy McDonald [16], Nicola McStravick [16], Chloe Millington [10], Camille Newby [21], Catharine Noble [17], Rachel Pereira [22], Alex Pinto [1], Louise Robertson [9], Abigail Robotham [21], Kathleen Ross [23], Kath Singleton [24], Rachel Skeath [10], Allyson Terry [25], Karen Van Wyk [26], Fiona White [26], Lucy White [4], Jo Wildgoose [27], Alison Woodall [28] and Anita MacDonald [1]

1. Birmingham Women's & Children's NHS Foundation Trust, Birmingham B4 6NH, UK; catherine.ashmore@nhs.net (C.A.); a.daly3@nhs.net (A.D.); alex.pinto@nhs.net (A.P.); anita.macdonald@nhs.net (A.M.)
2. Royal Hospital for Children, Glasgow G51 4TF, UK; sarah.adam@ggc.scot.nhs.uk (S.A.); Barbara.Cochrane@ggc.scot.nhs.uk (B.C.)
3. Royal Victoria Infirmary, Newcastle upon Tyne NE1 4LP, UK; sandraadams2@nhs.net
4. Sheffield Children's NHS Foundation Trust, Sheffield S10 2TH, UK; heather.allen10@nhs.net (H.A.); lucy.white19@nhs.net (L.W.)
5. Cardiff and Vale UHB, Cardiff CF14 4XW, UK; Sarah.Bailey3@wales.nhs.uk
6. Wrightington, Wigan and Leigh NHS Foundation Trust, Wigan WN1 2NN, UK; janette.banks@nhs.net
7. University College London Hospitals NHS Foundation Trust, London WC1N 3BG, UK; harriet.churchill@nhs.net (H.C.); c.ellerton@nhs.net (C.E.)
8. Guy's and St Thomas' NHS Foundation Trust, London SE1 7EU, UK; Jennifer.Cook@gstt.nhs.uk (J.C.); Sarah.Firman@gstt.nhs.uk (S.F.)
9. University Hospitals Birmingham NHS Foundation Trust, Birmingham B15 2TH, UK; Clare.Dale@uhb.nhs.uk (C.D.); Louise.robertson@uhb.nhs.uk (L.R.)
10. Great Ormond Street Hospital for Children NHS Foundation Trust, London WC1N 3JH, UK; marjorie.dixon@gosh.nhs.uk (M.D.); Ide.Herlihy@gosh.nhs.uk (I.H.); Chloe.Millington@gosh.nhs.uk (C.M.); Rachel.skeath@gosh.nhs.uk (R.S.)
11. Royal Hospital for Sick Children, Edinburgh EH9 1LF, UK; carolyn.dunlop@nhslothian.scot.nhs.uk
12. University Hospital Southampton NHS Foundation Trust, Southampton SO16 6YD, UK; Anita.Emm@uhs.nhs.uk
13. The National Society for Phenylketonuria, Purley CR8 9DD, UK; Suzanne.ford@nspku.org
14. University Hospitals of Leicester NHS Trust, Leicester LE1 5WW, UK; moira.french@uhl-tr.nhs.uk
15. Evelina London Children's Hospital, London SE1 7EH, UK; Joanna.gribben@gstt.nhs.uk
16. Belfast Health & Social Care Trust, Belfast BT9 7AB, UK; Anne.Grimsley@belfasttrust.hscni.net (A.G.); joy.mcdonald@belfasttrust.hscni.net (J.M.); nicola.mcstravick@belfasttrust.hscni.net (N.M.)
17. Sheffield Teaching Hospitals NHS Foundation Trust, Sheffield S5 7AU, UK; melanie.hill13@nhs.net (M.H.); catharine.noble@nhs.net (C.N.)
18. Royal Liverpool and Broadgreen University Hospitals NHS Trust, Liverpool L7 8XP, UK; shirley.judd@rlbuht.nhs.uk
19. Ninewells Hospital, Dundee DD1 9SY, UK; karen.lang@nhs.net
20. Aneurin Bevan University Health Board Wales, Newport NP20 2UB, UK; Jo.males@wales.nhs.uk
21. Bristol Royal Hospital for Children, Bristol BS2 8BJ, UK; Camille.Newby@uhbw.nhs.uk (C.N.); Abigail.Robotham@uhbw.nhs.uk (A.R.)
22. Norfolk and Norwich University Hospital, Norwich NR4 7UY, UK; rachel.pereira@nnuh.nhs.uk
23. Royal Aberdeen Children's Hospital, Aberdeen AB25 2ZG, UK; Kathleenross@nhs.net
24. University Hospital of Wales, Cardiff CF14 4XW, UK; kath.singleton@wales.nhs.uk

[25] Alder Hey Children's NHS Foundation Trust, Liverpool L12 2AP, UK; allyson.terry@alderhey.nhs.uk
[26] Royal Manchester Children's Hospital, Manchester M13 9WL, UK; Karen.VanWyk@mft.nhs.uk (K.V.W.); fiona.white@mft.nhs.uk (F.W.)
[27] Bradford Teaching Hospitals NHS Foundation Trust, Bradford BD5 0NA, UK; joannewildgoose@hotmail.com
[28] Salford Royal NHS Foundation Trust, Salford M6 8HD, UK; Alison.Woodall@srft.nhs.uk
* Correspondence: evanss21@me.com; Tel.: +44-0121-333-8021

Received: 17 June 2020; Accepted: 22 July 2020; Published: 24 July 2020

Abstract: In phenylketonuria (PKU), variable dietary advice provided by health professionals and social media leads to uncertainty for patients/caregivers reliant on accurate, evidence based dietary information. Over four years, 112 consensus statements concerning the allocation of foods in a low phenylalanine diet for PKU were developed by the British Inherited Metabolic Disease Dietitians Group (BIMDG-DG) from 34 PKU treatment centres, utilising 10 rounds of Delphi consultation to gain a majority (≥75%) decision. A mean of 29 UK dietitians (range: 18–40) and 18 treatment centres (range: 13–23) contributed in each round. Statements encompassed all foods/food groups divided into four categories based on defined protein/phenylalanine content: (1) foods high in protein/phenylalanine (best avoided); (2) foods allowed without restriction including fruit/vegetables containing phenylalanine ≤75 mg/100 g and most foods containing protein ≤0.5 g/100 g; (3) foods that should be calculated/weighed as an exchange food if they contain protein exchange ingredients (categorized into foods with a protein content of: >0.1 g/100 g (milk/plant milks only), >0.5 g/100 g (bread/pasta/cereal/flours), >1 g/100 g (cook-in/table-top sauces/dressings), >1.5 g/100 g (soya sauces)); and (4) fruit/vegetables containing phenylalanine >75 mg/100 g allocated as part of the protein/phenylalanine exchange system. These statements have been endorsed and translated into practical dietary management advice by the medical advisory dietitians for the National Society for PKU (NSPKU).

Keywords: phenylketonuria (PKU); consensus; Delphi method; food labelling; phenylalanine; Phe; protein; exchanges

---

1. Introduction

Phenylketonuria (PKU) is a rare, inherited metabolic disorder (IMD) caused by phenylalanine hydroxylase deficiency, leading to an abnormal accumulation of blood phenylalanine. Without treatment, it causes severe and irreversible intellectual disability. However, national newborn screening programmes detect PKU, which enables treatment to commence in early infancy with outcomes associated with a broad range of normal general ability. In the UK, the only available treatment is a rigorous, life-long dietary restriction of natural protein (i.e., meat, eggs, fish, cheese, nuts, bread, flour, pasta) in order to control blood phenylalanine levels [1,2] within the target range, as pharmacological treatments are not reimbursed by the National Health Service. Individuals with PKU tolerate only a limited amount of natural protein, with the amount individually determined. It is estimated that 80% of those with classical PKU are prescribed less than 10 g/day of protein [3].

Whilst the goal of dietetic management is essentially the same, the method for allocating phenylalanine/protein intake varies between and sometimes within countries, with insufficient evidence to recommend any one method [4–9]. Broadly there are two different methods, each with its merits and drawbacks: (1) patients are allocated a daily amount of phenylalanine/protein from all food, which is calculated to provide the prescribed phenylalanine intake; and (2) a phenylalanine/protein exchange system whereby amounts of food are calculated for a defined amount of phenylalanine (ranging from 10–50 mg of phenylalanine for each exchange) or protein (e.g., 1 g protein = 1 exchange). One food exchange can be replaced with an alternative exchange of an equivalent phenylalanine/protein amount.

In the UK, natural protein is apportioned using a protein exchange system whereby one exchange is equivalent to the amount of food that is calculated/measured to provide 1 g protein or 50 mg of phenylalanine. Individual tolerance based on blood phenylalanine levels determines the number of exchanges allocated, but is typically between 3–10 g exchanges/day (or 150–500 mg/day phenylalanine). The diet is then supplemented with a phenylalanine-free/low-phenylalanine protein substitute, foods naturally low in phenylalanine/protein, and special low protein foods (SLPF) (e.g., bread, flour, pasta) approved by the Advisory Board for Borderline Substances (ACBS), and prescribed by the General Practitioner (GP). Overall dietary management is complex, and a high degree of patient/caregiver knowledge and application is required to effectively implement dietary care. Successful dietary treatment is hampered by inconsistent dietary advice associated with unclear and historical recommendations, lack of comprehensive food phenylalanine analysis, similar plant species containing a variable phenylalanine content per 100 g of food, and an overwhelming range of manufactured foods with unclear declarations of protein content. In the UK, national practical guidance about the implementation of dietary treatment in PKU has not been reviewed since 1993 [10].

Patients with PKU are cared for by multidisciplinary teams of health professionals working across many care settings within the UK. Almost every professional involved has an opinion about the practical implementation of dietary management. This causes wide variability in dietary advice given by practitioners, leading to uncertainty for patients and caregivers and inexperienced health professionals seeking advice from peers. In addition, patients/caregivers readily turn to social media for information that may be based on the erroneous interpretations of others. Multiple sources of discordant information are also available through apps and specialist manufacturers of dietary products. International dietary practices also vary, adding another tier of misunderstanding and ambiguity [4]. It is important that patients with PKU and their caregivers receive reliable and uniform dietary information from their health care professionals.

In 2016, the British Inherited Metabolic Disease Dietitians Group (BIMDG-DG) identified several controversial issues concerning the calculation of protein in a low phenylalanine diet [11]. This included EU legislation that foods containing a protein content of ≤0.5 g/100 g do not need to declare their specific protein content on food labels [12]. This has caused considerable confusion about which foods could be permitted within a phenylalanine restricted diet. There was also discrepancy about the allocation of fruits and vegetables according to their phenylalanine/protein content. This led to the development of 23 BIMDG-DG national consensus statements about the interpretation of protein food labelling and allocation of foods in a low phenylalanine diet [11].

Since the publication of the first report [11], the BIMDG-DG have continued to develop new statements about food allocation in a low phenylalanine diet utilising Delphi methodology. This time, all food categories have been systematically examined, and a comprehensive set of consensus guidelines for interpreting their protein content and allocating these foods within a low phenylalanine diet has been developed.

2. Materials and Methods

Over a 4-year period (November 2015 to September 2019), 112 consensus statements from UK dietitians working in PKU were developed regarding the inclusion, exclusion, and allocation of foods within a low phenylalanine diet *specifically for use by dietitians*. These were aimed at restrictions of protein of ≤10 g/day or phenylalanine ≤500 mg/day. These statements considered most commercially available foods as well as special low protein foods. The process included 10 rounds of Delphi consultation to gain a majority decision in a structured and systematic way [13]. It involved an independent facilitator collecting the opinion of clinical dietitians working with PKU. The facilitator issued proposed statements about food allocation, each time presenting any available research evidence along with the protein/phenylalanine content of foods. Questionnaire responses in each round were gathered and results discussed by telephone conference. Metabolic dietitians from the BIMDG voted on each statement. The process was repeated, with modification of the statements, until there was at least 75% agreement (an arbitrary figure chosen to represent a majority decision). Any unresolved statements at round one of the Delphi process were carried over to the next round of statements for further discussion. A more detailed description of this process has been reported previously [11].

All foods/food groups were systematically divided into nineteen subgroups with a mean of six statements per subgroup (range: 1–14): milk and milk replacements (n = 2); dairy products and alternatives (n = 12); breads and cereals (n = 4); spreads and dips (n = 5); sauces and soups (n = 13); pasta and rice (n = 2); potato and potato products (n = 6); fruit and vegetables (n = 7); meat and alternatives (n = 7); drinks (n = 8); sweet snacks (n = 13); savoury snacks (n = 10); sugars, sweeteners and syrups (n = 9); herbs and spices (n = 1); low protein special foods (n = 2); flours and starch (n = 5); baking ingredients (n = 4); gluten-free products (n = 1); and gelatine containing products (n = 1).

No ethical approval was required for this project as it is not considered research as defined by the UK Policy Framework for Health and Social Care Research [14]. Descriptive analysis was used to present the results.

3. Results

In the UK, there are nine paediatric and nine adult PKU specialist centres for PKU, with around 18 district general hospitals who share PKU care with the specialist centres.

In total, 93 BIMDG dietitians (57% (n = 53) paediatric; 32% (n = 30) adult; 11% (n = 10) working with both adults and children) from all represented UK PKU treatment centres (n = 34) responded to the consensus statements during the four year study period. In each round of the Delphi process, there were approximately 70 active BIMDG dietetic members potentially able to participate. The numbers of dietetic contributors varied in each round due to the movement of dietitians (maternity leave, retirement, illness, secondment, or specialty change). A median of 29 dietitians (range: 18–40) and 18 centres (range: 13–23) contributed to each round. All specialist UK PKU IMD centres were represented in each round except for one adult centre only able to participate in two of 10 rounds. All BIMDG dietitians received copies of the results for each round and the minutes of meetings. Fifty-eight percent (n = 65/112) of statements received 100% agreement, a further 26% (n = 29/112) received ≥90% agreement, and 15% (17/112) were between 79–89%.

Of the 112 statements, most had majority agreement within one round of the Delphi process. Only nine statements received less than 75% agreement on the first round of discussion (fruit/vegetables, low protein milks, coconut desserts, soya sauce, special low protein foods, vegetable crisps, eggs, cheese, and seeds). These statements were then modified and reconsidered in the next round. New statements for other foods continued to be added at each round. Consensus was agreed in seven of nine statements on the second round of discussion. Two statements (fruit/vegetables and low protein milks) required discussion, modification, and voting over three consecutive rounds to achieve consensus. No statements required deletion, only modification, as all foods/food groups were required to have a statement.

Each food subgroup was systematically discussed via the Delphi process and allocated into one of four categories (with further subcategories), based on their protein or phenylalanine content.

Category 1: Foods high in protein or phenylalanine that are best avoided (Table 1). This included two subcategories:

(a) Foods high in protein (generally containing protein >15 g/100 g).

(b) Foods containing aspartame and therefore phenylalanine.

Category 2: Foods allocated without restriction or measurement (defined as exchange-free foods) (Table 2). Generally, any foods with an upper protein content cut-off point ≤0.5 g/100 g or containing exchange-free ingredients were considered exchange-free. Exceptions included spices (with a higher protein content) that are used for flavouring purposes only and consumed in small quantities. Overall, this category consisted of four subcategories:

Table 1. Category 1: Foods high in protein or phenylalanine.

1. Foods High in Protein or Phenylalanine that Are Best Avoided	
(a) Foods high in protein (approximately >15 g/100 g)	• Meat, fish, eggs, nuts, cheeses, seeds, soya products, Quorn, goji berries, peanut butter, tofu, spreadable yeast extracts. • Exceptions: soft cheeses, soya cheese, baked products containing seeds or eggs as an ingredient, baked goods with eggs as an ingredient - these fall into category 3b and are used as part of the exchange system. • Eggs contain protein <15 g/100 g. Although they are used as part of the protein exchange system in baked goods, one hen's egg is high in protein and best avoided.
(b) Foods containing aspartame	• Aspartame containing food and drinks (e.g., fizzy drinks, fruit juice, fruit tea, milkshake powders/syrup, smoothies, squash, chewing/bubble gum, desserts, jelly, sweets, tabletop sweeteners).

Table 2. Category 2: Foods allocated without restriction.

2. Foods Allocated without Restriction or Measurement (Defined as Exchange-Free Foods)	
(a) Fruits and Vegetables containing phenylalanine ≤75 mg/100 g (except potatoes and vegetable crisps)	• Apples, apricots, avocado, bananas, banana chips, bilberries, blackberries, blueberries, candied angelica, candied peel, cherries, clementines, cranberries, currants, custard apples, damsons, dates, dragon fruit, fruit crisps (e.g., apple, pineapple), fruit pie filling, fruit mincemeat, fruit salad, glacé cherries, gooseberries, grapes, grapefruit, greengages, guavas, jackfruit, kiwi fruit, kumquats, lemons, limes, loganberries, lychees, mandarins, mango, medlars, melon, nectarines, olives, oranges, papaya (paw paw), peaches, pears, physalis, pineapple, plums, pomegranate, prickly pear, prunes, quince, raisins, raspberries, rhubarb, satsumas, Sharon fruit, star fruit, strawberries, sultanas, tamarillo, tangerines, watermelon. • Artichoke, aubergine, baby corn, beetroot, cabbage, capers, caperberries, carrots, cassava, celeriac, celery, chayote, chicory, courgette, cucumber, dudhi, eddoes, endive, fennel, garlic, gherkin, ginger, green beans (dwarf, French, runner), karela, kohlrabi, leeks, lettuce, marrow, mooli, mushrooms, okra, onion, pak choi, parsnips, peppers, pickled vegetables (e.g., onion, gherkins, red cabbage), plantain, pumpkin, radish, salad cress, samphire, squash (butternut, acorn, spaghetti), swede, sweet potato, tomato, turnip, watercress, water chestnuts. • Fruit and vegetable-based foods containing exchange-free fruits/vegetables and other exchange-free ingredients (e.g., frozen or canned fruit/vegetables, tomato puree/passata).
(b) Manufactured foods containing protein ≤0.5 g/100 g or exchange-free ingredients	• Sugar (brown, cane, caster, demerara, fruit, glucose, granulated, icing, molasses, muscovado, white). • Jam, honey, marmalade, syrup (agave, fruit, golden, maple, treacle). • Fats (oils, oil sprays, ghee, lard). • Baking ingredients (arrowroot, baking powder, bicarbonate of soda, cassava/tapioca flour, cornflour/maize starch, cream of tartar, sago). • Aspartame-free drinks (squash, fruit drinks, soft drinks, black/herbal tea and coffee). • Artificial sweeteners (except aspartame). • Condiments (mint jelly, mint sauce, salt, vinegar). • Fibres/gums (e.g., psyllium fibre/husks, xanthan gum). • Aspartame-free milkshake powders/syrups and custard powder containing exchange-free ingredients. • Plants & cereals (konnyaku, sago, tapioca, cassava crisps).
(c) Manufactured food containing protein >0.5 g/100 g but used in small amounts	• Fats (butter, margarine). • Herbs, spices, condiments (e.g., pepper). • Food colouring and flavourings/essences.
(d) Special low protein foods containing exchange-free ingredients or a phenylalanine content <25 mg/100 g	• Low protein: bread (sliced, rolls, baguettes), biscuits, breakfast cereals, cereal bars, cakes, chocolate, chocolate spread, cheese sauce, crackers, cake mix, dessert/custard mixes, egg replacer, fish substitute mixes, flour, pizza bases, pasta, rice, sausage/burger mixes. • Low protein milk replacement Prozero (Vitaflo). • Includes most UK ACBS prescribed low protein products.

(a) Fruits and vegetables containing phenylalanine ≤75 mg/100 g.

(b) Manufactured foods containing protein ≤0.5 g/100 g or exchange-free ingredients.

(c) Manufactured food containing protein >0.5 g/100 g, but used in small amounts so provide minimal contribution to protein intake.

(d) SLPF containing exchange-free ingredients or a phenylalanine content <25 mg/100 g that have been ACBS approved and are available on prescription for low protein diets.

Category 3: Manufactured foods/SLPF allocated as part of the protein/phenylalanine exchange system according to their protein/phenylalanine content per 100 g (Table 3). One protein/phenylalanine exchange is the amount of food that is calculated/measured to provide either 1 g of protein or 50 mg of phenylalanine from its food analysis. This category included any SLPF and manufactured foods with a phenylalanine/protein content above the upper cut-off point (>0.5 g/100 g) and containing protein exchange ingredients (e.g., milk, wheat, flour, rice, egg, and soya). This section was divided into four subcategories with different protein/phenylalanine upper cut-off points based on the food portion size that would typically be consumed or if they contained a high proportion of exchange-free fruit or vegetables (e.g., commercial cooking sauces):

Table 3. Category 3: Manufactured foods allocated as part of an exchange system based on protein or phenylalanine content.

3. Manufactured Foods Allocated as Part of the Protein Exchange System According to Their Protein/Phenylalanine Content per 100 g	
(a) Liquid plant and animal milks with protein >0.1 g/100 g or 0.1 g/100 mL or specialist low protein milks with phenylalanine >5 mg/100 mL	• Animal milks (e.g., cow, goat, sheep), full fat, semi-skimmed, skimmed, condensed. • Plant milks (e.g., coconut, oat, almond, soya). Includes coffee with these added (e.g., lattes, cappuccino, frappuccino, macchiato, coffee pods/sachets). • Low protein milk replacements–Dalia 6.4 mg Phe/100 mL (Taranis), Lattis 12 mg Phe/100 mL (Mevalia), Loprofin 10 mg Phe/100 mL and SnoPro 8.7 mg Phe/100 mL (Nutricia).
(b) Foods containing protein >0.5 g/100 g or specialist low protein foods containing phenylalanine >25 mg/100 g and containing exchange ingredients	• Bread and bread products, biscuits and cakes made from regular flour, butter, cheese spread, cream, cream cheese, chocolate spreads, cocoa powder, breakfast cereals, cereal grains, cereal bars, cereal products (pancakes, waffles, stuffing, Yorkshire pudding), chocolate, coconut based desserts and products, corn/rice based snacks, crackers, cream, dairy desserts (custard, instant, fromage frais, mousse), dips (sweet & savoury), drinking chocolate, flour and flour products, free-from and vegan/plant cheeses, fondant icing, fruit bars, fudge, gelatine containing foods, gluten-free foods, gravy, herb/spice rubs and coatings, hummus, ice cream (dairy/non-dairy), ice lollies, icing/frosting, jelly, legumes/pulses (baked beans, lentils), lemon curd, liquorice, marshmallows/mallows, marzipan, milk based sauces, milkshake powders/syrups, mustard, nut spread, pasta/noodles, pesto, plant/vegetable spreads, popcorn, potato crisps, pretzels, pot noodles, puddings/desserts, rice, rice/oat cakes, soft cheese, sorbets, soups, stock cubes, sweets, tapenade, toffee, tofu, vegan meat/fish or egg alternatives, vegetable crisps, vegetable soups, yoghurt (dairy/non-dairy). • Note: yogurts, dairy desserts and coconut-based puddings with a protein content ≤0.5 g/100 g should be limited to 1 per day. • Low protein ACBS prescription products: Promin potato pots, Promin potato cakes, Taranis fish substitute.
(c) Commercial sauces and tabletop sauces containing protein >1 g/100 g and containing exchange ingredients	• Cook-in, pour-over or liquid sauces (curry, sweet & sour, tomato, vegetable), oil-based dressings (mayonnaise, salad cream, vinaigrette), table top sauces (brown, chilli, chutney, horseradish, mint, pickles, tartare, tomato ketchup). • Cake decorations/sprinkles.
(d) Soya sauces containing protein >1.5 g/100 g	• Most soya sauces have protein >1.5 g/100 g.

(a) Liquid plant milks and animal milks that contain protein >0.1 g/100 mL or specialist low protein milks with a phenylalanine content >5 mg/100 mL that have been approved by the UK ACBS and available on prescription for low protein diets. The upper protein/phenylalanine cut-off is set at a low amount due to potential high daily volumes that may be consumed.

(b) Manufactured foods containing protein >0.5 g/100 g or SLPF containing phenylalanine >25 mg/100 g and containing exchange ingredients. This group contained most of the manufactured foods. It also included some SLPF approved by the UK ACBS and available on prescription for low protein diets.

(c) Commercial sauces and tabletop sauces containing protein >1 g/100 g and exchange ingredients. This subcategory mainly consisted of commercial sauces containing vegetables, which have a lower phenylalanine content per 1 g of protein than cereals or animal foods [15]. This group also included cake decorations, as the amounts consumed are small.

(d) Soya sauces containing protein >1.5 g/100 g. This subcategory allowed soya sauce with a higher amount of protein because the amount consumed is small and only a few brands contained protein less than this amount.

Category 4: Fruit/vegetables containing phenylalanine >75 mg/100 g allocated as part of the protein/phenylalanine exchange system (Table 4). This included vegetable crisps prepared from exchange-free vegetables that had a higher phenylalanine content and were discussed in Evans et al. [11]. Potatoes have a lower phenylalanine content than 75 mg/100 g but are calculated/measured as exchange foods due to the amount consumed daily in the UK diet.

Table 4. Category 4: Exchange fruit and vegetables (phenylalanine >75 mg/100 g).

4. Exchange Fruit/Vegetables Containing Phenylalanine >75 mg/100 g Allocated as Part of the Protein Exchange System According to Their Phenylalanine/Protein Content per 100 g	
Fruit & Vegetables with phenylalanine content 75–99 mg/100 g	• Figs • Asparagus, bamboo shoots, beansprouts, broccoli, brussels sprouts, cauliflower, mange tout, sugar snap peas, whole hearts of palm. • A standard portion size of 60 g is used for 1 phenylalanine exchange.
Fruit & Vegetables with phenylalanine content >100 mg/100 g	• Passionfruit • Broad beans, chestnuts, choi sum, corn on the cob, kale, mixed vegetables, peas and petit pois, romanesco, rocket, spinach, spring greens, sweetcorn kernels, sweet potato fries with coating, vine leaves, yams. • Phenylalanine content is used to determine amount for 1 phenylalanine exchange.
Potatoes	• All potatoes and potato products.
Vegetable crisps	• All vegetable crisps (except cassava).

All statements were accepted and endorsed by the NSPKU. The statements were then translated into easier guidance for patients and carers [16]. A detailed list of all statements is provided in Supplementary Materials.

4. Discussion

This paper reports the results of four years of in-depth national discussions amongst experienced UK metabolic dietitians using Delphi methodology to gain consensus on the suitability and allocation of foods in a low phenylalanine diet, for patients tolerating a natural protein intake of ≤10 g/day. Having uniform national recommendations across all UK centres treating PKU should enable health professionals and support groups to provide consistent information to patients with PKU. Securing consensus amongst health professionals was challenging, but essential, as there were many differing opinions leading to disparate patient information and unfounded dietary practices. The Delphi methodology was systematic, impartial, and consistent, involving representation from all major UK PKU centres. All terms of reference were agreed in advance and an impartial facilitator ensured the process was conducted transparently and without bias [13]. Eighty-four percent (94/112) of statements received at least 90% agreement.

Historically in the UK, exceptionally low protein foods such as sugar, jam, honey, and vegetable oils have been permitted as exchange-free and there is no evidence to suggest this advice should change.

There was much discussion about the allocation of protein cut off points for different food groups and categorisation of foods/drinks within each group considering the impact on blood phenylalanine control if foods were not calculated/measured as part of the phenylalanine exchange system. It was clear that not all food subgroups could be considered in the same way, dependent on the weight of food that would be consumed and the role of each food within the diet. Most manufactured foods were defined as exchange-free only if they had a protein content ≤0.5 g/100 g of food. However, even within this subcategory, some foods required additional rules to prevent over consumption when protein content was close to the upper cut off point, and the amount consumed would exceed 100 g in one portion. This applied to dairy-free yoghurts and desserts based on plant milks that contained a protein content of around 0.5 g/100 g.

Since the 1960s, UK health professionals have given most fruits and vegetables (except potatoes) with a phenylalanine content ≤75 mg/100 g as exchange-free in a low phenylalanine diet. This guidance was reconsidered when developing the consensus statements, as evidence from a series of studies indicated that this maximum cut off did not adversely impact blood phenylalanine control in PKU [17–20]. This was also consistent with the 2017 PKU European Guidelines [21] and in turn, this influenced other consensus statements. For example, many cook-in/pour over sauces are primarily made from exchange-free vegetables and other exchange-free ingredients such as starches and seasonings only. It was therefore decided to calculate/measure vegetable sauces as part of the phenylalanine exchange system only if the sauce contained exchange ingredients (such as cream, flour) and a protein content >1 g/100 g. Soya sauce was considered exchange-free if the protein content was ≤1.5 g/100 g. A higher cut off point was given for soya sauce, as the amount used in recipes is generally small.

Plant milks and special milk replacements were a group that also required more specific definition of protein/phenylalanine cut off points due to potentially high daily volumes being consumed, thereby contributing a significant amount of protein/phenylalanine. A stringent exchange-free upper limit of ≤0.1 g protein/100 mL was set for regular and plant milks, and an upper limit of phenylalanine of ≤5 mg/100mL for low protein special milks available on ACBS prescription. This was because the volumes consumed may be high when taking this as a regular hot or cold drink, 'milk' shake or 'latte'. Only a small number of plant milks (some coconut milks) and low protein special milks (protein-free only) can therefore be given without calculation/measurement in the diet.

Currently, around 10% of SLPF contain more phenylalanine than the upper cut off point (≤25 mg/100 g) due to added ingredients such as milk, seeds, and rice flour [22]. It is hoped that these consensus statements will encourage manufacturers of SLPF to develop new low protein special foods with a phenylalanine content of ≤25 mg/100 g, so that they can be eaten without measurement or restriction. There seems to be little value in having SLPF that must still be limited and controlled within the diet. All SLPF should have clear labelling identifying their phenylalanine content (per 100 g of raw ingredients and per 100 g after preparation) with the full list of ingredients given, and any protein containing ingredients identified in bold on the ingredients list [23].

There were study strengths and limitations. Although dietitians represented both paediatric and adult care, they had differing opinions on the stringency of dietary guidance needed. Whilst this added to discussions, it also increased the challenge of producing statements that would meet the needs of the majority of the PKU population. Although all BIMDG dietitians had the opportunity to comment on each statement and vote in each round of the Delphi process, due to career changes or other circumstances such as maternity leave, the dietitians responding in each round were not necessarily the same each time. Even so, a representative from almost all major IMD specialty treatment centres (in both paediatric and adult care) was represented in each round. Overall, 100% consensus was reached for 64 statements. It was considered impractical to aim for 100% agreement for all statements, but a consensus cut off of 75% was agreed upon as it represented the majority of opinion. It may have been valuable to seek patient/carer opinion on each statement. Pragmatically, it was considered important as a first step to gain dietetic consensus as this was considered a substantial barrier to consistency of care, before translating into practical guidance for patients.

It is acknowledged that internationally, different systems are used to calculate/measure protein/phenylalanine in the PKU diet, each with its own inherent weaknesses [4–9]. Using upper protein/phenylalanine cut-off points has the disadvantage of having to measure/calculate foods as part of an exchange system if they contain protein marginally over the cut-off, whilst eating foods as exchange-free if they are just under the cut-off point, but it does give direct guidance and allows many foods to be eaten without measurement or calculation. Overall, these statements are designed to provide broad guidance for dietitians with limited experience in treating PKU on how to safely manage individuals with PKU. It is also important that they understand the foundation for the statements. The proposed guidance should be used in conjunction with individual tailored advice considering patient food likes, aversions, and level of understanding.

The statements were designed to be used by dietitians and heath care professionals. However, since completion of the statements, the NSPKU medical advisory dietitians have developed a dietary information guide using the statements as a basis for practical advice [16]. The next step will be to evaluate the adherence to these statements by both health professionals and patients with PKU and their caregivers.

5. Conclusions

The development and publication of UK consensus statements for food labelling and protein allocation in the PKU diet is an important step in harmonising dietary advice and effecting consistency of care for patients with PKU. Developing these statements in partnership with BIMDG dietitians using Delphi methodology has ensured that all dietitians have had the opportunity to participate in the development in an impartial, transparent, and consistent process. In the UK, this is the first time that such extensive agreement has been reached amongst specialist dietitians, with the results being simplified and implemented into patient care. It is also hoped that these guidelines will be respected and adopted by manufacturers of SLPF to ensure patients with PKU can gain full advantage from consuming low protein foods in a low phenylalanine diet.

Supplementary Materials: The following are available online at http://www.mdpi.com/2072-6643/12/8/2205/s1, Low protein labelling consensus statements for people with PKU on ≤10 g/day of natural protein.

Author Contributions: Conceptualization, all authors; Data curation, S.E.; Formal analysis, S.E. and A.M.; Investigation, all authors; Methodology, all authors; Project administration, S.E.; Resources, S.E. and A.M.; Supervision, A.M.; Validation, all authors; Visualization, S.E. and A.M.; Writing—original draft, S.E. and A.M.; Writing—review and editing, all authors. All authors have read and agreed to the published version of the manuscript.

Funding: No funding was required for this project. All work was conducted as part of routine service by all dietitians involved.

Acknowledgments: We would also like to thank the following dietetic staff for their contributions: Jane Ash (University Hospital of Wales), Assunta Aversano (Great Ormond Street Hospital for Children NHS Foundation Trust), Nicole Bickerdyke (Great Ormond Street Hospital for Children NHS Foundation Trust), Gillian Caine (Mid Yorkshire Hospitals NHS Trust), Emma Cameron (Bristol Royal Hospital for Children), Rachel Carruthers (University College London Hospitals NHS Foundation Trust), Sarah Cawtherley (Great Ormond Street Hospital for Children NHS Foundation Trust), Satnam Chahal (Birmingham Children's Hospital), Heidi Chan (Guy's and St Thomas' NHS Foundation Trust), Anne Clark (Children's University Hospital Dublin), Emma Coates, Karen Dines (Belfast Health and Social Care Trust), Sarah Donald (Cambridge University Hospitals NHS Foundation Trust), Carla Fitzcary (Guy's and St Thomas' NHS Foundation Trust), Francine Freedman (University College London Hospitals NHS Foundation Trust), Lisa Gaff (Cambridge University Hospitals NHS Foundation Trust), Cerys Gingell (Nottingham University Hospitals NHS Trust), Diane Green (Salford Royal NHS Foundation Trust), Louise Gubb (Royal Berkshire Hospital), Paula Hallam (Great Ormond Street Hospital for Children NHS Foundation Trust), Una Hendroff (Mater Misericordiae University Hospital Dublin), Rachel Hoban (Birmingham Children's Hospital), Sarah Howe (University Hospitals Birmingham NHS Foundation Trust), Rachel Hudghton (Leicester Royal Infirmary), Inderdip Hunjan (Bradford Teaching Hospitals NHS Foundation Trust), Kit Kaalund (University College London Hospitals NHS Foundation Trust), Eimear Kelleher (Mater Misericordiae University Hospital Dublin), Farzana Khan (Bradford Teaching Hospitals NHS Foundation Trust), Steve Kitchen (Birmingham Children's Hospital), Sharan Lowry (Sheffield Children's NHS Foundation Trust), Georgina Martin (Cambridge

University Hospitals NHS Foundation Trust), Avril Micciche (Guy's and St Thomas' NHS Foundation Trust), Claire Nicol (Royal Victoria Infirmary Newcastle Upon Tyne), Pat Portnoi (NSPKU), Janet Purves (Royal Hospital for Sick Children Edinburgh), Carla Rennie (Royal Hospital for Children, Glasgow), Sarah Ripley (Salford Royal NHS Foundation Trust), Alice Runnacles (Royal Victoria Infirmary Newcastle Upon Tyne), Gurpreet Sagoo (Great Ormond Street Hospital for Children NHS Foundation Trust), Emma Simpson (Royal Manchester Children's Hospital), Arlene Slabbert (Guy's and St Thomas' NHS Foundation Trust), Jacky Stafford (Great Ormond Street Hospital for Children NHS Foundation Trust), Abigail Swancott (Guy's and St Thomas' NHS Foundation Trust), Ruth Thom (Belfast Health and Social Care Trust), Phillipa Thomas (Bristol Royal Hospital for Children), Lyndsey Tomlinson (University Hospitals Birmingham NHS Foundation Trust), Alison Tooke (Nottingham University Hospitals NHS Trust), and Rychelle Winstone (Guy's and St Thomas' NHS Foundation Trust).

Conflicts of Interest: B.C. received financial support from Nutricia and Vitaflo to attend study days and conferences; J.C. received financial support from Nutricia and Vitaflo to attend study days; C.D. received honoraria from Nutricia; A.D. received research funding from Vitaflo and financial support from Nutricia and Vitaflo to attend study days and conferences; C.E. received honoraria and financial support from Vitaflo and Nutricia to attend study events and conferences and participation in advisory boards for Nutricia Metabolics and Meta Healthcare; S.E. received research funding from Nutricia, financial support from Nutricia and Vitaflo to attend study days and conferences; S.F. (Sarah Firman). received financial support from Nutricia and Vitaflo to attend study days and conferences; S.F. (Suzanne Ford). received funding for conferences and study days from Nutricia, Vitaflo, and Mevalia, advisory board participation for Nutricia and Biomarin; M.F. received financial support from Nutricia and Vitaflo to attend study days and conferences; A.M. received research funding and honoraria from Nutricia, Vitaflo International, and Merck Serono, member of European Nutrition Expert Panel (Merck Serono international), member of Sapropterin Advisory Board (Merck Serono international), member of the Advisory Board Element (Danone-Nutricia); J.M. (Joy McDonald). received financial support from Nutricia and Vitaflo to attend study days and conferences; N.M. received financial support from Nutricia and Vitaflo to attend study days and conferences; C.N. (Camille Newby) received financial support from Nutricia and Vitaflo to attend study days and conferences; A.P. received an educational grant from Cambrooke Therapeutics and grants from Vitaflo, Nutricia, Merck Serono, Biomarin, and Mevalia to attend scientific meetings; A.R. received financial support from Nutricia and Vitaflo to attend study days and conferences; L.R. received financial support from Nutricia and Vitaflo to attend study days and conferences and is a member of the Nutricia Metabolics Advisory Board; R.S. received financial support from Nutricia and Vitaflo to attend study days and conferences; K.V. is a member of the Nutricia Paediatric Metabolic Advisory Board; F.W. received honoraria and financial support from Vitaflo and Nutricia to attend study events and conferences.

References

1. MacDonald, A. Inborn errors in metabolism. In *Paediatric Nutrition in Practice*, 2nd ed.; Koletzko, B., Ed.; Karger: Basel, Switzerland, 2015; pp. 226–233.
2. MacDonald, A.; White, F. Amino Acid Disorders. In *Clinical Paediatric Dietetics*, 4th ed.; Shaw, V., Ed.; Wiley Blackwell: Oxford, UK, 2015; pp. 391–455.
3. Ford, S.; O'Driscoll, M.; MacDonald, A. Living with Phenylketonuria: Lessons from the PKU community. *Mol. Genet. Metab. Rep.* **2018**, *17*, 57–63. [CrossRef] [PubMed]
4. Ahring, K.; Bélanger-Quintana, A.; Dokoupil, K.; Ozel, H.G.; Lammardo, A.M.; MacDonald, A.; Motzfeldt, K.; Nowacka, M.; Robert, M.; van Rijn, M. Dietary management practices in phenylketonuria across European centres. *Clin. Nutr.* **2009**, *28*, 231–236. [CrossRef] [PubMed]
5. Pinto, A.; Adams, S.; Ahring, K.; Allen, H.; Almeida, M.F.; Garcia-Arenas, D.; Arslan, N.; Assoun, M.; Altınok, Y.A.; Barrio-Carreras, D.; et al. Weaning practices in phenylketonuria vary between health professionals in Europe. *Mol. Genet. Metab. Rep.* **2019**, *18*, 39–44. [CrossRef] [PubMed]
6. Rohde, C.; Thiele, A.G.; Och, U.; Schönherr, K.; Meyer, U.; Rosenbaum-Fabian, S.; Maddalon, C.; Matzken, S.; Blessing, H.; Lang, F.; et al. Effect of dietary regime on metabolic control in phenylketonuria: Is exact calculation of phenylalanine intake really necessary? *Mol. Genet. Metab. Rep.* **2015**, *5*, 36–41. [CrossRef] [PubMed]
7. Bernstein, L.; Burns, C.; Sailer-Hammons, M.; Kurtz, A.; Rohr, F. Multiclinic Observations on the Simplified Diet in PKU. *J. Nutr. Metab.* **2017**, *2017*, 4083293. [CrossRef] [PubMed]
8. Singh, R.H.; Rohr, F.; Frazier, D.; Cunningham, A.; Mofidi, S.; Ogata, B.; Splett, P.L.; Moseley, K.; Huntington, K.; Acosta, P.B.; et al. Recommendations for the nutrition management of phenylalanine hydroxylase deficiency. *Genet. Med.* **2014**, *16*, 121–131. [CrossRef] [PubMed]
9. MacDonald, A.; Van Wegberg, A.M.; Ahring, K.; Beblo, S.; Bélanger-Quintana, A.; Burlina, A.; Campistol, J.; Coşkun, T.; Feillet, F.; Giżewska, M.; et al. PKU dietary handbook to accompany PKU guidelines. *Orphanet. J. Rare Dis.* **2020**, *15*, 171. [CrossRef] [PubMed]

10. Medical Research Council. Recommendations on the dietary management of phenylketonuria. Report of Medical Research Council Working Party on Phenylketonuria. *Arch. Dis. Child* **1993**, *68*, 426–427. [CrossRef] [PubMed]
11. Evans, S.; Ford, S.; Adam, S.; Adams, S.; Ash, J.; Ashmore, C.; Caine, G.; Carruthers, R.; Cawtherley, S.; Chahal, S.; et al. Development of national consensus statements on food labelling interpretation and protein allocation in a low phenylalanine diet for PKU. *Orphanet. J. Rare Dis.* **2019**, *14*, 2. [CrossRef] [PubMed]
12. European Parliament. Regulation (EU) No 1169/2011 of the European Parliament and the Council of the European Union of 25 October 2011 on the setting of tolerances for nutrient values declared on a label. *Off. J. Eur. Union* **2011**, *L304*, 18–63.
13. Dahlgaard-Park, S.M. *The SAGE Encyclopedia of Quality and the Service Economy: Delphi Method*; SAGE Publications, Inc.: Thousand Oaks, CA, USA, 2015; Available online: http://sk.sagepub.com/reference/the-sage-encyclopedia-of-quality-and-the-service-economy (accessed on 17 June 2020).
14. Health Research Authority and Medical Research Council. Is My Study Research? Decision Tool. National Health Service (NHS). 2020; Available online: http://www.hra-decisiontools.org.uk/research/ (accessed on 17 June 2020).
15. Weetch, E.; MacDonald, A. The determination of phenylalanine content of foods suitable for phenylketonuria. *J. Hum. Nutr. Diet.* **2006**, *19*, 229–236. [CrossRef] [PubMed]
16. NSPKU. Dietary Information Booklet for the Treatment of Phenylketonuria 2020. Available online: www.nspku.org/documents (accessed on 17 June 2020).
17. MacDonald, A.; Rylance, G.; Davies, P.; Asplin, D.; Hall, S.K.; Booth, I.W. Free use of fruits and vegetables in phenylketonuria. *J. Inherit. Metab. Dis.* **2003**, *26*, 327–338. [CrossRef] [PubMed]
18. Rohde, C.; Mütze, U.; Weigel, J.F.; Ceglarek, U.; Thiery, J.; Kiess, W.; Beblo, S. Unrestricted consumption of fruits and vegetables in phenylketonuria: No major impact on metabolic control. *Eur. J. Clin. Nutr.* **2012**, *66*, 633–638. [CrossRef] [PubMed]
19. Rohde, C.; Mütze, U.; Schulz, S.; Thiele, A.G.; Ceglarek, U.; Thiery, J.; Mueller, A.S.; Kiess, W.; Beblo, S. Unrestricted fruits and vegetables in the PKU diet: A 1-year follow-up. *Eur. J. Clin. Nutr.* **2014**, *68*, 401–403. [CrossRef] [PubMed]
20. Zimmermann, M.; Jacobs, P.; Fingerhut, R.; Torresani, T.; Thöny, B.; Blau, N.; Baumgartner, M.R.; Rohrbach, M. Positive effect of a simplified diet on blood phenylalanine control in different phenylketonuria variants, characterized by newborn BH4 loading test and PAH analysis. *Mol. Genet. Metab.* **2012**, *106*, 264–268. [CrossRef] [PubMed]
21. Van Wegberg, A.M.; MacDonald, A.; Ahring, K.; Bélanger-Quintana, A.; Blau, N.; Bosch, A.M.; Burlina, A.; Campistol, J.; Feillet, F.; Giżewska, M.; et al. The complete European guidelines on phenylketonuria: Diagnosis and treatment. *Orphanet. J. Rare Dis.* **2017**, *12*, 162. [CrossRef] [PubMed]
22. Wood, G.; Evans, S.; Pointon-Bell, K.; Rocha, J.; MacDonald, A. Special low protein foods in the UK: An examination of their macronutrient composition in comparison to regular foods. *Nutrients* **2020**, *12*, 1893. [CrossRef] [PubMed]
23. Wood, G.; Evans, S.; Pointon-Bell, K.; MacDonald, A. Special low protein foods: The challenges in obtaining accurate information about the availability of special low protein foods (SLPFs) and their nutritional composition. *Netw. Health Dig.* **2020**, *2020*, 36–38.

© 2020 by the authors. Licensee MDPI, Basel, Switzerland. This article is an open access article distributed under the terms and conditions of the Creative Commons Attribution (CC BY) license (http://creativecommons.org/licenses/by/4.0/).

nutrients

MDPI

Article

Preliminary Investigation to Review If a Glycomacropeptide Compared to L-Amino Acid Protein Substitute Alters the Pre- and Postprandial Amino Acid Profile in Children with Phenylketonuria

Anne Daly [1,*], Sharon Evans [1], Alex Pinto [1], Richard Jackson [2], Catherine Ashmore [1], Júlio César Rocha [3,4] and Anita MacDonald [1]

1 Dietetic Department, Birmingham Children's Hospital, Steelhouse Lane, Birmingham B4 6NH, UK; evanss21@me.com (S.E.); alex.pinto@nhs.net (A.P.); catherine.ashmore@nhs.net (C.A.); anita.macdonald@nhs.net (A.M.)
2 Liverpool Clinical Trials Centre, University of Liverpool, Brownlow Hill, Liverpool L69 3GL, UK; r.j.jackson@liverpool.ac.uk
3 Nutrition and Metabolism, NOVA Medical School, Faculdade de Ciências Médicas, Universidade Nova de Lisboa, 1169-056 Lisboa, Portugal; rochajc@nms.unl.pt
4 Centre for Health Technology and Services Research (CINTESIS), 4200-450 Porto, Portugal
* Correspondence: a.daly3@nhs.net

Received: 28 July 2020; Accepted: 11 August 2020; Published: 14 August 2020

Abstract: In Phenylketonuria (PKU), the peptide structure of the protein substitute (PS), casein glycomacropeptide (CGMP), is supplemented with amino acids (CGMP-AA). CGMP may slow the rate of amino acid (AA) absorption compared with traditional phenylalanine-free amino acids (Phe-free AA), which may improve nitrogen utilization, decrease urea production, and alter insulin response. Aim: In children with PKU, to compare pre and postprandial AA concentrations when taking one of three PS's: Phe-free AA, CGMP-AA 1 or 2. Methods: 43 children (24 boys, 19 girls), median age 9 years (range 5–16 years) were studied; 11 took CGMP-AA1, 18 CGMP-AA2, and 14 Phe-free AA. Early morning fasting pre and 2 h postprandial blood samples were collected for quantitative AA on one occasion. A breakfast with allocated 20 g protein equivalent from PS was given post fasting blood sample. Results: There was a significant increase in postprandial AA for all individual AAs with all three PS. Postprandial AA histidine ($p < 0.001$), leucine ($p < 0.001$), and tyrosine ($p < 0.001$) were higher in CGMP-AA2 than CGMP-AA1, and leucine ($p < 0.001$), threonine ($p < 0.001$), and tyrosine ($p = 0.003$) higher in GCMP-AA2 than Phe-free AA. This was reflective of the AA composition of the three different PS's. Conclusions: In PKU, the AA composition of CGMP-AA influences 2 h postprandial AA composition, suggesting that a PS derived from CGMP-AA may be absorbed similarly to Phe-free AA, but this requires further investigation.

Keywords: phenylketonuria; PKU; glycomacropeptide; amino acid; absorption

---

1. Introduction

Protein substitutes are an essential source of synthetic protein in the dietary treatment of classical phenylketonuria (PKU). Protein is the second major constituent in the body, critical for growth and supporting a wide range of metabolic and cellular functions. Amino acids (AA) are engaged in a dynamic process of protein synthesis and degradation. In PKU, it is critical that the AA profile of protein substitutes are carefully developed, with a balance of AAs that meet WHO 2007 [1] minimal AA requirements [2,3]. Furthermore, there is evidence that modification of the large neutral amino acid (LNAA) profile (including tyrosine, leucine, isoleucine, methionine, valine, histidine, threonine and tryptophan) will enhance brain AA concentrations. In mice fed LNAA, an improvement in brain neurophysiology with increased brain serotonin and norepinephrine, and lower phenylalanine concentrations was observed [4]. It is also important that the composition and protein source of protein substitutes help support physiological AA absorption [5].

In PKU, protein substitutes are either manufactured using artificial AA without phenylalanine or based on caseinglycomacropeptide (CGMP), a by-product of the cheese making process [6]. Artificial AAs are generated from plant-based materials converted into sugars, fermented, and purified [7]. In contrast, CGMP modified for use in PKU is a mix of bioactive glycopeptides containing residual phenylalanine with the addition of essential and semi essential artificial AAs (CGMP-AA). Some authors [8,9] describe CGMP as an intact protein. However, proteins are large macromolecules made up of one or more polypeptide chains, whereas CGMP is a 64 macropeptide and therefore classification as an intact protein source is a misrepresentation [10]. Although there is knowledge about the absorption of AAs and intact protein, little is known about the absorption properties of CGMP-AA.

It is well recognized that the kinetic and biochemical properties of natural proteins change depending on the quality of protein, which is determined by its AA pattern, the speed of digestion, absorption and release of AAs into the circulation [11]. Protein metabolism has been extensively studied [12,13]. It is established that whey protein is rapidly absorbed, and the release of some individual AAs (leucine, isoleucine) influences anabolic and hormonal pathways [11,14,15]. In contrast, the appearance of plasma AAs following a meal with casein is slower, with protein synthesis increased and breakdown inhibited [11,16,17]. AAs are directly available for absorption by the small intestine, and so are quickly absorbed, potentially resulting in their transient imbalance, and altering their bioavailability. AA antagonism, the presence of high concentrations of specific AAs, may alter the AA equilibrium, impairing the absorption of other AAs and limiting absorption and metabolism [18]

In PKU, several authors [19–22] have demonstrated improved utilization of AAs when protein substitutes are taken in divided doses throughout the day. However, this still does not produce a normal physiological response; protein substitutes are usually taken as an addition and not as an integral part of a meal. In an animal study, the metabolic and biological influences of CGMP were compared with Phe-free AA, with CGMP giving a more physiological response decreasing metabolic stress and immunity [23]. In subjects with PKU, van Calcar et al. [24] suggested CGMP improved protein synthesis and nitrogen retention compared to Phe-free AA. However, peptides may be absorbed more rapidly than AA [14,25], thereby questioning the rate of delivery of AAs from CGMP-AA.

In this pilot, parallel study in children with PKU, we aimed to investigate if there were any differences following an overnight fast in the pre and postprandial AA absorption after taking one breakfast dose of Phe-free AA compared with two different CGMP-AA formulations with varying AA compositions.

2. Ethical Permission

The South Birmingham Research Ethics committee granted a favorable ethical opinion. The study was registered 13/WM/0435 IRAS (integrated research application system) 129497. Written informed consent was obtained for all subjects from at least one caregiver with parental responsibility and written assent obtained from the subject if appropriate for their age and level of understanding.

3. Materials and Methods

3.1. Inclusion Criteria

Entry into the study included: children diagnosed with PKU by newborn screening, aged 5–16 years and treated with diet only. Children had to be adherent with diet and protein substitute intake, with 70% of routine blood phenylalanine concentrations within European PKU Guideline target range [26] for six months before study enrolment. Target blood phenylalanine ranges were 120 to 360 µmol/L for children aged 5 up to 12 years and 120 to 600 µmol/L for 12 years and older [26].

3.2. Study Design

Pre and postprandial AA absorption was measured on one occasion after six months of taking either Phe-free AA, or one of two CGMP formulations (CGM-AA1 or CGMP-AA2). Children attended the hospital after an overnight fast (minimal fasting time 10 h). CGMP-AA1 had been taken for six months by 11 children as part of a pilot study the results have previously been published [27]. Following the results of the pilot study, a further 18 children were recruited and took CGMP-AA2, which was a modification of CGMP-AA1. Nineteen children remained on Phe-free AA. All children had fasting capillary finger pricks (0.5 mL), for quantitative plasma AAs. Children then took 20 g protein equivalent from Phe-free AA, CGMP-AA1 or CGMP-AA2, followed by a breakfast providing less than one third of their phenylalanine/natural protein allowance (median 2 g natural protein (100 mg phenylalanine), range 1–6). After 120 min post protein substitute and breakfast, a second capillary sample was taken for AAs.

3.3. Protein Substitutes (PHE-FREE AA and CGMP1, CGMP2)

The AA profile and nutritional composition of the three different protein substitutes (provided by Vitaflo International) are given in Table 1. All the children in the Phe-free AA group took the same liquid pouch (PKU Cooler 20). For each 20 g protein equivalent, Phe-free AA provided 124 kcal, 9.4 g carbohydrate, and 0.7 g fat, and CGMP-AA1 and CGMP-AA2, 120 kcal, 6.5 g carbohydrate, and 1.5 g fat. CGMP-AA2 had increased amounts of tyrosine, leucine, histidine, and tryptophan, and less methionine, lysine, glycine, and aspartic acid than CGMP-AA1. Except for threonine (higher in CGMP-AA1 and 2), glycine and methionine (higher in CGMP-AA1), and leucine (higher in CGMP-AA2), all the other AAs were slightly but not significantly higher in the Phe-free AA. Glutamine was naturally present in CGMP-AA, but not added to Phe-free AA. The energy content of the three products was similar, although the carbohydrate content was 30% higher in the Phe-free AA, and fat content 53% higher in the two CGMP-AA products, but overall fat intake was low from all three protein substitutes.

The single dose of Phe-free AA, CGMP-AA1, and CGMP-AA2 given in this study provided 20 g protein equivalent. The CGMP-AA1 and CGMP-AA2 also provided an additional 36 mg phenylalanine for each 20 g protein equivalent. The children chose either Phe-free AA or CGMP-AA, depending on their taste preference.

Table 1. Nutritional composition of CGMP-AA 1, CGMP-AA2, and PHE-FREE AA protein substitutes.

Protein Substitute		**CGMP-AA1**	**CGMP-AA2**	**PHE-FREE AA**
Nutrients	**Units**	**Per 20 g PE Sachet**	**Per 20 g PE Sachet**	**Per 20 g PE Pouch**
Calories	Kcal	120	120	124
Protein equivalent	g	20	20	20
Total Carbohydrate	g	6.5	6.5	9.4
Sugars	g	2.2	2.2	7.8
Total Fat	g	1.5	1.5	0.7
Docosahexaenoic Acid	mg	84	84	134
Arachidonic Acid	mg	-	-	-
Fiber	g	0.1	0.1	-
Comprehensive amino acid profile				
L-amino acids		**CGMP-AA1**	**CGMP-AA2**	**PHE-FREE AA**
		20 g PE	**20 g PE**	**20 g PE**

Table 1. *Cont.*

L-Alanine	g	0.76	0.83	0.92
L-Arginine	g	1.00	0.96	1.5
L-Aspartic Acid	g	2.04	1.31	2.37
L-Cystine	g	0.01	0.24	0.61
L-Glutamine	g	2.49	2.70	-
Glycine	g	2.77	1.20	2.35
L-Histidine	g	0.42	0.70	0.92
L-Isoleucine	g	1.37	1.35	1.62
L-Leucine	g	1.30	3.00	2.54
L-Lysine	g	1.07	0.80	1.67
L-Methionine	g	0.54	0.28	0.45
L-Phenylalanine	g	0.03	0.03	-
L-Proline	g	1.51	1.52	1.69
L-Serine	g	0.98	0.96	1.04
L-Threonine	g	2.17	2.20	1.62
L-Tryptophan	g	0.17	0.40	0.5
L-Tyrosine	g	1.01	2.24	2.38
Taurine	g	-	-	0.04
L-Valine	g	1.13	1.09	1.86

CGMP-AA 1/CGMP-AA 2: casein glycomacropeptide formula 1 and 2; PHE-FREE AA: Phenylalanine-free L-amino acid (PKU Cooler 20, Vitaflo International).

3.4. *Measurement of Quantitative Plasma Amino Acids*

Capillary blood samples were collected into a Sarstedt tube and analyzed by ion exchange HPLC with postcolumn derivatization and spectrophotometric detection (Biochrom, Harvard Bioscience, Holliston, MA, USA). Prior to analysis, separated lithium-heparinized plasma was deproteinized 1:1 with 8% sulphosalicylic acid containing an internal standard, S-2-amino-ethyl-L-cysteine hydrochloride (Sigma, Merck, St. Louis, MO, USA). Quantitative amino acids (QAA) were analyzed except tryptophan and asparagine, which are not reported by our laboratory. Nonproteinogenic AA ornithine, citrulline, and taurine were included in the analysis. The individual pre- and postprandial AAs were quantitated (QAA) and the total AAs, total large neutral amino acids (LNAA), total essential amino acids (EAA), and total branched chain amino acids (BCAA) were calculated from these results. We report total AAs, LNAAs, BCAA, and EAAs, together with individual AAs.

3.5. *Statistics*

Descriptive statistics are reported as medians with associated interquartile ranges. Differences in AAs at baseline and follow-up are assessed using a paired *t*-test. Differences between the three treatment groups are performed using linear regression with differences at follow-up adjusting for baseline covariate values. All analyses are performed in the statistical package R (Version 3.3).

4. Results

4.1. *Subjects*

Forty-three (41 European and 2 Asian origin) children with PKU, with a median age of 9 years (range 5–16) were recruited and participated in the study. The number of children in each group was: Phe-free AA, n = 14 (8 boys and 6 girls), CGMP-AA1, n = 11 (5 boys and 6 girls), and CGMP-AA2, n = 18 (11 boys and 7 girls). The median age (range) in the groups were: CGMP-AA1, 8.3 years (6–16), CGMP-AA2, 8.4 years (5–14) and Phe-free AA, 12.9 years (5–15). There was significant difference in age between CGMP-AA2 and the Phe-free AA group ($p = 0.001$). The median phenylalanine concentration for 12 months pre-study (all the children were taking Phe-free-AA) was 288 μmol/L (140–600).

In all three groups, the median daily dose of protein equivalent from protein substitute was 60 g/day (range 40–80 g), and the median amount of prescribed natural protein was 5.5 g/day (range

3–30 g) or 275 mg phenylalanine (range 150–1500 mg). The majority had classical PKU, except two children who were mild according to their untreated blood phenylalanine levels at diagnosis and dietary phenylalanine tolerance.

4.2. Quantitative Plasma Amino Acid Results

Individual Amino Acids

Significant pre and postprandial differences for most individual AA were observed within each group (Table 2).

Preprandial valine was significantly lower with CGMP-AA2 than CGMP-AA1 ($p = 0.031$) and CGMP-AA2 vs. Phe-free AA ($p < 0.001$). For CGMP-AA1 vs. Phe-free AA, there were no significant preprandial changes.

Postprandial CGMP-AA1 vs. CGMP-AA2: histidine ($p < 0.001$), leucine ($p < 0.001$) and tyrosine ($p < 0.001$) were significantly higher for CGMP-AA2, while methionine ($p < 0.001$) significantly lower compared with CGMP-AA1.

Postprandial CGMP-AA1 vs. Phe-free AA: histidine ($p = 0.005$) and tyrosine ($p = 0.005$) were significantly higher in Phe-free AA, but isoleucine ($p = 0.008$), methionine ($p < 0.001$), threonine ($p = 0.001$) significantly lower. For CGMP-AA2 v Phe-free AA, leucine ($p < 0.001$), threonine ($p < 0.001$), and tyrosine ($p = 0.003$) were all significantly higher in CGMP-AA2.

Changes in the pre and postprandial AA concentrations between the groups appeared to be mainly a reflection of the different amino acid compositions of the three protein substitutes, being most evident between CGMP-AA1 and CGMP-AA2. CGMP-AA2 had higher amounts of histidine, leucine, and tyrosine, and lower methionine and valine compared to CGMP-AA1.

4.3. Total Amino Acids, LNAA, BCAA, and EAA

There were similar significant pre and postprandial changes within the groups for total AAs, LNAA, BCAA and EAA (Table 3(a–d), Figure 1a–d). No significant pre or postprandial changes were observed between any of the three groups when comparing total AA, LNAA, BCAA, or EAAs.

Table 2. Median (range) fasting pre and postprandial individual amino acids for CGPM-AA1, CGMP-AA2, and PHE-FREE AA protein substitutes.

Median L-Amino Acids µmol/L (range)	CGMP-AA1 (*n* = 11)		CGMP-AA 2 (*n* = 18)		PHE-FREE AA (*n* = 14)		*p* Value
	Pre-Prandial	Post-Prandial	Pre-Prandial	Post-Prandial	Pre-Prandial	Post-Prandial	
Alanine	320 ^ (171–425)	482 ^,** (266–685)	277 ^^,§ (182–566)	457 ^^ (304–701)	356 ^^^,§ (153–463)	412 ^^^,** (319–612)	^ 0.007, ^^ <0.0001, ^^^ <0.002 § 0.012, ** 0.036
Arginine	44 ^ (27–56)	77 ^ (28–138)	38 ^^ (14–56)	94 ^^ (23–130)	38 ^^^ (16–57)	74 ^^^ (33–168)	^ 0.003, ^^ <0.0001, ^^^ 0.0001
Aspartic acid	21 T (12–35)	19 (12–36)	17 T (11–28)	15 (10–31)	21 (6–31)	18 (11–42)	T 0.03
Citrulline	39 T (30–51)	39 (28–55)	29 T (23–46)	33 (24–49)	37 (16–48)	35 (17–47)	T 0.02
Cystine	36 ^,T (18–45)	28 ^ (12–44)	16 ^^,T,§ (4–35)	21 ^^,§§ (5–35)	30 § (19–57)	31 §§ (20–171)	^ 0.02, ^^ 0.001 T 0.003, § <0.001, §§ <0.001
Glutamine	495 ^ (406–542)	476 ^,TT (258–578)	472 ^^ (350–606)	513 ^^,TT (396–675)	461 (339–583)	455 (345–613)	^ 0.001, ^^ 0.001 TT 0.035
Glutamic acid	62 ^ (40–79)	71 ^ (45–104)	63 ^^ (31–133)	70 ^^ (41–155)	75 (22–110)	78 (33–111)	^ 0.04, ^^ 0.02
Glycine	323 ^ (227–429)	476 ^,TT,** (291–767)	306 ^^ (195–446)	342 ^^,TT (227–526)	303 ^^^ (189–380)	368 ^^^,** (199–642)	^ 0.003, ^^ <0.0001, ^^^ 0.004 TT <0.001, ** 0.001
Histidine	66 (56–83)	62 TT,** (39–83)	75 ^^ (62–119)	107 ^^,TT (70–151)	81 ^^^ (46–99)	99 ^^^,** (53–181)	^^ <0.0001, ^^^ 0.003 TT <0.001, ** 0.005
Isoleucine	51^ (39–71)	172 ^,** (97–270)	43 ^^ (34–62)	168 ^^ (96–230)	55 ^^^ (30–72)	130 ^^^,** (68–311)	^ 0.001, ^^ <0.0001, ^^^ 0.0001 ** 0.008
Leucine	92 ^ (73–116)	195 ^,TT (99–296)	87 ^^ (72–130)	356 ^^,TT,§§ (227–440)	108 ^^^ (66–127)	199 ^^^,§§ (108–474)	^ 0.001, ^^ <0.0001, ^^^ 0.0001 TT <0.001, §§ <0.001
Lysine	147 ^ (102–179)	210 ^ (128–314)	130 ^^,§ (94–172)	188 ^^ (96–332)	155 ^^^,§ (109–193)	216 ^^^ (148–385)	^ 0.002, ^^ <0.0001, ^^^ 0.0001 § 0.04
Methionine	20 ^ (13–23)	48 ^,TT,** (28–82)	18 ^^ (15–27)	30 ^^,TT (15–42)	19 ^^^ (14–27)	29 ^^^,** (19–52)	^ 0.001, ^^ <0.0001, ^^^ 0.002 TT <0.001, ** <0.001
Ornithine	66 ^ (53–114)	87 ^ (55–175)	72 ^^ (37–96)	91 ^^ (69–136)	83 ^^^ (37–99)	99 ^^^ (51–141)	^ 0.004, ^^ 0.0004, ^^^ 0.004

Table 2. *Cont.*

Median L-Amino Acids µmol/L (range)	CGMP-AA1 (n = 11)		CGMP-AA 2 (n = 18)		PHE-FREE AA (n = 14)		p Value
	Pre-Prandial	Post-Prandial	Pre-Prandial	Post-Prandial	Pre-Prandial	Post-Prandial	
Proline	116 ^ (93–183)	282 ^ (174–522)	106 ^^ (80–284)	275 ^^ (198–530)	141 ^^^ (83–322)	258 ^^^ (161–447)	^0.002, ^^ <0.0001, ^^^ 0.0001
Serine	156 ^ (123–227)	202 ^ (131–321)	146 ^^ (119–196)	193 ^^ (119–251)	155 (71–228)	169 (132–325)	^ 0.008, ^^ <0.0001
Taurine	54 (35–73)	53 (36–326)	55 (26–92)	47 (33–74)	56 (39–98)	56 (47–100)	
Threonine	161 ^ (90–241)	307 ^,** (200–614)	141 ^^ (80–215)	331 ^^,§§ (240–440)	129 ^^^ (67–193)	214 ^^^,**,§§ (122–358)	^ 0.002, ^^ <0.0001, ^^^ 0.0001 ** 0.001, §§ <0.001
Tyrosine	37 ^ (29–68)	80 ^,TT,** (34–149)	49 ^^ (36–84)	181 ^^,TT,§§ (126–327)	47 ^^^ (31–73)	136 ^^^,**,§§ (52–206)	^ 0.002, ^^ <0.0001, ^^^ 0.0001 TT <0.001,** 0.005, §§ 0.003
Valine	214 ^,T (157–278)	345 ^ (206–580)	168 ^^,§,T (131–243)	363 ^^ (248–538)	227 ^^^,§ (136–345)	407 ^^^ (240–628)	^ 0.002, ^^ <0.0001,^^^ 0.0001 T 0.031, § <0.001

Key: ^, ^^, ^^^ significant differences within CGMP1,2 and Phe-free AA. Preprandial changes: T = CGMP1 v2, § = CGMP2 vPhe-free AA. Postprandial changes TT = CGMP1v2, ** = GGMP1 vs. Phe-free AA, §§ = CGMP2 vs. Phe free AA. p = significant value

Table 3. (**a**) Median pre and postprandial total amino acids results for CGMP-AA1, CGMP-AA2, and PHE-FREE AA protein substitutes. (**b**) Median pre and postprandial EAA for CGMP-AA1, CGMP-AA2, and PHE-FREE AA protein substitutes. (**c**) Median pre and postprandial BCAA for CGMP-AA1, CGMP-AA2, and PHE-FREE AA protein substitutes. (**d**) Median pre and postprandial LNAA for CGMP-AA1, CGMP-AA2, and PHE-FREE AA protein substitutes.

(a)							
	CGMP-AA1 *n*= 11		**CGMP-AA2 *n* = 18**		**PHE-FREE AA *n* = 14**		***p* Value**
Total AA µmol/L (range)	**Pre-Prandial**	**Post-Prandial**	**PRE-PRANDIAL**	**Pre-Prandial**	**Pre-Prandial**	**Pre-Prandial**	
	753 * (219–5257)	1473 * (237–5659)	1375 ** (317–8344)	3249 ** (291–9139)	1067 *** (336–8513)	1922 *** (396–9064)	* 0.0003, ** <0.0001, *** <0.0001
Amino acids: alanine, arginine, aspartic acid, cystine, glutamine, glutamic acid, glycine, histidine, isoleucine, leucine, lysine, methionine, proline, serine, threonine, tyrosine, valine (excluding phenylalanine, tryptophan, citrulline)							
(b)							
EAA µmol/L (range)	**Pre-Prandial**	**Pre-Prandial**	**Pre-Prandial**	**Pre-Prandial**	**Pre-Prandial**	**Pre-Prandial**	
	92 * (20–214)	195 * (48–345)	87 ** (18–168)	188 ** (30–363)	108 *** (17–223)	199 *** (30–415)	* 0.03 ** 0.02 *** 0.02
Amino acids: histidine, isoleucine, leucine, lysine, methionine, threonine, valine (excluding phenylalanine, tryptophan)							
(c)							
BCAA µmol/L (range)	**Pre-Prandial**	**Pre-Prandial**	**Pre-Prandial**	**Pre-Prandial**	**Pre-Prandial**	**Pre-Prandial**	
	92 (51–214)	195 (172–345)	87 (43–168)	356 (168–363)	98 (46–223)	214 (131–415)	ns
Amino acids: isoleucine, leucine, valine							
(d)							
LNAA µmol/L (range)	**Pre-Prandial**	**Pre-Prandial**	**Pre-Prandial**	**Pre-Prandial**	**Pre-Prandial**	**Pre-Prandial**	
	66 * (20–214)	172 * (48–345)	75 ** (18–167)	180 ** (30–363)	67 *** (17–223)	138 *** (30–415)	* 0.03, ** 0.02, *** 0.02
Amino acids: histidine, isoleucine, leucine, methionine, threonine, tyrosine, valine							

CGMP-AA 1, CGMP-AA 2: caseinglycomacropeptide formula 1 and 2, PHE-FREE AA: Phenylalanine-free L-amino acids.*, **, *** p = significant value, ns not significant.

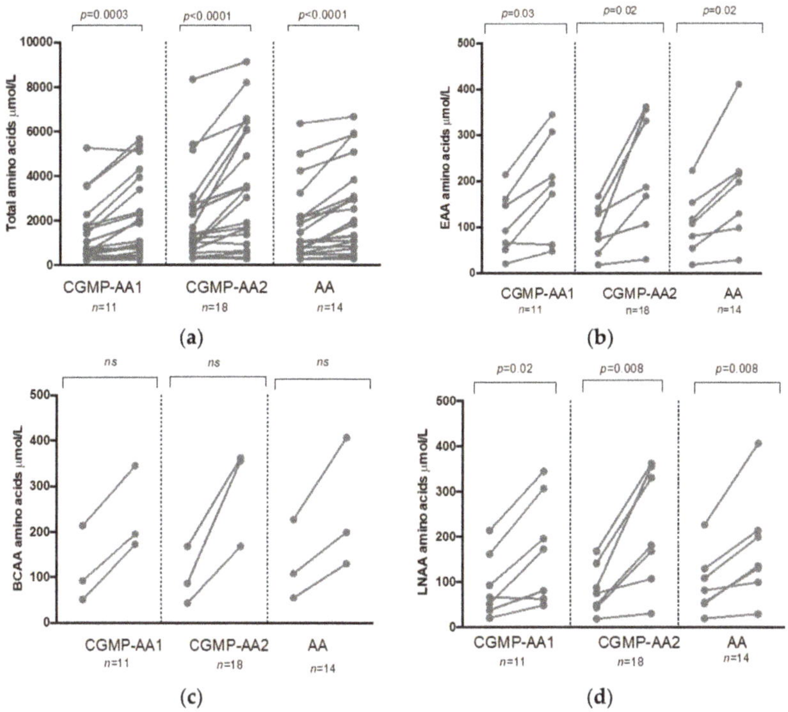

Figure 1. (**a**) Total median pre and postprandial amino acid concentrations for total amino acids ($n = 17$) for CGMP-AA1, CGMP-AA2, and PHE-FREE AA protein substitutes. (**b**). Total median pre and postprandial amino acid concentrations for EAA ($n = 7$) for CGMP-AA1, CGMP-AA 2, and PHE-FREE AA protein substitutes. (**c**) Total median pre and postprandial amino acid concentrations for BCAA ($n = 3$) for CGMP-AA1, CGMP-AA2, and PHE-FREE AA protein substitutes. (**d**) Total median pre and postprandial amino acid concentrations for LNAA ($n = 8$) for CGMP-AA1, CGMP-AA2, and PHE-FREE AA protein substitutes.

5. Discussion

This pilot study showed that that the postprandial AA concentrations largely reflected the AA profile of each of the protein substitutes used. CGMP-AA2 contained higher amounts of tyrosine, histidine, and leucine, and lower amounts of methionine and valine compared to CGMP-AA1. These changes were mirrored in the higher postprandial peaks of tyrosine, histidine, and leucine observed between CGMP-AA1 vs. CGMP-AA2. Although there was no postprandial change between the groups for valine, preprandial levels were lower between CGMP-AA1 vs. CGMP-AA2 and CGMP-AA2 vs. Phe-free AA. This is not easily explained physiologically, but may reflect a chance finding, or changes as a result of the competition between the AAs.

It is interesting to speculate on the postprandial changes, as it seems the more AA added to the protein substitute, the higher the AA concentrations were when measured at 120 min. The physiological consequence of these higher AAs is unknown. Postprandial tyrosine was significantly higher in CGMP-AA2 compared with CGMP-AA1 as a direct response of adding extra tyrosine. Norepinephrine is derived from tyrosine and is a principal brain neurotransmitter and so the provision of adequate tyrosine is essential to produce this monaminergic neurotransmitter, which is of clinical significance. Ney et al. [28] measured fasting tyrosine and tryptophan concentrations in subjects taking Phe-free AA compared to CGMP-AA and found their concentrations were 50% higher with Phe-free AA. Gut serotonin levels and microbiome-derived compounds made from tyrosine and tryptophan, although not significantly different, were higher in the CGMP-AA group, suggesting an improved bioavailability of tyrosine and tryptophan [29].

In our study, although individual AAs changed significantly within groups, no significant differences were observed between groups for total AAs, LNAAs, BCAAs, and EAAs. After 120 min, AA concentrations had increased significantly above fasting levels with a 56% increase in CGMP-AA1, 73% increase in CGMP-AA2 and a 42% increase in the Phe-free AA group. The total AA concentration per 20 g protein equivalent for CGMP-AA1 was 21 g, CGMP-AA2, 22 g, and Phe-free AA 24 g. It seems unlikely that the peptide-based CGMP-AA offered any advantage in minimizing the kinetic release of AA, although postprandial bloods were not measured in the first hour post consumption.

This exploratory investigation was a crude assessment to explore if there were any kinetic differences between AAs and a peptide based CGMP-AA with a different AA profile. In a crossover study in eleven adults with PKU, MacLeod et al. [30] measured postprandial AAs after 180 min following a breakfast with Phe-free AA or CGMP-AA. CGMP-AA was consumed as GMP foods rather than drinks. In the CGMP-AA group, postprandial threonine and isoleucine were significantly higher, and total AA concentrations just reached a significant difference compared to the Phe-free AA group. The authors suggested that based on the higher concentrations of insulin and total plasma AAs in the GMP group, CGMP-AA had an improved AA absorption profile compared to Phe-free AA. Although the preprandial breakfast was isocaloric, the AA composition of both products was not stated and the protein substitute in the form of a food versus a liquid may alter the absorption of AAs. Similarly, a non-physiological response causing a rapid rise in insulin concentrations may not be ideal. What remains unknown both in this and our own study is at what point maximum and nadir concentrations for AAs were reached, and neither study measured concentrations over 240 min or used a whole protein source as a comparison from which maximum and minimum AA concentrations could be compared.

Ahring et al. [31] compared two groups of protein substitutes: group 1, CGMP only versus Phe-free AA (different protein sources but the same AA composition) and group 2, CGMP-AA (CGMP with added AAs) versus Phe-free AA, (different protein sources and the same AA composition). The AA profile was different between group 1 and 2. Measurements were made over 240 min. They reported no differences in the absorption of total AAs between the two groups, suggesting that CGMP made no impact on the absorption rate of AAs. However, some individual AAs changed significantly between the groups. For example, the tyrosine amounts in the different product consumed were markedly different; group 1, 0.05 g and group 2, 10.81 g. At 30 min post ingestion, plasma tyrosine concentrations in group 2 were double those in group 1, reflecting changes in the AA profile of the protein substitute rather than the source of protein.

Gropper et al. [32] and others [33–35] have demonstrated that the type and quality of protein influences kinetic absorption. Healthy volunteers ingested one of three protein sources: AAs, a mixture of 75% AA with 25% natural protein, or whole protein. After 150 min, the only AA profile significantly higher than baseline was the group ingesting whole protein. In the two other groups, peak AA concentrations were reached before this time point. These studies showed that peak AA concentrations from a Phe-free AA or Phe-free AA combined with an intact protein source were more rapidly absorbed compared with an intact protein source only.

Both the time of arrival and pattern of AAs in the systemic circulation are important for effective protein synthesis. For this to occur efficiently all essential AAs must be presented to the tissues in appropriate amounts simultaneously. In PKU, the delivery of AAs to tissues is accelerated compared to a diet based on mixed proteins [12]. Glutamine is the most abundant free AA in the body, with a wide range of diverse molecular actions [36]. Its primary source is skeletal muscle. Both BCAA and lysine, by different mechanisms, act as precursors for glutamine synthesis, and leucine can stimulate the release of glutamine and alanine from muscles. Threonine in high concentrations can decrease glutamine formation [37]. The importance of understanding the delivery of AAs from protein substitutes and their effect on molecular pathways is crucial to long-term health outcomes for patients reliant on protein substitutes for their main source of nitrogen.

There are limitations to our findings. This was a pilot study with the aim to explore AA absorption from protein substitutes with different AA compositions. The liquid Phe-free AA preparation contained 30% more carbohydrate and 50% less fat than the CGMP-AA preparations. Breakfast provided similar food choices and the protein content was controlled, providing no more than 30% of their natural/phenylalanine daily allowance; however, we did not standardize the breakfast for all subjects, nor did we measure the AA concentrations every 30 min or over the course of 240 min as recommended by others [11,37], thereby missing the peak and baseline values. Similarly, we did not collect any other supporting data such as insulin and glucose concentrations to review the effect of insulinotropic AAs between the different protein sources. Our AA analysis did not measure tryptophan. We did not compare children with PKU with a control group taking a standard breakfast only without protein substitute.

6. Conclusions

In conclusion, the delivery, timing, and ratios of AAs are essential to maximize nitrogen utilization and biochemical functions. This pilot investigation compared three protein substitutes with different AA compositions and two different protein sources. It appeared that the AA composition rather than protein source was more important in determining postprandial plasma AAs. Further detailed work is needed to understand the kinetic and functional roles of protein substitute based on different protein sources and their metabolic impact

Author Contributions: Conceptualization, A.M., A.D.; methodology, A.M., A.D.; software, R.J., A.D.; validation, A.M., A.D., S.E., A.P., C.A., J.C.R.; formal analysis, A.D., R.J.; investigation, A.D.; writing—original draft preparation, A.D., A.M. Writing review and editing, A.D., S.E., A.P., R.J., J.C.R., A.M.; supervision, A.M. Funding acquisition, funding was part of PhD project from Vitaflo International. All authors have read and agreed to the published version of the manuscript.

Funding: This research was funded by Vitaflo International as part of PhD grant.

Conflicts of Interest: A.D. research funding from Vitaflo International, financial support from Nutricia, Mevalia and Vitaflo International to attend study days and conferences. S.E. research funding from Nutricia, financial support from Nutricia and Vitaflo International to attend study days and conferences. A.P. educational grant from Cambrooke Therapeutics and grants from Vitaflo International, Nutricia, Merck Serono, Biomarin and Mevalia to attend scientific meetings. C.A. received honoraria from Nutricia and Vitaflo International to attend study days and conferences. J.C.R. member of the European Nutritionist Expert Panel (Biomarin), the Advisory Board for Applied Pharma Research and Nutricia, and received honoraria as a speaker from APR, Merck Serono, Biomarin, Nutricia, Vitaflo, Cambrooke, PIAM and Lifediet. A.M. research funding and honoraria from Nutricia, Vitaflo International & Merck Serono, Member of European Nutrition Expert Panel (Merck Serono international), member of Sapropterin Advisory Board (Merck Serono international), member of the Advisory Board Element (Danone–Nutricia). The funders had no role in the design of the study; in the collection, analyses, or interpretation of data; in the writing of the manuscript, or in the decision to publish the results.

References

1. World Health Organisation; Food and Agriculture Organization of the United Nations; United Nations University. Protein and Amino Acid Requirements in Human Nutition. In *Report of a Joint WHO/FAO/UNU Expert Consultation*; WHO Technical Report Series 935; WHO: Geneva, Switzerland, 2007; pp. 1–265.
2. Macdonald, A.; Daly, A.; Davies, P.; Asplin, D.; Hall, K.; Rylance, G.; Chakrapani, A. Protein substitutes for PKU: What's new? *J. Inherit. Metab. Dis.* **2004**, *27*, 363–371. [CrossRef] [PubMed]
3. Woolf, L.I.; Griffiths, R.; Moncrieff, A. Treatment of phenylketonuria with a diet low in phenylalanine. *Br. Med. J.* **1995**, *1*, 57–64. [CrossRef] [PubMed]
4. van Vliet, D.; Bruinenberg, V.M.; Mazzola, P.N.; van Faassen, M.H.; de Blaauw, P.; Kema, I.P.; Fokkema, M.R.H.; van Anholt, R.D.; van der Zee, E.A.; van Spronsen, F.J. Large Neutral Amino Acid Supplementation Exerts Its Effect through Three Synergistic Mechanisms: Proof of Principle in Phenylketonuria Mice. *PLoS ONE* **2015**, *10*, e0143833. [CrossRef] [PubMed]

5. MacDonald, A.; Singh, R.H.; Rocha, J.C.; van Spronsen, F.J. Optimising amino acid absorption: Essential to improve nitrogen balance and metabolic control in phenylketonuria. *Nutr. Res. Rev.* **2019**, *32*, 70–78. [CrossRef]
6. Neelima, S.R.; Rajput, Y.S.; Mann, B. Chemical and functional properties of glycomacropeptide (GMP) and its role in the detection of cheese whey adulteration in milk: A review. *Dairy Sci. Technol.* **2013**, *93*, 21–43. [CrossRef]
7. Usuda, Y.; Hara, Y.; Kojima, H. Toward Sustainable Amino Acid Production. *Adv. Biochem. Eng. Biotechnol.* **2017**, *159*, 289–304.
8. Macleod, E.L.; Ney, D.M. Nutritional Management of Phenylketonuria. *Ann. Nestle Eng.* **2010**, *68*, 58–69. [CrossRef] [PubMed]
9. Ney, D.M.; Gleason, S.T.; van Calcar, S.C.; MacLeod, E.L.; Nelson, K.L.; Etzel, M.R.; Rice, G.M.; Wolff, J.A. Nutritional management of PKU with glycomacropeptide from cheese whey. *J. Inherit. Metab. Dis.* **2009**, *32*, 32–39. [CrossRef]
10. O'Riordan, N.; Kane, M.; Joshi, L.; Hickey, R.M. Structural and functional characteristics of bovine milk protein glycosylation. *Glycobiology* **2014**, *24*, 220–236. [CrossRef]
11. Boirie, Y.; Dangin, M.; Gachon, P.; Vasson, M.P.; Maubois, J.L.; Beaufrere, B. Slow and fast dietary proteins differently modulate postprandial protein accretion. *Proc. Natl. Acad. Sci. USA* **1997**, *94*, 14930–14935. [CrossRef]
12. Dangin, M.; Boirie, Y.; Garcia-Rodenas, C.; Gachon, P.; Fauquant, J.; Callier, P.; Ballevre, O.; Beufrere, B. The digestion rate of protein is an independent regulating factor of postprandial protein retention. *Am. J. Physiol. Endocrinol. Metab.* **2001**, *280*, E340–348. [CrossRef] [PubMed]
13. Fouillet, H.; Mariotti, F.; Gaudichon, C.; Bos, C.; Tome, D. Peripheral and splanchnic metabolism of dietary nitrogen are differently affected by the protein source in humans as assessed by compartmental modeling. *J. Nutr.* **2002**, *132*, 125–133. [CrossRef] [PubMed]
14. Gorissen, S.H.M.; Trommelen, J.; Kouw, I.W.K.; Holwerda, A.M.; Pennings, B.; Groen, B.B.L.; Wall, B.T.; Churchward, T.A.-V.; Horstman, A.M.H.; Koopman, R.; et al. Protein Type, Protein Dose, and Age Modulate Dietary Protein Digestion and Phenylalanine Absorption Kinetics and Plasma Phenylalanine Availability in Humans. *J. Nutr.* **2020**, *150*, 2041–2050. [CrossRef] [PubMed]
15. Park, S.; Church, D.D.; Azhar, G.; Schutzler, S.E.; Ferrando, A.A.; Wolfe, R.R. Anabolic response to essential amino acid plus whey protein composition is greater than whey protein alone in young healthy adults. *J. Int. Soc. Sports Nutr.* **2020**, *17*, 9. [CrossRef] [PubMed]
16. Luiking, Y.C.; Deutz, N.E.; Jakel, M.; Soeters, P.B. Casein and soy protein meals differentially affect whole-body and splanchnic protein metabolism in healthy humans. *J. Nutr.* **2005**, *135*, 1080–1087. [CrossRef] [PubMed]
17. Mulet-Cabero, A.I.; Torcello-Gomez, A.; Saha, S.; Mackie, A.R.; Wilde, P.J.; Brodkorb, A. Impact of caseins and whey proteins ratio and lipid content on in vitro digestion and ex vivo absorption. *Food Chem.* **2020**, *319*, 126514. [CrossRef]
18. Roth, E.; Druml, W. Plasma amino acid imbalance: Dangerous in chronic diseases? *Curr. Opin. Clin. Nutr. Metab. Care* **2011**, *14*, 67–74. [CrossRef]
19. Macdonald, A.; Rylance, G.; Davies, P.; Asplin, D.; Hall, S.K.; Booth, I.W. Administration of protein substitute and quality of control in phenylketonuria: A randomized study. *J. Inherit. Metab. Dis.* **2003**, *26*, 319–326. [CrossRef]
20. Mönch, E.; Herrmann, M.E.; Brösicke, H.; Schöffer, A.; Keller, M. Utilisation of amino acid mixtures in adolescents with phenylketonuria. *Eur. J. Nucl. Med. Mol. Imaging* **1996**, *155*, S115–S120. [CrossRef]
21. Weigel, C.; Rauh, M.; Kiener, C.; Rascher, W.; Knerr, I. Effects of Various Dietary Amino Acid Preparations for Phenylketonuric Patients on the Metabolic Profiles along with Postprandial Insulin and Ghrelin Responses. *Ann. Nutr. Metab.* **2007**, *51*, 352–358. [CrossRef]
22. Macdonald, A.; Rylance, G.; Hall, S.K.; Asplin, D.; Booth, I.W. Factors affecting the variation in plasma phenylalanine in patients with phenylketonuria on diet. *Arch. Dis. Child.* **1996**, *74*, 412–417. [CrossRef] [PubMed]
23. Solverson, P.; Murali, S.G.; Brinkman, A.S.; Nelson, D.W.; Clayton, M.K.; Yen, C.-L.E.; Ney, D.M. Glycomacropeptide, a low-phenylalanine protein isolated from cheese whey, supports growth and attenuates metabolic stress in the murine model of phenylketonuria. *Am. J. Physiol. Metab.* **2012**, *302*, E885–E895. [CrossRef] [PubMed]

24. Van Calcar, S.C.; MacLeod, E.L.; Gleason, S.T.; Etzel, M.R.; Clayton, M.K.; Wolff, J.; Ney, D.M. Improved nutritional management of phenylketonuria by using a diet containing glycomacropeptide compared with amino acids. *Am. J. Clin. Nutr.* **2009**, *89*, 1068–1077. [CrossRef] [PubMed]
25. Silk, D.B.A.; Fairclough, P.D.; Clark, M.L.; Hegarty, J.E.; Addison, J.M.; Burston, D.; Clegg, K.M.; Matthews, D.M. Use of a Peptide Rather Than Free Amino Acid Nitrogen Source in Chemically Defined "Elemental" Diets. *J. Parenter. Enter. Nutr.* **1980**, *4*, 548–553. [CrossRef] [PubMed]
26. Van Wegberg, A.M.J.; Macdonald, A.; Ahring, K.; Bélanger-Quintana, A.; Blau, N.; Bosch, A.M.; Burlina, A.B.; Campistol, J.; Feillet, F.; Gizewska, M.; et al. The complete European guidelines on phenylketonuria: Diagnosis and treatment. *Orphanet J. Rare Dis.* **2017**, *12*, 162. [CrossRef]
27. Daly, A.; Evans, S.; Chahal, S.; Santra, S.; Macdonald, A. Glycomacropeptide in children with phenylketonuria: Does its phenylalanine content affect blood phenylalanine control? *J. Hum. Nutr. Diet.* **2017**, *30*, 515–523. [CrossRef]
28. Ney, D.M.; Murali, S.G.; Stroup, B.M.; Nair, N.; Sawin, E.A.; Rohr, F.; Levy, H.L. Metabolomic changes demonstrate reduced bioavailability of tyrosine and altered metabolism of tryptophan via the kynurenine pathway with ingestion of medical foods in phenylketonuria. *Mol. Genet. Metab.* **2017**, *121*, 96–103. [CrossRef]
29. Pinto, A.; Almeida, M.F.; Ramos, P.C.; Rocha, S.; Guimas, A.; Ribeiro, R.; Martins, E.; Bandeira, A.; Macdonald, A.; Rocha, J.C. Nutritional status in patients with phenylketonuria using glycomacropeptide as their major protein source. *Eur. J. Clin. Nutr.* **2017**, *71*, 1230–1234. [CrossRef]
30. MacLeod, E.L.; Clayton, M.K.; Van Calcar, S.C.; Ney, D.M. Breakfast with glycomacropeptide compared with amino acids suppresses plasma ghrelin levels in individuals with phenylketonuria. *Mol. Genet. Metab.* **2010**, *100*, 303–308. [CrossRef]
31. Ahring, K.K.; Lund, A.M.; Jensen, E.; Jensen, T.G.; Brøndum-Nielsen, K.; Pedersen, M.; Bardow, A.; Holst, J.J.; Rehfeld, J.F.; Møller, L.B. Comparison of Glycomacropeptide with Phenylalanine Free-Synthetic Amino Acids in Test Meals to PKU Patients: No Significant Differences in Biomarkers, Including Plasma Phe Levels. *J. Nutr. Metab.* **2018**, *2018*, 6352919. [CrossRef]
32. Gropper, S.S.; Gropper, D.M.; Acosta, P.B. Plasma Amino Acid Response to Ingestion of L-Amino Acids and Whole Protein. *J. Pediatr. Gastroenterol. Nutr.* **1993**, *16*, 143–150. [CrossRef] [PubMed]
33. Adibi, S.A.; Morse, E.L. Intestinal transport of dipeptides in man: Relative importance of hydrolysis and intact absorption. *J. Clin. Investig.* **1971**, *50*, 2266–2275. [CrossRef] [PubMed]
34. Adibi, S.A.; Allen, E.R. Impaired Jejunal Absorption Rates of Essential Amino Acids Induced by Either Dietary Caloric or Protein Deprivation in Man. *Gastroenterol.* **1970**, *59*, 404–413. [CrossRef]
35. Marrs, T.C.; Addison, J.M.; Burston, D.; Matthews, D.M. Changes in plasma amino acid concentrations in man after ingestion of an amino acid mixture simulating casein, and a tryptic hydrolysate of casein. *Br. J. Nutr.* **1975**, *34*, 259–265. [CrossRef]
36. Curi, R.; Lagranha, C.; Doi, S.; Sellitti, D.; Procopio, J.; Pithon-Curi, T.; Corless, M.; Newsholme, P. Molecular mechanisms of glutamine action. *J. Cell. Physiol.* **2005**, *204*, 392–401. [CrossRef]
37. Nilsson, M.; Holst, J.J.; Björck, I.M. Metabolic effects of amino acid mixtures and whey protein in healthy subjects: Studies using glucose-equivalent drinks. *Am. J. Clin. Nutr.* **2007**, *85*, 996–1004. [CrossRef]

© 2020 by the authors. Licensee MDPI, Basel, Switzerland. This article is an open access article distributed under the terms and conditions of the Creative Commons Attribution (CC BY) license (http://creativecommons.org/licenses/by/4.0/).

nutrients

MDPI

Article

An Observational Study Evaluating the Introduction of a Prolonged-Release Protein Substitute to the Dietary Management of Children with Phenylketonuria

Anita MacDonald *, Catherine Ashmore, Anne Daly, Alex Pinto and Sharon Evans

Dietetic Department, Birmingham Women's and Children's Hospital, Steelhouse Lane, Birmingham B4 6NH, UK; catherine.ashmore@nhs.net (C.A.); a.daly3@nhs.net (A.D.); alex.pinto@nhs.net (A.P.); evanss21@me.com (S.E.)

* Correspondence: anita.macdonald@nhs.net; Tel.: +44-121-333-8021

Received: 9 August 2020; Accepted: 31 August 2020; Published: 3 September 2020

Abstract: Dietary restriction of phenylalanine combined with a protein substitute prevents intellectual disability in patients with phenylketonuria (PKU). However, current protein substitutes are associated with low adherence owing to unpalatability and burdensome administration regimens. This prospective, observational acceptability study in children with PKU assessed the use of a prolonged-release protein substitute designed with an ethyl cellulose and arginate coating masking the bitter taste, smell and reducing the osmolarity of free amino acids. The study product was mixed with the subject's food or drink and replaced ≥1 dose per day of the subject's usual protein substitute for 7 days. Seven of 13 subjects were able to take their prescribed dose over the 7 day period. Most subjects mixed the test protein substitute with food or fruit juice. Reduced blood phenylalanine levels ($n = 5$) and improved phenylalanine/tyrosine ratio ($n = 4$) were recorded from baseline to Day 7, respectively. Four subjects reported fewer gastrointestinal symptoms compared to baseline. There were no cases of diarrhoea, constipation, bloating, nausea or vomiting. No adverse reactions were reported. In conclusion, the novel prolonged-release protein substitute was taken in a different way to a typical protein substitute and enabled satisfactory blood phenylalanine control. The study product was well tolerated; subjects experienced fewer gastrointestinal symptoms than with their previous treatment. Although the results of this pilot study provide reassuring data, longer-term studies evaluating adherence and blood phenylalanine control are necessary.

Keywords: phenylketonuria; diet therapy; phenylalanine; protein substitute; gastrointestinal symptoms; prolonged release

1. Introduction

Phenylketonuria (PKU) is a rare metabolic disorder caused by a deficiency of phenylalanine hydroxylase (PAH), the enzyme that catalyses the hydroxylation of phenylalanine to tyrosine, and which leads to irreversible intellectual impairment in untreated children [1]. While there is no cure for PKU, the dietary restriction of phenylalanine has been highly successful in preventing intellectual disability and achieving near-normal intellect. However, current dietary treatments are associated with some major issues, such as low adherence attributed to unpalatable and burdensome dietary supplements and subtle but chronic neuropsychological impairments despite early intervention, particularly in adulthood, including mood and psychiatric issues [2–5].

Lifelong dietary management of PKU involves severe restriction of phenylalanine plus supplementation with protein substitutes, usually consisting of phenylalanine-free amino acids [6]. Protein substitutes provide essential and nonessential amino acids and commonly include micronutrients that would otherwise be lacking in a low-phenylalanine diet [6,7]. Since the 1960s, when the first manufactured protein substitutes were introduced, improvements have been made to their nutritional composition, presentation, taste and acceptability [7,8]. Currently, protein substitutes are available in a variety of forms, including powders, gels, liquids and tablets [9], and are traditionally administered as high-volume hyperosmolar drinks (if diluted with too little water, they may cause abdominal pain, diarrhoea or constipation [10]).

Generally, many patients have a poor acceptance of protein substitutes and parents struggle to ensure that their children take them as prescribed [11,12]. Poor adherence to a low-phenylalanine diet and protein substitute increases with age and is associated with worsening of blood phenylalanine control [11,13,14]. Furthermore, children may take up to an hour to take their full dose of protein substitute, with some failing to take the prescribed quantity [15]. Although reasons for poor adherence to protein substitutes are multiple, the bitter taste and aftertaste of synthetic amino acids are frequently reported as important factors [9,16]. Manufacturers have tried to improve the taste by lowering the quantities of unpalatable sulphurous and dicarboxylic amino acids and adding flavourings and sweeteners, but the results have been suboptimal [17], with minimal impact on aftertaste [18]. Commonly, children complain of breath odour following consumption of protein substitutes [19]. Although reports of adding protein substitute to food are few [20], this is perceived to have low acceptance [21].

Synthetic amino acids bypass degradation by proteases in the digestive process, resulting in blood levels that are higher, peak faster and decrease more quickly than when compared with natural protein. Therefore, it is recommended to take synthetic protein substitutes in small, frequent dosages in equally distributed amounts [8]. If protein substitutes are taken less frequently, there may be wide variations in blood phenylalanine levels over 24 h, which is associated with reduced blood phenylalanine control in PKU [8,22]. Protein substitutes with the ability to delay absorption of phenylalanine and tyrosine, mimicking physiological absorption kinetics, are expected to improve the rate of protein accretion, minimizing fluctuations in quantitative plasma amino-acid levels [8]. However, despite previous efforts, there has been little success in developing a slow-release protein substitute that mirrors the physiological absorption kinetics of intact natural proteins.

There is a need for more palatable and physiological protein substitutes that are both effective at reducing 24 h variability in phenylalanine levels and are accepted by patients. Here, we report outcomes from an observational study assessing the introduction of a prolonged-release protein substitute to the diets of children with PKU [16,23].

2. Materials and Methods

2.1. Study Design

This was a prospective, observational acceptability study performed in children with PKU aged 3–16 years who attended a single clinic at Birmingham Women's and Children's Hospital, Birmingham, United Kingdom. Caregivers of eligible children were identified and sent a study information sheet. Research dietitians discussed the study details with interested parents and patients on their request. This study was conducted according to the requirements stated by the UK Advisory Committee for Borderline Substances (ACBS). This is a committee that recommends dietary products to be reimbursed by the National Health Service (NHS). The study product met the criteria of a Type 2 product: *"a formulation which was broadly similar in composition to existing products already on the market."* According to ACBS guidelines, *"All acceptability studies must be for at least 1 week and at least 15 patients must complete the study. Where nutritional products are intended for use in very rare conditions, such as inherited metabolic disorders, fewer patients are acceptable"*. Thirteen patients were enrolled for 7 days of treatment and were able to complete the study, but only 54% (7 of 13) were able to take 100% of the prescribed product and were included in the final analyses. These studies conformed to the principles of good clinical practice.

2.2. Inclusion and Exclusion Criteria

Inclusion criteria included: male or female; aged 3–16 years; proven diagnosis of PKU requiring a protein substitute; already taking a protein substitute; and willing to take the study product for 7 days.

Exclusion criteria included: presence of serious concurrent illness; chief investigator's uncertainty about the willingness or ability of the patient to comply with protocol requirements; participation in any other study involving investigational or marketed products within two weeks prior to study entry; and children who received antibiotics over the two weeks prior to the study.

2.3. Study Product

The study product, PKU GOLIKE PLUS 3–16, (APR Applied Pharma Research, Switzerland) is a protein substitute for oral use in the form of off-white/light yellow granules, consisting of a prolonged-release amino-acid mixture with vitamins and minerals and other nutrients (i.e., carnitine, taurine, choline and inositol). The study product was developed with a coating able to overcome practical issues associated with free amino acids, such as bitter taste, smell, aftertaste and osmolarity. The coating consisted of ethyl cellulose plus alginates encasing granules of amino acids (without phenylalanine). The study product could be mixed with food or fruit juice or taken as granules. It was gluten and lactose-free and contained no added fat, with a nutritional profile suitable for patients aged 3–16 years (Table 1).

The study product was introduced into the standard therapeutic diets of enrolled children for 7 days by replacing at least one dose per day of the patient's usual protein substitute, according to ACBS requirements. Considering the short study duration, all the protein substitute requirements were not replaced as patients had little time to adapt to a different type of protein substitute given in another format.

Table 1. Nutritional declaration for the study product (PKU GOLIKE PLUS 3–16) [1].

Component	Per 100 g	Per Sachet of 24 g
Energy	280 kcal/1187 kJ	67 kcal/286 kJ
Fat	0 g	0 g
of which saturated	0 g	0 g
Carbohydrate	4.3 g	1.0 g
of which sugars	0 g	0 g
Fibre	7.1 g	1.7 g
Protein equivalent [1]	62.2 g	15 g
Salt	0.06 g	0.015 g
	Amino Acids	
L-serine	2.5 g	0.6 g
L-threonine	3.8 g	0.9 g
L-leucine	8.6 g	2.1 g
Glycine	3.8 g	0.9 g
L-alanine	2.3 g	0.5 g
L-arginine	3.0 g	0.7 g
L-cysteine	1.5 g	0.4 g
L-glutamine	15.0 g	3.6 g
L-histidine	2.1 g	0.5 g
L-aspartic acid	4.5 g	1.1 g
L-proline	4.5 g	1.1 g
L-isoleucine	4.1 g	1.0 g
L-lysine	5.3 g	1.3 g
L-tryptophan	1.5 g	0.4 g
L-valine	3.8 g	0.9 g
L-methionine	1.0 g	0.3 g
L-tyrosine	7.5 g	1.8 g

Table 1. *Cont.*

Component	Per 100 g	Per Sachet of 24 g
	Vitamins	
Vitamin A (RE)	1295 mcg	311 mcg
Vitamin D	25 mcg	6.0 mcg
Vitamin E (αTE)	13 mg	3.2 mg
Vitamin K	100 mcg	24 mcg
Vitamin C	135 mg	32.31 mg
Thiamine	2.0 mg	0.5 mg
Riboflavin	1.9 mg	0.5 mg
Niacin	27 mg	6.4 mg
Vitamin B6	2.6 mg	0.6 mg
Folic acid	267 mcg	64.1 mcg
Vitamin B12	4.2 mcg	1.0 mcg
Biotin	54 mcg	13 mcg
Pantothenic acid	11 mg	2.6 mg
	Minerals	
Potassium	1250 mg	300 mg
Calcium	1339 mg	321 mg
Magnesium	304 mg	72.9 mg
Phosphorus	1060 mg	254 mg
Chloride	0.75 mg	0.18 mg
Sodium	25 mg	5.9 mg
Iron	23 mg	5.6 mg
Zinc	14 mg	3.4 mg
Copper	1.4 mg	0.3 mg
Manganese	2.5 mg	0.6 mg
Selenium	58 mcg	14 mcg
Chromium	46 mcg	11 mcg
Molybdenum	88 mcg	21 mcg
Iodine	225 mcg	54.0 mcg
	Other Nutrients	
Carnitine	0.08 g	0.02 g
Taurine	0.21 g	0.05 g
Choline	321 mg	77.1 mg
Inositol	214 mg	51.4 mg

[1] 1 g of protein equivalent = 1.2 g of amino acids. The protein content is provided by the amino acids. Ingredients: L-glutamine, L-leucine, L-tyrosine, L-lysine acetate, glazing agent: ethyl cellulose; calcium hydrogen phosphate dihydrate, maltodextrin, L-aspartic acid, L-proline, L-isoleucine, L-threonine, glycine, L-valine, potassium bicarbonate, L-arginine, L-serine, L-alanine, L-histidine, L-cysteine, L-tryptophan, L-methionine, choline bitartrate, magnesium oxide, iron, maize starch, ferric pyrophosphate, glazing agent: sunflower lecithin, stabiliser: sodium alginate; inositol, taurine, L-ascorbic acid, L-carnitine, zinc sulphate, nicotinamide, DL-alpha tocopheryl acetate, chromium chloride hexahydrate, sodium molybdate, manganese gluconate, calcium-d-pantothenate, cupric gluconate, retinyl palmitate, pyridoxine hydrochloride, thiamine hydrochloride, riboflavin, cholecalciferol, folic acid, potassium iodide, phytomenadione, sodium selenite, D-biotin, cyanocobalamin.

Patients followed their usual low-phenylalanine diets during the study. Any foods containing protein up to 0.5 g/100 g and fruits and vegetables containing phenylalanine up to 75 mg/100 g were given without measurement. We aimed to maintain the same total protein equivalent intake for each patient whilst taking the prolonged-release protein substitute.

2.4. Preparation of Study Product

The research dietitians explained the study product's characteristics and its theoretical advantages to the caregivers and patients. Verbal and written information was given about how to administer the study product, i.e., granules could be taken either in liquids or semisolid foods with a creamy consistency (fruit smoothies, low-protein vegetable soup, fruit or vegetable purees, low-protein puddings or desserts). Each caregiver and patient were also given a practical demonstration on how to mix the study product. The type of food or drink that the study product was mixed with was selected by the patient.

2.5. Assessments

Subject demographics were recorded at study baseline. Prior to treatment initiation, information was recorded regarding the current protein substitute (dose, type, palatability and presence of any gastrointestinal side effects).

During treatment, parents/caregivers and subjects completed daily questionnaires to record treatment preparation/administration and any problems, including adverse events or gastrointestinal side effects. An additional questionnaire was completed at the end of treatment.

All patients had known adherence with their usual protein substitute prior to entering this study as well as routine blood spot phenylalanine monitoring, with three retrospective blood spots available in addition to blood tests taken during the study.

A fasting blood spot for phenylalanine was taken at home by parents/caregivers both before and at the end of the treatment period. Children aged ≤12 years aimed to maintain blood phenylalanine between 120–360 μmol/L; children aged >12 years aimed to maintain levels between 120–600 μmol/L. Early-morning fasted blood spots were collected on filter cards (Perkin Elmer 226, Greenville, SC, USA, UK Standard NBS). Blood specimens were sent via first-class post to the laboratory at Birmingham Children's Hospital. All cards had a standard thickness and the blood phenylalanine concentrations were calculated on a 3.2 mm punch by Waters Xevo TQD tandem mass spectrometer (Elstree, Herts, UK).

2.6. Ethical Permission

The Solihull Research Ethics Committee granted a favourable ethical opinion (REC reference: 19/WM/0151, IRAS project ID: 256519). Written consent was obtained for all subjects from at least one caregiver with parental responsibility and written assent obtained from the subject if appropriate for their age and level of understanding.

2.7. Statistical Analysis

Descriptive statistics were used to examine demographics and disorder characteristics, protein substitute use, adverse events and plasma phenylalanine and tyrosine levels.

3. Results

3.1. Subjects

A total of 13 subjects were enrolled into the acceptability study (12 with classical PKU; 1 with moderate PKU); mean age was 11.6 years (range 7 to 16 years). All subjects were diagnosed with PKU via newborn screening and started a low-phenylalanine diet from the time of screening. None were treated with sapropterin as an adjunctive therapy. At the start of the study, the subjects routinely took either one or two different types/brands of protein substitute daily (eight subjects took a single protein substitute daily; five subjects took two different types). Seven of the subjects routinely took a protein substitute derived from casein glycomacropeptide (CGMP), and three subjects took this as their sole source of protein substitute, with four subjects taking CGMP with an amino acid substitute. In total, five subjects usually took both a liquid and a powdered protein substitute, five took a liquid substitute only, and three took a powdered substitute only. The median (range) dose of protein equivalent from protein substitute was 1.4 (1.0–3.1) g/kg/day.

3.2. Substudy Cohort Results

The substudy cohort consisted of subjects who were able to take the entire prescribed dose of the study product (n = 7; mean age 10.9 years) (Table 2). The mean percentage of the prescribed dose consumed over 7 days for these patients was 98% (range 92–100%). Subjects that consumed less than

90% of the prescribed dose during the 7 days of treatment are not included in the substudy cohort analyses. Subjects were prescribed either 15 g/day (*n* = 6) or 20 g/day (*n* = 1) protein equivalent of the study product to replace one dose of the total protein substitute intake. The study product provided 20–27.3% of the subject's usual protein equivalent intake per day; the remaining protein equivalent intake was from each subjects' usual protein substitute. Three subjects (Subjects 1, 2 and 6) were 5 g of protein equivalent short of their pre-study dose, due to differences in sizes of protein substitute sachets/pouches.

Table 2. Baseline demographics and treatment in the substudy cohort.

Baseline Demographics							
Subject	1	2	3	4	5	6	7
Age (years)	11	11	12	9	7	11	15
Weight (kg)	60.3	53.3	45.9	25.8	26.6	45.6	55.8
Height (cm)	152.3	147.5	155.8	124.5	119.7	154.8	174.4
PKU classification	Classical	Classical	Moderate	Classical	Classical	Classical	Classical
Blood Phe on diagnosis (μmol/L)	1700	1680	900	1390	1590	2520	2690
Gender	Female	Female	Male	Male	Male	Male	Male
Ethnicity	Pakistani	Pakistani	White European	White British	Mixed race	White British	White British
Diet and Protein-Substitute Profile							
Natural protein allowance (g/day)	4.0	4.0	7.0	3.0	6.5	7.5	18.0
Protein equivalent from usual protein substitutes g/day (g/kg/day)	60.0 (1.0)	60.0 (1.1)	60.0 (1.3)	80.0 (3.1)	60.0 (2.3)	80.0 (1.7)	60.0 (1.2)
Number of different protein substitutes/day	2	2	2	1	2	2	1
Number of doses/day	3	3	4	4	4	4	3
Study Product Treatment Schedule and Preparation							
Daily dose (g)	24.0	24.0	24.0	32.0	24.0	24.0	24.0
Protein equivalent from study product (g)	15.0	15.0	15.0	20.0	15.0	15.0	15.0
Total daily protein equivalent from all protein substitutes (g) [1]	55.0	55.0	55.0	80.0	60.0	75.0	55.0
Protein equivalent from study product (% of daily intake)	27.3	27.3	27.3	25.0	25.0	20.0	27.3
Method of administration	In fruit juice	In fruit juice	Food and drinks	In fruit juice	In fruit juice and food	Fruit smoothie	Smoothie
Timing of administration of study product	Evening	Evening	Evening	Evening	Morning midday and bedtime	Morning or evening	Morning or evening
Comments on study product	Left some bits behind on cup	Last bit was hard to take	No comments	No comments	No comments	Required blender to mix	No comments

[1] Subject 1, Subject 2 and Subject 6 were 5 g of protein short of their pre-study dose due to differences in sizes of protein substitute sachets/pouches when incorporating the study product with existing protein-substitute diet plan. Abbreviations: Phe: phenylalanine; PKU: phenylketonuria.

3.3. *Administration of Study Product*

The study product was mixed with a variety of different drinks and foods, but the preferred method was to prepare it as a fruit smoothie or mixed with fruit juice (Table 2). All subjects or their parents found the study product easy to prepare. Seven of 13 children commented that the product had little taste or that it tasted "okay". The reason why children (*n* = 6 of 13) were unable to take the entire prescribed dose of the study product was primarily because of texture (described as bitty and gritty) and one child described the amount of powder as "a little overpowering" when added to food.

3.4. *Adherence with Study Product*

All but one of the subjects included in the substudy cohort took 100% of the prescribed study product on at least one day during the study period; five subjects took 100% of the dose every day.

The lowest percentage daily intake of the study product was 75%, which occurred once throughout the study (Day 1, Subject 2).

3.5. Blood Phenylalanine and Tyrosine Control

Blood phenylalanine and tyrosine control was satisfactory over the study period, with lower blood phenylalanine levels recorded in five of the seven subjects (Subject 1: −40%, Subject 2: −7%, Subject 3: NA, Subject 4: −33%, Subject 6: −7%, Subject 7: −50%) (Table 3 and Figure 1). One subject had blood sample labelling issues, so blood phenylalanine was not reported whilst on the study product (Subject 3). The blood phenylalanine level increased in one child (Subject 5: +21%) but remained within the recommended target range. Tyrosine baseline and Day 7 data were available for five subjects: tyrosine levels increased in three subjects and were lower in two subjects. The phenylalanine/tyrosine ratio improved in four of five subjects.

Table 3. Phenylalanine and tyrosine blood levels of the substudy patient cohort during the study.

Subject	1	2	3	4	5	6	7
Target Phe levels (μmol/L)	120–360	120–360	120–360	120–360	120–360	120–360	120–600
			Phenylalanine Levels (μmol/L)				
Baseline	500	270	60	500	290	430	520
Day 7	460	250	NA [1]	385	350	400	260
			Tyrosine Levels (μmol/L)				
Baseline	110	120	50	30	60	50	70
Day 7	150	160	NA [1]	NA [1]	40	60	50
			Phenylalanine/Tyrosine Ratio				
Baseline	4.5	2.3	1.2	16.7	4.8	8.6	7.4
Day 7	3.1	1.6	NC [1]	NC [1]	8.8	6.7	5.2

[1] Sample labelling issue. NA: not available; NC: not calculable.

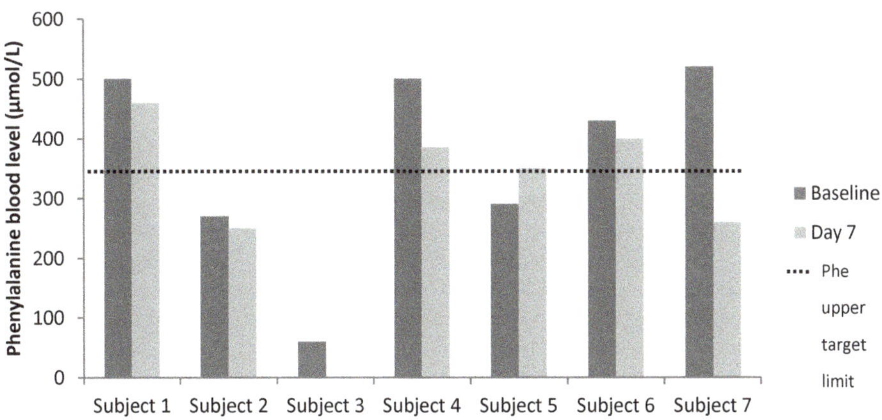

Figure 1. Phenylalanine blood levels of the substudy patient cohort over the study period. Phe target range was 120–360 μmol/L for patients ≤12 years and 120–600 μmol/L for patients >12 years. (Note: Sample labelling issue for Subject 3 on Day 7, so no result recorded). Abbreviation: Phe: phenylalanine.

3.6. Gastrointestinal Symptoms

The study product was well tolerated by all seven subjects for the entire study period. Four subjects reported fewer gastrointestinal symptoms, with less burping, flatulence and regurgitation whilst taking the study product. In one subject, 'severe' burping, flatulence and regurgitation was recorded at baseline and reduced to 'mild' with the study product. There were no cases of mild, moderate or severe diarrhoea, constipation, bloating/distension, nausea or vomiting during the study. There was one case of moderate abdominal discomfort and pain (attributed to menstruation).

3.7. Adverse Events

No adverse reactions were reported.

4. Discussion

In this study, we assessed the introduction of a novel protein substitute with a coating that supports a more physiological release of amino acids than existing substitutes and also masks their bitter taste, aftertaste and smell [16]. The study design followed the guidelines of the UK ACBS with the objective of evaluating the acceptability of the new protein substitute with a limited number of subjects and within a short evaluation time. Despite a short time period for adaptation to the study product, subjects did not report that taste and smell were an issue. During the study, most of the subjects either showed improved blood phenylalanine and tyrosine control or at least maintained them within their target range. This is an encouraging preliminary outcome.

Many patients with PKU are food neophobic [24,25]: and are very suspicious of anything different. Consequently, this short evaluation study was particularly challenging. The study product was dissimilar from current protein substitutes, which patients were well used to taking. Each child understood the potential advantages of taking a prolonged-release protein in terms of impact on blood phenylalanine control and product taste and was very open to trying this new product. However, none had previously experienced taking a protein substitute added to their usual food or drink, so this was an unfamiliar concept to them. The abrupt introduction of a new substitute was overwhelming to some children. Understandably, it is likely to take time to adapt to change, and a slower, more gradual introduction may have been more acceptable. Thereby, a program of slow and systematic introduction, with consistent support given by the PKU team, is essential. Additionally, accompanying educational messages may significantly influence motivation and persistence when introducing any new protein substitute [26]. Generally, patients with PKU and their caregivers welcome new treatments, changes in treatment strategies, new foods, and different presentations of protein substitutes if it improves acceptability and tolerance of treatment [27].

Most of the subjects added the study product to fruit smoothies or juice. This form of administration was quick and similar to their current method of taking protein substitute. This study group has always treated protein substitute differently to food, consumed it immediately before or after meals. If the study evaluation had been extended, they might have eventually accepted mixing the new protein substitute with food or drinks. Due to the study product's coating, the taste and smell were masked, enabling the protein substitute to be taken as an 'add-on' to meals and snacks without affecting the taste of the original food. This type of protein substitute may be particularly useful for patients returning to a low-phenylalanine diet, who have unpleasant memories associated with the smell and taste of protein substitute. It may also be helpful during pregnancy for women struggling with nausea and vomiting associated with the taste and smell of protein substitutes [28]. The excipients used for the coating of the study product are considered safe for pregnancy and lactation [29].

The study product was well tolerated over the treatment period. Protein substitutes are known to have a high osmolarity and therefore may cause gastrointestinal upsets, including abdominal pain, diarrhoea and constipation [12,15,30]. In this cohort, there were no cases of diarrhoea, constipation, bloating/distension, nausea or vomiting during the study. Four subjects reported improvements in the severity of burping, flatulence and regurgitation that they had experienced while receiving their previous treatment regimen. Importantly, there were no adverse reactions reported during the study period. The use of an ethyl cellulose and alginate coating on the study product resulted in granules of amino acids that were stable in the stomach with a gradual disintegration during small-intestine transit [29,31]. This may explain the favourable gastrointestinal tolerability in this study. Both excipients (ethyl cellulose and alginate) are widely used in pharmaceutical technology and recognised as safe for use in medical foods [16]. It would be important to observe whether benefits to gastrointestinal tolerance could be observed longer-term, particularly if the study product provided a higher proportion of protein-substitute intake.

In recent years, several new protein-substitute compositions and formulations have been developed with the aim of improving adherence, which remains suboptimal especially in adolescents and adults with PKU [32]. However, all contain free L-amino acids with absorption kinetics that are more rapid than intact protein sources, causing a lower biological and functional efficacy [33–35]. The introduction of a prolonged-release protein substitute with the ability to prolong absorption of synthetic amino acids is expected to minimise blood phenylalanine level fluctuations and improve phenylalanine control and other metabolic markers in PKU [8,23], but longer-term studies are required to examine the impact on blood phenylalanine control and 24 h blood phenylalanine variability.

There were limitations to this study evaluation. It represented a small patient population with a limited treatment period. Only around 20 to 30% of the usual protein substitute requirements were substituted with the study product due to the short time of adaptation. There were limited blood phenylalanine control data. Additional, more extensive studies performed in a larger population and over a longer timeframe will provide more evidence regarding the adherence and tolerability of this protein substitute, together with impact on metabolic control in patients with PKU.

5. Conclusions

In children with PKU, partial replacement of standard protein-substitute with a prolonged-release protein substitute was achievable. The study product was mixed with food or fruit juice. Subjects maintained satisfactory blood phenylalanine control and were able to take their protein substitute in a different way to their usual practice. The prolonged-release protein substitute was well tolerated, with subjects experiencing fewer gastrointestinal symptoms than with their previous treatment regimen.

Author Contributions: Conceptualization, A.M.; methodology, A.M., A.D., S.E.; software, A.M.; validation, A.M.; formal analysis, A.M., C.A., A.D., and A.P.; investigation, A.M.; resources, A.M.; data curation, A.M.; writing—original draft preparation, A.M., C.A., A.D., A.P., and S.E.; writing—review and editing, A.M., C.A., A.D., A.P., and S.E.; visualization, A.M.; supervision, A.M.; project administration, A.M.; funding acquisition, A.M. All authors have read and agreed to the published version of the manuscript.

Funding: This research was funded by APR Applied Pharma Research SA.

Acknowledgments: The authors thank Sam Coates for their assistance in preparing the manuscript for publication and Patrizia Marzorati at APR Applied Pharma Research S.A. for her review and comments.

Conflicts of Interest: A.M. is an advisory board member for ELEMENT Danone-Nutricia, Arla, and Applied Pharma Research, and has received research funding and honoraria from Nutricia, and Vitaflo International. S.E. has received research funding from Nutricia, and financial support and honoraria from Nutricia and Vitaflo International to attend/speak at study days and conferences. C.A., and A.D. declare no conflict of interest. A.P. has received an educational grant from Cambrooke Therapeutics, research funding from Vitaflo, Nutricia, and Biomarin. Biomarin, Mevalia, Vitaflo and Nutricia have provided research funding to attend scientific meetings.

References

1. Van Wegberg, A.M.J.; MacDonald, A.; Ahring, K.; Belanger-Quintana, A.; Blau, N.; Bosch, A.M.; Burlina, A.; Campistol, J.; Feillet, F.; Gizewska, M.; et al. The complete European guidelines on phenylketonuria: Diagnosis and treatment. *Orphanet J. Rare Dis.* **2017**, *12*, 162. [CrossRef] [PubMed]
2. Hood, A.; Grange, D.K.; Christ, S.E.; Steiner, R.; White, D.A. Variability in phenylalanine control predicts IQ and executive abilities in children with phenylketonuria. *Mol. Genet. Metab.* **2014**, *111*, 445–451. [CrossRef] [PubMed]
3. DeRoche, K.; Welsh, M. Twenty-five years of research on neurocognitive outcomes in early-treated phenylketonuria: Intelligence and executive function. *Dev. Neuropsychol.* **2008**, *33*, 474–504. [CrossRef] [PubMed]

4. Simon, E.; Schwarz, M.; Roos, J.; Dragano, N.; Geraedts, M.; Siegrist, J.; Kamp, G.; Wendel, U. Evaluation of quality of life and description of the sociodemographic state in adolescent and young adult patients with phenylketonuria (PKU). *Health Qual. Life Outcomes* **2008**, *6*, 25. [CrossRef]
5. Al Hafid, N.; Christodoulou, J. Phenylketonuria: A review of current and future treatments. *Transl. Pediatr.* **2015**, *4*, 304–317.
6. Van Spronsen, F.J.; van Wegberg, A.M.; Ahring, K.; Belanger-Quintana, A.; Blau, N.; Bosch, A.M.; Burlina, A.; Campistol, J.; Feillet, F.; Gizewska, M.; et al. Key European guidelines for the diagnosis and management of patients with phenylketonuria. *Lancet Diabetes Endocrinol.* **2017**, *5*, 743–756. [CrossRef]
7. Macleod, E.L.; Ney, D.M. Nutritional Management of Phenylketonuria. *Ann. Nestle* **2010**, *68*, 58–69. [CrossRef]
8. MacDonald, A.; Singh, R.H.; Rocha, J.C.; van Spronsen, F.J. Optimising amino acid absorption: Essential to improve nitrogen balance and metabolic control in phenylketonuria. *Nutr. Res. Rev.* **2019**, *32*, 70–78. [CrossRef]
9. MaCdonald, A.; van Rijn, M.; Feillet, F.; Lund, A.M.; Bernstein, L.; Bosch, A.M.; Gizewska, M.; van Spronsen, F.J. Adherence issues in inherited metabolic disorders treated by low natural protein diets. *Ann. Nutr. Metab.* **2012**, *61*, 289–295. [CrossRef]
10. Shaw, V. (Ed.) *Clinical Paediatric Dietetics*, 4th ed.; Wiley-Blackwell: Oxford, UK, 2014; p. 864.
11. MacDonald, A. Diet and compliance in phenylketonuria. *Eur. J. Pediatr.* **2000**, *159* (Suppl. S2), S136–S141. [CrossRef]
12. Ford, S.; O'Driscoll, M.; MacDonald, A. Living with Phenylketonuria: Lessons from the PKU community. *Mol. Genet. Metab. Rep.* **2018**, *17*, 57–63. [CrossRef] [PubMed]
13. Walter, J.H.; White, F.J. Blood phenylalanine control in adolescents with phenylketonuria. *Int. J. Adolesc. Med. Health* **2004**, *16*, 41–45. [CrossRef] [PubMed]
14. Green, B.; Browne, R.; Firman, S.; Hill, M.; Rahman, Y.; Kaalund Hansen, K.; Adam, S.; Skeath, R.; Hallam, P.; Herlihy, I.; et al. Nutritional and metabolic characteristics of UK adult phenylketonuria patients with varying dietary adherence. *Nutrients* **2019**, *11*, 2459. [CrossRef] [PubMed]
15. MacDonald, A.; Harris, G.; Rylance, G.; Asplin, D.; Booth, I.w. Abnormal feeding behaviours in phenylketonuria. *J. Hum. Nutr. Diet.* **1997**, *10*, 163–170. [CrossRef]
16. Giarratana, N.; Gallina, G.; Panzeri, V.; Frangi, A.; Canobbio, A.; Reiner, G. A new Phe-free protein substitute engineered to allow a physiological absorption of free amino acids for phenylketonuria. *J. Inborn Errors Metab.* **2018**, *6*, 1–9. [CrossRef]
17. Pena, M.J.; de Almeida, M.F.; van Dam, E.; Ahring, K.; Belanger-Quintana, A.; Dokoupil, K.; Gokmen-Ozel, H.; Lammardo, A.M.; MacDonald, A.; Robert, M.; et al. Protein substitutes for phenylketonuria in Europe: Access and nutritional composition. *Eur. J. Clin. Nutr.* **2016**, *70*, 785–789. [CrossRef]
18. Paul, D.B.; Brosco, J.P. *The PKU Paradox. A Short History of a Genetic Disease*; Johns Hopkins University Press: Baltimore, MD, USA, 2013.
19. Tiele, A.; Daly, A.; Hattersley, J.; Pinto, A.; Evans, S.; Ashmore, C.; MacDonald, A.; Covington, J.A. Investigation of paediatric PKU breath malodour, comparing glycomacropeptide with phenylalanine free L-amino acid supplements. *J. Breath Res.* **2019**, *14*, 016001. [CrossRef]
20. Lim, K.; van Calcar, S.C.; Nelson, K.L.; Gleason, S.T.; Ney, D.M. Acceptable low-phenylalanine foods and beverages can be made with glycomacropeptide from cheese whey for individuals with PKU. *Mol. Genet. Metab.* **2007**, *92*, 176–178. [CrossRef]
21. MacDonald, A.; Gokmen-Ozel, H.; van Rijn, M.; Burgard, P. The reality of dietary compliance in the management of phenylketonuria. *J. Inherit. Metab. Dis.* **2010**, *33*, 665–670. [CrossRef]
22. Duran, G.P.; Rohr, F.J.; Slonim, A.; Guttler, F.; Levy, H.L. Necessity of complete intake of phenylalanine-free amino acid mixture for metabolic control of phenylketonuria. *J. Am. Diet. Assoc.* **1999**, *99*, 1559–1563. [CrossRef]
23. Scheinin, M.; Barassi, A.; Junnila, J.; Lovro, Z.; Reiner, G.; Sarkkinen, E.; MacDonald, A. Amino acid plasma profiles from a prolonged-release protein substitute for phenylketonuria: A randomized, single-dose, four-way crossover trial in healthy volunteers. *Nutrients* **2020**, *12*, 1653. [CrossRef] [PubMed]
24. Evans, S.; Daly, A.; Chahal, S.; MacDonald, J.; MacDonald, A. Food acceptance and neophobia in children with phenylketonuria: A prospective controlled study. *J. Hum. Nutr. Diet.* **2016**, *29*, 427–433. [CrossRef] [PubMed]

25. Evans, S.; Daly, A.; Chahal, S.; Ashmore, C.; MacDonald, J.; MacDonald, A. The influence of parental food preference and neophobia on children with phenylketonuria (PKU). *Mol. Genet. Metab. Rep.* **2018**, *14*, 10–14. [CrossRef] [PubMed]
26. Bernstein, L.E.; Helm, J.R.; Rocha, J.C.; Almeida, M.F.; Feillet, F.; Link, R.M.; Gizewska, M. Nutrition education tools used in phenylketonuria: Clinician, parent and patient perspectives from three international surveys. *J. Hum. Nutr. Diet.* **2014**, *27* (Suppl. S2), 4–11. [CrossRef]
27. Giovannini, M.; Verduci, E.; Salvatici, E.; Paci, S.; Riva, E. Phenylketonuria: Nutritional advances and challenges. *Nutr. Metab.* **2012**, *9*, 7. [CrossRef] [PubMed]
28. Ford, S.; O'Driscoll, M.; MacDonald, A. Reproductive experience of women living with phenylketonuria. *Mol. Genet. Metab. Rep.* **2018**, *17*, 64–68. [CrossRef] [PubMed]
29. Wasilewska, K.; Winnicka, K. Ethylcellulose—A pharmaceutical excipient with multidirectional application in drug dosage forms development. *Materials (Basel)* **2019**, *12*, 3386. [CrossRef]
30. PBMK, A. *Nutritional Management of Patients with Inherited Disorders of Aromatic Amino Acid Metabolism*; Jones and Bartlett: Boston, MA, USA, 2010.
31. Arica, B.; Calis, S.; Atilla, P.; Durlu, N.T.; Cakar, N.; Kas, H.S.; Hincal, A.A. In vitro and in vivo studies of ibuprofen-loaded biodegradable alginate beads. *J. Microencapsul.* **2005**, *22*, 153–165. [CrossRef]
32. Rocha, J.C.; MacDonald, A. Dietary intervention in the management of phenylketonuria: Current perspectives. *Pediatr. Health Med. Ther.* **2016**, *7*, 155–163. [CrossRef]
33. Lacroix, M.; Bos, C.; Leonil, J.; Airinei, G.; Luengo, C.; Dare, S.; Benamouzig, R.; Fouillet, H.; Fauquant, J.; Tome, D.; et al. Compared with casein or total milk protein, digestion of milk soluble proteins is too rapid to sustain the anabolic postprandial amino acid requirement. *Am. J. Clin. Nutr.* **2006**, *84*, 1070–1079. [CrossRef]
34. Bujko, J.; Schreurs, V.V.; Nolles, J.A.; Verreijen, A.M.; Koopmanschap, R.E.; Verstegen, M.W. Application of a [13CO2] breath test to study short-term amino acid catabolism during the postprandial phase of a meal. *Br. J. Nutr.* **2007**, *97*, 891–897. [CrossRef] [PubMed]
35. Metges, C.C.; Barth, C.A. Metabolic consequences of a high dietary-protein intake in adulthood: Assessment of the available evidence. *J. Nutr.* **2000**, *130*, 886–889. [CrossRef] [PubMed]

© 2020 by the authors. Licensee MDPI, Basel, Switzerland. This article is an open access article distributed under the terms and conditions of the Creative Commons Attribution (CC BY) license (http://creativecommons.org/licenses/by/4.0/).

nutrients

MDPI

Article

The Impact of the Use of Glycomacropeptide on Satiety and Dietary Intake in Phenylketonuria

Anne Daly [1,*], Sharon Evans [1], Alex Pinto [1], Richard Jackson [2], Catherine Ashmore [1], Júlio César Rocha [3,4] and Anita MacDonald [1]

1. Dietetic Department, Birmingham Children's Hospital, Steelhouse Lane, Birmingham B4 6NH, UK; evanss21@me.com (S.E.); alex.pinto@nhs.net (A.P.); catherine.ashmore@nhs.net (C.A.); anita.macdonald@nhs.net (A.M.)
2. Liverpool Clinical Trials Centre, University of Liverpool, Brownlow Hill, Liverpool L69 3GL, UK; r.j.jackson@liverpool.ac.uk
3. Nutrition and Metabolism, NOVA Medical School, Faculdade de Ciências Médicas, Universidade Nova de Lisboa, 1169–056 Lisboa, Portugal; rochajc@nms.unl.pt
4. Centre for Health and Technology and Services Research (CINTESIS), 4200–450 Porto, Portugal

* Correspondence: a.daly3@nhs.net

Received: 29 July 2020; Accepted: 1 September 2020; Published: 4 September 2020

Abstract: Protein is the most satiating macronutrient, increasing secretion of gastrointestinal hormones and diet induced thermogenesis. In phenylketonuria (PKU), natural protein is restricted with approximately 80% of intake supplied by a synthetic protein source, which may alter satiety response. Casein glycomacropeptide (CGMP-AA), a carbohydrate containing peptide and alternative protein substitute to amino acids (AA), may enhance satiety mediated by its bioactive properties. Aim: In a three-year longitudinal; prospective study, the effect of AA and two different amounts of CGMP-AA (CGMP-AA only (CGMP100) and a combination of CGMP-AA and AA (CGMP50) on satiety, weight and body mass index (BMI) were compared. Methods: 48 children with PKU completed the study. Median ages of children were: CGMP100; (n = 13), 9.2 years; CGMP50; (n = 16), 7.3 years; and AA (n = 19), 11.1 years. Semi-quantitative dietary assessments and anthropometry (weight, height and BMI) were measured every three months. Results: The macronutrient contribution to total energy intake from protein, carbohydrate and fat was similar across the groups. Adjusting for age and gender, no differences in energy intake, weight, BMI, incidence of overweight or obesity was apparent between the groups. Conclusion: In this three-year longitudinal study, there was no indication to support a relationship between CGMP and satiety, as evidenced by decreased energy intake, thereby preventing overweight or obesity. Satiety is a complex multi-system process that is not fully understood.

Keywords: phenylketonuria; PKU; glycomacropeptide; satiety

1. Introduction

Phenylketonuria (PKU), due to phenylalanine hydroxylase deficiency, leads to accumulation of phenylalanine and irreversible brain damage if untreated [1,2]. A lifelong phenylalanine/protein restricted diet is essential, and most patients with classical PKU tolerate ≤ 500 mg/day of phenylalanine (equivalent to ≤10 g/day protein). Meats, fish, eggs and cheese are avoided with foods such as potatoes, cereals and peas given in restricted and calculated amounts; special low protein foods together with some fruits and vegetables (containing phenylalanine ≤ 75 mg/100 g) are given without restriction. Therefore, the diet is lacking in high quality protein and protein intake is supplemented with a minimal/phenylalanine-free protein substitute, usually supplying up to 80% of protein requirements. It is unclear if this synthetic source of protein alters satiety. Satiety, which is a sense of fullness after eating, is important in regulating food intake [3,4]. In general nutrition, there is evidence that the amount and type of dietary protein alters appetite and may influence weight regulation [5,6].

Protein substitutes for PKU are obtained from either casein glycomacropeptide (CGMP) or phenylalanine-free amino acids (AAs). CGMP and AAs are compositionally different. Amino acid-based protein substitutes are composed of free L-amino acids only, whereas CGMP is a glycosylated peptide containing varying amounts of oligosaccharides, mostly sialic acid (N-acetylneuraminic acid), galactosamine and galactose [7]. CGMP has prebiotic, antimicrobial and immunomodulatory effects [8,9] and is prevalent in bovine milk. It constitutes 20–25% of the total protein in whey products [10]. Whey protein has been shown to induce satiety to a greater extent than other protein sources such as casein, soya and egg albumin [11,12]. There is some evidence that CGMP influences hormone responses affecting satiety [13–15]. However, human studies investigating the effect of CGMP on food intake and satiety have resulted in mixed findings [16–19].

Gut absorption of protein is thought to modulate satiety, although the influence of the protein source and individual amino acids on the control of food intake is not completely understood. It involves complex pathways that affect vagus-mediated signals, satiety related hormones and their metabolites (including the peptide ghrelin, cholecystokinin (CCK), glucagon-like peptide-1 (GLP-1), glucose dependent insulinotropic polypeptide (GIP) and peptide tyrosine–tyrosine (PYY)) [11,15,20–23]. Some blood amino acids, particularly leucine, lysine, tryptophan, isoleucine and threonine, are linked to satiety responses [24–27]. CGMP is a small peptide and likely to be quickly absorbed [22,28,29]. A sharp rise in the mean circulating amino-acid concentrations following protein ingestion has been associated with a reduction in appetite [30,31]. Korompokis et al. [32] investigated the absorption kinetics of protein and the impact of amino acids on appetite and satiety related hormones. In a randomized cross over study using liquid preloads with a similar energy density but variable energy from protein, carbohydrate and fat, no postprandial kinetic effect of amino acids on appetite was shown. They concluded protein intake affected the amino acid profile but was not related to appetite regulation.

Equally, it is well established that amino acids bypass degradation by proteases and compared to whole proteins are absorbed faster [33–36]. In our own center, we reviewed plasma amino acid concentrations both fasting and after 2 h following a breakfast meal and 20 g protein equivalent from protein substitute either based on AA or CGMP. Although there were significant differences for individual amino acids, probably related to the amino acid composition of the protein substitutes, postprandial total amino acid concentrations were not different between the protein substitutes [37]. Until the absorption kinetics of CGMP (CGMP with added rate limiting amino acids) has been reported, its influence on satiety remains speculative. Additionally, CGMP and AA supplements modified for PKU usually contain added carbohydrate, which may also influence amino acid kinetics and hormone response [28].

If CGMP did increase satiety in PKU as suggested [38], it may bring important benefits, potentially helping control obesity, commonly reported in PKU [38,39]. In this study, we compared energy intake, weight and body mass index (BMI) over a three-year period in a group of children with PKU taking either CGMP at two different concentrations or amino acid supplements. We hypothesized that, if CGMP influenced satiety, then energy intake should be lower and weight and BMI changes altered between the groups.

Ethical Permission

The South Birmingham Research Ethics committee granted a favorable ethical opinion, referenced 13/WM/0435 and IRAS (integrated research application system) number 129497. Written informed consent was obtained for all subjects from at least one caregiver with parental responsibility and written assent obtained from the subject if appropriate for their age and level of understanding.

2. Methods and Materials

2.1. Subjects

In a three-year long-term prospective study, 50 children (28 boys, 22 girls) with PKU were recruited. Their median recruitment age was 9.2 years (range 5–16 years). Forty-seven children were European and three were of Pakistani origin. Inclusion criteria included: diagnosed by newborn screening, aged 5–16 years, not treated with sapropterin dihydrochloride and adherent to protein substitute. Seventy percent of routine blood phenylalanine concentrations were within phenylalanine target range for six months before study enrolment. Target blood phenylalanine concentrations for children aged 5–11 years was <360 µmol/L and for 12 years and older <600 µmol/L as recommended by the European PKU guidelines [40].

2.2. Protein Substitutes

Two types of protein substitute were studied: one based on phenylalanine-free AA (liquid pouches or powders) and the other on a powdered CGMP-AA supplement (a study product, made by Vitaflo, International Ltd, UK) (see Table 1 for nutritional analysis comparisons). The CGMP contained a residual amount of phenylalanine (36 mg/20 g protein equivalent). It was supplemented with essential and semi-essential amino acids, to provide a balanced amino acid profile to sustain nitrogen requirements and so the term CGMP-AA is used.

Table 1. The nutrient composition of CGMP-AA compared with conventional AA.

Protein Substitute		CGMP-AA	Phe-Free AA *
Nutrients	**Units**	**Per 20 g PE**	**Per 20 g PE**
Calories	Kcal	120	124
Protein equivalent	g	20	20
Total Carbohydrate	g	6.5	9.4
Sugars	g	2.2	7.8
Total Fat	g	1.5	0.7
Docosahexaenoic acid	mg	84	134
Salt	g	0.53	0.43
Vitamin A	µg RE	283	278
Vitamin D	µg	4.5	10
Vitamin E	mg αTE	6.5	5.2
Vitamin C	mg	38	36
Vitamin K	µg	35	34
Thiamine	mg	0.68	0.70
Riboflavin	mg	0.78	0.77
Niacin	mg	8.4	8.4
Vitamin B6	mg	1.0	0.9
Folic Acid	µg	136	134
Vitamin B12	µg	1.6	1.6
Biotin	µg	63.9	63
Pantothenic acid	mg	2.7	2.6
Choline	mg	204	201
Sodium	mmol	9.0	7.3
Potassium	mmol	6.8	7.9
Chloride	mmol	0.2	3.9
Calcium	mg	407	400

Table 1. *Cont.*

Protein Substitute		CGMP-AA	Phe-Free AA *
Nutrients	**Units**	**Per 20 g PE**	**Per 20 g PE**
Phosphorus	mmol	12	11.4
Magnesium	mg	128	125
Iron	mg	7.3	7.3
Copper	µg	748	730
Zinc	mg	7.3	7.3
Manganese	mg	1.1	1
Iodine	µg	85.7	84
Molybdenum	µg	49	48
Selenium	µg	29.9	29
Chromium	µg	29.9	29
Amino acids			
L-Alanine	g	0.78	0.92
L-Arginine	g	0.95	1.5
L-Aspartic Acid	g	1.12	2.37
L-Cystine	g	0.01	0.61
L-Glutamine [1]	g	2.57	-
Glycine	g	1.2	2.35
L-Histidine	g	0.7	0.92
L-Isoleucine	g	1.35	1.62
L-Leucine	g	3.00	2.54
L-Lysine	g	0.80	1.67
L-Methionine	g	0.28	0.45
L-Phenylalanine	g	0.03	0
L-Proline	g	1.52	1.69
L-Serine	g	0.96	1.04
L-Threonine	g	2.20	1.62
L-Tryptophan	g	0.40	0.5
L-Tyrosine	g	2.24	2.38
L-Valine	g	1.09	1.86

CGMP-AA, casein glycomacropeptide; AA, phenylalanine-free amino acid; PE, protein equivalent; * based on liquid Phe-free amino acids (Vitaflo International Ltd); [1] glutamine content varies according to if the AA protein substitute is a liquid or powder.

2.3. Study Design

The primary aim of this three-year single center, longitudinal study was to compare the efficacy of CGMP-AA compared to AA on bone health in a group of 50 PKU children (this will be reported separately). The secondary aim was to study energy intake, with particular reference to protein intake (from protein substitutes and phenylalanine exchanges), weight and BMI changes between the two groups, exploring the theory that CGMP-AA is actively related to satiety. Following the findings from a pilot study [41], it was clear that not all children in the CGMP-AA group were able to tolerate their entire protein substitute from CGMP-AA, due to its phenylalanine content. Therefore, within this group, there was a further subdivision: CGMP100, those taking all their substitute from CGMP-AA and those taking a combination of CGMP-AA and AA, named CGMP50. A group of children remained on AA only.

2.4. Selection into AA or CGMP-AA Group

The children chose AA or CGMP-AA, depending on their taste preference. They remained on this protein substitute for the three-year duration of the study.

2.5. Dietary Assessment

A three-day semi-quantitative dietary assessment was completed once every three months. The diet diaries were all checked during a face-to-face interview with caregivers by one of two trained metabolic dietitians. Protein containing foods were weighed. A picture book of pre-weighed foods was used to help caregivers estimate the portion size of other foods such as low protein pasta, fruit and vegetables. Twice a year, average food portions were weighed, e.g., low protein pasta, bread and low protein sausages. All children were observed eating at least one meal annually.

2.6. Nutritional Analysis

The dietary assessments were analyzed using Nutritics Nutritional Software (v5.093). The following macronutrients were analyzed: daily energy (Kcal), protein (g), carbohydrate (unrefined and refined) (g) and fat (g). For each subject, the annual median macronutrient value was calculated, and the median value for all subjects in each group was determined, giving the median of the median value for each macronutrient. The results were compared with age and gender specific UK dietary reference values or estimated average requirement (EAR) for energy (UK Scientific Advisory Committee on Nutrition (SCAN)) [42]. The nutrient contribution from CGMP-AA and AA were included in this analysis as well as all special low protein foods.

2.7. Anthropometric Measurements

Weight, height and BMI were measured once every three months by one of two metabolic dietitians. Height was measured with a Harpenden stadiometer (Holtain Ltd, Crymych, UK) and weight on calibrated digital scales (Seca UK model 875); they were measured to the nearest 0.1 cm and 0.1 kg, respectively.

2.8. Blood Phenylalanine Levels

Trained parents/caregivers collected weekly early morning fasted blood spots on filter cards, Perkin Elmer 226 (UK Standard NBS). Blood specimens were sent via first class post to the laboratory at Birmingham Children's Hospital. All the cards had a standard thickness and the blood phenylalanine concentrations were calculated on a 3.2-mm punch by MS/MS tandem mass spectrometry.

2.9. Statistical Analysis

Continuous data are represented as median (IQR) and categorical data were summarized as frequencies of counts and associated percentages. Analyses of study endpoints were performed using Analysis of Covariance (ANCOVA) techniques, which analyzes the data at three years follow-up while adjusting for baseline values. As there was a difference in age between participants between the two groups, all models were adjusted for patient age as well as gender. The impact of CGMP compared to AA was evaluated by comparison of CGMP100 and CGMP50. Descriptive statistics are reported as medians and differences at baseline and follow up were assessed using a paired t test. Analysis was performed using R (Version 3).

3. Results

Of the 50 children recruited, 48 completed the study: CGMP group, $n = 29$; AA group, $n = 19$. The CGMP group was divided into CGMP100, $n = 13$ (45%), and CGMP50, $n = 16$, (55%). The median ages at enrolment were: CGMP100, 9.2 years; CGMP50, 7.3 years; and AA, 11.1 years. There was a significant difference in age between the AA and CGMP50 ($p = 0.005$) and between CGMP50 and CGMP100 ($p = 0.04$).

Prior to starting the CGMP-AA, all patients were prescribed AA as their source of protein substitute. Six subjects took powdered amino acids (XP Maxamum (Nutricia Ltd., Trowbridge, UK), $n = 1$; PKU Anamix first spoon (Nutricia Ltd.), $n = 3$; PKU gel (Vitaflo International Ltd.), $n = 2$) and 44 subjects took liquid pouches (PKU Lophlex LQ (Nutricia Ltd.), $n = 3$; PKU Cooler (Vitaflo International Ltd., Liverpool, UK), $n = 41$). The AA group remained on their usual protein substitute (PKU Lophlex LQ (Nutricia Ltd.), $n = 1$; PKU Cooler (Vitaflo International Ltd) $n = 14$) or a powdered preparation

(PKU gel (Vitaflo International Ltd.), $n = 4$) throughout the study. The median (range) phenylalanine concentrations at baseline were not statistically different: CGMP100, 255 μmol/L (170–360); CGMP50, 290 μmol/L (220–430); and AA, 315 μmol/L (215–600). The majority had classical PKU, except two children who were mild based on untreated blood phenylalanine levels at diagnosis and dietary phenylalanine tolerance.

The median daily dose of protein equivalent from protein substitute was 60 g/day (range 40–80 g), and the median amount of prescribed natural protein was 5.5 g protein/day in all groups (range 3–30 g) or 275 mg/day phenylalanine (range 150–1500 mg).

3.1. Subject Withdrawal

One boy and one girl (aged 12 and 11 years, respectively) in the CGMP-AA group were excluded from the study, as both were unable to adhere with the study protocol. One failed to return blood phenylalanine samples and both had poor adherence to their phenylalanine restricted diet.

3.2. Nutritional Intake

Change in Nutrient Intake

In all the groups, the energy intake expressed as a percentage of EAR decreased over the three years. In the AA group: baseline, 106% (77–177), year 3, 95% (80–138); CGMP50, baseline, 105% (90–120), year 3, 100% (88–144); and CGMP100, baseline, 104% (85–126), year 3, 101% (87–118) (Table 2).

ANCOVA analysis adjusting for age, gender, energy intake (Kcal/d) and EAR showed that the difference in three-year EAR between AA and CGMP50 was not statistically significant ($p = 0.717$) and neither was the difference between AA and CGMP100 ($p = 0.673$). Further longitudinal analysis showed no significant differences between the groups over the three years for energy intake.

Table 2. Median daily energy intake (range) and %EAR (range) for AA, CGMP50 and CGMP100 from baseline to year 3.

	Median Energy (Kcal/Day)			**Median % EAR**		
Year	**AA (Range)** $n = 19$	**CGMP50 (Range)** $n = 16$	**CGMP100 (Range)** $n = 13$	**AA (Range)** $n = 19$	**CGMP50 (Range)** $n = 16$	**CGMP100 (Range)** $n = 13$
Baseline	1950 (1138–2999)	1701 (1466–2494)	1831 (1612–2591)	106 (77–177)	105 (90–120)	104 (85–126)
1	1957 (1151–3191)	1793 (1560–2681)	1917 (1642–2708)	102 (77–140)	99 (85–132)	96 (73–124)
2	1976 (1517–2669)	1828 (1512–2814)	1965 (1258–2810)	99 (72–130)	105 (90–158)	107 (85–152)
3	2120 (1111–3387)	1966 (1523–2405)	2064 (1672–3144)	95 (80–138)	100 (88–144)	101 (87–118)

AA, amino acid; CGMP, casein glycomacropeptide; EAR, estimated average requirement; CGMP50, patients taking a combination of CGMP-AA and AA; CGMP100, patients taking all their protein substitute from CGMP-AA.

Over the three-year period, the median percentage energy contribution from carbohydrate, protein and fat was not significantly different among the three groups (Table 3).

The median percentage energy contribution from protein (including protein substitute and natural protein from food) was similar for all three groups. Protein provided a median of 15% (75 g) of the total energy intake, and natural protein intake supplied a median intake of 10–16% (8–12 g/day) of the total protein intake. A small but significant difference was noted for the contribution of protein substitute between CGMP50 (88%) and CGMP100 (85%) ($p = 0.01$) as well as between CGMP50 (88%) and AA (85%) ($p = 0.007$). Although the protein substitutes still provided >85% of the total protein intake in all groups, there were significant differences in the youngest age group (CGMP50) and in those children unable to tolerate extra phenylalanine from CGMP-AA (Table 3).

Table 4 describes the total protein intake from all foods including fruits and vegetables that are usually allowed without restriction in a UK diet. The actual intake was higher compared to the "prescribed" or allocated phenylalanine exchanges (due to the small amounts of protein present in foods given without restriction). There was a significant increase in the actual intake between baseline and year 3 in AA ($p = 0.002$) and CGMP50 ($p = 0.02$) groups. The median protein intake from food was significantly different between and within the groups. CGMP50 had the lowest natural protein intake, with significant differences between CGMP50 and both AA and CGMP100 (Table 4).

These differences reflect protein tolerance, being lower in the youngest age group and those unable to tolerate the extra phenylalanine from CGMP-AA to meet all their protein substitute requirements.

Table 3. Median three-year percentage (range) energy contribution for protein, carbohydrate and fat from food and protein substitute in AA, CGMP50 and CGMP100.

	Median % Energy from Protein			**Median % Energy from Carbohydrate**			**Median % Energy from Fat**		
	AA $n = 19$	**CGMP50** $n = 16$	**CGMP100** $n = 13$	**AA** $n = 19$	**CGMP50** $n = 16$	**CGMP100** $n = 13$	**AA** $n = 19$	**CGMP50** $n = 16$	**CGMP100** $n = 13$
Baseline (Range)	15 (10–30)	15 (9–26)	15 (11–26)	57 (45–68)	56 (43–70)	57 (45–69)	26 (15–40)	26 (17–37)	28 (17–33)
Year 1–3 (Range)	15 (9–27)	15* (10–22)	16* (12–23)	58 (42–70)	57 (46–67)	58 (48–67)	27 (16–39)	26 (18–37)	27 (18–35)

* $p = 0.02$. AA, amino acid; CGMP, casein glycomacropeptide; CGMP50, patients taking a combination of CGMP-AA and AA; CGMP100, patients taking all their protein substitute from CGMP-AA.

Table 4. Median natural protein intake (g/day) (range) from food sources only and median percentage intake from protein equivalent from substitute from baseline to year 3 in AA, CGMP50 and CGMP100.

Natural Protein Intake from Food (g/d)				
Year	**AA** $n = 19$	**CGMP50** $n = 16$	**CGMP100** $n = 13$	p **Value**
Baseline (Range)	11.5 * (5–32)	8 *,§ (4–23)	12 § (5–36)	*,§ $p = 0.001$
1 (Range)	12 * (5–33)	8 *,§ (4–24)	12 § (5–34)	* $p = 0.0001$, § $p = 0.001$
2 (Range)	13 * (5–33)	9 *,§ (3–23)	10 § (5–44)	* $p = 0.014$, § $p = 0.02$
3 (Range)	12 * (6–46)	9 *,§ (3–24)	13 § (6–36)	* <0.0001, § $p = 0.0006$

Table 4. *Cont.*

Median % protein intake from protein substitute equivalent (range)				
Year	**AA**	**CGMP50**	**CGMP100**	***p* Value**
Baseline (Range)	85 (68–92)	87 (72–93)	85 (58–92)	
1 (Range)	86 (72–92)	88 (71–93)	83 (65–91)	
2 (Range)	86 (65–93)	87 (70–91)	86 (59–93)	
3 (Range)	84 (68–90)	88 (75–93)	84 (68–91)	
Year 1–3 (Range)	85 * (65–93)	88 *,§ (70–93)	85 § (59–93)	* $p = 0.01$, § $p = 0.007$

AA, amino acid; CGMP, casein glycomacropeptide; CGMP50, patients taking a combination of CGMP-AA and AA; CGMP100, patients taking all their protein substitute from * significant difference between AA and CGMP50; § significant difference between CGMP50 and CGMP 100.

3.3. Phenylalanine Control

There was a significant increase in blood phenylalanine levels between baseline and year 3 for AA ($p = 0.02$) and CGMP50 ($p = 0.04$), although all groups had median blood phenylalanine control within target (Table 5). This increase is an expected finding due to children reaching teenage years, when dietary adherence is known to deteriorate [43,44].

Table 5. Median (range) blood phenylalanine concentrations from baseline to year 3 in AA, CGMP50 and CGMP100.

	Baseline Median Phe µmol/L			Year 3 Median Phe µmol/L		
	AA µmol/L *n* = 19	**CGMP50 µmol/L *n* = 16**	**CGMP100 µmol/L *n* = 13**	**AA µmol/L *n* = 19**	**CGMP50 µmol/L *n* = 16**	**CGMP100 µmol/L *n* = 13**
Phe (Range)	315* (140–600)	255 § (170–360)	290 (200–710)	360 * (210–830)	290 § (220–430)	320 (250–895)

* $p = 0.02$, § $p = 0.04$. AA, amino acid; CGMP, casein glycomacropeptide; CGMP50, patients taking a combination of CGMP-AA and AA; CGMP100, patients taking all their protein substitute from CGMP-AA.

3.4. Anthropometric Data

For weight and BMI measurements, ANCOVA was applied adjusting for age, gender, weight and BMI over the three-year duration (Table 6).

Table 6. Change in weight and BMI Z scores from baseline to year 3 in the AA, CGMP50 and CGMP100 groups applying ANCOVA adjusting for variables in age and gender differences.

BMI Z Score	Baseline	36 m	Differences
AA	−0.15	0.63	0.3
CGMP50	0.17	0.85	0.5
CGMP100	−0.11	0.95	0.6
Wt Z score	**Baseline**	**36 m**	**Differences**
AA	−0.25	0.74	0.2
CGMP50	0.28	0.91	0.2
CGMP100	0.02	0.97	0.4

BMI: body mass index; Wt, weight; AA, amino acid; CGMP, casein glycomacropeptide; CGMP50, patients taking a combination of CGMP-AA and AA; CGMP100, patients taking all their protein substitute from CGMP-AA.

Weight and BMI Z Scores

There was no statistical difference in three-year weight Z scores between AA and CGMP50 ($p = 0.7$), AA and CGMP100 ($p = 0.7$) and CGMP50 and CGMP100 ($p = 0.95$). Similarly, for BMI Z scores there was no statistical difference in the three-year BMI Z scores between AA and CGMP50 ($p = 0.784$), AA and CGMP100 ($p = 0.553$) and CGMP50 and CGMP100 ($p = 407$).

Using the World Health organization (WHO) definition of obesity (BMI equivalent to two standard deviations over the reference median), obesity rates between years 1 and 3 remained unchanged in the AA group (26%, $n = 5/19$), increased in the CGMP50 group from 0% to 19% ($n = 3/16$) and remained at 0% in the CGMP100 group over the three years. Overweight, defined as one standard deviation over the reference median, decreased in the AA group from 37% ($n = 7/19$) to 26%, ($n = 5/19$), remained unchanged in the CGMP50 group at 19% ($n = 3/16$) and increased in the CGMP100 group from 15% ($n = 2/13$) to 46% ($n = 6/13$).

4. Discussion

This three-year longitudinal study in PKU systematically reviewed the macronutrient intake and anthropometry of children taking AA compared to CGMP-AA, with CGMP-AA provided at two different CGMP concentrations. Although satiety was not directly measured through satiety visual analog scales and hormone concentrations, the hypothesis that CGMP would enhance satiety leading to lower energy intake and slower weight gain was not observed. EAR decreased in all three groups over the three years, the most consistent decrease was in the AA group, suggesting the relationship between satiety and CGMP-AA was less convincing. Adjusting for age and gender, no differences in weight or BMI were apparent between the AA and CGMP-AA groups. Although there was no obesity present in the CGMP100 group over the three years of study, the rate of overweight increased from 15% to 45%. Our data suggest that the use of CGMP50 or CGMP100 did not reduce energy intake and thereby appetite.

Protein substitutes based on 60 g of protein equivalent from CGMP-AA provided a typical intake of approximately 45 g/day of CGMP. We cannot evaluate from this study if this quantity may have had a significant impact on hormones such as ghrelin and GLP-1, which are important in controlling satiety. It has been shown that whey protein given to healthy volunteers at 10% of the energy intake suppressed subjective hunger but made no impact on actual energy intake. At a higher protein intake of 25% of energy intake, it increased insulin, and active GLP-1 and incretin hormones were higher, but made no difference to overall satiety and energy intake [5]. In non-PKU human studies, consensus is strong that CGMP has no effect on food intake or satiety [16,45], and weight loss after long-term consumption of CGMP has not been demonstrated [17]. Overall, subjective feelings of appetite indicate that CGMP is not critical for whey-induced satiety or energy intake. In our study, natural protein intake was vegetable based, which remained consistent over the study period for each of the three groups and was not a consideration influencing satiety.

Indirectly, CGMP may influence satiety by its prebiotic properties but a direct relationship on appetite has not been studied [46]. Microbiota play a key role in regulating energy balance and satiety, by the interaction of gut hormones and pathways such as leptin-melanocortin, a critical system controlling appetite and energy balance [47,48]. CGMP can change microbiota. In mice fed CGMP, there was a reduction in Desulfovibrio bacteria, an increase in short chain fatty acids and reduced inflammatory markers [9]. Three products, lactose, commercial CGMP (70% GMP, <2% lactose) and semi-purified CGMP (51% GMP, 4% lactose), were tested on fecal high and low diversity microbiota from healthy and elderly subjects. Both CGMP products resulted in a healthy microbiota, being more pronounced in the lower lactose CGMP preparation [49]. How this translates to satiety remains unanswered.

Only one study has examined the effect of CGMP-AA and satiety in subjects with PKU. MacLeod et al. [50] gave 11 subjects (eight adults and three children) a standard breakfast with either CGMP-AA or phenylalanine-free AA. Plasma insulin, ghrelin and amino acids were measured 180 min after breakfast. Postprandial ghrelin was significantly lowered associated with fullness in the

CGMP group, with total plasma amino acids and insulin concentrations only just reaching significance between the groups. This short four-day study only measured one incretin hormone at one time point. Studies on satiety ideally need to employ a wide variation in time intervals, capturing differences in appetite and plasma amino acid profiles [51]. The carbohydrate intake was higher in the CGMP-AA breakfast, which may sustain insulin concentrations over the short study period. Results from the satiety questionnaire show 60% of the adult group were overweight. It also included only three children. Overall, these limitations render any direct interpretation of the effect of CGMP-AA on satiety challenging.

There were limitations to our study. Firstly, dietary assessments regardless of method are inaccurate, although face-to-face interviews and periodic weighed food intakes were conducted to minimize these difficulties. Satiety visual analog scales post protein substitute consumption were not conducted, although their value is subjective particularly in children and obese subjects. The AA group took different amino acid preparations but 15 of 19 consumed low energy liquid preparations. Age varied between the groups but was statistically accounted for; the influence of growth and exercise on appetite was unmeasured. No biochemical hormone markers were assessed. Both insulin and ghrelin are endocrine mediators of food intake. However, insulin is an anabolic hormone and ghrelin has growth hormone functions, further complicating any clear relationship with satiety in children who are growing and reaching adolescence. Children were not randomized to one of the three protein substitute groups; choice of group was dependent on protein substitute acceptance and blood phenylalanine control when taking CGMP-AA. One further limitation was the lack of a non-PKU control group; although dietary composition would have been different, a comparison of energy intake, weight and BMI would have been useful. Despite these limitations, no obvious impact on satiety was found between our two study groups. There are inherent difficulties in studying appetite given the behavioral and environmental factors that counterbalance the physiological regulators of appetite.

5. Conclusions

In this three-year longitudinal study in children with PKU, CGMP-AA when compared to AA did not appear to influence energy intake, weight gain or BMI and by implication satiety. In PKU, there is little understanding of the optimal dietary composition needed to control appetite preventing overweight and obesity. All macronutrients have unique physiological properties that influence metabolic pathways. The impact of CGMP-AA on satiety, particularly the amounts, timing of ingestion and its effect when combined with other foods related to satiety signals remains to be fully explored.

Author Contributions: Conceptualization, A.M. and A.D.; methodology, A.M. and A.D.; software, R.J. and A.D.; validation, A.M., A.D., S.E., A.P., C.A. and J.C.R.; formal analysis, A.D. and R.J.; investigation, A.D. and A.M.D.; writing—original draft preparation, A.D. and A.M.; writing—review and editing, A.D., S.E., A.P., R.J., J.C.R. and A.M.; and supervision, A.M.D. Funding was part of PhD project from Vitaflo International. All authors have read and agreed to the published version of the manuscript.

Funding: This research was funded by Vitaflo International as part of Ph.D. grant.

Conflicts of Interest: A.D. research funding from Vitaflo International, financial support from Nutricia, Mevalia and Vitaflo International to attend study days and conferences. S.E. research funding from Nutricia, financial support from Nutricia and Vitaflo International to attend study days and conferences A.P. has received an educational grant from Cambrooke Therapeutics and grants from Vitaflo International, Nutricia, Merck Serono, Biomarin and Mevalia to attend scientific meetings. C.A. received honoraria from Nutricia and Vitaflo International to attend study days and conferences. J.C.R. member of the European Nutritionist Expert Panel (Biomarin), the Advisory Board for Applied Pharma Research and Nutricia, and received honoraria as a speaker from APR, Merck Serono, Biomarin, Nutricia, Vitaflo, Cambrooke, P.I.A.M. and Lifediet. AMD research funding and honoraria from Nutricia, Vitaflo International and Merck Serono, Member of European Nutrition Expert Panel (Merck Serono international), member of Sapropterin Advisory Board (Merck Serono international), member of the Advisory Board Element (Danone-Nutricia). The funders had no role in the design of the study; in the collection, analyses, or interpretation of data; in the writing of the manuscript, or in the decision to publish the results.

References

1. Bickel, H. The first treatment of phenylketonuria. *Eur. J. Pediatr.* **1996**, *155* (Supp. l), 2–3. [CrossRef]
2. Bickel, H.; Gerrard, J.; Hickmans, E.M. Influence of phenylalanine intake on phenylketonuria. *Lancet* **1953**, *265*, 812–813. [CrossRef]
3. Li, T.; Zhao, M.; Raza, A.; Guo, J.; He, T.; Zou, T.; Song, H. The effect of taste and taste perception on satiation/satiety: A review. *Food Funct.* **2020**, *11*, 2838–2847. [CrossRef] [PubMed]
4. Tome, D.; Schwarz, J.; Darcel, N.; Fromentin, G. Protein, amino acids, vagus nerve signaling, and the brain. *Am. J. Clin. Nutr.* **2009**, *90*, 838S–843S. [CrossRef] [PubMed]
5. Veldhorst, M.A.; Nieuwenhuizen, A.G.; Hochstenbach-Waelen, A.; Westerterp, K.R.; Engelen, M.P.K.J.; Brummer, R.-J.M.; Deutz, N.E.P.; Westerterp-Plantenga, M.S. A breakfast with alpha-lactalbumin, gelatin, or gelatin + TRP lowers energy intake at lunch compared with a breakfast with casein, soy, whey, or whey-GMP. *Clin. Nutr.* **2009**, *28*, 147–155. [CrossRef]
6. Sousa, G.T.; Lira, F.S.; Rosa, J.C.; de Oliveira, E.P.; Oyama, L.M.; Santos, R.V.; Pimentel, G.D. Dietary whey protein lessens several risk factors for metabolic diseases: A review. *Lipids Health Dis.* **2012**, *11*, 67. [CrossRef]
7. Walstra, P.; Wouters, J.T.M.; Geurts, T.J. *Dairy Science and Technology*; CRC Press: Boca Raton, FL, USA; Taylor & Francis Group, LLC: Boca Raton, FL, USA, 2006; p. 768.
8. O'Riordan, N.; Kane, M.; Joshi, L.; Hickey, R.M. Structural and functional characteristics of bovine milk protein glycosylation. *Glycobiology* **2014**, *24*, 220–236. [CrossRef]
9. Sawin, E.A.; De Wolfe, T.J.; Aktas, B.; Stroup, B.M.; Murali, S.G.; Steele, J.L.; Ney, D.M. Glycomacropeptide is a prebiotic that reduces Desulfovibrio bacteria, increases cecal short-chain fatty acids, and is anti-inflammatory in mice. *Am. J. Physiol. Gastrointest Liver Physiol.* **2015**, *309*, G590–G601. [CrossRef]
10. Sharma, R.; Rajput, Y.S.; Mann, B. Chemical and functional properties of glycomacropeptide (GMP) and its role in the detection of cheese whey adulteration in milk: A review. *Dairy Sci. Technol.* **2013**, *93*, 21–43.
11. Hall, W.L.; Millward, D.J.; Long, S.J.; Morgan, L.M. Casein and whey exert different effects on plasma amino acid profiles, gastrointestinal hormone secretion and appetite. *Br. J. Nutr.* **2003**, *89*, 239–248. [CrossRef]
12. Veldhorst, M.A.; Nieuwenhuizen, A.G.; Hochstenbach-Waelen, A.; van Vught, A.J.; Westerterp, K.R.; Engelen, M.P.K.J.; Brummer, R.-J.M.; Deutz, N.E.P.; Westerterp-Plantenga, M.S. Dose-dependent satiating effect of whey relative to casein or soy. *Physiol. Behav.* **2009**, *96*, 675–682. [CrossRef]
13. Hoefle, A.S.; Bangert, A.M.; Rist, M.J.; Gedrich, K.; Lee, Y.M.; Skurk, T.; Danier, J.; Schwarzenbolz, U.; Daniel, H. Postprandial metabolic responses to ingestion of bovine glycomacropeptide compared to a whey protein isolate in prediabetic volunteers. *Eur. J. Nutr.* **2019**, *58*, 2067–2077. [CrossRef]
14. Royle, P.J.; McIntosh, G.H.; Clifton, P.M. Whey protein isolate and glycomacropeptide decrease weight gain and alter body composition in male Wistar rats. *Br. J. Nutr.* **2008**, *100*, 88–93. [CrossRef]
15. Cordova-Davalos, L.E.; Jimenez, M.; Salinas, E. Glycomacropeptide Bioactivity and Health: A Review Highlighting Action Mechanisms and Signaling Pathways. *Nutrients* **2019**, *11*, 598. [CrossRef]
16. Gustafson, D.R.; McMahon, D.J.; Morrey, J.; Nan, R. Appetite is not influenced by a unique milk peptide: Caseinomacropeptide (CMP). *Appetite* **2001**, *36*, 157–163. [CrossRef]
17. Keogh, J.B.; Clifton, P. The effect of meal replacements high in glycomacropeptide on weight loss and markers of cardiovascular disease risk. *Am. J. Clin. Nutr.* **2008**, *87*, 1602–1605. [CrossRef]
18. Lam, S.M.; Moughan, P.J.; Awati, A.; Morton, H.R. The influence of whey protein and glycomacropeptide on satiety in adult humans. *Physiol. Behav.* **2009**, *96*, 162–168.
19. Keogh, J.B.; Clifton, P.M. Weight loss using meal replacements high in glycomacropeptide-enriched whey powder. *Aust. J. Dairy Technol.* **2009**, *64*, 32–33.
20. Veldhorst, M.A.; Nieuwenhuizen, A.G.; Hochstenbach-Waelen, A.; Westerterp, K.R.; Engelen, M.P.; Brummer, R.-J.M.; Deutz, N.E.P.; Westerterp-Plantenga, M.S. Comparison of the effects of a high- and normal-casein breakfast on satiety, 'satiety' hormones, plasma amino acids and subsequent energy intake. *Br. J. Nutr.* **2009**, *101*, 295–303. [CrossRef]
21. Bowen, J.; Noakes, M.; Clifton, P.M. Appetite regulatory hormone responses to various dietary proteins differ by body mass index status despite similar reductions in ad libitum energy intake. *J. Clin. Endocrinol. Metab.* **2006**, *91*, 2913–2919. [CrossRef]

22. Batterham, R.L.; Heffron, H.; Kapoor, S.; Chivers, J.E.; Chandarana, K.; Herzog, H.; Le Roux, C.W.; Thomas, E.L.; Bell, J.D.; Withers, D.J. Critical role for peptide YY in protein-mediated satiation and body-weight regulation. *Cell Metab.* **2006**, *4*, 223–233. [CrossRef]
23. Pena, M.J.; Guerreiro, S.G.; Rocha, J.C.; Morais, T.; Pereira, S.S.; Monteiro, M.P.; Borges, N. Acute effect of an amino acid mixture in the rat glycemic profile. *J. Cell. Biochem.* **2019**, *120*, 13056–13065. [CrossRef]
24. Adams, R.L.; Broughton, K.S. Insulinotropic Effects of Whey: Mechanisms of Action, Recent Clinical Trials, and Clinical Applications. *Ann. Nutr. Metab.* **2016**, *69*, 56–63. [CrossRef]
25. Dangin, M.; Boirie, Y.; Garcia-Rodenas, C.; Gachon, P.; Fauquant, J.; Callier, P.; Ballevre, O.; Beaufrere, B. The digestion rate of protein is an independent regulating factor of postprandial protein retention. *Am. J. Physiol. Endocrinol. Metab.* **2001**, *280*, E340–E348. [CrossRef]
26. Poppitt, S.D.; Strik, C.M.; McArdle, B.H.; McGill, A.T.; Hall, R.S. Evidence of enhanced serum amino acid profile but not appetite suppression by dietary glycomacropeptide (GMP): A comparison of dairy whey proteins. *J. Am. Coll. Nutr.* **2013**, *32*, 177–186. [CrossRef]
27. Meyer-Gerspach, A.C.; Hafliger, S.; Meili, J.; Doody, A.; Rehfeld, J.F.; Drewe, J.; Beglinger, C.; Wolnerhanssen, B. Effect of L-tryptophan and L-leucine on gut hormone secretion, appetite feelings and gastric emptying rates in lean and non-diabetic obese participants: A randomized, double-blind, parallel-group trial. *PLoS ONE* **2016**, *11*, e0166758. [CrossRef]
28. Chungchunlam, S.M.; Henare, S.J.; Ganesh, S.; Moughan, P.J. Dietary whey protein influences plasma satiety-related hormones and plasma amino acids in normal-weight adult women. *Eur. J. Clin. Nutr.* **2015**, *69*, 179–186. [CrossRef]
29. Lacroix, M.; Bos, C.; Leonil, J.; Airinei, G.; Luengo, C.; Dare, S.; Benamouzig, R.; Fouillet, H.; Fauquant, J.; Tome, D.; et al. Compared with casein or total milk protein, digestion of milk soluble proteins is too rapid to sustain the anabolic postprandial amino acid requirement. *Am. J. Clin. Nutr.* **2006**, *84*, 1070–1079. [CrossRef]
30. Westerterp-Plantenga, M.S.; Nieuwenhuizen, A.; Tome, D.; Soenen, S.; Westerterp, K.R. Dietary protein, weight loss, and weight maintenance. *Annu. Rev. Nutr.* **2009**, *29*, 21–41. [CrossRef]
31. Mellinkoff, S.M.; Frankland, M.; Boyle, D.; Greipel, M. Relationship between serum amino acid concentration and fluctuations in appetite. *J. Appl. Physiol.* **1956**, *8*, 535–538. [CrossRef]
32. Korompokis, K.; Ostman, E.; Dougkas, A. The impact of liquid preloads varying in macronutrient content on postprandial kinetics of amino acids relative to appetite in healthy adults. *Appetite* **2016**, *107*, 511–520. [CrossRef] [PubMed]
33. Gropper, S.S.; Acosta, P.B. Effect of simultaneous ingestion of L-amino acids and whole protein on plasma amino acid and urea nitrogen concentrations in humans. *JPEN J. Parenter. Enter. Nutr.* **1991**, *15*, 48–53. [CrossRef]
34. Ha, E.; Zemel, M.B. Functional properties of whey, whey components, and essential amino acids: Mechanisms underlying health benefits for active people (review). *J. Nutr. Biochem.* **2003**, *14*, 251–258. [CrossRef]
35. Bos, C.; Metges, C.C.; Gaudichon, C.; Petzke, K.J.; Pueyo, M.E.; Morens, C.; Everwand, J.; Benamouzig, R.; Tome, D. Postprandial kinetics of dietary amino acids are the main determinant of their metabolism after soy or milk protein ingestion in humans. *J. Nutr.* **2003**, *133*, 1308–1315. [CrossRef] [PubMed]
36. Weigel, C.; Rauh, M.; Kiener, C.; Rascher, W.; Knerr, I. Effects of various dietary amino acid preparations for phenylketonuric patients on the metabolic profiles along with postprandial insulin and ghrelin responses. *Ann. Nutr. Metab.* **2007**, *51*, 352–358. [CrossRef]
37. Daly, A.; Evans, S.; Pinto, A.; Jackson, R.; Ashmore, C.; Rocha, J.C.; MacDonald, A. Preliminary Investigation to Review If a Glycomacropeptide Compared to L-Amino Acid Protein Substitute Alters the Pre- and Postprandial Amino Acid Profile in Children with Phenylketonuria. *Nutrients* **2020**, *12*, 2443. [CrossRef]
38. Belanger-Quintana, A.; Martinez-Pardo, M. Physical development in patients with phenylketonuria on dietary treatment: A retrospective study. *Mol. Genet. Metab.* **2011**, *104*, 480–484. [CrossRef]
39. Burrage, L.C.; McConnell, J.; Haesler, R.; O'Riordan, M.A.; Sutton, V.R.; Kerr, D.S.; McCandless, S.E. High prevalence of overweight and obesity in females with phenylketonuria. *Mol. Genet. Metab.* **2012**, *107*, 43–48. [CrossRef]
40. Van Wegberg, A.M.J.; MacDonald, A.; Ahring, K.; Belanger-Quintana, A.; Blau, N.; Bosch, A.M.; Burlina, A.; Campistol, J.; Feillet, F.; Gizewska, M.; et al. The complete European guidelines on phenylketonuria: Diagnosis and treatment. *Orphanet J. Rare Dis.* **2017**, *12*, 162. [CrossRef]

41. Daly, A.; Evans, S.; Chahal, S.; Santra, S.; MacDonald, A. Glycomacropeptide in children with PKU: Does its phenylalalnine content affect blood phenylalanine control? *J. Hum. Nutr. Diet.* **2017**, *30*, 515–523. [CrossRef]
42. Scientific Advisory Committee on Nutrition. *SACN Carbohydrates and Health Report*; Public Health England: London, UK, 2015; p. 384.
43. MacDonald, A. Diet and compliance in phenylketonuria. *Eur. J. Pediatr.* **2000**, *159* (Suppl. S2), S136–S141. [CrossRef]
44. MacDonald, A.; Gokmen-Ozel, H.; van Rijn, M.; Burgard, P. The reality of dietary compliance in the management of phenylketonuria. *J. Inherit. Metab. Dis.* **2010**, *33*, 665–670. [CrossRef]
45. Burton-Freeman, B.M. Glycomacropeptide (GMP) is not critical to whey-induced satiety, but may have a unique role in energy intake regulation through cholecystokinin (CCK). *Physiol. Behav.* **2008**, *93*, 379–387. [CrossRef]
46. Singh, R.K.; Chang, H.W.; Yan, D.; Lee, K.M.; Ucmak, D.; Wong, K.; Abrouk, M.; Farahnik, B.; Nakamura, M.; Zhu, T.H.; et al. Influence of diet on the gut microbiome and implications for human health. *J. Transl Med.* **2017**, *15*, 73. [CrossRef] [PubMed]
47. Crowley, V.E. Overview of human obesity and central mechanisms regulating energy homeostasis. *Ann. Clin. Biochem.* **2008**, *45*, 245–255. [CrossRef]
48. Forte, N.; Fernandez-Rilo, A.C.; Palomba, L.; Di Marzo, V.; Cristino, L. Obesity affects the microbiota-gut-brain axis and the regulation thereof by endocannabinoids and related mediators. *Int. J. Mol. Sci.* **2020**, *21*, 1554. [CrossRef]
49. Ntemiri, A.; Chonchuir, F.N.; O'Callaghan, T.F.; Stanton, C.; Ross, R.P.; O'Toole, P.W. Glycomacropeptide sustains microbiota diversity and promotes specific taxa in an artificial colon model of elderly gut microbiota. *J. Agric. Food Chem.* **2017**, *65*, 1836–1846. [CrossRef]
50. MacLeod, E.L.; Clayton, M.K.; van Calcar, S.C.; Ney, D.M. Breakfast with glycomacropeptide compared with amino acids suppresses plasma ghrelin levels in individuals with phenylketonuria. *Mol. Genet. Metab.* **2010**, *100*, 303–308. [CrossRef]
51. Chungchunlam, S.M.; Moughan, P.J.; Henare, S.J.; Ganesh, S. Effect of time of consumption of preloads on measures of satiety in healthy normal weight women. *Appetite* **2012**, *59*, 281–288. [CrossRef]

© 2020 by the authors. Licensee MDPI, Basel, Switzerland. This article is an open access article distributed under the terms and conditions of the Creative Commons Attribution (CC BY) license (http://creativecommons.org/licenses/by/4.0/).

nutrients

MDPI

Article

A 3 Year Longitudinal Prospective Review Examining the Dietary Profile and Contribution Made by Special Low Protein Foods to Energy and Macronutrient Intake in Children with Phenylketonuria

Anne Daly [1,*], Sharon Evans [1], Alex Pinto [1], Catherine Ashmore [1], Júlio César Rocha [2,3] and Anita MacDonald [1]

1 Dietetic Department, Birmingham Children's Hospital, Steelhouse Lane, Birmingham B4 6NH, UK; evanss21@me.com (S.E.); alex.pinto@nhs.net (A.P.); catherine.ashmore@nhs.net (C.A.); anita.macdonald@nhs.net (A.M.)
2 Nutrition and Metabolism, NOVA Medical School, Faculdade de Ciências Médicas, Universidade Nova de Lisboa, 1169-056 Lisboa, Portugal; rochajc@nms.unl.pt
3 Centre for Health Technology and Services Research (CINTESIS), 4200-450 Porto, Portugal
* Correspondence: a.daly3@nhs.net

Received: 19 September 2020; Accepted: 12 October 2020; Published: 15 October 2020

Abstract: The nutritional composition of special low protein foods (SLPFs) is controlled under EU legislation for 'Foods for Special Medical Purposes (FSMP)'. They are designed to meet the energy needs of patients unable to eat a normal protein containing diet. In phenylketonuria (PKU), the macronutrient contribution of SLPFs has been inadequately examined. Aim: A 3-year longitudinal prospective study investigating the contribution of SLPFs to the macronutrient intake of children with early treated PKU. Methods: 48 children (27 boys) with a mean recruitment age of 9.3 y were studied. Semi-quantitative dietary assessments and food frequency questionnaires (FFQ) were collected three to four times/year for 3 years. Results: The mean energy intake provided by SLPFs was 33% (SD ± 8), and this figure was 42% (SD ± 13) for normal food and 21% (SD ± 5) for protein substitutes (PS). SLPFs supplied a mean intake of 40% carbohydrate (SD ± 10), 51% starch (SD ± 18), 21% sugar (SD ± 8), and 38% fat (SD ± 13). Fibre intake met 83% of the Scientific Advisory Committee on Nutrition (SACN) reference value, with 50% coming from SLPFs with added gums and hydrocolloids. Low protein bread, pasta and milk provided the highest energy contribution, and the intake of sweet SLPFs (e.g., biscuits, cakes, and chocolate) was minimal. Children averaged three portions fruit/vegetable daily, and children aged ≥ 12 y had irregular meal patterns. Conclusion: SLPFs provide essential energy in phenylalanine restricted diets. Optimising the nutritional quality of SLPFs deserves more attention.

Keywords: phenylketonuria; PKU; glycomacropeptide; special low protein foods; macronutrient intake; protein substitute

1. Introduction

In phenylketonuria (PKU), deficiency or reduced activity of the phenylalanine hydroxylase enzyme (PAH) limits the conversion of phenylalanine to tyrosine. Without intervention, intellectual disability, significant delays in developmental milestones, hyperactive behaviour with autistic features, and seizures may occur. High levels of brain phenylalanine are probably the main cause of neurotoxicity [1] by interfering with cerebral protein synthesis [2], increasing myelin turnover and inhibiting neurotransmitter synthesis [3,4]. In children with classical PKU, their only effective management option is a severely restricted low protein/phenylalanine diet that aims to lower blood phenylalanine levels to within a strict target range [5]. Although alternative treatments, such as sapropterin and pegvaliase (PEGylated recombinant *Anabaena variabilis* phenylalanine ammonia lyase (PAL)), have been licensed, they are only suitable for subsections of the PKU population, and access is restricted in some countries. Thereby, outcome in PKU is dependent on the early introduction of dietary treatment and the quality of lifelong blood phenylalanine control, which in turn is determined by the ability to adhere to dietary treatment.

A low protein diet aims to prevent long-term phenylalanine toxicity, with most patients tolerating < 10 g/day of natural protein [6]. All high biological protein foods (e.g., meat, eggs, fish, ordinary bread, pasta and flour) are not allowed [7]. The diet is supplemented with a minimal/free phenylalanine synthetic protein (protein substitute), which in classical PKU provides approximately 80% of total protein intake [8]. In the UK, the protein substitute is given with an allocated daily amount of measured phenylalanine from a range of regular foods (e.g., potatoes, peas, cereals) to meet essential requirements. Low protein foods are given without restriction and include low protein regular foods (containing protein ≤ 0.5 g/100 g), fruit and vegetables (containing phenylalanine ≤ 75 mg/100 g), and special low protein foods (SLPFs) (containing phenylalanine ≤ 25 mg/100 g) [5,9].

In a low phenylalanine diet, SLPFs are essential safe foods, satisfying satiety, offering choice and replicating some normality in a lifelong restricted diet [7]. They are categorised as 'Foods for Special Medical Purposes' (FSMPs) and are defined as specialist foods for the dietary management of patients with a medical condition who are unable to achieve a suitable nutritional intake from regular foods. They are described as 'evidence based nutritional solutions for disorder related conditions' [10]. They are 'highly regulated' by European Union law, with *'Delegated Regulation (EU) No 2016/128'* setting in place policies on composition and labelling. FSMPs should be used only under medical/health professional supervision and must be labelled according to their intended use.

SLPFs replace basic food items such as milk, bread, and pasta. They help optimise growth, provide energy to prevent catabolism, and avoid consequential raised blood phenylalanine [11]. A wide range of SLPFs may be key in helping patients sustain their dietary treatment for life, although access to SLPFs varies across Europe (European Society for Phenylketonuria (ESPKU)) [12]. Pena et al. [13] showed that the availability of SLPFs in different countries in Europe ranges from 73 SLPFs in Portugal to 256 in Italy, while no information was available for some countries. Only a few specialist companies manufacture SLPFs, and due to the constraints of a phenylalanine restricted diet, their nutritional composition consists mainly of carbohydrate and fat. Their taste and aesthetic properties are prioritised over their nutritional composition, although producing acceptable, high-quality SLPFs from isolated food starches with good organoleptic properties, texture and colour is challenging. While the variety and availability of SLPFs has improved, the choice, ease of access and quality in comparison to regular foods are all still narrow, inflexible and inadequate. There is only limited data about the nutritional contribution made by SLPFs to a phenylalanine restricted diet, or the types of SLPFs that are consumed.

In a longitudinal 3-year prospective study, the aim was to evaluate the contribution of SLPFs to the energy and macronutrient intake of a group of well-controlled children with PKU on dietary treatment only. As a secondary aim, we examined the dietary patterns of children when using SLPFs routinely in their diets.

2. Materials and Methods

In total, 50 children (28 boys, 22 girls) with PKU were recruited. Of these children, 47 were European and 3 were of Pakistani origin. Study inclusion criteria: diagnosed by newborn screening; aged 5–16 years; not treated with sapropterin dihydrochloride; and 70% of blood phenylalanine concentrations within target range for 6 months before study enrolment. The target blood phenylalanine ranges for children aged 5–12 years were 120 to ≤360 μmol/L, and for 12 years and older they were 120 to ≤600 μmol/L, as recommended by the European PKU guidelines [5]. Based on untreated blood phenylalanine levels at newborn screening (<1000 μmol/L) and dietary phenylalanine tolerance (>750 mg/day), the majority of participants had classical PKU, except two children with mild PKU.

2.1. Protein Substitute Intake

At enrolment, the protein substitute sources were casein glycomacropeptide (CGMP-AA), $n = 31$, and amino acid supplements (AA), $n = 19$. The AAs were either ready-to-drink liquid pouches providing 10, 15 or 20 g protein equivalent or powders made up with water to a semi-solid consistency. The CGMP-AA was a powder mixed with 120 mL water. It contained 20 g of protein equivalent, with a residual amount of phenylalanine (36 mg/20 g protein equivalent) and it was given as a drink.

Dietary intake was assessed by 2 types of dietary assessment technique:

- Three-day recorded food diary: caregivers were instructed on how to record food intake using scales, household measures or from a pictorial handbook with measured food portion sizes. A three-day semi-quantitative dietary assessment was completed, with an annual mean of 4 (range 3–6) assessments per child for a period of 3 years. Assessments were checked via face-to-face interviews by one of two trained metabolic dietitians. Portion weights of SLPFs were provided by manufacturers' information or estimated from the 'low protein' portion size picture book. At least once annually, children were observed eating one meal at home, with portion sizes weighed and checked.
- Food frequency questionnaire (FFQ): the FFQ, specifically designed for patients with PKU, contained a series of questions on the consumption of both SLPFs and regular foods, estimating the portion sizes eaten and frequency of consumption of each food item. Foods were grouped into dairy products, cereals, fats, sugar and sweet foods, drinks, fruit and vegetables, and 'meat, fish, eggs', with 'free from' or SLPF alternatives for each category. The FFQ diaries were completed at the same time as the 3-day diet diaries, with a mean of 3 (range 3–6) FFQs completed yearly, constructing a comprehensive database on the actual consumption of special low protein and regular foods.

All the dietary assessments were analysed using Nutritics Nutritional Software (v5.093) [14]. The results were compared with age and gender specific UK dietary reference values and estimated average requirements for energy (EAR) (UK Scientific Advisory Committee on Nutrition (SACN) [15]. The database included the nutrient analysis of protein substitutes and SLPFs, using nutritional information supplied by the manufacturers.

The contribution of SLPFs to macronutrient intake was calculated by age for children ≤11 years and those ≥12 years. This age range was chosen to match with the age-dependent upper target blood phenylalanine concentrations given by the European PKU Guidelines [5]. At the point that children reached ≥12 years of age, they were then transferred to the older age group. The following macronutrients were analysed: energy (Kcal), protein (g), carbohydrate (g), starch (g), sugar (g), fat (g) and fibre (g). For each subject for each year, the mean contributions of each macronutrient from protein substitute, SLPFs and regular foods were calculated, and the mean value for all subjects is presented. The annual mean total energy intake (Kcal/day) and % EAR has been compared with UK dietary reference values or EAR for energy [15]. From the FFQ, the regularity of meals (breakfast, midday and evening meal), frequency of snacks and drinks, and the amount of foods consumed each week were also estimated. The FFQ was used to calculate the number of portions in grams of each food from the different food groups. Using this data, the mean number of grams of food eaten each week was calculated. This data complemented the dietary assessment analysis, showing the foods that were consumed regularly, but also highlighting any foods that might have been omitted from the 3-day dietary assessments. This data was not used to estimate energy or nutrient intake and was not statistically analysed, but showed the typical weekly pattern of foods consumed, and how meals were structured.

2.2. Anthropometric Measurements

Weight, height and BMI were measured once every 3 months by one of two metabolic dietitians. Height was measured with a Harpenden stadiometer (Holtain Ltd., Crymych, UK) and weight on calibrated digital scales (Seca, Medical Measuring Systems and Scales, Birmingham, UK model 875); both were measured to the nearest 0.1 cm or kg, respectively.

2.3. Blood Phenylalanine Levels

Trained parents/caregivers collected weekly early morning fasted blood spots on filter cards, Perkin Elmer 226 (UK Standard NBS). Blood specimens were sent via first class post to the laboratory at Birmingham Children's Hospital. All the cards had a standard thickness, and the blood phenylalanine concentrations were calculated on a 3.2 mm punch by MS/MS tandem mass spectrometry. At enrolment, the median blood phenylalanine concentrations for the previous 12 months were collected and referred to as the enrolment blood phenylalanine concentration.

2.4. Statistical Analysis

Mann Whitney nonparametric unpaired t tests comparing two unmatched groups of data were used to compare macronutrient differences (energy, protein, carbohydrate, starch, sugar, fat and fibre) between children ≤11 years and those ≥12 years. The quantitative outcome measures have been summarised and descriptive statistics reported as means and differences assessed between the groups, with a statistically significant value of $p < 0.05$.

2.5. Ethical Permission

The South Birmingham Research Ethics (REC) committee granted a favourable ethical opinion, referenced REC13/WM/0435 and IRAS (Integrated research application system) ID 129497. Written informed consent was obtained for all subjects from at least one caregiver with parental responsibility, and written assent obtained from the subject if appropriate for their age and level of understanding.

3. Results

In total, 48 children (21 girls and 27 boys) completed the study. The mean age at enrolment was 9.3 years (5–16 years). There were 35 children aged ≤11 years and 13 aged ≥ 12 years.

3.1. Subject Withdrawal

One boy and one girl (aged 12 years) were excluded from the study as both failed to comply with the study protocol. One failed to return blood phenylalanine samples and both had poor dietary adherence.

3.2. Dietary Prescription

Over the study period the total mean daily dose of protein equivalent from protein substitute was 64 g/day (range 40–80 g) or 1.5 g/kg (1–2 g/kg) with the mean amount of prescribed natural protein 5.5 g protein/day (range 3–30 g) or 275 mg phenylalanine (range 150–1500 mg)/day. The protein substitute source was AA, $n = 19$ (liquid pouches (PKU Lophlex LQ, Nutricia Ltd. Trowbridge, UK. $n = 1$; PKU Cooler, Vitaflo International Ltd., Liverpool, UK. $n = 14$), or powder (PKU gel, Vitaflo International Ltd., $n = 4$)). In total 29 children took CGMP-AA (GMP study product, Vitaflo International Ltd.); $n = 13$ had their entire protein substitute requirement as CGMP-AA, and $n = 16$ took a combination of CGMP-AA and AA. The numbers of children taking AA products in combination with CGMP were liquid pouches $n = 15$, (PKU Lophlex, Nutricia, $n = 4$; PKU Cooler, Vitaflo International Ltd., $n = 11$), and for those taking powder (PKU gel, Vitaflo International Ltd.) $n = 1$.

3.3. Energy and Macronutrient Intake

Mean daily energy intake and % EAR are described in Table 1. For both age groups the energy as a percentage of EAR was age appropriate and within 5% of the EAR. In both groups the total percentage contribution of energy from carbohydrate, protein and fat was similar.

Table 1. Mean (standard deviation) total energy intake, percentage of EAR, and the contribution of carbohydrate, protein and fat as a percentage of mean total energy intakes in children ≤ 11 years and ≥ 12 years, and for the combined age groups over the 3-year study period.

Year 1 to 3 Mean Intake (SD)	≤11 years n = 35	≥12 years n = 13	All Children (≤11 ≥12 years) n = 48
Energy intake Kcal/day	1921 (255)	2224 (417)	2059 (394)
% EAR	105 (21)	95 (13)	99 (15)
Mean % (±SD) energy contribution from carbohydrate, protein and fat			
Carbohydrate	58 (4)	56 (7)	57 (5)
Protein	15 (3)	14 (4)	14 (4)
Fat	27 (4)	28 (7)	28 (5)

EAR, estimated average requirement. SD, standard deviation. EAR for energy was 2175 kcal (1422–2809) [15]. Calculated by taking the median (range) for the combined ages and gender. Kcal Kilocalories.

3.4. Contribution of SLPFs to Mean Macronutrient Intake

3.4.1. Energy Intake

The total mean energy intake in the combined age groups was 2059 Kcal/day, of which the percentage mean energy intake from SLPFs was 33% (SD ± 8), regular foods 42% (SD ± 13) and protein substitute 21% (SD ± 5) (Table 1). Of the total energy, 2% (SD ± 3) was provided by phenylalanine/natural protein containing foods (potatoes, crisps, and vegetables with a phenylalanine content ≥ 75 mg/100 g protein).

Table 2. Mean (standard deviation) contribution of calories and grams per day from carbohydrate (including starch and sugar), fat and protein to mean total energy intake for all children over the 3-year study period.

	All Children (≤11 ≥12 years) n = 48	
Macronutrient	**Mean kcals/day (±SD)**	**Mean g/day (±SD)**
Energy	2059 (394)	-
CHO	1176 (60)	294 (15)
Starch	687 (24)	174 (5)
Sugar	477 (20)	119 (5)
Fat	576 (36)	63 (4)
Protein	307 (5)	74 (2)

SD, standard deviation. CHO, carbohydrate.

The contribution of SLPFs to the mean daily intake of carbohydrate was 40% (SD ± 10), starch 51% (SD ± 18), sugar 21% (SD ± 8) and fat 38% (SD ± 13). The daily intake of sweet SLPFs (e.g., biscuits, cakes, and chocolate) was low, and overall contributed minimally to the energy, fat or carbohydrate intake. Table 2 describes the mean contribution of energy and grams per day from carbohydrate (starch and sugar), fat and protein to total daily energy intake.

3.4.2. Carbohydrate Intake

The mean intake of carbohydrate from SLPFs, regular foods and protein substitute for all children was 294 g/day, of which starch provided 174 g/day (59%) and sugar 119 g/day (40%) (Table 2).

SLPFs were the highest contributor to total carbohydrate intake, with a mean intake from bread of 50 g/day (17%), pasta of 39 g/day (13%) and low protein milk replacement of 9 g/day (3%). The highest contribution from regular foods to mean carbohydrate intake came from drinks (carbonated and cordials) at 20 g/day (7%), potatoes at 19 g/day (6%), fruit at 14 g/day (5%), and crisps and confectionary, both providing 12 g/day (4%). Protein substitute contributed 26 g/day (9%) to the total mean carbohydrate intake. Other foods making up the total carbohydrate are shown in Table 3. The only significant difference in the intake of SLPFs between children aged ≤11 and ≥12 years was for the low protein milk replacement, this being higher in the younger age group ($p < 0.0001$).

Table 3. Mean contribution of SLPFs, regular foods and protein substitute in grams to total carbohydrate, starch, sugar and protein intake for children with PKU.

Macronutrient	CHO (g)		Starch (g)		Sugar (g)		Fat (g)		Protein (g)		CHO (g)	Starch (g)	Sugar (g)	Fat (g)	Protein (g)
Age	≤11 years	≥12 years	≤11 years	≥12 years	≤11 years	≥12 years	≤11 years	≥12 years	≤11 years	≥12 years	All Ages				
Total macronutrient intake from all foods and PS	274 g	313 g	165 g	182 g	110 g	128 g	57 g	68 g	71 g	77 g	294 g	174 g	119 g	63 g	74 g
SLPFs															
Bread	43	56	34	40	3	4	6	7	-	-	50	37	4	7	-
Pasta/Rice	38	40	30	38	-	-	1	1	-	-	39	34	-	1	-
Milk substitute	14 §	3 §	2	1	7	3	7	2	-	-	9	3	5	5	-
Biscuits	5	4	3	2	2	1	2	2	-	-	5	3	2	2	-
Cereal bars	5	4	3	2	8	6	3	1	-	-	3	4	7	2	-
Cakes/puddings	6	4	4	2	4	3	5	3	-	-	5	5	3	4	-
Miscellaneous: burger, sausage, pizza, homemade dishes	11	14	3	5	3	5	2	4	-	-	6	3	4	3	
Total macronutrients from SLPFs	122 g	125 g	79 g	90 g	27 g	22 g	26 g	20 g	-	-	117 g	89 g	25 g	24 g	-
Regular foods															
Potato	16	22	17	20	2	4	3	4	2	3	19	19	3	3	2
Crisps	11	13	12	13	-	-	4	5	1	2	12	12	-	5	2
Cereals	9	6	7	4	1	1	-	-	1	1	7	5	1	-	1
Dairy	3	3	1	1	1	1	1	1	-	-	3	1	1	1	-
Vegetables	3	6	-	-	5	5	2	4	1	2	5	-	5	3	2
Fruit	14	13	-	-	15	13	-	-	1	1	14	-	14	-	1
Sweets/chocolate	8	15	2	5	5	12	2	4	-	1	12	4	10	3	-
Cereal bars	3	3	2	2	2	3	1	1	-	1	3	2	3	1	1
Cakes/puddings	5	3	3	3	3	3	1	1	-	-	4	3	3	1	-
Drinks	14	25	-	-	14	20	-	-	-	-	20	-	17	-	-
Butter/oils	-	-	-	-	-	-	11	14	-	-	-	-	-	13	-
Sauces, jam	24	31	24	29	13	19	2	4	-	1	31	17	16	4	1
Total macronutrients from regular foods	110 g	140 g	68 g	77 g	61 g	81 g	27 g	38 g	6 g	12 g	130 g	63 g	73 g	34 g	10 g
Protein substitute	27 g	25 g	13 g	9 g	18 g	17 g	3 g	4 g	65 g	64 g	26 g	11 g	18 g	4 g	64 g
Total intake from all food and PS	259 g	290 g	160 g	176 g	106 g	120 g	56 g	62 g	71 g	76 g	273 g	163 g	116 g	62 g	74 g
Total % intake	95%	93%	97%	97%	96%	94%	98%	91%	100%	99%	93%	94%	97%	98%	100%

§ $p \leq 0.0001$, CHO carbohydrate, PS protein substitute, SLPFs special low protein foods. NB figures do not add to 100% due to inaccurate measurement of CHO, starch and sugar from SLPF information.

The combined age groups had a mean daily starch intake of 174 g/day. The highest starch intakes were provided from low protein bread, 37 g/day (21%) and pasta/rice, 34 g/day (20%). Potato and crisps (phenylalanine containing foods) were the other main non-SLPF contributors to starch intake (Table 3). Protein substitutes provided 11 g/day (6%) of the total starch intake.

Sugar intake supplied by SLPFs was minimal. In the combined age groups, low protein bread provided a mean intake of 4 g/day (3%). Sugar from low protein milk provided 5 g/day (4%), with a higher intake in children ≤11 years of age. The regular foods contributing to sugar intake were sweet drinks, providing 17 g/day (14%), fruit, 14 g/day (12%), and confectionary, 10 g/day (8%). Protein substitutes provided 18 g/day or 15% of the total sugar intake. The total mean amount of free sugar, defined as sugars added to cooked or manufactured food, was 26 g/day (22%), and this came largely from sweet drinks, e.g., cola, lemonade, sweets, jams, honey and condiments such as tomato sauce. The daily amount of free sugar, particularly from sweet drinks, was higher in children aged ≥12 years. Free sugars represented 5% of total energy intake, in line with SACN recommendations [16]. Fruit provided a high sugar intake, but this was from natural rather than refined sugars.

3.4.3. Fat Intake

Fat intake provided a mean of 63 g/day for all children, with SLPFs supplying minimal fat intake. In the combined ages, bread supplied a mean fat intake of 7 g/day (11%), and in children ≤11 years milk contributed 5 g/day (8%). The highest fat sources were from butter and oils, 13 g/day (21%), potato crisps, 5 g/day (8%), and fried potatoes, 3 g/day (5%). The main sources of fat were saturated fat from oil/butter, fried potatoes and crisps. Protein substitutes provided a mean fat intake of 4 g/day (6%), with some containing essential fatty acids and/or long-chain polyunsaturated fatty acids (LCPUFAs).

3.4.4. Protein Intake

SLPFs made no significant contribution to protein intake, while the protein equivalent from protein substitute consistently provided a mean intake of 64 g/day, which was 86% (SD ± 9) of the total protein intake.

3.4.5. Fibre Intake

SLPFs provided approximately 50% (SD ± 23) of the mean daily fibre intake. Low protein bread and pasta provided higher fibre sources than potatoes, vegetables and fruits. In children aged ≤11 y, the total mean fibre intake was 18 g/day, providing 83% of the recommended intake (SACN) [16], with 9 g/day (50%) from SLPFs. Children aged ≥12 years consumed a mean fibre intake of 20 g/day, providing 82% of the recommended intake, of which 11 g/day (55%) came from SLPFs. The blend of fibre was limited to the fibre sources added to SLPFs, which was commonly derived from gums and hydrocolloids.

3.5. Median Blood Phenylalanine Concentrations throughout the 3-Year Study Period

Statistically, the phenylalanine concentrations were significantly different both within and between the groups from enrolment to year 3 (Table 4). This group of children were well controlled, with median blood phenylalanine levels within the European PKU guidelines [5]. There was no correlation between energy intake from SLPFs and phenylalanine concentrations.

Table 4. Median (range) phenylalanine concentrations in children aged ≤ 11 years and ≥ 12 years at enrolment and follow up at 3 years.

Median Phenylalanine µmol/L	≤ 11 years (n = 35)	≥ 12 years (n = 13)	p Value
Enrolment (range)	270 µmol/L *§ (140–470)	356 µmol/L * (230–600)	* $p = 0.003$
3 year follow up (range)	300 µmol/L **§ (200–730)	485 µmol/L ** (320–895)	** $p < 0.0001$ § $p = 0.02$

*, **, §—p-values between and within the groups.

4. Anthropometry

At the 3-year follow up, the median weight, height and BMI Z scores (range) were 1.0 (0.3–1.7), 0.3 (−0.01–0.6) and 1.1 (0.5–0.8) respectively (Table 5). Using the WHO [17] definitions of overweight and obesity, between enrolment and year 3, overweight (defined as BMI one standard deviation over the reference median) increased from 25% (n = 12/48) to 29% (n = 14/48), and obesity (BMI equivalent to two standard deviations over the reference median) increased from 10% (n = 5/48) to 17% (n = 8/48).

Table 5. Median (range) annual weight, height and BMI Z scores for all children at enrolment and follow up at year 3.

Follow up Duration year	Weight Z Score	Height Z Score	BMI Z Score
Enrolment	0.7 (−0.1–1.2)	0 (−0.2–0.5)	0.7 (0.0–1.2)
year 1	0.8 (0.3–1.4)	0.2 (−0.3–0.4	1.0 (0.3–1.5)
year 2	0.9 (0.4–1.7)	0.2 (−0.1–0.6)	1.0 (0.3–1.9)
year 3	1.0 (0.3–1.7)	0.3 (−0.01–0.6)	1.1 (0.5–1.8)

BMI: Body mass index.

Food Patterns from the Food Frequency Questionnaires

Children aged ≤11 years ate regular main meals, including breakfast, midday and evening meal, with a mean of two snacks per day. Children aged ≥12 years were more independent, and some would cook their own meals, usually based on pasta or bread. They had irregular meal patterns with less supervision around mealtimes. They commonly missed breakfast, eating snack foods for their midday meal particularly when in school and eating more food towards the evening after the school day had finished.

Over the 3-year study period, low protein milk replacement decreased in children aged ≤11 years from a mean of 1750 mL/week to 1300 mL/week, remaining consistent at 700 mL/week in children aged ≥12 years. In the younger age group, the mean low protein bread intake increased from 670 g/week to 750 g/week at year 3, based on an average bread slice weighing 30 g. This was equivalent to three to four slices/day. Low protein bread intake remained consistent in children aged ≥12 years of age, at 900 g/week (four to five slices/day). The mean intake of low protein pasta in children aged ≤11 years was 600 g/week based on an estimated cooked portion of 200 g (three portions/week). This increased in children aged ≥12 years from 700 to 900 g/week (four portions/week).

The overall daily fruit intake was low. In children aged ≤11 years, the mean intake was 1200 g/week (two portions/day), and 700 g/week (one portion/day) in children ≥12 years. Vegetable intake decreased over the 3 years in both groups, decreasing in children aged ≤11 years from a mean of 660 g/week to 560 g/week (approximately one portion/day) and from 900 g/week to 700 g/week (one to two portions/day) in children ≥ 12 years. Children aged ≤ 11 years consumed a mean intake of 1200 mL/week of sweet drinks (mainly from fizzy drinks), with an average serving of 200 mL (one sweet drink/day), whilst the older group drank 2300 mL/week (two sweet drinks/day).

The SLPFs that were regularly eaten by the entire group of children were bread, at 92% (n = 44/48), pasta, at 85% (n = 41/48), and low protein milk at 77% (n = 37/48).

5. Discussion

This study demonstrated that SPLFs were an essential energy source, providing over 30% of energy intake in children with PKU aged 5 to 16 years of age. The low protein staple foods bread and pasta made the largest consistent contribution to energy intake. There were few differences in SLPF intake between children ≤ 11 and ≥ 12 years of age, the exception being the younger children who consumed more low protein milk replacement. Protein substitutes provided 18 g/day (15%) of the total sugar intake. Concern has been expressed about the energy content of SLPF snack foods [13], but in this study, low protein cakes, biscuits and chocolate made a minimal contribution to daily energy intake. Instead, aspartame-free sweet drinks provided the highest intake of free sugars. Some may argue that the sugar content of protein substitutes is too high; however, in children it is important to provide a source of energy to ensure nitrogen is used efficiently.

Very few studies have examined the energy contribution made by SLPFs. An Italian study, in children with PKU aged 5 to 11 years, reported that SLPFs provided 47% of energy intake [18]. In a small German study, reporting on eight children aged 6 to 16 years of age, when on dietary treatment only, the SLPFs provided 39% of the energy intake [19], a higher energy intake than was observed in our study. Throughout Europe, SLPFs are available through a number of systems, including state national health schemes (either prescription or monthly financial family allowance), or in some countries patients/carers may be expected to make a complete or partial contribution to their purchase. It is unknown how these systems or patient acceptance impact the usage of SLPFs. In addition, it is also unknown how adherence to a phenylalanine-restricted diet and the overall quality of blood phenylalanine alters the usage of SLPFs.

In the UK, access to SPLFs is controlled. The National Society for PKU (NSPKU) provides age-defined guidance on the maximum monthly units of SLPFs that can be prescribed by community general practitioners (GPs). This is based on the assertion that SLPFs provide up to 50% of energy intake. It has been reported that both the NHS authority (e.g., the Clinical Commissioning Group or Health Board) and GPs have refused to prescribe or have limited the amounts of SLPFs that patients can access [6,20]. In some cases, patient requests for low protein cake mixes, or low protein cereals bars, have been rejected, even though our study indicates that these contribute a negligible energy intake. Both Cochrane et al. and Ford et al. [6,20,21] have described the stigma caregivers and patients encounter when obtaining SLPFs via their GP. Due to access issues, on occasion patients are without these foods, leading to anxiety about food insecurity, which has recently been reported in PKU [22]. Even in the early 1950s, when dietary treatment commenced, it was recognised that catabolism led to increased blood phenylalanine concentrations, and therefore an adequate energy intake supplied by SLPF is a necessity [23].

It has been suggested that the uncontrolled consumption of SLPFs may cause obesity [13,18], although the principle low protein foods eaten by our patient cohort were bread and pasta. However, as in other studies [10,24], our children consumed a low fat, high carbohydrate diet, leading to an imbalance in macronutrient composition. Despite energy intake only meeting recommendations, both overweight and obesity increased over the 3-year study. It has been observed from the age of one year that the energy provided by carbohydrate is higher than in healthy controls [24]. Clearly, a balanced diet prevents co-morbidities, such as metabolic syndrome, obesity, coronary heart disease and diabetes type II [17,25,26]. The type of carbohydrate is also an important health consideration. Insulin resistance measured by HOMA-IR (Homeostasis Model Assessment Insulin Resistance) has been shown to be higher in subjects with PKU [27], especially those who are overweight [27] or those with central obesity [28]. Similarly, the dietary glycaemic index and load was higher in children with PKU, suggesting a link between the quality of carbohydrate and peripheral insulin resistance [18]. Furthermore, lower total/LDL and higher triglyceride/HDL cholesterol ratios have been reported in children with PKU, suggesting an association between the quality of carbohydrate and triglyceride glucose index [28].

The starch sources from SLPFs, bread and pasta, eaten by children in our study were derived from starch isolated from wheat, maize and rice. Isolated starches are refined, having different physiologic properties compared to complex starch forms, and foods containing these may have a higher glycaemic index than those made from wheat flour [29,30]. However, a high intake of sugar from regular sweet drinks is also problematic. Many 'sugar free' drinks are unsuitable for children with PKU as they contain aspartame, a source of phenylalanine, limiting choice and increasing the glycaemic index of foods consumed. Importantly, the aspartame content of drinks may vary significantly, and the phenylalanine content is not identified on the food label [31].

Fibre sources may alter the gut microbiome, increasing the risk of chronic diseases such as inflammatory bowel disease and obesity [32]. The main fibre sources added to low protein bread and pasta were hydrocolloids. These are common additives in the food manufacturing industry, aiding texture and viscosity, but their role in gut health is limited [33,34]. Although the health benefits of hydrocolloids have been reported, there is little understanding of how these function in the intestine, or of their physiological benefits [35]. The fruit, vegetable and fibre intake of children in this study was less than the UK government's '5 a day' healthy eating recommendations, derived from World Health Organisation (WHO) and SACN recommendations [16,17,36]. Fruits and vegetables low in phenylalanine (≤75 mg/100 g), except potatoes, make a valuable contribution to the dietary intake, as they can be eaten ad libitum [37]. The free consumption of these fruits and vegetables does not impact metabolic control and should be encouraged in a low phenylalanine diet as a source of beneficial fibre. Cereal and wholegrain fibres, associated with a lower risk of cardio metabolic disease and colorectal cancer, and promoted by SACN [16], are precluded in a phenylalanine restricted diet. The challenge and responsibility of manufacturers making SLPFs is to provide a source of beneficial fibre maximising gut microbiome health.

There are limitations to this study. Firstly, all dietary assessment methods are open to misinterpretation. To minimise error, the standard weights of foods were collected regularly, and at least one meal was observed and food items weighed by the same dietitian. The food frequency questionnaire was not validated, although two dietary assessment tools were used, and mealtime portion sizes were observed and weighed by a dietetic researcher to help improve the quality of the dietary data collected. The nutritional composition, including the starch and sugar content, of SLPFs was not always available on food labels, and there were some discrepancies between food labels and manufacturers' websites [29]. It was not possible to accurately assess the salt intake from food labels or from the amounts added to food in cooking or at the table.

6. Conclusions

In children with PKU, dietary intake is based on a lower number of regular foods, offering limited variety. This study showed that SLPFs make an important contribution to energy intake in a phenylalanine restricted diet, with consistent dietary patterns over time demonstrating long-term dependence on essential foods, such as low protein bread, pasta and milk. The intake of sugar and fat from SLPFs was minimal. SLPFs should be unlimited to all patients on a phenylalanine-restricted diet, helping their ability to sustain their dietary restriction and reducing anxiety around food insecurity. Further improvements in the nutritional quality of the diet would aid in securing longer-term health benefits and adherence to a severe lifelong regimen.

Author Contributions: Conceptualisation, A.M., A.D.; methodology, A.M., A.D.; software, A.D.; validation, A.M., A.D., S.E., A.P., C.A., J.C.R.; formal analysis, A.D., investigation, A.D.; writing—original draft preparation, A.D., A.M. Writing review and editing, A.D., S.E., A.P., J.C.R., A.M.; supervision, A.M. Funding acquisition, funding was part of PhD project from Vitaflo International. All authors have read and agreed to the published version of the manuscript.

Funding: This research was funded by Vitaflo International as part of PhD grant.

Conflicts of Interest: A.D. received research funding from Vitaflo International, financial support from Nutricia, Mevalia and Vitaflo International to attend study days and conferences. SE received research funding from Nutricia, financial support from Nutricia and Vitaflo International to attend study days and conferences. A.P. received an educational grant from Cambrooke Therapeutics and grants from Vitaflo International, Nutricia, Merck Serono, Biomarin and Mevalia to attend scientific meetings. C.A. received honoraria from Nutricia and Vitaflo International to attend study days and conferences. J.C.R. is a member of the European Nutritionist Expert Panel (Biomarin), the Advisory Board for Applied Pharma Research and Nutricia, and received honoraria as a speaker from APR, Merck Serono, Biomarin, Nutricia, Vitaflo, Cambrooke, PIAM and Lifediet. A.M. received research funding and honoraria from Nutricia, Vitaflo International and Merck Serono, and is a Member of European Nutrition Expert Panel (Merck Serono international), a member of Sapropterin Advisory Board (Merck Serono international), and a member of the Advisory Board Element (Danone–Nutricia). The funders had no role in the design of the study; in the collection, analyses, or interpretation of data; in the writing of the manuscript, or in the decision to publish the results.

References

1. González, M.J.; Gassió, R.; Artuch, R.; Campistol, J. Impaired Neurotransmission in Early-treated Phenylketonuria Patients. *Semin. Pediatr. Neurol.* **2016**, *23*, 332–340. [CrossRef] [PubMed]
2. Van Vliet, D.; Bruinenberg, V.M.; Mazzola, P.N.; Van Faassen, M.H.J.R.; De Blaauw, P.; Kema, I.P.; Heiner-Fokkema, M.R.; Van Anholt, R.D.; Van Der Zee, Y.A.; Van Spronsen, F. Large Neutral Amino Acid Supplementation Exerts Its Effect through Three Synergistic Mechanisms: Proof of Principle in Phenylketonuria Mice. *PLoS ONE* **2015**, *10*, e0143833. [CrossRef] [PubMed]
3. Hoeksma, M.; Reijngoud, D.-J.; Pruim, J.; De Valk, H.W.; Paans, A.M.; Van Spronsen, F.J. Phenylketonuria: High plasma phenylalanine decreases cerebral protein synthesis. *Mol. Genet. Metab.* **2009**, *96*, 177–182. [CrossRef] [PubMed]
4. Surtees, R.; Blau, N. The neurochemistry of phenylketonuria. *Eur. J. Nucl. Med. Mol. Imaging* **2000**, *159*, S109–S113. [CrossRef] [PubMed]
5. Van Wegberg, A.M.J.; Macdonald, A.; Ahring, K.; Bélanger-Quintana, A.; Blau, N.; Bosch, A.M.; Burlina, A.; Campistol, J.; Feillet, F.; Giżewska, M.; et al. The complete European guidelines on phenylketonuria: Diagnosis and treatment. *Orphanet J. Rare Dis.* **2017**, *12*, 162. [CrossRef]
6. Ford, S.; O'Driscoll, M.; Macdonald, A. Living with Phenylketonuria: Lessons from the PKU community. *Mol. Genet. Metab. Rep.* **2018**, *17*, 57–63. [CrossRef]
7. Macdonald, A.; Van Wegberg, A.M.J.; Ahring, K.; Beblo, S.; Bélanger-Quintana, A.; Burlina, A.; Campistol, J.; Coşkun, T.; Feillet, F.; Giżewska, M.; et al. PKU dietary handbook to accompany PKU guidelines. *Orphanet J. Rare Dis.* **2020**, *15*, 171. [CrossRef]
8. MacDonald, A.; Chakrapani, A.; Hendriksz, C.; Daly, A.; Davies, P.; Asplin, D.; Hall, K.; Booth, I.W. Protein substitute dosage in PKU: How much do young patients need? *Arch. Dis. Child.* **2006**, *91*, 588–593. [CrossRef]
9. Evans, S.; Adam, S.; Adams, S.; Allen, H.; Ashmore, C.; Bailey, S.; Banks, J.; Churchill, H.; Cochrane, B.; Cook, J.; et al. Uniformity of Food Protein Interpretation Amongst Dietitians for Patients with Phenylketonuria (PKU): 2020 UK National Consensus Statements. *Nutrients* **2020**, *12*, 2205. [CrossRef]
10. European Union, Supplementing regulation (EU). No 609/2013 of the European Parliament and of the Council as regards the specific compositional and information requirements for food for special medical purposes. *Off. J. Eur. Union* **2016**, *L25/30*, 35–48.
11. Rocha, J.C.; Macdonald, A. Dietary intervention in the management of phenylketonuria: Current perspectives. *Pediatr. Health Med. Ther.* **2016**, *7*, 155–163. [CrossRef] [PubMed]
12. European Society of Phenylketonuria. Available online: http://www.espku.org/ (accessed on 16 September 2020).
13. Pena, M.J.; Almeida, M.F.; Van Dam, E.; Ahring, K.; Belanger-Quintana, A.; Dokoupil, K.; Gokmen-Ozel, H.; Lammardo, A.M.; Macdonald, A.; Robert, M.; et al. Special low protein foods for phenylketonuria: Availability in Europe and an examination of their nutritional profile. *Orphanet J. Rare Dis.* **2015**, *10*, 162. [CrossRef] [PubMed]
14. Nutritics, Research Edition, Version 5. *093*; Nutritics: Dublin, Ireland, 2019.
15. Public Health England. *Dietary Reference Values for Energy Scientific Advisory Committee on Nutrition*; Public Health England: London, UK, 2011; p. 228.
16. Scientific Advisory Committee on Nutrition. *Carbohydrates and Health*; Public Health England: London, UK, 2015.
17. World Health Organisation. *Diet, Nutrition and the Prevention of Chronic Diseases Report*; WHO: Geneva, Switerland, 2002.
18. Moretti, F.; Pellegrini, N.; Salvatici, E.; Rovelli, V.; Banderali, G.; Radaelli, G.; Scazzina, F.; Giovannini, M.; Verduci, E. Dietary glycemic index, glycemic load and metabolic profile in children with phenylketonuria. *Nutr. Metab. Cardiovasc. Dis.* **2017**, *27*, 176–182. [CrossRef] [PubMed]
19. Thiele, A.G.; Rohde, C.; Mütze, U.; Arélin, M.; Ceglarek, U.; Thiery, J.; Baerwald, C.; Kiess, W.; Beblo, S. The challenge of long-term tetrahydrobiopterin (BH4) therapy in phenylketonuria: Effects on metabolic control, nutritional habits and nutrient supply. *Mol. Genet. Metab. Rep.* **2015**, *4*, 62–67. [CrossRef]
20. Ford, S.; O'Driscoll, M.; Macdonald, A. Prescribing issues experienced by people living with phenylketonuria in the UK. *Mol. Genet. Metab. Rep.* **2019**, *21*, 100527. [CrossRef]

21. Cochrane, B.; Schwahn, B.; Galloway, P.; Robinson, P.; Gerasimidis, K. A questionnaire survey on the usage of low protein staple foods by people with phenylketonuria in Scotland. *J. Hum. Nutr. Diet.* **2014**, *27*, 533–541. [CrossRef]
22. Coakley, K.E.; Porter-Bolton, S.; Salvatore, M.L.; Blair, R.B.; Singh, R.H. Food insecurity in females with phenylketonuria. *JIMD Rep.* **2020**, *53*, 103–110. [CrossRef]
23. Bickel, H.; Gerrard, J.; Hickmans, E.M. The influence of phenylalanine intake on the chemistry and behaviour of a phenyl-ketonuric child. *Acta Paediatr.* **1954**, *43*, 64–77. [CrossRef]
24. Evans, S.; Daly, A.; Wildgoose, J.; Cochrane, B.; Chahal, S.; Ashmore, C.; Loveridge, N.; Macdonald, A. Growth, Protein and Energy Intake in Children with PKU Taking a Weaning Protein Substitute in the First Two Years of Life: A Case-Control Study. *Nutrients* **2019**, *11*, 552. [CrossRef]
25. Albertson, A.M.; Thompson, D.; Franko, D.L.; Holschuh, N.M.; Bauserman, R.; Barton, B. Prospective Associations among Cereal Intake in Childhood and Adiposity, Lipid Levels, and Physical Activity during Late Adolescence. *J. Am. Diet. Assoc.* **2009**, *109*, 1775–1780. [CrossRef]
26. Després, J. Is visceral obesity the cause of the metabolic syndrome? *Ann. Med.* **2006**, *38*, 52–63. [CrossRef] [PubMed]
27. Couce, M.L.; Guler, I.; Anca-Couce, A.; Lojo, M.; Mirás, A.; Leis, R.; Pérez-Muñuzuri, A.; Fraga, J.M.; Gude, F. New insights in growth of phenylketonuric patients. *Eur. J. Pediatr.* **2015**, *174*, 651–659. [CrossRef] [PubMed]
28. Rocha, J.C.; Van Spronsen, F.J.; Almeida, M.F.; Soares, G.; Quelhas, D.; Ramos, E.; Guimarães, J.T.; Borges, N. Dietary treatment in phenylketonuria does not lead to increased risk of obesity or metabolic syndrome. *Mol. Genet. Metab.* **2012**, *107*, 659–663. [CrossRef] [PubMed]
29. Wood, G.; Evans, S.; Pointon-Bell, K.; Rocha, J.C.; Macdonald, A. Special Low Protein Foods in the UK: An Examination of Their Macronutrient Composition in Comparison to Regular Foods. *Nutrients* **2020**, *12*, 1893. [CrossRef] [PubMed]
30. Soltanizadeh, N.; Mirmoghtadaie, L. Strategies Used in Production of Phenylalanine-Free Foods for PKU Management. *Compr. Rev. Food Sci. Food Saf.* **2014**, *13*, 287–299. [CrossRef]
31. Van Vliet, K.; Melis, E.S.; De Blaauw, P.; Van Dam, E.; Maatman, R.G.H.J.; Abeln, D.; Van Spronsen, F.J.; Heiner-Fokkema, M.R. Aspartame and Phe-Containing Degradation Products in Soft Drinks across Europe. *Nutrients* **2020**, *12*, 1887. [CrossRef] [PubMed]
32. Valdes, A.M.; Walter, J.; Segal, E.; Spector, T.D. Role of the gut microbiota in nutrition and health. *BMJ* **2018**, *361*, k2179. [CrossRef]
33. Gidley, M.J. Hydrocolloids in the digestive tract and related health implications. *Curr. Opin. Colloid Interface Sci.* **2013**, *18*, 371–378. [CrossRef]
34. Rutenberg, R.; Bernstein, S.; Paster, N.; Fallik, E.; Poverenov, E. Antimicrobial films based on cellulose-derived hydrocolloids. A synergetic effect of host–guest interactions on quality and functionality. *Colloids Surfaces B Biointerfaces* **2016**, *137*, 138–145. [CrossRef]
35. Edwards, C.A.; Garcia, A.L. *The Health Aspects of Hydrocolloids*; Handbook of Hydrocolloids; Phillips, G.O., Williams, A., Eds.; Woodhead Publishing: Sawston, UK, 2009; pp. 50–81.
36. Public Health England. *From Plate to Guide: What, Why and How for the Eatwell Model, H*; PHE Publications Gateway Number: 2016451; Public Health England: London, UK, 2016.
37. Mac Donald, A.; Rylance, G.; Davies, P.; Asplin, D.; Hall, S.K.; Booth, I.W. Free use of fruits and vegetables in phenylketonuria. *J. Inherit. Metab. Dis.* **2003**, *26*, 327–338. [CrossRef]

Publisher's Note: MDPI stays neutral with regard to jurisdictional claims in published maps and institutional affiliations.

© 2020 by the authors. Licensee MDPI, Basel, Switzerland. This article is an open access article distributed under the terms and conditions of the Creative Commons Attribution (CC BY) license (http://creativecommons.org/licenses/by/4.0/).

nutrients

MDPI

Article

Protein Labelling Accuracy for UK Patients with PKU Following a Low Protein Diet

Dilyana Kraleva [1], Sharon Evans [2], Alex Pinto [2], Anne Daly [2], Catherine Ashmore [2], Kiri Pointon-Bell [1], Júlio César Rocha [3,4] and Anita MacDonald [2,*]

1 Faculty of Health, Education & Life Sciences, Birmingham City University City South Campus, Westbourne Road, Edgbaston, Birmingham B15 3TN, UK; didincee@gmail.com (D.K.); Kiri.Pointon-Bell@bcu.ac.uk (K.P.-B.)

2 Birmingham Women's and Children's NHS Foundation Trust, Steelhouse Lane, Birmingham B4 6NH, UK; Sharon.morris6@nhs.net (S.E.); alex.pinto@nhs.net (A.P.); a.daly3@nhs.net (A.D.); Catherine.ashmore@nhs.net (C.A.)

3 Nutrition & Metabolism, NOVA Medical School, Faculdade de Ciências Médicas, Universidade Nova de Lisboa, Campo Mártires da Pátria, 130, 1169-056 Lisbon, Portugal; rochajc@nms.unl.pt

4 Center for Health Technology and Services Research (CINTESIS), R. Dr. Plácido da Costa, s/n, 4200-450 Porto, Portugal

* Correspondence: anita.macdonald@nhs.net

Received: 21 September 2020; Accepted: 27 October 2020; Published: 10 November 2020

Abstract: A phenylalanine (protein)-restricted diet is the primary treatment for phenylketonuria (PKU). Patients are dependent on food protein labelling to successfully manage their condition. We evaluated the accuracy of protein labelling on packaged manufactured foods from supermarket websites for foods that may be eaten as part of a phenylalanine-restricted diet. Protein labelling information was evaluated for 462 food items ("free from", $n = 159$, regular, $n = 303$), divided into 16 food groups using supermarket website data. Data collection included protein content per portion/100 g when food was "as sold", "cooked" or "prepared"; cooking methods, and preparation instructions. Labelling errors affecting protein content were observed in every food group, with overall protein labelling unclear in 55% ($n = 255/462$) of foods. There was misleading, omitted, or erroneous (MOE) information in 43% ($n = 68/159$) of "free from" foods compared with 62% ($n = 187/303$) of regular foods, with fewer inaccuracies in "free from" food labelling ($p = 0.007$). Protein analysis was available for uncooked weight only but not cooked weight for 58% ($n = 85/146$) of foods; 4% ($n = 17/462$) had misleading protein content. There was a high rate of incomplete, misleading, or inaccurate data affecting the interpretation of the protein content of food items on supermarket websites. This could adversely affect metabolic control of patients with PKU and warrants serious consideration.

Keywords: phenylketonuria; food labelling; protein content; free from; gluten free

1. Introduction

Phenylketonuria (PKU) is a rare, autosomal recessive inborn error of metabolism due to low or absent activity of the enzyme phenylalanine hydroxylase (PAH), required for degradation of phenylalanine to tyrosine. It causes elevated levels of phenylalanine in the blood and brain and if untreated, leads to severe, irreversible, intellectual disability [1]. Maintaining low blood phenylalanine levels within defined target ranges prevents phenylalanine toxicity [1]. Although it can be managed with a combined approach of dietary and pharmaceutical treatment, the only treatment option in the UK is a lifelong, phenylalanine-restricted diet [2]. Dietary management is stringent, requiring discipline and tenacity, and it is well established that many patients with PKU of all ages are unable to sustain satisfactory blood phenylalanine control [3,4]. Although there are multiple causes for unsatisfactory metabolic control, relatively small deviations from dietary prescription can adversely affect blood phenylalanine levels in patients with classical PKU [5].

Dietary treatment involves the avoidance of high-protein foods such as meat, fish, eggs, cheese, seeds, soya, and nuts and a limited intake of natural protein from foods such as cereals, potatoes, milk, and some vegetables. Any natural protein intake should be calculated, measured, and controlled and up to 80% of patients tolerate <10 g/day [3]. Fruit and vegetables with a phenylalanine content ≤75 mg/100 g, butter, oils, and sugars are given without restriction [2]. Dietary protein is supplemented with synthetic protein, either phenylalanine-free amino acids or low-phenylalanine glycomacropeptide, with added vitamins and minerals to meet nutritional protein requirements. Caregivers and people with PKU are trained in reading and interpreting the protein amounts on manufactured food labels. They are reliant on supermarkets and manufacturers to provide accurate and easily interpreted information about protein content on food labels. Almost every UK supermarket offers an online food delivery service and survey data suggest that 29% of people purchase food via online shopping [6], with online grocery shopping available in at least 60 countries worldwide [7]. Website supermarket shopping is popular for those with special dietary requirements, giving the opportunity to examine food labels prior to food purchase [8]. Patients with PKU and their families can browse the protein nutritional analysis of foods and examine information about food preparation, cooking, and reconstitution of foods. Some online supermarket websites are intuitive to dietary needs and can even create a specific dietary profile that will highlight products that should be avoided for food allergies, although the needs of patients with PKU are not considered [9].

On 25 October 2011, the European Parliament and Council adopted Regulation (EU) No 1169/2011 that issued legal standards for the labelling and information given to consumers by food manufacturers (called the "Food Information to Consumers (FIC) Regulation") [10]. This regulation has been applied since 2016. When pre-packaged foods are sold "online", it is regulated that the responsibility for providing mandatory food information (except the date of minimum durability or the "use by" date) sits with the owner of the online website (the responsible food business operator). Online pre-packaged mandatory food information should include information on the weight and volume of food (net quantity information), a list of ingredients, protein content per 100 g/100 mL, and instructions for use or cooking, if applicable. For non-packaged food, there are fewer rigid stipulations, but the food business operator is required to provide allergen information [11].

In PKU, if foods are eaten because of inaccurate or ambiguous website or food labelling information it may cause unexplained, poor blood phenylalanine control. In countries like the UK, there is high reliance on manufactured foods, so reliability of food labelling information is particularly important. Inadequate, misleading, or unclear information about protein content may deter caregivers/patients with PKU from purchasing specific foods. Therefore, it is in the best interests of manufacturers to supply suitable and trustworthy information that is easy to understand and accurate. The clarity of food labelling and food information on online supermarket websites remains unstudied for people with special dietary requirements such as PKU.

The aim of this study was to evaluate the accuracy of protein labelling on packaged manufactured foods from supermarket websites for foods that may be eaten as part of a phenylalanine-restricted diet. Patients with PKU may use some "free from foods", particularly gluten free, which may have a lower protein content than foods containing wheat or milk, so emphasis was placed on this group of foods.

2. Materials and Methods

From January 2019 to April 2020, 462 packaged food items were examined using descriptive information given by major UK supermarkets (Asda, Morrisons, Sainsbury's, Tesco, and Waitrose) available on their website. For each food, factors that may alter or affect the protein analysis were tabulated. The selection of foods was not random as this was conducted based on food popularity and common usage. Foods were chosen based on their potential suitability in a protein-restricted diet and mainly had a protein content <10 g/100 g. A selection of "free from" (all gluten-free) and regular foods were examined. Both commercial branded products and the supermarkets' own brands were included. Meat and fish products were avoided, although two types of regular cheese were included to compare information with "free from" varieties. Items were divided and analysed by the following food groups: bread and bread products, breakfast cereals, vegan cheese, cakes, sweet biscuits, pastries/tarts, crackers, chocolate, crisps, desserts, flours, gravies/sauces, pasta, vegetable foods, dried pot noodles and yoghurts. The supermarket websites accessed were required to give a product description and nutritional analysis for each food.

The following data were collected for each food item: product description, ingredients, preparation, cooking instructions and usage, net and portion size. Specific information collected about protein content included: protein content per portion as cooked/prepared, protein content per portion as sold; protein content per 100 g expressed as cooked/prepared, protein content per 100 g as sold; any misleading information about protein content per 100 g or per food portion (e.g., when a food protein content states 0.0 g per portion, but was >0.1 g/100 g or if the protein content for each portion was only described as <0.5 g with no other relevant information, or if there was a discrepancy between the ingredients listed and the protein content); reconstitution instructions for dry powders, including protein content per portion/100 g supplied when protein analysis was given after dry products had been reconstituted.

All information was transferred onto a database and coded according to the accuracy of information. All products were checked twice by two different dietitians for accuracy and to minimise any risk of bias. Statistical analysis was performed using Mann–Whitney unpaired *t*-tests to compare numbers of misleading, omitted, or erroneous (MOE) foods in "free from" and "regular" food groups and to compare types of MOE information between the two groups. Percentage error in "free from" and "regular" foods were also compared using Wilcoxon signed-rank tests.

3. Results

3.1. Accuracy of Product Description

The product description, ingredients, net weight, portion size, protein content per 100 g cooked and uncooked, and preparation and cooking instructions were checked for 462 foods from five supermarket websites (Asda, Morrisons, Sainsbury's, Tesco, and Waitrose). There were 159 "free from" foods and 303 "regular" foods (Table 1). All the "free from" foods were gluten free. Overall, 255 of 462 (55%) foods had information that was MOE from the website product information given by supermarkets, thereby affecting the interpretation of protein content for food items. There were fewer inaccuracies in "free from foods" (MOE, 68 of $n = 159$ foods, 43%), compared with regular foods (MOE, 187 of $n = 303$ foods, 62%) ($p = 0.007$, Wilcoxon signed-rank test) most notably for breads, bread products, and flours.

3.2. Types of Misleading Omitted, or Erroneous (MOE) Information in Food Product Information that Affected Protein Content

All types of MOE information are categorised by issue in Table 2. Some food items had more than one descriptive issue that affected interpretation of their protein content.

3.2.1. Food Label did not Distinguish if Protein Content was for Food Item when Cooked/Prepared or as Sold

Thirty two percent (n =146/462) of foods required further cooking or preparation. Of these 7% ($n = 10/146$) did not specify whether the protein analysis per 100 g/weight of food was for the cooked/prepared or 'as sold' product. When expressed per portion size this figure was 14% ($n = 20/146$).

Table 1. Number of individual foods with misleading, omitted, or erroneous (MOE) information that would affect the protein content calculation from information given on the supermarket websites.

Food Groups	"Free From/Gluten-Free" Foods (*n* = 159) Number of Foods Examined	"Free From/Gluten-Free" Foods (*n* = 159) Number of Foods with MOE Issues (%)	"Regular" Foods (*n* = 303) Number of Foods Examined	"Regular" Foods (*n* = 303) Number of Foods with MOE Issues (%)	*p* Value *
Biscuits	29	9 (31)	48	16 (33)	>0.99
Bread and bread products	33	11 (33)	35	22 (63)	0.02
Breakfast cereals	7	4 (57)	13	8 (62)	>0.99
Cakes	14	4 (28)	27	15 (56)	0.19
Cheese	1	0 (0)	2	2 (100)	ID
Chocolate	13	2 (15)	26	8 (31)	0.45
Crackers, crispbread, rice cakes	17	5 (29)	34	16 (47)	0.37
Crisps, pretzels	2	1 (50)	4	2 (50)	>0.99
Desserts, puddings, dessert mixes, ice cream	2	0 (0)	20	12 (60)	0.19
Flours, flour mixes	4	2 (50)	8	8 (100)	0.09
Gravies/sauces	8	7 (88)	16	16 (100)	0.33
Pastries/tarts/pancakes/waffles	11	8 (73)	16	12 (75)	>0.99
Pasta	13	13 (100)	22	22 (100)	>0.99
$ Pot noodles, meal pots	1	1 (100)	22	22 (100)	ID
Vegetable foods	1	1 (100)	4	4 (100)	ID
Yogurt	3	0 (0)	6	2 (33)	0.50
Total food numbers	159	68 (43)	303	187 (62)	

Abbreviations: MOE, foods with misleading, omitted, or erroneous information. * Mann–Whitney unpaired *t*-test; $ Pot noodles: a mix of dehydrated noodles, assorted dried vegetables and flavouring powder in a pot. They are prepared by adding boiling water. ID = insufficient data.

Table 2. Type of misleading, omitted, or erroneous (MOE) information that would affect protein content describing individual foods given on the supermarket websites.

Issue	All Foods *n* = 462 (%)	"Free From" Foods *n* = 159 (%)	"Regular" Foods *n* = 303 (%)	*p* Value *
Unspecified if protein content given per food portion is for cooked/prepared or weight as sold	20 (14) **	9 (22) **	11 (10) **	0.11
Unspecified if protein content given per 100 g is for cooked/prepared or weight as sold	10 (7) **	3 (2) **	7 (7) **	>0.99
Protein amount given is the same per 100 g and per portion (but one portion does not weigh 100 g)	1 (<1)	0 (0)	1 (<1)	0.47
Cooking/preparation instructions missing	12 (8) **	5 (12) **	7 (7) **	0.32
Protein content per 100 g cooked/prepared missing	85 (58) **	30 (73) **	55 (52) **	0.03
Protein content per 100 g uncooked/unprepared missing	48 (33) **	7 (17) **	41 (39) **	0.02
Protein content per portion size missing	47 (10)	16 (10)	31 (10)	0.96
Protein content per portion size cooked/prepared missing	51 (35) **	14 (34) **	37 (35) **	>0.99
Protein content per portion as sold missing but provided when cooked/prepared	38 (26) **	3 (7) **	35 (33) **	0.001
Weight of portion size missing	125 (27)	28 (18)	97 (32)	0.001
Missing net size	21 (5)	5 (3)	16 (5)	0.30
Protein content states 0 g per portion, even though contains protein > 0.1 g/100 g	7 (2)	6 (4)	1 (<1)	0.004
Protein content per portion described as <0.5 g	8 (2)	1 (1)	7 (2)	0.19
Protein content per 100 g described as <0.5 g protein	2 (<1)	0 (0)	2 (1)	0.31
Incorrect protein analysis	1 (<1)	0 (0)	1 (<1)	0.47
Missing protein analysis	2 (<1)	0 (0)	2 (1)	0.31
Protein content per portion only after prepared with milk	10 (7) **	0 (0)	10 (10) **	0.06
Protein content per 100 g only after prepared with milk	8 (5) **	0 (0)	8 (8) **	0.11
Missing ingredients list	9 (2)	1 (1)	8 (3)	0.14

* Mann–Whitney unpaired *t*-test. ** total number of foods requiring cooking/preparation *n* = 146; free from *n* = 41, regular *n* = 105.

3.2.2. Missing Information about Protein Content per 100 g (either Omitted Protein Content for Cooked/Prepared Weight or for Uncooked/Unprepared Weight)

Protein content was given for uncooked weight only but not cooked weight for 58% (*n* = 85/146) of foods requiring preparation. In contrast, 33% (*n* = 48/146) of foods gave protein value for cooked but not uncooked weight. These issues were more commonly "regular" foods than "free from" foods (p = 0.03 and 0.02 respectively).

3.2.3. Missing Information about Protein Content per Portion Size

Twenty-seven percent ($n = 125/462$) of foods did not give a weight for an estimated portion/serving size; these were more likely to be "regular" foods than "free from" foods ($p = 0.001$). Ten percent ($n = 47/462$) of foods did not give the protein analysis of the portion size. Protein content was omitted per cooked portion size in 35% ($n = 51/146$) of foods or omitted per portion as sold in 26% ($n = 38/146$).

3.2.4. Omitted Cooking Instructions

This information was omitted in 8% ($n = 12/146$) of food items.

3.2.5. Missing Net Size

This information was omitted in 5% of foods ($n = 21/462$).

3.2.6. Misleading/Incorrect Protein Content

Four percent ($n = 17/462$) of foods had misleading protein analysis. Either the protein content per portion size stated 0 g protein when the analysis per 100 g stated a protein content >0.0 g (more commonly in "regular" foods than "free from" foods; $p = 0.004$), or the protein content per 100 g or per portion stated <0.5 g but did not give a specific protein amount. One food had an incorrect protein analysis; this product contained 70% peas, and it was stated that it contained a protein content of only 0.5 g/100 g when it should have contained around 4 g/100 g.

3.2.7. Missing Protein Analysis

Two foods contained no protein analysis (a jelly and frozen potato product). These products were produced by the same manufacturer.

3.2.8. Preparation/Reconstitution Information

Twelve percent ($n = 18/146$) of foods requiring preparation gave the protein analysis only after a product had been reconstituted/prepared with milk even though milk was not part of the ingredients list. Consequently, this "theoretical" protein analysis portrayed these foods to be unsuitable in a low-protein diet. The protein content of the dry ingredients was not given. This was more likely to occur in "regular" foods than in "free from" foods although it did not reach statistical significance.

3.2.9. Omitted Ingredients List

Two percent ($n = 9/462$) of foods did not give an ingredients list.

3.3. *Frequency of Misleading, Omitted, or Erroneous (MOE) Food Information for Food Groups and for Individual Foods*

Thirty eight percent ($n = 96/255$) of foods with MOE information had one inaccuracy, 37% ($n = 95/255$) had two inaccuracies, and 16% ($n = 41/255$) had three inaccuracies regarding their information, which affected the interpretation of the food protein content. Bread and bread products, cake, and biscuits commonly had missing information about portion sizes. Pasta and vegetable products regularly had omitted information about the protein content for cooked or uncooked product (either per 100 g or per portion size). Pot noodles were particularly misleading; their protein content was commonly given per 100 g reconstituted weight rather than dry weight, but this was unclear. The protein content of "regular" custards, instant desserts, and some "regular" and "free from" cereals were only given after reconstitution with milk, and commonly had unclear portion sizes.

The frequency of MOE information and the number of problems for the same food items ("free from" and "regular" food items) that would affect their protein content given on the supermarket websites are presented in Figures 1 and 2.

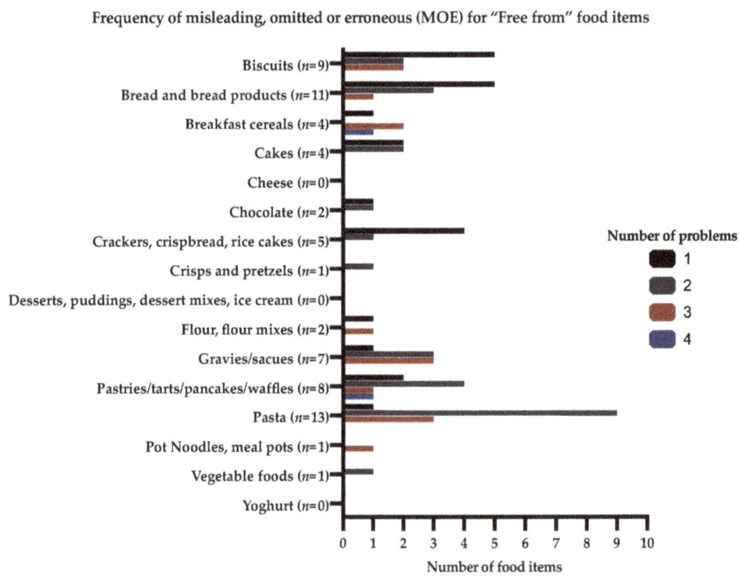

Figure 1. Frequency of misleading, omitted, or erroneous (MOE) information for "free from" food items that would affect their protein content given on the supermarket websites.

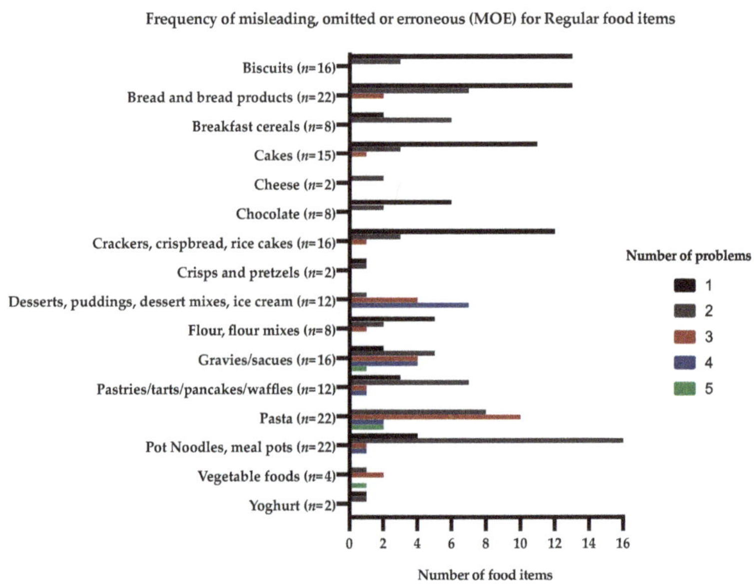

Figure 2. Frequency of misleading, omitted, or erroneous (MOE) information for regular food items that would affect their protein content given on the supermarket websites.

4. Discussion

This research indicates that interpreting the protein content for some common supermarket foods available via online websites is inadequate, unclear and even misleading for people with PKU. Information about the protein content per portion size was sometimes omitted or indeterminate, particularly for "regular" foods compared with "free from" foods. For "regular" dried products requiring reconstitution, the protein content was commonly given only after the product has been prepared with "added" cow's milk, which then increased the protein content of the food, rendering it unsuitable for most people with PKU. For other products consisting of dry ingredients, it was sometimes uncertain if protein labelling was for the dry product or after preparation. For products such as gluten-free biscuits, the protein content was stated as <0.5 g per portion only, even though the protein per 100 g was much higher, and the food item included protein-containing ingredients. Not all products identified net weight.

Food regulations, manufacturers, and online food business operators have not considered the impact of any inaccurate product information for people on very low-protein diets. Fortunately, mandatory FIC nutrition labelling for pre-packaged foods does include protein content, but it is listed only after energy, fat (including saturates), and carbohydrates (including sugar). For non–prepacked foods, there is no requirement in the EU FIC regulations for any nutrition information to be provided, but many manufacturers voluntarily declare the protein content. The FIC regulations states that food manufacturers are not required to do their own laboratory analysis for protein content and it is possible for a food business operator to calculate the values themselves (1) from the known or actual average values of the ingredients used or (2) from generally established and accepted data [12]. The accuracy of protein measurement by these methods is unknown and the definition of what is meant by "generally established and accepted" data is not given, so manufacturers could interpret this in different ways. It is also unknown how many food businesses estimate the protein content by using published protein values of similar foods rather than estimating individual foods by chemical analysis.

Some patients with PKU tolerate a minimal amount of protein (3 to 4 g/day) so accurate protein information is crucial [13–15]. In conflict, the FIC regulations apply protein tolerances to food labels on the basis that protein analysis is not precise due to natural variations in ingredient composition and changes in production. They appear unaware of the needs of patients on very low-protein diets. For foods containing protein <10 g/100 g, they state that the protein content may be within ±2 g; for foods containing protein 10–40 g/100 g, the protein content is ±20%; and for foods containing protein >40 g per 100 g, protein content is within ±8 g [16,17]. Additionally, rounding guidance suggested by the EU states that for food containing protein ≥10 g/100 g or 100 mL, the protein should be declared to the nearest 1 g (no decimals); protein between <10 g and >0.5 g/100 g or mL to the nearest 0.1 g; and protein at ≤0.5 g/100 g or mL as "0 g" or "<0.5 g." We identified eight foods, particularly "free from" items, that stated that the food portion contained <0.5 g protein, even though each portion could have contained 0.4 g protein (<0.5 g); this amount would need to be calculated in a very low-protein diet as it may impact on metabolic control. Some patients with PKU have unexplained fluctuating daily blood phenylalanine levels and some of this may be due to the approximate nature of food protein labelling [18].

The FIC regulations state that the nutrition declaration is required for the food as sold, but, instead and where appropriate, it can relate to the food as prepared, provided sufficiently detailed preparation instructions are given. It is therefore possible to include only the nutrition information "as prepared" for foods such as dehydrated powdered soup or desserts. This is deceptive and unsafe for people with PKU. Commonly, we found that nutrition labels for "regular" dessert mixes were calculated based on their preparation with cow's milk, and this should be avoided in PKU. The addition of milk substantially increased the protein content, even though many of the raw ingredients of dessert mixes were low in protein. By declaring protein content after preparation with other added ingredients, products appear unsuitable for patients on a low-protein diet, even though it may have been possible to consume the food product if it had been made up with a low-protein milk alternative. Additionally, giving the protein content of pot noodles after preparation is confusing and this has led to several incidents when caregivers/patients have miscalculated and underestimated their protein content [19].

It was common for supermarket websites to omit information about whether the protein analysis was associated with cooked or uncooked food. The protein content of a food product will vary depending on if it is dry cooked, fried, microwaved, or uncooked [20]. Commonly, foods such as potatoes have a high water content, and dry cooking results in moisture loss and a more concentrated protein amount [20]. These protein differences must be considered in a low-protein diet. The FIC regulations state that instructions on how to prepare and cook the food, including heating in a microwave oven, must be given on the label if they are needed [21–23]. If the food must be heated, the temperature of the oven and the cooking time should usually be stated, so it is sensible to give food analysis both for the "as sold" state and for "cooked", as recommended.

The protein content of a portion size was either omitted or the portion size was not quantified by weight for 172 of 462 foods (37%), particularly in "regular" foods, contributing to the difficulty in calculating the protein content of foods consumed in a phenylalanine-restricted diet. The FIC regulations state that the portion or consumption unit should be easily recognisable by the consumer, quantified on the label in close proximity to the nutrition declaration, and the number of portions or units contained in the package must be stated on the label. The "consumption unit" information requires improvement.

Legislation on food labelling gives instruction to producers and retailers; it also gives the consumers rights to basic information. We have shown that the information on the protein content of foods via supermarket websites is inaccurate and potentially harmful to those with PKU. Unlike allergies, there is little understanding of the essential role of a very low-protein diet and the harmful impact of poor control on patients' neuropsychological health [9]. This may also apply to other patients with inherited disorders of protein metabolism such as Maple Syrup Urine Disease or Tyrosinaemia type I or II. They also rely on accurate food labelling to manage their dietary treatment safely.

There appears to be no audit or regular assessment of supermarket websites to check accuracy of information that is provided to the consumer. Manufacturers should indicate on food labels how they have estimated protein content. We identified a packaged food product containing 70% peas and 30% carrots (with no other added ingredients), but it stated that it only contained protein 0.5 g/100 g, when it should have contained a protein amount of around 4 g/100 g (based on the established protein content of peas). For children to be given a food they enjoy in error, leads to additional psychological stress and guilt for the parents. We identified two other packaged products made by the same company without any protein analysis. These products were targeted at young children, so likely to be mistakenly eaten by a population very vulnerable to the impact of high blood phenylalanine concentrations.

This study did have some limitations. Although almost 500 foods were examined, matched numbers and types of "free from" foods were not compared with regular foods. However, as an overall group, "free from" foods website supermarket information gave more comprehensive data that would enable the consumer to assess the protein content of the product consumed. This was commonly due to the low availability of some "free from" foods, such as pot noodles or dessert pots. There was not an equal number of foods examined in all the different food groups, with small numbers of the following products examined: cheese, yoghurts, ice cream, and vegetable products. Food products in this study were not chosen by random, but commonly selected in order of popularity and usage, so it is accepted that there are limitations in product selection, especially with gluten-free or "free from" products that are usually purchased by patients with coeliac disease or food intolerance. Website information was not compared with product labelling on packages as purchased from the supermarket shelves, which may have identified further discrepancies.

5. Conclusions

Obtaining accurate information about the protein content of some foods from online supermarket website information is challenging. A high proportion of incomplete, misleading, or inaccurate data was identified that directly affected the interpretation of the protein content of food items. Inadequate protein food labelling is likely to contribute to the difficulties in maintaining good metabolic control in PKU. It is important that all dietitians, patients, and families of patients with PKU are aware of the food label limitations and potential problems.

Although food producers and business operators are expected to provide information to consumers that is clear and accurate, little attention is paid to the exactness of protein food labelling. The FIC regulations should be reconsidered, with more attention given to monitoring the accuracy of information provided by supermarket websites. Poor awareness of the impact and inattention to the factors that affect food protein content and carelessness about the accuracy of protein labelling can adversely affect the neurological health of people with PKU and deserves urgent consideration.

Author Contributions: Conceptualisation, D.K., A.M., and S.E.; methodology, D.K., A.M., and S.E.; formal analysis, D.K., S.E., A.P., and A.M.; investigation, D.K.; data curation, D.K.; writing—original draft preparation, D.K.; writing—review and editing, D.K., S.E., A.P., A.D., C.A., K.P.B., J.C.R., and A.M.; visualisation, D.K., S.E., A.P., A.D., C.A., K.P.-B., J.C.R., and A.M.; supervision, A.M. and K.P.-B.; project administration, A.M. and K.P.-B. All authors have read and agreed to the published version of the manuscript.

Funding: This research received no external funding.

Conflicts of Interest: The authors have no conflicts of interest to declare related to the paper's content.

References

1. Van Wegberg, A.M.J.; Macdonald, A.; Ahring, K.; Bélanger-Quintana, A.; Blau, N.; Bosch, A.M.; Burlina, A.; Campistol, J.; Feillet, F.; Giżewska, M.; et al. The complete European guidelines on phenylketonuria: Diagnosis and treatment. *Orphanet J. Rare Dis.* **2017**, *12*, 162. [CrossRef] [PubMed]
2. Macdonald, A.; Van Wegberg, A.M.J.; Ahring, K.; Beblo, S.; Bélanger-Quintana, A.; Burlina, A.; Campistol, J.; Coşkun, T.; Feillet, F.; Giżewska, M.; et al. PKU dietary handbook to accompany PKU guidelines. *Orphanet J. Rare Dis.* **2020**, *15*, 171. [CrossRef] [PubMed]
3. Ahring, K.; Bélanger-Quintana, A.; Dokoupil, K.; Gokmen-Ozel, H.; Lammardo, A.M.; Macdonald, A.; Motzfeldt, K.; Nowacka, M.; Robert, M.; Van Rijn, M. Blood phenylalanine control in phenylketonuria: A survey of 10 European centres. *Eur. J. Clin. Nutr.* **2011**, *65*, 275–278. [CrossRef] [PubMed]
4. Jurecki, E.; Cederbaum, S.; Kopesky, J.; Perry, K.; Rohr, F.; Sanchez-Valle, A.; Viau, K.; Sheinin, M.; Cohen-Pfeffer, J. Adherence to clinic recommendations among patients with phenylketonuria in the United States. *Mol. Genet. Metab.* **2017**, *120*, 190–197. [CrossRef] [PubMed]
5. Cazzorla, C.; Bensi, G.; Biasucci, G.; Leuzzi, V.; Manti, F.; Musumeci, A.; Papadia, F.; Stoppioni, V.; Tummolo, A.; Vendemiale, M.; et al. Living with phenylketonuria in adulthood: The Pku Attitude study. *Mol. Genet. Metab. Rep.* **2018**, *16*, 39–45. [CrossRef] [PubMed]
6. Sabanoglu, T. Statistica. 2019. Available online: https://wwwstatistacom/statistics/286116/food-and-groceries-online-purchasing-in-great-britain-by-demographic/ (accessed on 1 August 2020).
7. The Nielson Company. The Future of Grocery E-Commerce, Digital Technology and Changing Shopping Preferences around the World. Available online: https://www.nielsen.com/wp-content/uploads/sites/3/2019/04/nielsen-global-e-commerce-new-retail-report-april-2015.pdf (accessed on 1 August 2020).
8. Hess, R.; Visschers, V.H.M.; Siegrist, M. The role of health-related, motivational and sociodemographic aspects in predicting food label use: A comprehensive study. *Public Health Nutr.* **2012**, *15*, 407–414. [CrossRef] [PubMed]
9. Ford, S.; O'Driscoll, M.; Macdonald, A. Living with Phenylketonuria: Lessons from the PKU community. *Mol. Genet. Metab. Rep.* **2018**, *17*, 57–63. [CrossRef] [PubMed]
10. Regulation (EU) No 1169/2011 of the European Parliament and of the Council of 25 October 2011 on the Provision of Food Information to Consumers, Amending Regulations (EC) No 1924/2006 and (EC) No 1925/2006 of the European Parliament and of the Council, and repealing Commission Directive 87/250/EEC, Council Directive 90/496/EEC, Commission Directive 1999/10/EC, Directive 2000/13/EC of the European Parliament and of the Council, Commission Directives 2002/67/EC and 2008/5/EC and Commission Regulation (EC) No 608/2004. Available online: https://eur-lex.europa.eu/eli/reg/2011/1169/oj (accessed on 1 August 2020).
11. Food Standards Agency. Food Sold Online. Available online: https://www.food.gov.uk/sites/default/files/media/document/foodsoldonline-guidance-for-local-authorities.pdf (accessed on 1 August 2020).
12. Food Standards Agency. Nutrition Labelling. 2018. Available online: https://wwwfoodgovuk/business-guidance/nutrition-labelling (accessed on 1 August 2020).

13. MacDonald, A. Phenylketonuria. In *Clinical Paediatric Dietetics*, 5th ed.; Shaw, V., Ed.; Wiley Blackwell: Chichester, UK, 2020; pp. 513–531.
14. Evans, S.; Adam, S.; Adams, S.; Allen, H.; Ashmore, C.; Bailey, S.; Banks, J.; Churchill, H.; Cochrane, B.; Cook, J.; et al. Uniformity of Food Protein Interpretation Amongst Dietitians for Patients with Phenylketonuria (PKU): 2020 UK National Consensus Statements. *Nutrients* **2020**, *12*, 2205. [CrossRef] [PubMed]
15. Evans, S.; British Inherited Metabolic Diseases Group (BIMDG) Dietitians Group; Ford, S.; Adam, S.; Adams, S.; Ash, J.; Ashmore, C.; Caine, G.; Carruthers, R.; Cawtherley, S.; et al. Development of national consensus statements on food labelling interpretation and protein allocation in a low phenylalanine diet for PKU. *Orphanet J. Rare Dis.* **2019**, *14*, 2. [CrossRef] [PubMed]
16. Guidance Document for Competent Authorities for the Control of Compliance with EU Legislation on: Regulation (EU) No 1169/2011 of the European Parliament and of the Council of 25 October 2011 on the Provision of Food Information to Consumers, Amending Regulations (EC) No 1924/2006 and (EC) No 1925/2006 of the European Parliament and of the Council, and Repealing Commission Directive 87/250/EEC, Council Directive 90/496/EEC, Commission Directive 1999/10/EC, Directive 2000/13/EC of the European Parliament and of the Council, Commission Directives 2002/67/EC and 2008/5/EC and Commission Regulation (EC) No 608/2004 and Council Directive 90/496/EEC of 24 September 1990 on Nutrition Labelling of Foodstuffs and Directive 2002/46/EC of the European Parliament and of the Council of 10 June 2002 on the Approximation of the Laws of the Member States Relating to Food Supplements with Regard to the Setting of Tolerances for Nutrient Values Declared on a Label. Available online: https://ec.europa.eu/food/sites/food/files/safety/docs/labelling_nutrition-vitamins_minerals-guidance_tolerances_1212_en.pdf (accessed on 1 August 2020).
17. Department of Health. Technical guidance on food labelling, Population Health Division. 2016. Available online: https://assets.publishing.service.gov.uk/government/uploads/system/uploads/attachment_data/file/595961/Nutrition_Technical_Guidance.pdf (accessed on 1 August 2020).
18. Romani, C.; Manti, F.; Nardecchia, F.; Valentini, F.; Fallarino, N.; Carducci, C.; De Leo, S.; Macdonald, A.; Palermo, L.; Leuzzi, V. Adult cognitive outcomes in phenylketonuria: Explaining causes of variability beyond average Phe levels. *Orphanet J. Rare Dis.* **2019**, *14*, 273. [CrossRef] [PubMed]
19. MacDonald, A. (Birmingham Women's and Children's Hospital, Birmingham, West Midlands, UK). A Case of Poor Blood Phenylalanine Control Associated with Interpreting Protein Analysis from Food Labelling of Pot Noodles. Unpublished. Personal communication, 2020.
20. Finglas, P.; Pinchen, H.M.; Berry, R.; Church, S.M.; Dodhia, S.K.; Farron-Wilson, M.; Swan, G. *McCance and Widdowson's the Composition of Foods*, 7th ed.; Royal Society of Chemistry: Cambridge, UK, 2014.
21. European Commission. Mandatory Food Information. Available online: https://eceuropaeu/food/safety/labelling_nutrition/labelling_legislation/mandatory-food-information_en (accessed on 1 August 2020).
22. Gov UK. Food Labelling: Giving Food Information to Consumers. 2017. Available online: https://www.gov.uk/guidance/food-labelling-giving-food-information-to-consumers (accessed on 1 August 2020).
23. Food Standards Agency. Packaging and Labelling. 2018. Available online: https://www.food.gov.uk/business-guidance/packaging-and-labelling (accessed on 1 August 2020).

Publisher's Note: MDPI stays neutral with regard to jurisdictional claims in published maps and institutional affiliations.

© 2020 by the authors. Licensee MDPI, Basel, Switzerland. This article is an open access article distributed under the terms and conditions of the Creative Commons Attribution (CC BY) license (http://creativecommons.org/licenses/by/4.0/).

nutrients

Review

Protein Substitutes in PKU; Their Historical Evolution

Anne Daly *, Sharon Evans , Alex Pinto, Catherine Ashmore and Anita MacDonald

Birmingham Women's and Children's Hospital, NHS Foundation Trust, Birmingham B4 6NH, UK; evanss21@me.com (S.E.); alex.pinto@nhs.net (A.P.); catherine.ashmore@nhs.net (C.A.); anita.macdonald@nhs.net (A.M.)
* Correspondence: a.daly3@nhs.net

Abstract: Protein substitutes developed for phenylketonuria (PKU) are a synthetic source of protein commonly based on L-amino acids. They are essential in the treatment of phenylketonuria (PKU) and other amino acid disorders, allowing the antagonistic amino acid to be removed but with the safe provision of all other amino acids necessary for maintaining normal physiological function. They were first formulated by a chemist and used experimentally on a 2-year-old girl with PKU and their nutritional formulations and design have improved over time. Since 2008, a bioactive macropeptide has been used as a base for protein substitutes in PKU, with potential benefits of improved bone and gut health, nitrogen retention, and blood phenylalanine control. In 2018, animal studies showed that physiomimic technology coating the amino acids with a polymer allows a slow release of amino acids with an improved physiological profile. History has shown that in PKU, the protein substitute's efficacy is determined by its nutritional profile, amino acid composition, dose, timing, distribution, and an adequate energy intake. Protein substitutes are often given little importance, yet their pharmacological actions and clinical benefit are pivotal when managing PKU.

Keywords: phenylketonuria; protein substitute; amino acid; glycomacropeptide

Citation: Daly, A.; Evans, S.; Pinto, A.; Ashmore, C.; MacDonald, A. Protein Substitutes in PKU; Their Historical Evolution. *Nutrients* **2021**, *13*, 484. https://doi.org/10.3390/nu13020484

Academic Editor: Jamie I. Baum
Received: 29 December 2020
Accepted: 28 January 2021
Published: 2 February 2021

Publisher's Note: MDPI stays neutral with regard to jurisdictional claims in published maps and institutional affiliations.

Copyright: © 2021 by the authors. Licensee MDPI, Basel, Switzerland. This article is an open access article distributed under the terms and conditions of the Creative Commons Attribution (CC BY) license (https://creativecommons.org/licenses/by/4.0/).

1. Introduction

Amino acids are unique substrates providing nitrogen, hydrocarbon skeletons and sulphur [1]. They are essential precursors for the synthesis of proteins, peptides, and low molecular weight substances such as glutathione, dopamine, nitric oxide, and serotonin [1]. In phenylketonuria (PKU), dietary treatment was made feasible with the introduction of low/free phenylalanine synthetic proteins (protein substitutes), that have gradually advanced with time. In the 1950s, these were originally derived from protein hydrolysates, but in the 1970s, phenylalanine-free amino acids were introduced. Protein substitutes provide the building blocks of tissue proteins and their amino acids are essential for the synthesis of hormones, enzymes, and other cellular processes. Therefore, their composition and nutritional profile is fundamental, helping prevent neurological devastation, allowing normal growth and biosynthetic functions. The original technology for making protein substitutes was crude and limited but now precision manufacturing has improved their quality.

Although Følling [2] first identified phenylpyruvic acid in the urine of untreated children with PKU, it was Penrose who recognised that it was a genetic recessive disorder and named it phenylketonuria (PKU) [3]. He was also the first to try a dietary treatment based on fruit, sugar, olive oil, and vitamins, but this protein-free diet lacked essential phenylalanine and all other amino acids, resulting in malnutrition and so the treatment was abandoned [4]. Twenty years later, protein substitutes were introduced, but their central role in the management of PKU remains undervalued.

2. Early Studies

Følling and Penrose [5] both demonstrated that giving phenylalanine to a PKU subject increased the excretion of phenylpyruvic acid. The type of phenylalanine ingested as

D or L isomers had different effects on phenylpyruvic excretion, with L phenylalanine leading to a greater production of phenylpyruvic acid. Similarly, in non-PKU subjects, L phenylalanine was the preferred metabolised substrate, with D and DL isomers leading to small amounts of phenylpyruvic acid but an absence when the L form was given due to its complete metabolism. From these studies, they concluded that phenylpyruvic acid excreted in PKU patients was due to an incomplete breakdown of phenylalanine. Tyrosine when administered had no effect on urine phenylpyruvic acid excretion, concluding this was metabolised normally. It was not until 1944 that Bernheim [6] demonstrated that the main metabolic pathway for phenylalanine was by parahydroxylation of phenylalanine to tyrosine. In 1953, Jarvis showed that it was the inability to perform this hydroxylation that resulted in phenylketonuria [7].

Penrose and Quastel [5] conducted a series of feeding studies where they found that by lowering the natural protein intake by >50% resulted in an immediate reduction in urinary phenylpyruvic acid in a patient with PKU. However, after the second day of treatment, urine phenylpyruvic acid re-appeared and increased over subsequent days. The authors noted a weight loss over the same time and hypothesised that catabolism led to the production of phenylpyruvic acid.

In 1951, a positive ferric chloride screening test in a symptomatic 2-year-old girl from Birmingham, UK, preceded the first successful dietary treatment in PKU. She was only the third child to be tested with the ferric chloride test at Birmingham Children's Hospital [8]. In PKU, phenylpyruvic acid present in urine causes the characteristic greenish-blue colour reaction when a few drops of ferric chloride are added [9]. On presentation, she was unable to talk, walk, or engage with her surroundings; her mother waited for the doctor every morning outside the hospital laboratory as she refused to accept that there was no treatment for her daughters' condition. Louis Woolf designed the first successful protein substitute formulation used in PKU. He was a chemist with a commercial background and had used hydrolysed casein to produce amino acids as a treatment for malnutrition after the Second World War. Cost was a priority in the post war years and protein hydrolysates were readily available and cheaper than pure amino acids. In 1949, he suggested that supplementation of carbon treated casein hydrolysate with appropriate amounts of missing amino acids (including a source of phenylalanine to prevent deficiency) could treat PKU. He was unable to convince his medical colleagues at Great Ormond Street (GOS) Children's Hospital, London to try his proposed treatment. He recalls: "*At GOS, the suggestion floated like a lead balloon, I was told not unkindly that I should be devising new diagnostic tests, not dreaming up crazy treatments for conditions that everybody knew were untreatable*" [10]. In collaboration with Drs Bickel, Hickman, and Gerrard from Birmingham, a modification of Louis Woolf's protein substitute was given to the 2-year-old child with PKU [11].

3. The First Protein Substitute

Casein was hydrolysed by boiling it for several hours with concentrated hydrochloric or sulphuric acid to produce a thick black amino acid liquid. This solution was neutralised with sodium hydroxide and then purified by the addition of activated carbon and finally filtered to produce a clear solution of amino acids. This solution contained phenylalanine, which was removed by a second filtration method using activated charcoal. This removed the aromatic amino acids: phenylalanine, tryptophan, and tyrosine (although a small residual amount of phenylalanine was detectable). To nutritionally improve the protein substitute carbohydrate, fat, vitamins and minerals were added, together with tryptophan and tyrosine [11]. This unpalatable solution was then mixed with sugar, wheat starch, double cream and water and given as a formula to infants. In older children, it was either flavoured with tomatoes and given as a soup [12], made into a blancmange with sugar, margarine and wheat starch [13] or mixed with vegetable oil, and sugar and flavoured with artificial flavourings [14].

The production of the original formula was difficult and time consuming and had to be done in a cold room or it would deteriorate. The black charcoal covered everything and as the first formula was prepared, the sight of Dr. Bickel wrapped in layers of jumpers topped by a charcoal smudged lab coat became a common sight [8]. Woolf identified that a small amount of phenylalanine should be added to the formula as it was an essential amino acid [15]. He stressed the need for careful monitoring, and he was also the first to propose treatment for life in PKU [12,16,17].

Phenylalanine-free amino acids as a protein substitute for PKU were first tried in the USA in the 1950s [18], but had to be abandoned most likely due to the pure amino acid mixtures causing vomiting.

4. Amino Acid Requirements

In the early stages of making the protein substitute, the exact amino acid composition of the casein hydrolysate was unknown. Casein was low in sulphur containing amino acids and cysteine was also partly removed by hydrolysis and charcoal filtration. Bickel [19] suggested adding L cystine to the hydrolysate and Woolf proposed the addition of DL methionine [20].

The amounts of tryptophan and tyrosine added to the first protein substitutes were determined by amino acid requirements established in the early 1950s; this knowledge was pioneered by Rose [21–25], Holt and Snyderman [26,27]. It was estimated that an adult man required 1000 mg/day of L phenylalanine to maintain nitrogen equilibrium or 300 mg/day if tyrosine was provided [24]. Synderman [28] suggested that 90 mg per kg/day of L phenylalanine was needed by an infant, but this was reduced to 25 mg/kg per day if sufficient tyrosine was supplied. Other essential amino acid requirements were estimated from the work of Rose, Leverton [23,29] and Swendseid [30]. In Woolf's [12] original formula, 25 mg/kg/day of L tryptophan and 25 mg/kg/day of L methionine were given in addition to 50 mg/kg/day of L tyrosine, a surrogate essential amino acid in PKU (Table 1).

Table 1. The original composition of the first protein substitute designed by Louis Woolf (1958).

Product	Daily Intake
Casein hydrolysate	**24 g**
DL-tryptophan	**1 g**
L-tyrosine	**2 g**
DL-methionine	**1 g**
Sucrose	**90 g**
Cows milk	**0–200 mL**
Double cream	**85 mL**
Calcium hydrogen phosphate	**0.71 g**
Potassium chloride	**0.65 g**
Sodium chloride	**0.016 g**
Magnesium sulphate	**0.165 g**
Sodium citrate	**0.177 g**
Potassium iodide	**0.00013 g**
Citric acid	**0.08 g**
Water	**850 mL**

Table 1. *Cont.*

Product	Daily Intake
Vitamins and minerals	
Choline chloride	**100 mg**
Inositol	**216 mg**
Vitamin B12	**4 μg**
Aneurine hydrochloride (vitamin B1)	**0.5 mg**
Riboflavin	**0.5 mg**
Pyridoxine	**0.33 mg**
Nicotinamide	**3.33 mg**
Ascorbic acid	**40 mg**
α-Tocopherol	**0.33 mg**
Acetomenaphthone (vitamin K)	**0.5 mg**
Biotin	**0.17 mg**
Folic acid	**0.35 mg**
Vitamin A	**3000 iu**
Vitamin D	**500 iu**
Zinc sulphate	**0.0014 g**
Ferrous sulphate	**0.15 g**
Manganous sulphate	**0.0008 g**
Cupric sulphate	**0.003 g**

There were challenges when administering the artificial diet in PKU [11]. The first child to start treatment was admitted to hospital for 6 weeks. The musty smell associated with PKU disappeared, plasma and urinary phenylalanine concentrations returned to normal, and there was a negative ferric chloride test when the diet was commenced. However, the child lost weight and within 5 days of treatment, plasma tyrosine concentrations were un-recordable (with a change in hair pigmentation), and plasma phenylalanine was raised. Tyrosine (1.5 g/daily) was added, correcting the low plasma tyrosine concentrations and temporarily arrested weight loss. However, after a further 3 weeks, aminoaciduria was noted, and in the fifth week, blood phenylalanine increased and phenylpyruvic acid reappeared in the urine; this was associated with weight loss, vomiting, and the child was described as unwell. These observations were important, highlighting that tyrosine became an essential amino acid in PKU as a consequence of the biochemical block in converting phenylalanine to tyrosine. Aminoaciduria, a result of weight loss and catabolism due to phenylalanine deficiency, led to an increase in catabolism and a subsequent increase in blood phenylalanine. Adding a measured amount of phenylalanine back into the diet (typically 250–500 mg or equivalent to around 5–10 g protein/day) increased plasma and urine concentrations, but to levels significantly below pre-treatment concentrations. Laboratory analysis was laborious, each blood test was analysed in duplicate and the production of a chromatogram took 3 days of intensive labour [8].

After 6 months of treatment, this child made remarkable progress; followed by a cascade of successful case studies. Woolf [17] reported 3 cases, with a further publication of 10 cases in 1958 [12], in which children were treated from the age of three weeks to 5 years of age. Armstrong and Tyler [31] reported the treatment of five children, and Armstrong and Binkley in 1956 [32] followed the progress of an infant starting treatment at 40 days of age. All reported that a low phenylalanine diet, supplemented with a low phenylalanine protein hydrolysate corrected the major biochemical abnormalities.

It was also established that sufficient carbohydrate and fat (including a source of linoleic acid) was necessary to prevent protein catabolism [20,33–35]. Woolf reported that the daily intake of hydrolysate should be high correcting for the inefficient utilisation of the amino acids [12].

5. Commercial Protein Hydrolysate Preparations

Production of the hydrolysate moved from hospital laboratories to commercial production in late 1953/early 1954. In Europe, Cymogran 1954/5 (Allen and Hanbury, London, UK), XP Albumaid (1960) (Scientific Hospital Supplies, Liverpool, UK) and Minafen (designed for infants in 1955), (Trufood Ltd., Guildford, UK) were developed and the US produced Lofenalac (Mead Johnson, Chicago, IL, USA) in 1958. In the spirit of commercial interest, Trufood and Allen and Hanbury agreed to share production with one company making an infant substitute Minafen (Trufood)and the other (Allen and Hanbury) a preparation for older children Cymogran. Limited practical instructions were provided on how to reconstitute these formulas and families had to weigh the prescribed powder. The main difference between hospital and factory production was the use of ion exchange resins to separate phenylalanine, dispensing with the sodium hydroxide and carbon filtration. These synthetic filters consisted of microbeads from resin or polymers, allowing the separation and purification of the hydrolysed casein. These products were supplemented with variable amounts of vitamins, minerals, carbohydrate and fat.

6. The First UK PKU Guidelines

In 1960, the UK Ministry of Health [9] provided guidelines on screening and early detection of PKU, together with recommendations on optimal blood phenylalanine concentrations and provision of protein substitutes. They proposed screening by the ferric chloride test at 4–6 weeks of age (which was later replaced by the Guthrie method in 1969 [36]). To prevent phenylalanine deficiency, a target blood phenylalanine concentration slightly above normal was recommended (90–120 mmol/L), with blood phenylalanine monitoring done twice weekly until stability was achieved, and then weekly or monthly monitoring was required.

In infancy, a protein substitute, formulated and reconstituted similar to regular milk-based infant formula was recommended. A second protein substitute with a lower energy content was advocated for older children.

7. Nutritional Deficiencies with Early Protein Substitutes

In the early history of treating PKU by diet, there were concerns about 'over- treating' patients and maintaining very low phenylalanine blood concentrations. Nutritional deficiencies, malnutrition, and even death were linked to dietary treatment [37]. In the 1960s, severe skin rashes in babies on Minafen (Allen and Hanbury Ltd., London, UK) were reported [38–40]. Woolf [12] described a child with faltering growth and hair loss when acetyl DL tryptophan was accidentally given instead of DL tryptophan; stopping the acetyl derivative immediately reversed the symptoms. Studies in animals fed synthetic low phenylalanine diets [41] led to the addition of choline, riboflavin, folic acid, and vitamin E to the hydrolysate preparations. Two reports of folic acid deficiency were described [42,43], one child had megaloblastic anaemia due to folic acid deficiency exacerbated by vomiting and poor feeding and subsequently died. Hypoglycaemia was also reported in two cases [44].

8. Amino Acid Preparations

In the late 1960s, commercial amino acid formulas were made from pure crystalline amino acids by fermentation of bacteria. They were manufactured by the Japanese at an affordable cost [45]. The first product for PKU was Aminogram Food Supplement [46,47], which had several advantages, compared to hydrolysed formulas including improved taste and a lower daily volume, with an amino acid composition that could be easily adapted for the treatment of other aminoacidopathies such as maple syrup urine disease, homocystinuria, and tyrosinaemia type 1.

Manz [48] reported anorexia and vomiting in some infants given amino acid preparations. Metabolic acidosis was observed when the preparations contained amino acids in the form of hydrochloride salts or when the ratio of sulphur containing amino acids was too high, leading to higher urinary pH and increased renal net acid excretion. Modifications in the amino acid preparations normalised the renal net acid excretion and acidosis.

The early commercial preparations of L amino acid substitutes required separate supplementation with vitamins and minerals and careful monitoring of nutritional status was essential. These vitamin and mineral supplements were commonly deficient in molybdenum, chromium, selenium, and pantothenic acid [49,50].

L-amino acid substitutes nutritionally complete: In 1980, the first UK amino acid preparation supplemented with carbohydrate, vitamins, mineral, and trace elements and designed for children over 1 year of age with PKU was manufactured. It was flavoured for improved taste and palatability. In 1988, a similar product (but also with added taurine and carnitine), but formulated for children over the age of 8 years and suitable for maternal PKU, was introduced [51]. From the 1990s, further advances were made in the nutritional formulations, taste, and presentation of protein substitutes (Table 2). Although selenium supplementation was added to protein substitutes from the late 1980s, many countries were wary about adding selenium to protein substitutes due to concerns about its toxicity which had been responsible for deaths in man and animals and was referred to as the 'essential poison' [52]. Consequently, this led to reports of many cases of biochemical selenium deficiency [53,54].

Table 2. Introduction of protein substitutes.

Laboratory produced preparation	
1952	**Hydrolysed casein, powdered preparation, nutritionally incomplete, low phenylalanine** Addition of tyrosine, tryptophan and methionine, carbohydrate, fat, vitamins and minerals
Commercial produced preparations	
1954	**Hydrolysed casein, powdered preparation, nutritionally incomplete, low phenylalanine** Trufood (infant product), Allen and Handbury/Mead Johnson (older children)
1960	**L amino acid, powdered preparation, nutritionally incomplete phenylalanine free** Powell and Scholfield powdered preparation taken as a drink
1980	**L amino acid, powdered preparation with added carbohydrate and fat nutritionally complete, phenylalanine free** Powdered flavoured preparation taken as a drink
1988	**L amino acid, powdered preparation with added carbohydrate and fat nutritionally complete, phenylalanine free** Powdered preparation taken as a drink for >8 years and adults
1988	**L amino acid, powdered preparation, with added carbohydrate and fat, nutritionally complete designed for infants, phenylalanine free** Powdered preparation used for infants 0–12 months
2001	**L amino acid, powdered preparation with added carbohydrate and essential fatty acids, nutritionally complete, phenylalanine free** Powdered flavoured preparation taken as a drink
2001	**L amino acid powdered preparation, spoonable paste added carbohydrate, fat free, nutritionally complete, phenylalanine free** Powdered flavoured preparation taken as a spoonable paste
2002	**L amino acid tablets, nutritionally incomplete, phenylalanine free** Amino acid tablets
2003	**L amino acid powdered preparation, low carbohydrate, no fat, nutritionally complete, phenylalanine free** Powdered flavoured preparation taken as low volume drink or spoonable paste for children >8 years

Table 2. *Cont.*

2006	**L amino acid ready to drink preparation, low carbohydrate no fat, nutritionally complete, phenylalanine free** Ready to drink flavoured liquid
2008	**Casein macropeptide with L amino acids, essential fatty acids, nutritionally complete, low phenylalanine** Powdered preparation made into a low volume drink
2008	**Casein macropeptide with L amino acid nutritionally complete with essential fatty acids, low phenylalanine.** Powdered preparation make into a low volume drink
2018	**Slow release L amino acid preparation, carbohydrate and fat free, nutritionally complete, phenylalanine free** Micro tablets, made from L amino acids coated with ethyl cellulose and alginate which slowly release the L amino acids

It was also established that the fat intake of children with PKU was low [55] and n-3 long chain polyunsaturated fatty acid status was sub-optimal [56]. This led to the addition of essential fatty acids to protein substitute powders designed for children [57]. Around the same time, long chain polyunsaturated fatty acids were added to infant protein substitutes in 2000; in a double blind randomised study, infants received either a formula with or without a supplemented fat blend of long chain polyunsaturated fatty acids (LC-PUFA). The results clearly showed the benefit of supplementation [58] and led to the addition of LC-PUFA to other products designed for older children.

Over the years, there has been much endeavor to ensure that protein substitutes meet changing nutritional trends and accommodate nutritional requirements according to life stage. In 2011, the first phenylalanine-free infant formula containing a specific mixture of prebiotic oligosaccharides was introduced [59]. This helped maintain levels of bifidobacteria and lower stool pH in infants with PKU. There is concern about increasing obesity rates in the PKU and non PKU population, so many recent protein substitutes introduced for children, teenage, and adults with PKU have a lower carbohydrate and energy composition [60]. Impact on lowering obesity rates has not yet been proven.

The nutritional adaptation of protein substitutes in order to gain clinical benefit is an area likely to grow in the future. Recently, specific nutrient combinations (containing uridine monophosphate, docosahexaenoic acid, eicosapentaenoic acid, choline, phospholipids, folic acid, vitamins B12, B6, C, and E, and selenium) have been studied in PKU mice to examine the impact on synaptic deficits in PKU [61]. The specific nutrients are precursors and cofactors for the synthesis of phospholipids thought to be beneficial in improving the neurotransmitter/synaptic changes in PKU. This combination of nutrients has been shown to have a benefit on synapse formation, morphology, and function in mouse models of Alzheimer's disease so it may be an important nutritional adaptation of protein substitutes for older patients [62].

9. Choice of Protein Substitutes

The choice, composition, and presentation of protein substitutes have expanded at fast moving rates since the turn of the century. This time was associated with the evolution of pre-packaged and premeasured products, which has not only improved convenience but also accuracy, adherence, and ease of protein substitute prescription for clinicians.

Spoonable low volume protein substitutes: An innovative substitute designed for young children with PKU was produced in 2000, based on phenylalanine free amino acids and starch to which a small amount of water was added, forming a gel/paste that was similar in consistency to a weaning food [60]. This low volume, fat free, lower calorie, more concentrated amino acid substitute had the advantage of allowing transition onto a second stage product from the age of 6 months, in line with complementary feeding. It was presented in premeasured sachets (dispensing with the need for large tins of formula), was easy to prepare, with a good consistency and acceptable taste. Normal infant feeding behaviour, teething, and intercurrent infection can lead to its rejection in late infancy so perseverance and a consistent approach is needed by parents [63,64].

Ready to drink liquid protein substitutes: In 2005, 'ready to drink' flavoured phenylalanine free amino acid pouches were introduced. These small volume, lower energy protein substitutes were convenient and compact, allowing greater independence for children and teenagers. Patients were less self-conscious taking a liquid drink compared to a powdered preparation [65]. The pharmacological efficacy of these lower volume substitutes did not compromise nutritional biochemistry or phenylalanine concentrations, which remained the same or improved [60]. One potential problem was abdominal discomfort (constipation/diarrhoea), attributed to the hyperosmolar concentration of the lower volume protein substitutes [66,67], and like all concentrated amino acid products, they should be administered with additional water.

Protein substitute tablets: Amino acid tablets and modular systems were also introduced around 2000. Modular systems are when a combination of amino acid tablets, capsules, liquids, powder, or bars of amino acids are used to provide daily protein substitute requirements, allowing flexibility of choice. In a randomised crossover study, it was shown that subjects with PKU successfully took at least 40% of their protein substitute as tablets, with an improvement in adherence and significantly lower blood phenylalanine concentration [68]. The quantity of tablets to meet protein requirements was around 70/day, and they were not nutritionally complete, requiring extra supplementation with vitamins and minerals. They provide an alternative for older children and adults who struggle taking conventional protein substitute. Micro-tablets of amino acids have since been introduced.

Caseinglycomacropeptide with amino acids (CGMP-AA): CGMP-AA was introduced in the UK in 2017, although first used in the USA as a protein substitute for PKU in 2008 [69]. CGMP is purified from whey by anion exchange chromatography, but the final product does contain residual amounts of aromatic amino acids including phenylalanine [70]. CGMP-AA is different from amino acid substitutes; approximately 40% of the product is composed of amino acids, with the rest as a bioactive peptide; based on a macropeptide, they are associated with improved taste and palatability [71].

Slow-release protein substitute: A prolonged release product was first developed in 2014 [72] but there was little supporting published data demonstrating its effectiveness. In 2018, a slow release preparation containing amino acids coated with ethyl cellulose and alginate was introduced. Based on physiomimic technology, the bitter taste and smell of amino acids was improved, and as this product is not mixed with fluid, it does not have an osmolality. Most importantly, the technology prolongs the release of amino acids into the systemic circulation. Animal and human kinetic studies demonstrate a reduced peak concentration of amino acids. This new technology suggests a physiological absorption of amino acids similar to natural protein [73,74]. In a short-term observational study using prolonged amino acids in subjects with PKU, it was well tolerated, with fewer gastrointestinal symptoms and no change in blood phenylalanine concentrations [75].

10. Pharmacological Importance of Protein Substitutes

The amount of protein equivalent (g/kg) from protein substitutes affects blood phenylalanine control [76–79]. As early as 1961, an observational study performed by O'Daly [80] showed protein substitutes significantly lowered blood phenylalanine concentrations. Further studies have shown that phenylalanine tolerance is increased when total protein intake from a protein substitute is increased [79,81].

Protein substitutes have an important function at the blood brain barrier. Large neutral amino acids including phenylalanine compete for LAT1, a large neutral amino acid transporter allowing entry of amino acids into the brain [82]. Phenylalanine has a particularly high affinity for LAT1 and protein substitutes are the only source of competitive large neutral amino acids necessary to prevent excess phenylalanine entering the brain.

These pharmacological effects of ingesting an amino acid rich formula are frequently neglected and given little scientific credence, and yet they have a significant impact on phenylalanine metabolism and long-term physical and neurological outcome. The gut also controls the absorption of amino acids across the epithelial membrane. Phenylalanine is transported as a carrier mediated sodium dependent process which requires energy. Similar to the blood brain barrier, large neural amino acids are transported in the gut by LAT1, also known as SLC5A7, for which phenylalanine has a high affinity [83,84].

11. Protein Substitute Requirements

Human requirements for each amino acid are specific to age, metabolic demands (immune/neuromuscular), and growth rate (protein deposition) [76,85,86]. For protein synthesis to occur, all the amino acids should be available; absence of one leads to the cessation of synthesis [1]. Snyderman [87] reported that the complete withdrawal of phenylalanine from the diet in a normal infant led to the depression in several other amino acids, the most prominent being tyrosine, hence the importance of tyrosine supplementation. Woolf showed that the nitrogen content of the artificial substitute was not an exchange for natural protein; the hydrolysate contained less nitrogen, was rapidly absorbed from the gut with greater oxidation and urinary amino acid losses [88,89], therefore sufficient product was needed to meet nitrogen requirements.

A protein substitute intake that just meets minimum WHO requirements (WHO/FAO/UNU 2007) [90] may result in 'latent' catabolism, leading to body tissue breakdown, increasing phenylalanine concentrations. Protein utilisation is enhanced by a supply of carbohydrate and fat [91,92] further illustrated in a randomised controlled study in PKU subjects by MacDonald [93] and supported by Illsinger [94].

The European PKU Guidelines recommend that the total protein intake should supply 40% more than the FAO/WHO/UNU safe levels of protein intake [95]. However, this amount is arbitrary and unconfirmed by research [67]. A collaborative study [96] involving 63 European and Turkish IMD centres concluded that the amount of total protein prescribed by different European countries was not uniform. All centres gave higher protein equivalents than the recommended 2007 WHO/FAO/UNU [90] safe levels of protein intake with Western European centres prescribing less total protein then other European regions.

To maximise the utilisation of amino acids and minimise the variation in phenylalanine concentrations, protein substitutes should be taken frequently, a minimum of three times a day. MacDonald [97] demonstrated that the greater the amount of protein substitute consumed between waking and 4 p.m., the greater the decrease in phenylalanine concentrations. Likewise, when protein substitute was given 4 hourly for 24 h, there was a marked stablisation in phenylalanine concentrations, reducing phenylalanine variability [98].

Tyrosine, a precursor of catecholamine neurotransmitters (dopamine, norepinephrine, and epinephrine), thyroxine, and melanin, is an essential amino acid in PKU due to the limited or absent hydroxylation of phenylalanine. It is hydrophobic and the absolute quantities added to protein substitutes are not defined. Indicator amino acid oxidation studies [99] suggest tyrosine should provide 19 mg/kg/day, although current protein substitutes provide approximately 5 times above current recommendations. The importance of tyrosine was recognised by the Report of the Medical Research Council working party on PKU [100], which recommended that protein substitutes should be nutritionally complete and contain 100–120 mg/kg/day of tyrosine.

12. Protein Substitute Administration

In the early history, the practicalities of administering an acid based hydrolysed unflavoured product were particularly challenging. Bentovim [46] described the struggles families faced trying to persuade children to take the acid tasting formula: the large daily volume that needed to be consumed, regular vomiting, refusal to eat permitted food due to negative associations with the substitute, the bad smell, and lack of palatability. Furthermore, children experienced isolation and psychological difficulties particularly in the school years. This was one of the factors leading to diet cessation as early as 6 years [101–103]. An extract from the *Cork Examiner* describes the struggles faced by one family adapting to the news that their two children had been diagnosed with PKU and the dietary changes and challenges made to improve their neurological outcome [104].

Despite the advances in technology, almost all protein substitutes have a strong taste and odour and are associated with poor palatability and breath odour. They are a burden to patients as they must be consumed a minimum of three times daily and spread evenly throughout the day. Ford [66] reported 293 of 631 participants with PKU (39% of adults, 11% of children) either did not take protein substitute or took less than their prescribed amounts.

> Verbatim extract from study: *Our greatest struggle is getting our son taking his protein substitute. He refuses to take it and it can take up to 45 min for him to finish one with a lot of upset.*

Evans [105], in a case control study in PKU children of weaning age, highlighted the stress, anxiety, and struggles associated with protein substitute administration. Maternal anxiety regarding child rejection of protein substitute increased with time peaking at 12–24 months. Similarly, in 2016, MacDonald [106] reported in 114 children with PKU, dietary management was associated with a considerable time and financial burden for caregivers, with much time spent supervising protein substitute intake.

13. Conclusions

In PKU, the early pioneers understood the physiological importance of protein substitutes. They stressed the need for a balanced amino acid profile, for even administration throughout the day, together with an adequate energy intake and dietary treatment for life. Although these principals remain unchanged 70 years later, each decade has witnessed improvements in the delivery and nutritional composition of protein substitutes, which remain of fundamental importance in the treatment of PKU. Further changes are needed, to deliver improved taste and odour-free products, with the properties of natural protein delivering a stable chemical environment associated with optimal physiological function and patient tolerance.

An extract from the *Irish Cork Examiner* describes the struggles of a family diagnosed with PKU in 1959, the son aged 4 and the daughter $2\frac{1}{2}$ years old. This extract describes the determination, sacrifice, hardship, and success against the better judgement of expert advice.

25 October 1962. Phenylketonuria: A story of heartbreak and hope.

> *"Treatment might help your daughter" he said "but for your son detection has come too late." He would deteriorate so much that at a later stage institutional care was inevitable. No one had attempted treatment on a child over 2 years. But the specialist was willing to give my little girl a trial. I pleaded for both of them not knowing the terrible struggle this entailed.*

> *The boy was difficult, backward and had no speech while his sister could neither walk or talk and was unable to sit up alone and was extremely difficult to manage. Those first months of 1959 were a nightmare from which there was no awakening. The introduction of the unpalatable diet and the cessation of stews, broths and chocolate sundaes brought tears and tantrums. How I dreaded the ice cream vendors that first summer and the laughing lolly licking youngsters who stood on our corner. The synthetic protein (Minafen) was unpleasant to take, but I have found it can be disguised reasonably well in savouries and cookies.*

> *Each child is allowed approximately 270 mg of phenylalanine per day according to body weight. If the child were to have 1oz of porridge oats this would cover 241.5 mg of their daily allowance, whereas 1oz of tomatoes would only represent 5 mg, so planning meals for them was at one time a highly complicated business. Now I possess a simplified chart of foods with a low phenylalanine content and by drawing on almost every cook book in print for ideas I have complied my own Cook Book for Phenylketonurias. Special gluten free flour must be used for bread making and Kosher margarine replaces butter, bread*

making with wheat starch was a different matter "Neolite or just plain leather "was my husband's query at my first attempt. Meals for ourselves present a real problem. It is so difficult to take a hearty T bone steak or peach meringue while two pair of eyes watch with longing. Meals out are impossible as is home entertainment, but it has been so worthwhile, the little girl unable to talk or walk takes some chasing and her speech is coming slowly. Responsiveness and alertness have taken place in slow but sure degrees. My son now 7 has made remarkable progress benefiting from a normal education.'

Although detection of PKU and treatment soon after birth is essential for complete recovery we have proved beyond all doubt that much can still be done.

Author Contributions: Conceptualization, A.M., A.D.; writing original draft preparation, A.D., A.M. Writing review and editing, A.D., S.E., A.P., C.A. and A.M.; supervision, A.M. All authors have read and agreed to the published version of the manuscript.

Funding: This research received no external funding.

Acknowledgments: We would like to thank R. Cowen, P. Cowen, J. Cowen, P. Portnoi and C. Clothier for their time and contribution to the historical facts in the manuscript.

Conflicts of Interest: A.D. received research funding from Vitaflo International, financial support from Nutricia, Mevalia and Vitaflo International to attend study days and conferences. SE received research funding from Nutricia, financial support from Nutricia and Vitaflo International to attend study days and conferences. A.P.received an educational grant from Cambrooke Therapeutics and grants from Vitaflo International, Nutricia, Merck Serono, Biomarin and Mevalia to attend scientific meetings. C.A. received honoraria from Nutricia and Vitaflo International to attend study days and conferences. A.M. received research funding and honoraria from Nutricia, Vitaflo International and Merck Serono, and is a Member of European Nutrition Expert Panel (Merck Serono international), a member of Sapropterin Advisory Board (MerckSerono international), and a member of the Advisory Board Element (Danone–Nutricia).

References

1. Wu, G. Dietary protein intake and human health. *Food Funct.* **2016**, *7*, 1251–1265. [CrossRef] [PubMed]
2. Følliing, A. Uber Ausscheidung von Phenylbrenztraubensäure in den Harn als Stoffwechselanomalie in Verbindung mit Imbezillität. *Z. Physiol. Chem.* **1934**, *227*, 169–176. [CrossRef]
3. Penrose, L. Inheritance of phenylpyruvic amentia (phenylketonuria). *Lancet* **1935**, *226*, 192–194. [CrossRef]
4. Laxova, R. Lionel Sharples Penrose, 1898–1972: A personal memoir in celebration of the centenary of his birth. *Genetics* **1998**, *150*, 1333–1340.
5. Penrose, L.; Quastel, J.H. Metabolic studies in phenylketonuria. *Biochem. J.* **1937**, *31*, 266–274. [CrossRef]
6. Bernheim, M.L.; Bernheim, F. The production of a hydroxyphenyl compound from l-phenylalanine incubated with liver slices. *J. Biol. Chem.* **1944**, *152*, 481. [CrossRef]
7. Jervis, G. Phenylpyruvic oligophrenia deficiency of phenylalanine-oxidizing system. *Proc. Soc. Exp. Biol. Med.* **1953**, *82*, 514–515.
8. Green, A. *Sheila—Unlocking the Treatment for PKU*; Brewin Books: Reddtich, UK, 2020; p. 172.
9. Medical, Research, and Council. Treatment of Phenylketonuria. *Br. Med. J.* **1963**, *1*, 1691–1697. [CrossRef]
10. Woolf, L.I.; Adams, J. The Early History of PKU. *Int. J. Neonatal Screen.* **2020**, *6*, 59. [CrossRef]
11. Bickel, H.; Gerrard, J.; Hickmans, E.M. Influence of phenylalanine intake on phenylketonuria. *Lancet* **1953**, *265*, 812–813. [CrossRef]
12. Woolf, L.I.; Griffiths, R.; Moncrieff, A.; Coates, S.; Dillistone, F. The Dietary Treatment of Phenylketonuria. *Arch. Dis. Child.* **1958**, *33*, 31–45. [CrossRef] [PubMed]
13. Blainey, J.D.; Gulliford, R. Phenylalanine-restricted Diets in the Treatment of Phenylketonuria. *Arch. Dis. Child.* **1956**, *31*, 452–466. [CrossRef] [PubMed]
14. Hsia, D.Y.; Knox, W.; Quinn, K.V.; Paine, R.S. A one-year, controlled study of the effect of low-phenylalanine diet on phenylketonuria. *Pediatrics* **1958**, *21*, 178–202. [PubMed]
15. Alonso-Fernández, J.R.; Colón, C. Article Commentary: The contributions of Louis I Woolf to the treatment, early diagnosis and understanding of phenylketonuria. *J. Med. Screen.* **2009**, *16*, 205–211. [CrossRef] [PubMed]
16. Woolf, L.I. The Dietary Treatment of Phenylketonuria: Not Proven. *Dev. Med. Child Neurol.* **2008**, *9*, 244–245. [CrossRef]
17. Woolf, L.I.; Griffiths, R.; Moncrieff, A. Treatment of Phenylketonuria with a Diet Low in Phenylalanine. *BMJ* **1955**, *1*, 57–64. [CrossRef]

18. Armstrong, M.D.; Shaw, K.N. Studies on phenylketonuria. III. The metabolism of o-tyrosine. *J. Biol. Chem.* **1955**, *213*, 805–811. [CrossRef]
19. Bickel, H. The effects of a phenylalanine-free and phenylalanine-poor diet in phenylpyruvic oligophrenis. *Exp. Med. Surg.* **1954**, *12*, 114–117.
20. Woolf, L.I. Nutrition in relation to phenylketonuria. *Proc. Nutr. Soc.* **1962**, *21*, 21–29. [CrossRef]
21. Rose, W.C. Amino acid requirements of man. *Nutr. Rev.* **2009**, *34*, 307–309. [CrossRef]
22. Rose, W.C.; Coon, M.J.; Lambert, G.F.; Howe, E. The amino acid requirements of man. VIII. The metabolic availability of the optical isomers of acetyltryptophan. *J. Biol. Chem.* **1955**, *212*, 201–205. [CrossRef]
23. Rose, W.C.; Leach, B.; Coon, M.J.; Lambert, G.F. The amino acid requirements of man. IX. The phenylalanine requirement. *J. Biol. Chem.* **1955**, *213*, 913–922. [CrossRef]
24. Rose, W.C.; Wixom, R.L. The amino acid requirements of man. XIV. The sparing effect of tyrosine on the phenylalanine requirement. *J. Biol. Chem.* **1955**, *217*, 95–101. [CrossRef]
25. Rose, W.C.; Wixom, R.L. The amino acid requirements of man. XIII. The sparing effect of cystine on the methionine requirement. *J. Biol. Chem.* **1955**, *216*, 753–773.
26. Holt, L.E.; Snyderman, S.E. Report to the Council the Amino Acid Requirements of Infants. *JAMA* **1961**, *175*, 100–103. [CrossRef] [PubMed]
27. Holt, L.; Snyderman, S. Protein and amino acid requirements of infants and children. *Nutr. Abstr. Rev.* **1965**, *35*, 1–13.
28. Snyderman, S.E.; Holt, L.E.; Smellie, F.; Boyer, A.; Westall, R.G. The Essential Amino Acid Requirements of Infants: Valine. *Arch. Pediatr. Adolesc. Med.* **1959**, *97*, 186–191. [CrossRef]
29. Leverton, R.M.; Johnson, N.; Ellison, J.; Geschwender, D.; Schmidt, F. The quantitative amino acid requirements of young women. IV. Phenylalanine, with and without tyrosine. *J. Nutr.* **1956**, *58*, 341–353. [CrossRef]
30. Swendseid, M.E.; Williams, I.; Dunn, M.S. Amino Acid Requirements of Young Women Based on Nitrogen Balance Data. *J. Nutr.* **1956**, *58*, 495–505. [CrossRef]
31. Armstrong, M.D.; Tyler, F.H. Studies on phenylketonuria. i. restricted phenylalanine intake in phenylketonuria. *J. Clin. Investig.* **1955**, *34*, 565–580. [CrossRef]
32. Armstrong, M.D.; Binkley, E.L., Jr. Studies on phenylketonuria. V. Observations on a newborn infant with phenylketonuria. *Proc. Soc. Exp. Biol. Med.* **1956**, *93*, 418–420. [CrossRef] [PubMed]
33. Brimblecombe, F.; Stoneman, M.; Maliphant, R. Dietary treatment of an infant with phenylketonuria. *Lancet* **1959**, *273*, 609–611. [CrossRef]
34. Munro, H.N.; Naismith, D.J. The influence of energy intake on protein metabolism. *Biochem. J.* **1953**, *54*, 191–197. [CrossRef] [PubMed]
35. Munro, H.N.; Thomson, W.S. Influence of glucose on amino acid metabolism. *Metabolism* **1953**, *2*, 354–361.
36. Guthrie, R.; Susi, A. A simple phenylalanine method for detecting phenylketonuria in large populations of newborn infants. *Pediatrics* **1963**, *32*, 338–343.
37. Holt, K.S. Difficulties and Dangers in the Management of Phenylketonuria. *Acta Paediatr.* **1963**, *52*, 417–423. [CrossRef]
38. Moncrieff, A.; Wilkinson, R.H. Further Experiences in the Treatment of Phenylketonuria. *BMJ* **1961**, *1*, 763–767. [CrossRef]
39. Umbarger, B. Phenylketonuria—Treating the Disease and Feeding the Child. *Arch. Pediatr. Adolesc. Med.* **1960**, *100*, 908. [CrossRef]
40. Brimblecombe, F.S.W.; Blainey, J.D.; Stoneman, M.E.R.; Wood, B.S.B. Dietary and Biochemical Control of Phenylketonuria. *BMJ* **1961**, *2*, 793–798. [CrossRef]
41. Wilson, K.M.; Clayton, B.E. Importance of Choline during Growth, with Particular Reference to Synthetic Diets in Phenylketonuria. *Arch. Dis. Child.* **1962**, *37*, 565–577. [CrossRef]
42. Arakawa, T.; Hirata, K.; Ohara, K.; Fujii, M.; Takahashi, Y.; Kato, H. Folic Acid Deficiency Due to a Low-phenylalanine Diet. *Tohoku J. Exp. Med.* **1965**, *87*, 296–300. [CrossRef] [PubMed]
43. Royston, N.J.W.; Parry, T.E. Megaloblastic Anaemia Complicating Dietary Treatment of Phenylketonuria in Infancy. *Arch. Dis. Child.* **1962**, *37*, 430–435. [CrossRef] [PubMed]
44. Dodge, P.R.; Mangall, E.L.; Crawford, J.D.; Knapp, J.; Paine, R.S. Hypoglycemia complicating treatment of phenylketonuria with a phenylalanine-deficient diet; report of two cases. *N. Engl. J. Med.* **1959**, *260*, 1104–1111. [CrossRef] [PubMed]
45. Kumagai, H. Microbial Production of Amino Acids in Japan. In *History of Modern Biotechnology I*; Fiechter, A., Ed.; Springer: Berlin/Heidelberg, Germany, 2000; pp. 71–85.
46. Bentovim, A.; Clayton, B.E.; Francis, D.E.M.; Shepherd, J.; Wolff, O.H. Use of an Amino Acid Mixture in Treatment of Phenylketonuria. *Arch. Dis. Child.* **1970**, *45*, 640–650. [CrossRef]
47. Smith, I.; Clayton, B.; Wolff, O. New variant of phenylketonuria with progressive neurological illness unresponsive to phenylalanine restriction. *Lancet* **1975**, *305*, 1108–1111. [CrossRef]
48. Manz, F.; Schmidt, H.; Schärer, K.; Bickel, H. Acid-Base Status in Dietary Treatment of Phenylketonuria. *Pediatr. Res.* **1977**, *11*, 1084–1087. [CrossRef]
49. Alexander, F.W.; Clayton, B.E.; Delves, H.T. Mineral and Trace-metal Balances in Children Receiving Normal and Synthetic Diets. *Qjm: Int. J. Med.* **1974**, *43*, 89–111. [CrossRef]
50. Lawson, M.S.; Clayton, B.; Delves, H.T.; Mitchell, J.D. Evaluation of a new mineral and trace metal supplement for use with synthetic diets. *Arch. Dis. Child.* **1977**, *52*, 62–67. [CrossRef]

51. Wardley, B.L.; Taitz, L.S. Clinical trial of a concentrated amino acid formula for older patients with phenylketonuria (Maxamum XP). *Eur. J. Clin. Nutr.* **1988**, *42*, 81–86.
52. Jukes, T.H. Selenium, an "essential poison". *J. Appl. Biochem.* **1983**, *5*, 233–234.
53. Greeves, L.G.; Carson, D.J.; Craig, B.G.; McMaster, D. Potentially Life-Threatening Cardiac Dysrhythmia in a Child with Selenium Deficiency and Phenylketonuria. *Acta Paediatr.* **2008**, *79*, 1259–1262. [CrossRef] [PubMed]
54. Lombeck, I.; Kasperek, K.; Feinendegen, L.; Bremer, H. Serum-selenium concentrations in patients with maple-syrup-urine disease and phenylketonuria under dieto-therapy. *Clin. Chim. Acta* **1975**, *64*, 57–61. [CrossRef]
55. Rose, H.J.; White, F.; Macdonald, A.; Rutherford, P.J.; Favre, E. Fat intakes of children with PKU on low phenylalanine diets. *J. Hum. Nutr. Diet.* **2005**, *18*, 395–400. [CrossRef] [PubMed]
56. Lohner, S.; Fekete, K.; Decsi, T. Lower n-3 long-chain polyunsaturated fatty acid values in patients with phenylketonuria: A systematic review and meta-analysis. *Nutr. Res.* **2013**, *33*, 513–520. [CrossRef]
57. Cleary, M.; Feillet, F.; White, F.; Vidailhet, M.; Macdonald, A.; Grimsley, A.; Maurin, N.; De Baulny, H.O.; Rutherford, P.J. Randomised controlled trial of essential fatty acid supplementation in phenylketonuria. *Eur. J. Clin. Nutr.* **2006**, *60*, 915–920. [CrossRef]
58. Agostoni, C.; Harvie, A.; McCulloch, D.L.; Demellweek, C.; Cockburn, F.; Giovannini, M.; Murray, G.; Harkness, R.A.; Riva, E. A randomized trial of long-chain polyunsaturated fatty acid supplementation in infants with phenylketonuria. *Dev. Med. Child Neurol.* **2006**, *48*, 207–212. [CrossRef]
59. Macdonald, A.; Cochrane, B.; Wopereis, H.; Loveridge, N. Specific prebiotics in a formula for infants with Phenylketonuria. *Mol. Genet. Metab.* **2011**, *104*, S55–S59. [CrossRef]
60. Macdonald, A.; Lilburn, M.; Cochrane, B.; Davies, P.; Daly, A.; Asplin, D.; Hall, S.K.; Cousins, A.; Chakrapani, A.; Robinson, P.; et al. A new, low-volume protein substitute for teenagers and adults with phenylketonuria. *J. Inherit. Metab. Dis.* **2004**, *27*, 127–135. [CrossRef]
61. Bruinenberg, V.M.; Van Vliet, D.; Van Der Goot, E.; Counotte, D.S.; Kuhn, M.; Van Spronsen, F.J.; Van Der Zee, E.A. Long-term dietary intervention with low Phe and/or a specific nutrient combination improve certain aspects of brain functioning in phenylketonuria (PKU). *PLoS ONE* **2019**, *14*, e0213391. [CrossRef]
62. De Wilde, M.C.; Penke, B.; van der Beek, E.M.; Kuipers, A.A.; Kamphuis, P.J.; Broersen, L.M. Neuroprotective effects of a specific multi-nutrient intervention against Abeta42-induced toxicity in rats. *J. Alzheimers Dis.* **2011**, *27*, 327–339. [CrossRef]
63. Evans, S.; Daly, A.; Chahal, S.; Macdonald, J. Food acceptance and neophobia in children with phenylketonuria: A prospective controlled study. *J. Hum. Nutr. Diet.* **2016**, *29*, 427–433. [CrossRef] [PubMed]
64. Evans, S.; Daly, A.; Macdonald, J.; Pinto, A.; Macdonald, A. Fifteen years of using a second stage protein substitute for weaning in phenylketonuria: A retrospective study. *J. Hum. Nutr. Diet.* **2017**, *31*, 349–356. [CrossRef] [PubMed]
65. Macdonald, A.; Lilburn, M.M.; Davies, P.; Evans, S.; Daly, A.; Hall, S.K.; Hendriksz, C.; Chakrapani, A.; Lee, P. 'Ready to drink' protein substitute is easier is for people with phenylketonuria. *J. Inherit. Metab. Dis.* **2006**, *29*, 526–531. [CrossRef] [PubMed]
66. Ford, S.; O'Driscoll, M.; Macdonald, A. Living with Phenylketonuria: Lessons from the PKU community. *Mol. Genet. Metab. Rep.* **2018**, *17*, 57–63. [CrossRef]
67. MacDonald, A.; Van Wegberg, A.M.J.; Ahring, K.; Beblo, S.; Bélanger-Quintana, A.; Burlina, A.; Campistol, J.; Coşkun, T.; Feillet, F.; Giżewska, M.; et al. PKU dietary handbook to accompany PKU guidelines. *Orphanet J. Rare Dis.* **2020**, *15*, 171. [CrossRef]
68. Macdonald, A.; Ferguson, C.; Rylance, G.; Morris, A.A.M.; Asplin, D.; Hall, S.K.; Booth, I.W. Are tablets a practical source of protein substitute in phenylketonuria? *Arch. Dis. Child.* **2003**, *88*, 327–329. [CrossRef]
69. Ney, D.M.; Hull, A.K.; Van Calcar, S.C.; Liu, X.; Etzel, M.R. Dietary Glycomacropeptide Supports Growth and Reduces the Concentrations of Phenylalanine in Plasma and Brain in a Murine Model of Phenylketonuria. *J. Nutr.* **2008**, *138*, 316–322. [CrossRef]
70. Nakano, T.; Silva-Hernandez, E.; Ikawa, N.; Ozimek, L. Purification of kappa-casien glycomacropeptide from sweet whey with undetectable level of phenylalanine. *Biotechnol. Prog.* **2002**, *18*, 409–412. [CrossRef]
71. Lim, K.; Van Calcar, S.C.; Nelson, K.L.; Gleason, S.T.; Ney, D.M. Acceptable low-phenylalanine foods and beverages can be made with glycomacropeptide from cheese whey for individuals with PKU. *Mol. Genet. Metab.* **2007**, *92*, 176–178. [CrossRef]
72. Giovannini, M.; Riva, E.; Salvatici, E.; Cefalo, G.; Radaelli, G. Randomized Controlled Trial of a Protein Substitute with Prolonged Release on the Protein Status of Children with Phenylketonuria. *J. Am. Coll. Nutr.* **2014**, *33*, 103–110. [CrossRef]
73. Scheinin, M.; Barassi, A.; Junnila, J.J.T.; Lovró, Z.; Reiner, G.; Sarkkinen, E.; Macdonald, A. Amino Acid Plasma Profiles from a Prolonged-Release Protein Substitute for Phenylketonuria: A Randomized, Single-Dose, Four-Way Crossover Trial in Healthy Volunteers. *Nutrients* **2020**, *12*, 1653. [CrossRef] [PubMed]
74. Giarratana, N.; Gallina, G.; Panzeri, V.; Frangi, A.; Canobbio, A.; Reiner, G. A New Phe-Free Protein Substitute Engineered to Allow a Physiological Absorption of Free Amino Acids for Phenylketonuria. *J. Inborn Errors Metab. Screen.* **2018**, *6*, 1–9. [CrossRef]
75. Macdonald, A.; Ashmore, C.; Daly, A.; Pinto, A.; Evans, S. An Observational Study Evaluating the Introduction of a Prolonged-Release Protein Substitute to the Dietary Management of Children with Phenylketonuria. *Nutrients* **2020**, *12*, 2686. [CrossRef] [PubMed]
76. Acosta, P.B.; Yannicelli, S. Protein intake affects phenylalanine requirements and growth of infants with phenylketonuria. *Acta Paediatr.* **1994**, *83*, 66–67. [CrossRef] [PubMed]

77. Duran, G.P.; Rohr, F.J.; Slonim, A.; Güttler, F.; Levy, H.L. Necessity of Complete Intake of Phenylalanine-free Amino Acid Mixture for Metabolic Control of Phenylketonuria. *J. Am. Diet. Assoc.* **1999**, *99*, 1559–1563. [CrossRef]
78. Kindt, E.; Lunde, H.A.; Gjessing, L.R.; Halvorsen, S.; Lie, S.O. Fasting Plasma Amino Acid Concentrations in PKU Children on Two Different Levels of Protein Intake. *Acta Paediatr.* **1988**, *77*, 60–66. [CrossRef] [PubMed]
79. Macdonald, A.; Chakrapani, A.; Hendriksz, C.; Daly, A.; Davies, P.; Asplin, D.; Hall, K.; Booth, I.W. Protein substitute dosage in PKU: How much do young patients need? *Arch. Dis. Child.* **2005**, *91*, 588–593. [CrossRef]
80. O'Daly, S. Phenylketonuria treated with a high phenylalanine intake and casein-hydrolysate/aminoacid mixtures. *Lancet* **1961**, *277*, 1379–1383. [CrossRef]
81. Acosta, P.B. Recommendations for protein and energy intakes by patients with phenylketonuria. *Eur. J. Nucl. Med. Mol. Imaging* **1996**, *155*, S121–S124. [CrossRef]
82. Van Vliet, D.; Ruinenberg, V.M.; Mazzola, P.N.; van Faassen, M.H.; de Blaauw, P.; Kema, I.P.; Heiner-Fokkema, M.R.; van Anholt, R.D.; van der Zee, E.A.; Spronsen, F.J. Large Neutral Amino Acid Supplementation Exerts Its Effect through Three Synergistic Mechanisms: Proof of Principle in Phenylketonuria Mice. *PLoS ONE* **2015**, *10*, e0143833. [CrossRef]
83. Adibi, S.A.; Morse, E.L. Intestinal transport of dipeptides in man: Relative importance of hydrolysis and intact absorption. *J. Clin. Investig.* **1971**, *50*, 2266–2275. [CrossRef] [PubMed]
84. Hidalgo, I.J.; Borchardt, R.T. Transport of a large neutral amino acid (phenylalanine) in a human intestinal epithelial cell line: Caco-2. *Biochim. Biophys. Acta Biomembr.* **1990**, *1028*, 25–30. [CrossRef]
85. Kindt, E.; Motzfeldt, K.; Halvorsen, S.; Lie, S. Protein requirements in infants and children: A longitudinal study of children treated for phenylketonuria. *Am. J. Clin. Nutr.* **1983**, *37*, 778–785. [CrossRef] [PubMed]
86. Przyrembel, H. Recommendations for protein and amino acid intake in phenylketonuria patients. *Eur. J. Nucl. Med. Mol. Imaging* **1996**, *155*, S130–S131. [CrossRef]
87. Snyderman, S.E.; Pratt, E.L.; Cheung, M.W.; Norton, P.; Holt, L.E.; Hansen, A.E.; Panos, T.C. The Phenylalanine Requirement of the Normal Infant. *J. Nutr.* **1955**, *56*, 253–263. [CrossRef]
88. Gropper, S.S.; Gropper, D.M.; Acosta, P.B. Plasma Amino Acid Response to Ingestion of L-Amino Acids and Whole Protein. *J. Pediatr. Gastroenterol. Nutr.* **1993**, *16*, 143–150. [CrossRef]
89. Jones, B.J.; Lees, R.; Andrews, J.; Frost, P.; Silk, D.B. Comparison of an elemental and polymeric enteral diet in patients with normal gastrointestinal function. *Gut* **1983**, *24*, 78–84. [CrossRef]
90. World Health Organisation. *Protein and Amino Acid Requirements in Human Nutition*; Report of a Joint WHO/FAO/UNU Expert Consultation (WHO Technical Report Series 935), in United Nations University; WHO: Geneva, Switzerland, 2007; pp. 1–265.
91. Reeds, P.J.; Garlick, P.J. Protein and Amino Acid Requirements and the Composition of Complementary Foods. *J. Nutr.* **2003**, *133*, 2953S–2961S. [CrossRef]
92. Viteri, F.E. INCAP Studies of Energy, Amino Acids, and Protein. *Food Nutr. Bull.* **2010**, *31*, 42–53. [CrossRef]
93. Macdonald, A.; Rylance, G.; Hall, S.K.; Asplin, D.; Booth, I.W. Factors affecting the variation in plasma phenylalanine in patients with phenylketonuria on diet. *Arch. Dis. Child.* **1996**, *74*, 412–417. [CrossRef]
94. Illsinger, S.; Meyer, U.; Vaske, B.; Das, A.M.; Lücke, T. Branched chain amino acids as a parameter for catabolism in treated phenylketonuria. *Amino Acids* **2004**, *28*, 45–50. [CrossRef] [PubMed]
95. Van Wegberg, A.M.J.; Macdonald, A.; Ahring, K.; Bélanger-Quintana, A.; Blau, N.; Bosch, A.M.; Burlina, A.; Campistol, J.; Feillet, F.; Giżewska, M.; et al. The complete European guidelines on phenylketonuria: Diagnosis and treatment. *Orphanet J. Rare Dis.* **2017**, *12*, 1–56. [CrossRef] [PubMed]
96. Aguiar, A.; Ahring, K.; Almeida, M.; Assoun, M.; Quintana, A.B.; Bigot, S.; Bihet, G.; Malmberg, K.B.; Burlina, A.; Bushueva, T.; et al. Practices in prescribing protein substitutes for PKU in Europe: No uniformity of approach. *Mol. Genet. Metab.* **2015**, *115*, 17–22. [CrossRef] [PubMed]
97. Macdonald, A.; Rylance, G.W.; Asplin, D.; Hall, S.K.; Booth, I.W. Does a single plasma phenylalanine predict quality of control in phenylketonuria? *Arch. Dis. Child.* **1998**, *78*, 122–126. [CrossRef] [PubMed]
98. Macdonald, A.; Rylance, G.; Davies, P.; Asplin, D.; Hall, S.K.; Booth, I.W. Administration of protein substitute and quality of control in phenylketonuria: A randomized study. *J. Inherit. Metab. Dis.* **2003**, *26*, 319–326. [CrossRef] [PubMed]
99. Elango, R.; Ball, R.O.; Pencharz, P.B. Amino acid requirements in humans: With a special emphasis on the metabolic availability of amino acids. *Amino Acids* **2009**, *37*, 19–27. [CrossRef]
100. Medical, Research, and Council. Phenylketonuria due to phenylalanine hydroxylase deficiency: An unfolding story. Medical Research Council Working Party on Phenylketonuria. *BMJ* **1993**, *306*, 115–119. [CrossRef]
101. Blaskovics, M.; Nelson, T.L. Phenylketonuria and its variations. A review of recent developments. *Calif. Med.* **1971**, *115*, 42–57.
102. Fisch, R.O.; Torres, F.; Gravem, H.J.; Greenwood, C.S.; Anderson, J.A. Twelve years of clinical experience with phenylketonuria: A statistical evaluation of symptoms, growth, mental development, electroencephalographic records, serum phenylalanine levels, and results of dietary management. *Neurology* **1969**, *19*, 659. [CrossRef]
103. Koff, E.; Kammerer, B.; Boyle, P.; Pueschel, S.M. Intelligence and phenylketonuria: Effects of diet termination. *J. Pediatr.* **1979**, *94*, 534–537. [CrossRef]
104. Cowan, M. *Phenylketonuria: A Story of Heartbreak and Hope*; Cork Examiner: Cork, Ireland, 1962.

105. Evans, S.; Daly, A.; Wildgoose, J.; Cochrane, B.; Chahal, S.; Ashmore, C.; Loveridge, N.; Macdonald, A. How Does Feeding Development and Progression onto Solid Foods in PKU Compare with Non-PKU Children During Weaning? *Nutrients* **2019**, *11*, 529. [CrossRef] [PubMed]
106. Macdonald, A.; Smith, T.A.; De Silva, S.; Alam, V.; Van Loon, J.M. The personal burden for caregivers of children with phenylketonuria: A cross-sectional study investigating time burden and costs in the UK. *Mol. Genet. Metab. Rep.* **2016**, *9*, 1–5. [CrossRef] [PubMed]

nutrients
MDPI

Article

Accidental Consumption of Aspartame in Phenylketonuria: Patient Experiences

Ella Newbould [1], Alex Pinto [2], Sharon Evans [2], Suzanne Ford [3], Mike O'Driscoll [4], Catherine Ashmore [2], Anne Daly [2] and Anita MacDonald [2,*]

1 Faculty of Health, Education & Life Sciences, Birmingham City University: City South Campus, Westbourne Road, Edgbaston, Birmingham B15 3TN, UK; ella.newbould@mail.bcu.ac.uk
2 Birmingham Women's and Children's NHS Foundation Trust, Steelhouse Lane, Birmingham B4 6NH, UK; alex.pinto@nhs.net (A.P.); sharon.morris6@nhs.net (S.E.); catherine.ashmore@nhs.net (C.A.); a.daly3@nhs.net (A.D.)
3 National Society for Phenylketonuria, Purley CR8 9DD, UK; suzanne.ford@nspku.org
4 School of Health and Education, Middlesex University, Room WG41A (Williams Building), The Burroughs Hendon, London NW4 4BT, UK; m.odriscoll@mdx.ac.uk
* Correspondence: anita.macdonald@nhs.net

Abstract: Aspartame is a phenylalanine containing sweetener, added to foods and drinks, which is avoided in phenylketonuria (PKU). However, the amount of phenylalanine provided by aspartame is unidentifiable from food and drinks labels. We performed a cross-sectional online survey aiming to examine the accidental aspartame consumption in PKU. 206 questionnaires (58% female) were completed. 55% of respondents (n = 114) were adults with PKU or their parent/carers and 45% (n = 92) were parents/carers of children with PKU. 74% (n = 152/206) had consumed food/drinks containing aspartame. Repeated accidental aspartame consumption was common and more frequent in children ($p < 0.0001$). The aspartame containing food/drinks accidentally consumed were fizzy drinks (68%, n = 103/152), fruit squash (40%, n = 61/152), chewing gum (30%, n = 46/152), flavoured water (25%, n = 38/152), ready to drink fruit squash cartons (23%, n = 35/152) and sports drinks (21%, n = 32/152). The main reasons described for accidental consumption, were manufacturers' changing recipes (81%, n = 123/152), inability to check the ingredients in pubs/restaurants/vending machines (59%, n = 89/152) or forgetting to check the label (32%, n = 49/152). 23% (n= 48/206) had been prescribed medicines containing aspartame and 75% (n = 36/48) said that medicines were not checked by medics when prescribed. 85% (n = 164/192) considered the sugar tax made accidental aspartame consumption more likely. Some of the difficulties for patients were aspartame identification in drinks consumed in restaurants, pubs, vending machines (77%, n = 158/206); similarities in appearance of aspartame and non-aspartame products (62%, n = 127/206); time consuming shopping/checking labels (56%, n = 115/206); and unclear labelling (55%, n = 114/206). These issues caused anxiety for the person with PKU (52%, n = 106/206), anxiety for parent/caregivers (46%, n = 95/206), guilt for parent/carers (42%, n = 87/206) and social isolation (42%, n = 87/206). It is important to understand the impact of aspartame and legislation such as the sugar tax on people with PKU. Policy makers and industry should ensure that the quality of life of people with rare conditions such as PKU is not compromised through their action.

Keywords: phenylketonuria; phenylalanine; aspartame; sugar tax

Citation: Newbould, E.; Pinto, A.; Evans, S.; Ford, S.; O'Driscoll, M.; Ashmore, C.; Daly, A.; MacDonald, A. Accidental Consumption of Aspartame in Phenylketonuria: Patient Experiences. *Nutrients* **2021**, *13*, 707. https://doi.org/10.3390/nu13020707

Academic Editor: Shanon L. Casperson

Received: 18 January 2021
Accepted: 19 February 2021
Published: 23 February 2021

Publisher's Note: MDPI stays neutral with regard to jurisdictional claims in published maps and institutional affiliations.

Copyright: © 2021 by the authors. Licensee MDPI, Basel, Switzerland. This article is an open access article distributed under the terms and conditions of the Creative Commons Attribution (CC BY) license (https://creativecommons.org/licenses/by/4.0/).

1. Introduction

Aspartame, a non-nutritive sweetener, is one of the most widely used artificial sweeteners and accounts for 62% of the artificial sweetener market [1]. It is a synthetic dipeptide known as N-L-alpha-aspartyl-L-phenylalanine methyl ester ($C_{14}H_{18}N_2O_5$) and was accidentally discovered in 1965 [2,3]. Aspartame is completely hydrolysed to phenylalanine (50%), aspartic acid (40%) and methanol (10%) in the intestinal lumen and is rapidly metabolised by esterases and peptidases [4,5]. It is around 200 times sweeter than sucrose and it is estimated that it is added to >6000 foods and drinks [6,7]. Aspartame is approved in more than 90 countries and its safety has been evaluated by the Joint FAO/WHO Expert Committee on Food Additives (JECFA), as well as by numerous national food safety authorities, including the US Food and Drug Administration (FDA) and the European Food Safety Authority (EFSA) [8–10]. Aspartame can be safely consumed by healthy individuals, but it has long been recognised as a hazard to individuals with phenylketonuria (PKU) and therefore, it should be avoided [11]. The amount of phenylalanine in aspartame containing foods and drinks is not declared on ingredient labels and its impact on metabolic control in patients with PKU is not well established [12–14].

PKU, an autosomal recessive inherited condition, is caused by mutations in the gene encoding phenylalanine hydroxylase. It is estimated to affect 0.45 million individuals worldwide, with a global prevalence of 1:23,930 live births [15]. A rigorous lifelong low-phenylalanine diet is the principal treatment option. It requires the avoidance of high protein foods such as meat, fish, eggs, lentils, nuts, soya, bread, pasta and cheese. Daily dietary phenylalanine intake is calculated, measured and continually controlled according to individual tolerance. Eighty per cent of patients tolerate <500 mg/day (10 g natural protein/day) in order to avoid elevated blood phenylalanine levels. Phenylalanine tolerance does vary between patients depending upon the severity of their disorder and the use of pharmaceutical treatment options such as sapropterin (synthetic tetrahydrobiopterin (BH4), or pegvaliase (phenylalanine ammonium lyase). Sapropterin, an oral drug, is effective in a subset of BH4 responsive patients with PKU and is usually given as an adjunct to dietary treatment [16]. Pegvaliase, delivered by subcutaneous injection, is only licensed for adults with blood phenylalanine levels above the European PKU guidelines target range [17,18]. Neither pharmaceutical treatment option is available via the National Health Service in England.

The additional scrutiny of checking all food ingredient labels for aspartame in food, drinks and drugs intensifies the complexity of management [19]. Aspartame is added to a wide variety of foods: low calorie sweeteners, soft drinks (including fizzy drinks, fruit squashes/cordials), iced tea, flavoured mineral water, energy drinks, dessert mixes, frozen desserts, syrups/dessert sauces, mints, jelly, chewing gum, fruit yogurt, ice lollies, and ice creams. It is also added to around 600 pharmaceutical products (both medically prescribed and over the counter) including chewable multivitamins and cough medications. According to European law, foods containing aspartame must declare it is added either by name or E number (E951) [20]. However, it is not mandatory for manufacturers to state the amount of aspartame added to foods, rendering it impossible for people with PKU to estimate the phenylalanine intake from this source.

A further concern in the UK is the Soft Drinks Industry Levy (SDIL) which was introduced by Her Majesty's Revenue and Customs (HMRC) in 2018 [21]. It is commonly referred to as the "sugar tax". This was devised in response to national concerns about rising childhood and teenage obesity and was designed to encourage manufacturers to reduce the added sugar content of their drinks. It is a two-tier levy system: including a standard tax rate applied to drinks with a sugar content between 5 g and <8 g per 100 mL and a higher tax rate applied to drinks with a sugar content ≥8 g per 100 mL. This "sugar tax" has been highly effective with at least 50% of manufacturers reducing the sugar content of their products [22] but it has also led to many manufacturers replacing sugar with artificial sweeteners such as aspartame, potentially marginalising the dietary choices of patients with PKU. A recent equality risk assessment conducted by the HMRC examining the SDIL, stated that they were unaware of any evidence to suggest that the existing warning on food labels about the presence of aspartame in soft drinks was inadequate for people with PKU [23].

It is important to understand the impact of added aspartame to foods, drinks and medications on people with PKU. This paper aims to examine the frequency of accidental aspartame consumption, the reasons for this, and the challenges associated with avoiding aspartame in PKU.

2. Materials and Methods

2.1. Study Design

We performed a cross sectional online survey. Patients with PKU and/or parents/caregivers of a person with PKU were invited to take part in this study. Respondents were excluded if they did not reside in the UK. The questionnaire was built in the Online Surveys platform (https://www.onlinesurveys.ac.uk (accessed on 1 April 2020)) and placed on the UK National Society for Phenylketonuria (NSPKU) website, with additional promotion on the NSPKU Twitter and Facebook accounts between April and July 2020.

This non-validated questionnaire contained 23 questions; 10 multiple choice (6 of which invited additional comments), 8 multiple response, 3 Likert scale and 2 open-ended questions. A group of experienced research dietitians from Birmingham Women's and Children's Hospital (A.P., S.E., A.M.), a colleague at the NSPKU (S.F.) and an expert in survey methodology (M.O.) helped develop the survey with a student dietitian from Birmingham City University (E.N.). The questionnaire was also reviewed by lay people to ensure its readability.

2.2. Data Collected

Demographic information was collected about the type of respondent (patient or parent/caregiver of patients aged $\geq$18 y or <18 y), gender of the person with PKU and confirmation of residency in the UK. Respondents answered questions about any known consumption of foods, drinks and medications containing aspartame, the frequency this had occurred, the reason behind this accidental ingestion and any symptoms this had caused. They were also asked about their knowledge and impact of the sugar tax with respect to the aspartame content of foods and drinks, and the ease of identifying aspartame on food, drinks and medication labels in addition to other challenges of identifying aspartame in products.

Overall themes explored in the survey were: accidental consumption of aspartame in food and drinks, accidental consumption of aspartame in medications, the sugar tax, drinks choice in different venues, label checking, and the effect of aspartame addition on the person with PKU and their family.

2.3. Statistics

Quantitative data analysis (inferential and descriptive statistics) was carried out with Statistical Package for the Social Sciences (SPSS) version 25 (SPSS Inc., Chicago, IL, USA). Multiple response questions were analysed with descriptive statistics only. Statistical significance was set at $p < 0.05$.

Qualitative data analyses of open-ended responses were carried out in NVIVO v.12 PRO. The whole survey dataset was imported into NVIVO so that the coding of open-ended responses could be broken down by survey questions including demographic questions. All open-ended responses were analysed thematically.

2.4. Ethics

Ethical approval to perform this study (approval number 6085, project title "The accidental consumption of aspartame in PKU: The experiences of patients and their caregivers") was given by Birmingham City University ethics committee. Adults with PKU and parents/carers of children and adults with PKU gave their consent at the beginning of the online questionnaire. Potential respondents were also advised that data from the survey may be published in an anonymized form. If names or hospitals were mentioned in verbatim abstracts, these were removed from results presented in this manuscript.

3. Results

There were 206 wholly or partially completed questionnaires. Fifty-five per cent (n = 114) of respondents were adults (18 or over) with PKU or parent/carers of adults with PKU and 45% (n = 92) were the parent or carers of children with PKU.

All respondents were normally residents in the UK. The PKU population described by the respondents were: 58% (n = 119) female; 41% (n = 85) male, 1 respondent was 'non-binary' and 1 preferred not to say.

3.1. Accidental Consumption of Aspartame in Food and Drink

Seventy-four per cent of participants (n = 152/206) said that people with PKU had consumed aspartame in a food or drink; 20% (n = 42/206) said they had not and 6% (n = 12/206) said they did not know.

Of those who had consumed aspartame by accident/error, just under half (47%, n = 72/152) said this occurred one to three times; 17% (n = 26/152) said 4 to 6 times and 6% (n = 9/152) said that it had occurred 7 to 9 times in the last 3 years. One in ten respondents (11%, n = 16/152) said that accidental consumption had occurred 10 times or more. Just under one fifth (19%, n = 29/152) of respondents could not recount how often accidental consumption had happened. Repeated accidental consumption of aspartame was more frequent in adults with PKU than for children ($p < 0.0001$). In the last 3 years, aspartame had been consumed accidentally 1 to 3 times in 79% (n = 42/53) of children and 43% (n = 30/70) of adults. In contrast, accidental consumption of 4 to 6 times occurred in 31% (n = 22/70) of adults compared to only 8% (n = 4/53) in children. Females (79%) with PKU were more likely to report having consumed aspartame than males (67%) ($p = 0.008$, Fisher's exact test). Eleven per cent (n = 8/74) of females had 7 to 9 incidents, compared to 0% (0/48) of males; 18% (n = 13/74) of females had 10 or more incidents, 3 times the proportion of males at 6% (n = 3/48). Patients that answered "don't know" were excluded.

The main reasons for accidental consumption of aspartame were manufacturers' changing product recipes (81%, n = 123/152), inability to check the ingredients e.g., drinks purchased in a pub or restaurant or from a vending machine (59%, n = 89/152), forgetting to check the label (32%, n = 49/152), and picking the wrong product from a shelf when shopping (29%, n = 44/152). Other reasons described by the respondents included: served the wrong drink in a bar or restaurant, (n = 22), unclear labelling (n = 16), not realising a product contained aspartame (n = 11), child unsupervised (n = 6), or other undefined reason (n = 4).

Examples of the verbatim quotes for the 5 most common themes for accidental aspartame consumption.

- "Drinks that were previously free from aspartame and fine to drink had their recipe changed without seemingly advertising the change. This meant that it was only on consumption and tasting the difference from how it used to be that the ingredients were checked, and aspartame was found."
- "I don't know how many times I have consumed aspartame, but I know I have. In a crowded bar it is hard to request a specific brand name and it is not possible to read a label on a multi dispensing tap such as that used by bar staff to add coke or tonic to a drink."
- "I have never seen a lolly with aspartame in before, so I didn't check it from the ice-cream man—I checked it only after she had eaten it."
- "Both my girls have autism. They do not understand consequences and are unable to challenge/ask people if the drinks contain aspartame, therefore they will just drink what is given to them. They have also picked up the wrong bottles of coke as the packaging is not much different at all".
- "Aspartame isn't required to be listed on alcoholic drinks, therefore it's hard to know if it's present or not."

3.2. *Foods/Drinks Involved in Accidental Aspartame Consumption*

The food or drinks containing aspartame most reported to be accidentally consumed were fizzy drinks e.g., Coca Cola/lemonade/Irn Bru (68%, n = 103/152), fruit squash/cordials e.g., Robinsons Summerfruit squash (40%, n = 61/152), chewing gum (30%, n = 46/152), flavoured water (25%, n = 38/152), ready to drink cartons or bottles of juice/squash e.g., Strawberry Ribena (23%, n = 35/152), sports drinks e.g., Lucozade/Powerade (21%, n = 32/152), alcoholic drinks (19%, n = 29/152), sweets (14%, n = 21/152), jelly (9%, n = 14/152), tonic water (7%, n = 11/152), mints (7%, n = 11/152), iced slush drinks (7%, n = 10/152), energy drinks e.g., Red Bull (5%, n = 7/152), and table top sweetener e.g., Half-Spoon (3%, n = 4/152).

3.3. *Aspartame Consumption of Medically Prescribed and over the Counter Medications*

Twenty-three per cent (n = 48/206) of responders said that people with PKU had been prescribed medicines by their doctors that contained aspartame. This was more likely to occur in children (30%, n = 28/92) than adults (18%, n = 20/114).

Seventy-five per cent (n = 36/48) said that medicines were not checked by doctors/pharmacists for aspartame, but it was identified by the person with PKU or their carer. Twenty-five per cent (n = 12/48) said they had been advised that it was better to take the medicine and not worry about the aspartame content. Four per cent (n = 2/48) of respondents said the amount of phenylalanine from aspartame was checked and the number of phenylalanine exchanges adjusted accordingly. Thirteen per cent (n = 6/48) gave an "other" response including: 'was given a replacement medication only after they requested for this to happen', 'they accepted the medicine even though they knew it contained aspartame', were 'refused an alternative medication', and 'health professionals (dispensing the medication) were unaware of aspartame or PKU'. Although most respondents managed to access an alternative suitable medication, it depended on the patient or carer first identifying that aspartame was on the list of ingredients on the original medication.

Most respondents (88%, n = 182/206) were aware that some over-the-counter medicines contained aspartame, but 20% (n = 37/182) had consumed aspartame from this source.

Some verbatim extracts about the experiences associated with aspartame in medications are given below.

- "I checked the ingredients and found the medicine contained aspartame and had a written warning about phenylalanine. I called the doctor who couldn't think of a different medicine so was told to go to hospital with my child to receive "better care."
- "Happens a lot. There have been times when I've had to visit several chemists to finally get a variation without aspartame. I've also asked the GP to issue a script for an alternative medicine. It's always down to the patient to check and Drs and pharmacists are unaware."
- "Always been told it's best to take the medication and get better then worry about levels afterwards."
- "We checked, and it only had a small amount of aspartame and he was very poorly and he needed to have it."

3.4. *"Sugar Tax"*

Most respondents (93%, n =192/206) were aware of the sugar tax. Many respondents (85%, n = 164/192) considered the sugar tax made accidental aspartame consumption more likely (either much more likely, 59% (n = 114/192) or slightly more likely, 26% (n = 50/192)). Eleven per cent (n = 21/192) thought that the sugar tax made no difference to the likelihood of accidental consumption of aspartame and just over 3% (n = 6/192) thought that the sugar tax had make it less likely.

Eighty-nine per cent (n = 170/192) thought the sugar tax led to fewer choices of drinks and more than two-thirds (68%, n = 130/192) considered that drink costs increased. More than four in 10 respondents said that the sugar tax had caused increased stress for the person with PKU and 27%, (n = 52/192) reported greater social isolation. Fifteen per cent

(n = 29/192) of respondents thought that the tax had led to worse blood phenylalanine control for people with PKU. Only 5% (n = 10/192) thought the tax had no effect. 'Other responses' were commonly expressions of anger, being disheartened or depressed about the situation as the sugar tax increased the burden of dietary treatment even more.

Some examples of verbatim quotes given to the open question responses about the impact of the sugar tax:

- "Drinks are something we can share and enjoy. Drinks that we could enjoy, experiment with, taste and talk about are now becoming less accessible and it has a really big impact on us. Sugar is actually one of the few things that we can ingest without fear of brain damage, and mental and physical damage."
- "My daughter is aware of the higher cost of the non-aspartame products so will often choose to go without; thinking about the extra expense to us as parents."
- "It has made an already difficult diet even harder to follow and people just think you are unhealthy choosing sugar versions and a faddy diet."
- "Soul destroying for a person to check every food label/every morsel they put into their mouths".

3.5. Choice of Drinks in Different Venues

Respondents stated their dissatisfaction with the supply of drinks in different venues (Table 1). This was highest in relation to leisure/sports centres (67%); followed by fast food chains (62%) and restaurants (60%). Forty-nine per cent (n = 81/167) were dissatisfied (fairly or extremely) with the choice of drinks in hospitals, when people with PKU attended their clinics. Museums, airports, petrol stations and other people's homes had some of the lowest dissatisfaction scores but even for these venues, dissatisfaction is high in absolute terms (i.e., there is low satisfaction across all venues and high proportions are neutral on most venues).

Table 1. Satisfaction with the range of drinks across various venues.

Venue	Extremely Dissatisfied	Fairly Dissatisfied	Neither Dissatisfied nor Satisfied	Fairly Satisfied	Extremely Satisfied
Leisure Centre/Sports Centre (n = 165)	n = 42 (25%)	n = 68 (41%)	n = 20 (12%)	n = 30 (18%)	n = 5 (3%)
Hospital Clinics (n = 167)	n = 41 (25%)	n = 40 (24%)	n = 32 (19%)	n = 47 (28%)	n = 7 (4%)
Fast Food Chains (n = 193)	n = 47 (24%)	n = 73 (38%)	n = 23 (12%)	n = 41 (21%)	n = 9 (5%)
Pubs/Bars (n = 188)	n = 42 (22%)	n = 69 (37%)	n = 13 (7%)	n = 59 (31%)	n = 5 (3%)
Restaurants (n = 195)	n = 41 (21%)	n = 76 (39%)	n = 26 (13%)	n = 46 (24%)	n = 6 (3%)
Schools (n = 123)	n = 24 (20%)	n = 42 (34%)	n = 24 (20%)	n = 26 (21%)	n = 7 (6%)
Tourist Attraction e.g., Alton Towers (n = 170)	n = 32 (19%)	n = 65 (38%)	n = 26 (15%)	n = 40 (24%)	n = 7 (4%)
Motorway Cafes (n = 170)	n = 31 (18%)	n = 58 (34%)	n = 28 (16%)	n = 42 (25%)	n = 11 (6%)
Cafes (n = 197)	n = 34 (17%)	n = 63 (32%)	n = 32 (16%)	n = 63 (32%)	n = 5 (3%)
Hotels (n = 170)	n = 29 (17%)	n = 55 (32%)	n = 36 (21%)	n = 43 (25%)	n = 7 (4%)
Workplace (n = 113)	n = 19 (17%)	n = 30 (27%)	n = 25 (22%)	n = 29 (26%)	n = 10 (9%)
College (n = 64)	n = 10 (16%)	n = 23 (36%)	n = 17 (27%)	n = 13 (20%)	n = 1 (2%)
Airports (n = 165)	n = 24 (15%)	n = 45 (27%)	n = 38 (23%)	n = 43 (26%)	n = 15 (9%)
Nurseries (n = 73)	n = 10 (14%)	n = 21 (29%)	n = 22 (30%)	n = 17 (23%)	n = 3 (4%)
Petrol Stations (n = 180)	n = 23 (13%)	n = 49 (27%)	n = 25 (14%)	n = 64 (36%)	n = 19 (11%)
University (n = 65)	n = 7 (11%)	n = 24 (37%)	n = 18 (28%)	n = 13 (20%)	n = 3 (5%)
Other People's Homes e.g., Friends/Family (n = 198)	n = 12 (6%)	n = 61 (31%)	n = 42 (21%)	n = 65 (33%)	n = 18 (9%)

Abbreviations: n: number of respondents. This varies considerably and is low for some venues such as 'university' and 'nurseries' because these are used predominantly by particular demographic groups and those that did not use them chose 'not applicable' and did not rate.

3.6. Label Checking

Respondents checked labels for food, drinks and medicines most of the time (Table 2). Drink labels (96%, n = 196/205) were checked either most of the time or always which is higher when compared with food labels (81%, n = 165/203).

Table 2. Proportion of respondents who check food, drinks and medicine labels.

	Not at All	Rarely	Sometimes	Most of the Time	Always
Food (n = 203)	n = 1 (<1%)	n = 13 (6%)	n = 24 (12%)	n = 51 (25%)	n = 114 (56%)
Drinks (n = 205)	n = 0 (0%)	n = 0 (0%)	n = 9 (4%)	n = 61 (30%)	n = 135 (66%)
Medicines (n = 203)	n = 8 (4%)	n = 18 (9%)	n = 18 (9%)	n = 30 (15%)	n = 129 (64%)

Food, drinks and medication labels were always checked more often by parents/caregivers for children. For food, 44% (n = 49/111) of adults or carers of adults always checked labels compared with 71% (n = 65/92) of parents/carers of children; for drinks, 56% (n = 63/113) of adults or carers of adults always checked labels compared with 78% (n = 72/92) of parents/carers of children; for medicine, 46% (n = 52/112) of adults or carers of adults always checked labels compared with 85% (n = 77/91) of parents/carers of children. On average, both adults or carers of adults and carers of children with PKU all checked food, drinks and medicines labels for aspartame 'most of the time' but carers of children significantly more so (Table 3).

Table 3. Mean levels of checking food, drink and medicine labels by age group.

	Adult (18 or Over) with PKU or Parent/Carer of Adult with PKU (Mean), n = 114	Parent or Carer of Child with PKU (Mean), n = 92	Total (Mean), n = 206	Mann Whitney Test p Value
Food	4.07	4.58	4.30	$p < 0.001$
Drinks	4.50	4.75	4.61	$p < 0.001$
Medicines	3.86	4.74	4.25	$p < 0.001$

Abbreviations: PKU, Phenylketonuria; n: number of respondents. The mean values relate to a scale of 1 to 5 (1 = Not at all; 2 = Rarely; 3 = Sometimes; 4 = Most of the time; 5 = Always).

3.7. Ease of Identifying Aspartame on the Ingredient Label

A high proportion of respondents reported it was very easy or fairly easy to identify aspartame on ingredient labels, 63% (n = 130/205) for food and 65% (n = 133/205) for drinks compared to those who had difficulty, 22% (n = 45/205) for food and 23% (n = 48/205) for drinks (Table 4). Ease of identification of aspartame on medicines was lower with 46% (n = 86/189) reporting it was very easy/fairly easy and 40% (n = 76/189) finding it difficult. The number remaining neutral was similar for food, drinks and medication.

Table 4. Perceived ease of label checking by product type.

	Very Difficult	Fairly Difficult	Neither Difficult nor Easy	Fairly Easy	Very Easy
Food (n = 205)	n = 11 (5%)	n= 34 (17%)	n = 30 (15%)	n = 89 (43%)	n = 41 (20%)
Drinks (n = 205)	n = 8 (4%)	n = 40 (20%)	n = 24 (12%)	n = 80 (39%)	n = 53 (26%)
Medicines (n = 189)	n = 19 (10%)	n = 57 (30%)	n = 27 (14%)	n = 57 (30%)	n = 29 (15%)

3.8. Challenges in Identifying Products which Contain Aspartame

The biggest challenges identified by respondents are presented in Table 5 in detail.

Table 5. Challenges in identifying products which contain aspartame.

Challenges Faced in Identifying If a Food, Drink or Medicine Contains Aspartame	Percentage Responses (%)	Number of Respondents Per Total Sample (n = 206)
Difficulties in Identifying Aspartame in Food or Drinks Consumed in Restaurants, Pubs, Cafes, Vending Machines	77	158
Similarities in Appearance of Non-Aspartame and Aspartame Containing Products	62	127
Time Taken to Identify if a Product Contains Aspartame	56	115
Unclear Labelling	55	114
Easy to Make Mistakes	44	91
Unable to Read the Writing on Food Labels (Writing too Small, too Shiny)	42	87
Lack of Knowledge about which Products Contain Aspartame	20	42
Have no Challenges	4	8
Other	3	7
Don't Know	<1	1

Perhaps unsurprisingly, the highest single response category in open-ended responses about the challenges in identifying aspartame is related to product labelling. This was mentioned by nearly half of those who responded to this question.

Verbatim quotes about the challenges relating to identifying aspartame from labels on foods and drinks:

- "Writing is often too small on supermarket products. Ingredients section often very full of text so hard to spot aspartame especially if you are rushing."
- "Sometimes I find it tricky to identify aspartame in products due to weird E numbers that I have no idea about. Clear labelling of aspartame needs to be on all consumable products."
- "If eating out often, the restaurant staff are reluctant to check labels or are unsure about ingredients. Catering size products are not easy for staff to find info. Details of ingredients might only be listed on the outer packaging which may have been discarded."

Respondents also mentioned that there was no prominent warning about the presence of aspartame or that this information was not consistently in the same place on packaging.

- "You are checking the label for aspartame, but the warning is not always in the same place".
- "The warning text is very small. It inhibits my son's independence as it's unrealistic to expect a child to check for labelling that is so hard to see. After the sugar tax, packaging changed and removed the easy visual clues that you could rely on to indicate that the product had aspartame. As an example, there is now a Coca Cola in a red can which has aspartame in it. There are frequently types with aspartame in and some without with virtually the same packaging, you have to check everything, and this is stressful". Many people commented about the time it takes to check labels.
- "We are really careful when we check labels, but it takes time, and it is difficult sometimes".

- "I can easily identify aspartame in products with labels, but it is time consuming and annoying. I worry that other caregivers, e.g., grandparents, would not be able to. There is no labelling in restaurants, so we err on the side of caution and only order what we know does not contain aspartame."

Some respondents suggested that the warning on packaging should be at least as prominent as allergen warnings.

- "Should be written in bold/special box like allergens."
- "Aspartame should be highlighted in a different colour or bold writing as they do for peanut allergies."

Overall, 74% (n = 152/206) of respondents thought that it would be helpful (fairly or extremely) if manufacturers listed the phenylalanine content of food, drink or medicines on the label. Only 6% (n = 13/206) were neutral and 20% (n = 41/206) thought it would be fairly or extremely unhelpful.

3.9. Effect of Aspartame on People with PKU and Parents/Carers

Table 6 gives the percentage of patients that reported each of the stated effects of aspartame on patients and parents/caregivers managing PKU.

Table 6. Reported effects of aspartame on the person with PKU and parent/carer.

Effects of Aspartame	Percentage Responses (%)	Number of Respondents Per Total Sample (n = 206)
Limits Suitable Drinks in Restaurants/Pubs/Cafes	86	178
Increases Time taken to do Food Shopping	80	164
Causes Anxiety for Person with PKU	52	106
Causes Anxiety for Parent/Carer	46	95
Causes Guilt for Parent/Carer	42	87
Causes Social Isolation	42	87
Person with PKU unable to buy Food or Drinks from Shops, Causing Loss of Independence	40	83
Have to Keep Food Products Separate in the House between PKU and Non-PKU Products	36	75
Causes Person with PKU to Feel Unwell	33	68
Causes Guilt for Person with PKU	33	67
Has no Effect	5	11
Other	4	8

Abbreviations: PKU: Phenylketonuria.

Coding of the open-ended responses about the effect of aspartame showed that the top four themes were: feelings of being different, lack of choice, stress or concern and the additional time required to check all labels. These issues are illustrated in the following verbatim quotes.

- "It's very isolating for our son. He feels people see him as fussy until we have to explain and even then, they don't seem to understand."
- "Feel bad when I can't find suitable drinks for my children with PKU, whereas my children without PKU can drink whatever they want."
- "The PKU diet is heavily restricted and time consuming. Aspartame adds another level of restriction and extra time is necessary to check everything before you can buy or eat it."

- "Checking for aspartame increases stress and anxiety especially when eating out which is supposed to be a nice/happy experience."

4. Discussion

This is the first UK survey to examine the impact of aspartame in food, drinks and medications on people with PKU and their caregivers. We found that repeated accidental aspartame consumption is common, particularly in adults with PKU. Many respondents acknowledged there may be occasions in which aspartame has been inadvertently ingested and there were many concerns about the inability to identify its presence in pre-mixed alcoholic drinks and draft soft drinks in restaurants and bars.

The most unintentionally consumed aspartame containing items included fizzy drinks, fruit squash, cordials, flavoured water, sports drinks and chewing gums. Changes to product recipes, selecting the wrong product when shopping, packaging similarities between aspartame and non-aspartame containing products, unclear labelling, and difficulties identifying aspartame in drinks purchased from restaurants and pubs were commonly identified challenges. This suggests the need for: mandatory ingredient lists for all drinks and foods in restaurants, cafes, bars, and vending machines; distinct front of package labelling when a product recipe has changed; and clear labelling when there are several products within a brand range with some containing aspartame and others not (e.g., Ribena, Fanta, Tango, Robinsons). There should also be mandatory visible "first glance" disclosure of aspartame on packaging. Recently Dutch researchers demonstrated there was wide variability in the aspartame content of soft drinks, particularly the same brand of soft drinks bought in different countries. They have urged European legislators to enforce manufacturers to declare the amount of phenylalanine obtained from aspartame on food and drink labels, so that individuals with PKU are aware of the phenylalanine content of foods and drinks [24]. This 'call for action' is supported by NSPKU Medical Advisory Panel of dietitians [25].

Accidental aspartame consumption due to medications occurred in almost a quarter of respondents. Respondents felt there was little awareness or concern about the presence of aspartame in medications amongst medical professionals when they prescribed medication for PKU. Generally, reminders to check prescriptions for aspartame came from patients/parents' instruction rather than the GP or pharmacist. Aspartame is commonly used as a sugar replacement in antibiotics, chewable tablets and sugar-free liquids. The European PKU guidelines [11,17] recommend that for immediate and short-term treatment of infections, if only aspartame containing medicines are available, it may be better to use these until aspartame-free medication is sourced rather than leave a person with PKU without treatment (for a concurrent illness) as blood phenylalanine levels will rise with infection. However, for chronic long-term use of medications, it is better to find alternative aspartame free medications. Aspartame can be identified from the list of excipients in the medication instruction leaflet or the EMC summary of product characteristics. The amount of estimated phenylalanine in a drug may also be listed and can vary from 1 to 25 mg per dose of medication. There is usually no aspartame warning on the outside packaging of medication and there is no legal obligation to include this [26]. However, it is considered important to have mandatory legislation to identify aspartame on the outer packaging for people with PKU, otherwise it is challenging to recognize its presence at the point of prescription or purchase, and it can be a cause of frustration, inconvenience and distress for carers or people with PKU.

The impact of aspartame in food and drinks on inhibiting socialisation, increasing the incumbrance of dietary management and decreasing autonomy for children and teenagers is evident. Respondents were particularly dissatisfied with the choice of suitable drinks at many venues including fast-food restaurants, leisure centres, tourist venues and even hospital clinics. Respondents were angry that waiters/waitresses or sales vendors convey little understanding or empathy. They were displeased with the lack of aspartame free soft drinks at their hospital, as they considered this to be one location that above all others should demonstrate understanding of their condition. For NHS England hospital trusts, the Commissioning for Quality and Innovation (CQUIN) offer a financial incentive if they provide healthier food and drinks. This includes that 80% of drinks provided/sold must not be sugary. If a hospital trust adheres to the CQUIN for healthy food for NHS staff, visitors, and patients, they receive additional funding worth 0.1% of the trust's overall budget [27]. Unfortunately, there are no exceptions for vulnerable groups who are unable to tolerate aspartame for medical reasons.

There was much anger and despondency concerning the sugar tax by the respondents to this survey. Although the sugar tax has been implemented to reduce national overweight/obesity, it will not necessarily change unhealthy lifestyle practices. Overall people with PKU and their caregivers felt marginalised by this government policy. The sugar tax has led to diminished choice of favourite branded drinks and increased the cost of sugar containing drinks. For many adults, most available soft drinks in bars now contain aspartame, so the freedom of choice and the ability to enjoy a drink with friends has been withdrawn, which is hard to endure when there are so many other dietary restrictions to contend with. Many people with PKU have a functional approach to food; they eat for necessity rather than pleasure. However, drinking 'normal' branded drinks brought normality and choice. Almost 60% of respondents considered that the sugar tax led to more dietary errors and 33% felt fatigued or unwell with aspartame consumption, although no other information was collected about symptoms. Sugar is one of the few foods that is protein free and can be eaten without adversely affecting blood phenylalanine control in PKU. Giving adequate energy intake from very low protein sources is essential to meet energy requirements and to minimise catabolism that can lead to poor blood phenylalanine control [11], so sugar is not an 'unhealthy' food for people with PKU when eaten in moderation. Although it is unlikely there will be any reversal of the sugar tax, and it is expected to be extended to other foods, it is disappointing there is little consideration about the impact of the sugar tax on PKU by Public Health England or HMRC. Promoting healthy eating and exercise habits in the general population should be the key to solving obesity rather than focusing on one food component. Taking a balanced approach, offering many healthy choices without compromising the aspartame-free options for people with PKU would be a better policy.

Confusion and regular recipe changes with the addition of aspartame to manufactured foods/drinks affect a child's ability to self-manage their diet. For foods such as fresh meat, fruit and vegetables there is clear guidance on whether these are either permitted or forbidden in a low phenylalanine diet; but the ingredients, particularly in popular manufactured sweetened products, may change without notice, adding aspartame, with no clear warning to the consumer. It is, therefore, difficult to give pragmatic advice about suitable foods and drinks. Aspartame may be added to many children's foods such as ice lollies, soft drinks and iced 'slush' drinks that may be purchased from an ice cream van or local shop. Consequently, an adult with dietary knowledge should always check the suitability of these foods and the continual checking of food labels is time consuming and endless.

It is incomprehensible that alcoholic beverages with added sweeteners with an alcohol by volume content of 1.2% or more, do not have to declare the type of sweetener on the label. Moreover, legally no nutrition information needs to be supplied on the label of alcohol although appropriate allergen information and relevant quantitative ingredient information should be given [28]. This renders it unmanageable for people with PKU to be confident that any alcoholic drinks with unnamed sweeteners are safe for consumption. Fortunately, it is likely this situation will improve in the next 2 years. A memorandum of understanding (Self-regulatory proposal from the European alcoholic beverages sectors on the provision of nutrition information and ingredients listing) was presented as a joint voluntary commitment to the EU Health Commissioner in June 2019. It committed that by the end of 2022 the list of ingredients on alcohol will be provided according to the

EU 1169/2011 law. This law asserts that aspartame should be identified on the list of ingredients and it must state that it contains a source of phenylalanine [29].

There are several limitations to this study. The participants were not randomly selected and individuals without internet access may have been unable to participate. The survey was also promoted on the NSPKU Twitter and Facebook page, meaning participants were more likely to be NSPKU members who may be more proactive and informed about PKU. Therefore, the survey population may not be representative of the entire PKU population for which it is estimated that there are around 2000 UK patients in hospital follow up. Some surveys were completed by caregivers on behalf of patients with PKU and therefore responses to some questions may have been the caregiver's opinion rather than the actual experiences of those with PKU. It may be that aspartame was consumed more often but respondents did not realise this. Some respondents were unable to remember how many times they had consumed aspartame over the three-year period. The survey was not validated and therefore has not been checked for reliability, however expert opinion was used to develop it. This study should be repeated and expanded in the future using a validated survey that is piloted and carefully applied by health professionals to further improve the accuracy of the data collected.

5. Conclusions

It is important that health care professionals and policy makers understand the impact of aspartame and policies affecting the increased use of aspartame such as the sugar tax on the lives of people with PKU. Aspartame addition to food and drinks introduces social constraints, impacts on metabolic control as well as providing a source of frustration, guilt and distress to people with PKU and their carers. It is difficult to adhere to the PKU diet when all ingredients are not readily declared on labels at the point of purchase or issue. This applies to food, drinks, and medicines. It is essential that that industry gives clear and 'front of package' labelling about aspartame presence and the amount of phenylalanine that the product contains. Manufacturers should also consider using alternative sweeteners that would be a suitable option for people with PKU.

Author Contributions: Conceptualisation, A.M., E.N., A.P. and S.E.; methodology, A.M., E.N., A.P. and S.E.; formal analysis, M.O., E.N., A.M. and A.P.; writing—original draft preparation, A.M., E.N., A.P. and A.D.; writing—review and editing, A.M., E.N., A.P., S.E., S.F., M.O., C.A., A.D., and supervision, A.M., A.P. and S.E. All authors have read and agreed to the published version of the manuscript.

Funding: This research received no external funding.

Institutional Review Board Statement: The study was conducted according to the guidelines of the Declaration of Helsinki, and approved by the Birmingham City University ethics committee with approval number 6085 (project title "The accidental consumption of aspartame in PKU: The experiences of patients and their caregivers").

Informed Consent Statement: Informed consent was given by all subjects when filling in the questionnaire.

Data Availability Statement: The data will be made available from the authors upon reasonable request.

Acknowledgments: We would like to acknowledge and thank all the patients and families that have taken their time to fill in this survey.

Conflicts of Interest: The authors declare no conflict of interest.

References

1. Fry, J. The world market for intense sweeteners. *World Rev. Nutr. Diet* **1999**, *85*, 201–211. [CrossRef] [PubMed]
2. Mazur, R.H. *Aspartic Acid-Based Sweeteners*; AVI Publishing: Westport, CT, USA, 1974.
3. Mazur, R.H. *Discovery of Aspartame*; Marcel Dekker: New York, NY, USA, 1984.

4. Mourad, I.M.; Noor, N.A. Aspartame (a widely used artificial sweetener) and oxidative stress in the rat cerebral cortex. *Int. J. Pharm. Biomed. Sci.* **2011**, *2*, 4–10.
5. Yang, Q. Gain weight by "going diet?" Artificial sweeteners and the neurobiology of sugar cravings: Neuroscience 2010. *Yale J. Biol. Med.* **2010**, *83*, 101–108. [PubMed]
6. Grembecka, M.; Szefer, P. Differentiation of Confectionery Products Based on Mineral Composition. *Food Anal. Methods* **2012**, *5*, 250–259. [CrossRef]
7. Mallikarjun, S.; Sieburth, R.M. Aspartame and Risk of Cancer: A Meta-analytic Review. *Arch. Environ. Occup. Health* **2015**, *70*, 133–141. [CrossRef] [PubMed]
8. JECFA (Joint FAO/WHO Expert Committee on Food Additives). *Toxicological Evaluation of Certain Food Additives: Aspartame*; JECFA: Geneva, Switzerland, 1981.
9. FDA (US Food and Drug Administration). *Food Additives Permitted for Direct Addition to Food for Human Consumption: Aspartame*; Register, F., Ed.; FDA: Silver Spring, MD, USA, 1996; pp. 33654–33656.
10. EFSA Panel on Food Additives Nutrient Sources added to Food. Scientific Opinion on the re-evaluation of aspartame (E 951) as a food additive. *EFSA J.* **2013**, *11*, 3496. [CrossRef]
11. MacDonald, A.; van Wegberg, A.M.J.; Ahring, K.; Beblo, S.; Bélanger-Quintana, A.; Burlina, A.; Campistol, J.; Coşkun, T.; Feillet, F.; Giżewska, M.; et al. PKU dietary handbook to accompany PKU guidelines. *Orphanet J. Rare Dis.* **2020**, *15*, 171. [CrossRef] [PubMed]
12. Da Silva, L.C.; Carvalho, T.S.; da Silva, F.B.; Pires, R.F.; Giugliani, R.; Pereira, M.L. Aspartame loading test in PKU heterozygous individuals bearing severe and moderate mutations. *Clin. Genet.* **2000**, *58*, 86–88. [CrossRef] [PubMed]
13. Stegink, L.D.; Filer, L.J., Jr.; Baker, G.L.; McDonnell, J.E. Effect of an abuse dose of aspartame upon plasma and erythrocyte levels of amino acids in phenylketonuric heterozygous and normal adults. *J. Nutr.* **1980**, *110*, 2216–2224. [CrossRef]
14. Stegink, L.D.; Koch, R.; Blaskovics, M.E.; Filer, L.J., Jr.; Baker, G.L.; McDonnell, J.E. Plasma phenylalanine levels in phenylketonuric heterozygous and normal adults administered aspartame at 34 mg/kg body weight. *Toxicology* **1981**, *20*, 81–90. [CrossRef]
15. Hillert, A.; Anikster, Y.; Belanger-Quintana, A.; Burlina, A.; Burton, B.K.; Carducci, C.; Chiesa, A.E.; Christodoulou, J.; Đorđević, M.; Desviat, L.R.; et al. The Genetic Landscape and Epidemiology of Phenylketonuria. *Am. J. Hum. Genet.* **2020**, *107*, 234–250. [CrossRef]
16. Scala, I.; Concolino, D.; Della Casa, R.; Nastasi, A.; Ungaro, C.; Paladino, S.; Capaldo, B.; Ruoppolo, M.; Daniele, A.; Bonapace, G.; et al. Long-term follow-up of patients with phenylketonuria treated with tetrahydrobiopterin: A seven years experience. *Orphanet J. Rare Dis.* **2015**, *10*, 14. [CrossRef]
17. Van Wegberg, A.M.J.; MacDonald, A.; Ahring, K.; Bélanger-Quintana, A.; Blau, N.; Bosch, A.M.; Burlina, A.; Campistol, J.; Feillet, F.; Giżewska, M.; et al. The complete European guidelines on phenylketonuria: Diagnosis and treatment. *Orphanet J. Rare Dis.* **2017**, *12*, 162. [CrossRef]
18. EMA—Palynziq Product Information. Available online: https://www.ema.europa.eu/en/documents/product-information/palynziq-epar-product-information_en.pdf (accessed on 16 February 2021).
19. Ford, S.; O'Driscoll, M.; MacDonald, A. Living with Phenylketonuria: Lessons from the PKU community. *Mol. Genet. Metab. Rep.* **2018**, *17*, 57–63. [CrossRef]
20. European Parliament. Regulation (EU) No 1169/2011 of the European Parliament and of the Council of 25 October 2011 on the provision of food information to consumers, amending Regulations (EC) No 1924/2006 and (EC) No 1925/2006 of the European Parliament and of the Council, and repealing Commission Directive 87/250/EEC, Council Directive 90/496/EEC, Commission Directive 1999/10/EC, Directive 2000/13/EC of the European Parliament and of the Council, Commission Directives 2002/67/EC and 2008/5/EC and Commission Regulation (EC) No 608/2004 Text with EEA relevance. Official Journal of the European Union. Available online: https://eur-lex.europa.eu/LexUriServ/LexUriServ.do?uri=OJ:L:2011:304:0018:0063:en:PDF (accessed on 15 January 2021).
21. *The Soft Drinks Industry Levy (Enforcement) Regulations 2018*; UK Statutory Instruments No 264; Legislation UK: London, UK, 2018.
22. Rimmer, A. Don't scrap the sugar tax, doctors tell Johnson. *BMJ* **2019**, *367*, l7051. [CrossRef] [PubMed]
23. HMRC. Available online: https://www.gov.uk/government/publications/soft-drinks-industry-levy--2/soft-drinks-industry-levy-initial-equality-impact-assessment (accessed on 28 December 2020).
24. Van Vliet, K.; Melis, E.S.; de Blaauw, P.; van Dam, E.; Maatman, R.; Abeln, D.; van Spronsen, F.J.; Heiner-Fokkema, M.R. Aspartame and Phe-Containing Degradation Products in Soft Drinks across Europe. *Nutrients* **2020**, *12*, 1887. [CrossRef]
25. NSPKU Statement on Aspartame. Available online: https://www.nspku.org/download/statement-on-aspartame/ (accessed on 15 January 2021).
26. CMPH (Committee for Medicinal Products for Human Use). *Information for the Package Leaflet Regarding Aspartame and Phenylalanine Used as Excipients in Medicinal Products for Human Use*; EMA/CHMP/134648/2015 corr. 1; European Medicine's Agency: Amsterdam, The Netherlands, 2017.
27. *NHS Staff Health & Wellbeing: CQUIN 2017-19 Indicator 1 Implementation Support*; NHS: London, UK, 2021.
28. Food Labelling Food Standards Agency. Available online: https://labellingtraining.food.gov.uk/module10/overview_1.html (accessed on 15 January 2021).
29. European Commission Labelling of Alcoholic Beverages in the EU: Some Facts. Available online: https://ec.europa.eu/food/safety/labelling_nutrition/labelling_legislation/alcohol_en (accessed on 12 January 2021).

nutrients

MDPI

Review

Protein Substitute Requirements of Patients with Phenylketonuria on BH4 Treatment: A Systematic Review and Meta-Analysis

Fatma Ilgaz [1], Cyril Marsaux [2,*], Alex Pinto [3], Rani Singh [4], Carmen Rohde [5], Erdem Karabulut [6], Hülya Gökmen-Özel [1], Mirjam Kuhn [2] and Anita MacDonald [3]

1 Department of Nutrition and Dietetics, Faculty of Health Sciences, Hacettepe University, 06100 Ankara, Turkey; fatma.celik@hacettepe.edu.tr (F.I.); hgokmen@hacettepe.edu.tr (H.G.-Ö.)
2 Danone Nutricia Research, 3584CT Utrecht, The Netherlands; mirjam.kuhn@danone.com
3 Department of Dietetics, Birmingham Women's and Children's Hospital, Birmingham B4 6NH, UK; alex.pinto@nhs.net (A.P.); Anita.Macdonald@nhs.net (A.M.)
4 Metabolic Genetics Nutrition Program, Department of Human Genetics, Emory University, Atlanta, GA 30322, USA; rsingh@emory.edu
5 Department of Paediatrics of the University Clinics Leipzig, University of Leipzig, 04103 Leipzig, Germany; Carmen.Rohde@medizin.uni-leipzig.de
6 Department of Biostatistics, Faculty of Medicine, Hacettepe University, 06100 Ankara, Turkey; ekarabul@hacettepe.edu.tr
* Correspondence: cyril.marsaux@danone.com

Citation: Ilgaz, F.; Marsaux, C.; Pinto, A.; Singh, R.; Rohde, C.; Karabulut, E.; Gökmen-Özel, H.; Kuhn, M.; MacDonald, A. Protein Substitute Requirements of Patients with Phenylketonuria on BH4 Treatment: A Systematic Review and Meta-Analysis. *Nutrients* **2021**, *13*, 1040. https://doi.org/10.3390/nu13031040

Academic Editor: Stéphane Walrand

Received: 19 February 2021
Accepted: 19 March 2021
Published: 23 March 2021

Publisher's Note: MDPI stays neutral with regard to jurisdictional claims in published maps and institutional affiliations.

Copyright: © 2021 by the authors. Licensee MDPI, Basel, Switzerland. This article is an open access article distributed under the terms and conditions of the Creative Commons Attribution (CC BY) license (https://creativecommons.org/licenses/by/4.0/).

Abstract: The traditional treatment for phenylketonuria (PKU) is a phenylalanine (Phe)-restricted diet, supplemented with a Phe-free/low-Phe protein substitute. Pharmaceutical treatment with synthetic tetrahydrobiopterin (BH4), an enzyme cofactor, allows a patient subgroup to relax their diet. However, dietary protocols guiding the adjustments of protein equivalent intake from protein substitute with BH4 treatment are lacking. We systematically reviewed protein substitute usage with long-term BH4 therapy. Electronic databases were searched for articles published between January 2000 and March 2020. Eighteen studies (306 PKU patients) were eligible. Meta-analyses demonstrated a significant increase in Phe and natural protein intakes and a significant decrease in protein equivalent intake from protein substitute with cofactor therapy. Protein substitute could be discontinued in 51% of responsive patients, but was still required in 49%, despite improvement in Phe tolerance. Normal growth was maintained, but micronutrient deficiency was observed with BH4 treatment. A systematic protocol to increase natural protein intake while reducing protein substitute dose should be followed to ensure protein and micronutrient requirements are met and sustained. We propose recommendations to guide healthcare professionals when adjusting dietary prescriptions of PKU patients on BH4. Studies investigating new therapeutic options in PKU should systematically collect data on protein substitute and natural protein intakes, as well as other nutritional factors.

Keywords: phenylalanine hydroxylase deficiency; hyperphenylalaninemia; PKU; protein substitute; medical formula; amino acid mixture; tetrahydrobiopterin; sapropterin; BH4

1. Introduction

Phenylketonuria (PKU) is an inborn error of phenylalanine (Phe) metabolism caused by deficiency of the Phe hydroxylase enzyme (PAH; EC 1.14.16.1), which catalyzes the conversion of Phe to tyrosine, with the help of the cofactor tetrahydrobiopterin (BH4) [1]. PKU is a rare disorder affecting approximately 1 in 24,000 newborns globally [2], although incidence varies greatly across ethnicities and geographic regions. Infants are usually diagnosed via newborn screening in the first 2 weeks of life and commence treatment if blood Phe levels exceed 360 μmol/L [3]. Untreated, PKU may cause severe neurological impairment with profound intellectual disability [1,3,4].

The traditional treatment for PKU is a Phe-restricted diet, which aims to avoid excessive accumulation of Phe to prevent adverse neurocognitive and psychological outcomes, while also meeting requirements for growth and development [3,5,6]. Phe tolerance, the maximum amount that can be eaten whilst maintaining blood Phe levels in the therapeutic range, varies between patients; it is influenced by the residual PAH activity and therefore the severity of PKU [3], and up to 80% of patients tolerate less than 10 g/day natural protein [7]. Therefore, a low-Phe diet requires supplementation with a Phe-free or low-Phe protein substitute, i.e., a protein replacement formula, based on either free L-amino acids (AA), or casein glycomacropeptide (cGMP) supplemented with free AA. Most protein substitutes contain additional tyrosine, micronutrients, essential fatty acids, and long-chain polyunsaturated fatty acids [6]. Protein substitutes are not only necessary to meet age-appropriate protein requirements for growth and to provide tyrosine [3,6], they also improve Phe tolerance and optimize metabolic control by suppressing blood Phe levels [6,8–10]. This is particularly important during illness and trauma, where protein substitutes have a protective role by counter-acting protein catabolism [6].

Although successful, dietary treatment of PKU constitutes a substantial burden for patients and their families. The difficulties to adhere life-long to this restrictive diet, as well as to maintain blood Phe levels within the recommended range, have called for new therapies to improve patients' quality of life [11]. Over the last 12 years, pharmaceutical adjunct therapies have been licensed including treatment with sapropterin dihydrochloride (a synthetic form of BH4) [12] and enzyme substitution therapy with pegvaliase (pegylated recombinant Phe ammonia lyase, PEG-PAL) [13]. Sapropterin therapy is prescribed to BH4-responsive patients with PKU; pegvaliase is only licensed for adults ($\geq$16 y in Europe) with blood Phe levels $\geq$600 µmol/L. Both pharmaceutical treatments may be used as monotherapies or in combination with Phe restriction. Kure et al. were among the first to report that oral administration of BH4 to some individuals with mild hyperphenylalaninemia led to a significant reduction in blood Phe levels [14]. Since then, it has been suggested that 20–50% of patients with PKU respond to sapropterin [15–19]. The basis of responsiveness may be associated with different molecular mechanisms. Increased liver BH4 concentrations may stimulate the activity of a partially active mutant PAH enzyme [20], as some mutations can decrease the enzyme affinity for its cofactor [21,22], or may act as a chemical chaperone to stabilize mutant PAH [22,23]. Potential responsiveness to BH4 may be predicted from a patient's *PAH* genotype and/or BH4 loading tests [3,24–26]. It varies according to metabolic phenotype—milder forms of PKU are more likely to respond, whereas patients with classic PKU are less likely to do so [2].

In responders, the BH4-induced decrease in blood Phe concentrations usually enables an increase in Phe/natural protein tolerance and, thereby, some relaxation of the Phe-restricted diet with lowering or cessation of protein substitute use. However, Phe tolerance is also affected by other factors including severity of PKU, patient's age, dosage of protein substitute, growth rate, and target blood Phe concentrations [3,27]. Additionally, it has been shown that some adolescents and young adults with PKU are able to tolerate more natural protein than prescribed when challenged [28]. This supports a periodic re-evaluation of Phe tolerance in all patients including responders to BH4 therapy.

The ultimate goals of BH4 treatment are to (1) allow dietary Phe relaxation and (2) obtain good metabolic control. If either objective is not achieved and sustained long term, continuation of BH4 treatment should be reconsidered. Protein substitutes are a major supplier of nutrients, not only of protein, but also of vitamins and minerals, leading to concerns about the impact on nutritional status of patients taking BH4 when they are stopped [29,30]. This highlights the importance of a systematic and gradual approach when considering reduction of protein substitute, while maximizing natural protein intake in patients on BH4 treatment, in order to avoid impairment of metabolic control and maintain nutritional status. To date, few dietary protocols are available to guide such adjustments [31].

Therefore, the present systematic review aimed to investigate the usage of protein substitute with BH4 therapy and to define criteria for continued protein substitute administration with BH4.

2. Methods

2.1. Terminology

In this manuscript, "BH4" refers to both the earlier synthetic BH4 formulation (6R-BH4; Schircks Laboratories) (mainly used in studies before 2009) and the later formulation sapropterin dihydrochloride (Kuvan®; Merck Serono or BioMarin Pharmaceutical Inc.). "Protein substitute" refers to the Phe-free/low-Phe protein replacement formula. Other names for protein substitute include synthetic protein, amino acid mixture (AAM), AA supplement, casein glycomacropeptide (cGMP or GMP-AA), and (special) medical food/formula. In contrast, we use "natural protein" as a synonym for intact protein.

2.2. Literature Search

Using the ProQuest platform, we performed a systematic literature search in a total of 92 electronic databases (including Medline, Embase, SciSearch and BIOSIS Previews) for any articles published in English between 1 January 2000 and 2 March 2020. The full list of electronic databases searched can be found here: https://dialog.com/commercial-databases/, accessed on 2 March 2020. The year 2000 was chosen as the starting date because responsiveness to BH4 was first reported by Kure and colleagues in 1999 [14] and BH4 was not used in PKU management until later. The following search string was used: Ti,ab(("phenyl ketonuri*" OR phenylketonuri* OR PKU OR "phenylalanine deficiency" OR "phenylalanine hydroxylase deficiency" OR "PAH deficiency" OR hyperphenylalaninemia OR hyperphenylalaninaemia OR HPA) AND (biopterin OR BH4 OR thb OR tetrahydrobiopterin OR sapropterin OR kuvan* OR biopten*)).

The Embase database includes many conference abstracts. At the time of the search, Embase covered the International Congress of Inborn Errors of Metabolism (ICIEM) as well as the annual symposia of the Society for the Study of Inborn Errors of Metabolism (SSIEM) and the annual meetings of the Society for Inherited Metabolic Disorders (SIMD) from 2009 until 2018. Therefore, in addition to the database search, electronic copies of the abstract books were retrieved and screened manually for SIMD 2019 and SSIEM 2019.

The Preferred Reporting Items for Systematic Reviews and Meta-Analyses (PRISMA) guidelines [32] were followed and the protocol published on the PROSPERO international prospective register of systematic reviews (CRD42020177311; www.crd.york.ac.uk/PROSPERO, accessed on 30 September 2020).

2.3. Study Selection

The PICO (population, intervention, comparison, outcomes) method was applied to formulate the review question, as well as to determine the eligibility criteria. All retrospective and prospective longitudinal studies, randomized controlled trials, and cross-sectional and case–control studies conducted in patients with hyperphenylalaninemia or PKU were included. Conference abstracts were also considered. To be eligible, studies had to include a minimum of 5 long-term BH4-responsive patients (no age restriction and no restriction regarding methodology for assessment of BH4 responsiveness), first treated with a Phe-restricted diet and protein substitute (PS), and subsequently being treated with BH4 for a minimum of 3 consecutive months from first dose received. Preclinical studies (in vitro and in vivo studies conducted on cell cultures or animals), case reports (with <5 BH4 responders), theses, non-original research (such as expert opinions and reviews), and studies without any information on protein substitute use were excluded. Patients with a diagnosis of BH4 deficiency or maternal PKU, or who were treated with pegvaliase were excluded. Patients who had never been prescribed protein substitute, who were treated with BH4 for <3 months, who had been found to be long-term non-responders, or who were not adherent with their treatment were removed from the analyses when known.

Two reviewers (F.I. and C.M.) screened titles and abstracts independently according to eligibility criteria. The full texts of all potentially relevant articles were reviewed. Conference abstracts without full text were kept if they (or the associated poster when available) contained sufficient information on the primary outcomes. Disagreements were resolved through discussion with all authors.

2.4. Outcome Measures

The primary outcomes were prescribed or self-reported intakes of protein substitute, natural protein, total protein, and dietary Phe. Secondary outcomes were nutritional status (i.e., micronutrient and fatty acid blood concentrations or dietary intakes), growth, metabolic control (e.g., blood Phe levels), authors' definition of protein adequacy (e.g., Food and Agriculture Organization of the United Nations/World Health Organization/United Nations University (FAO/WHO/UNU) safe levels of protein intake or national recommendations/reference amounts for protein intake), and authors' protocols for change in protein substitute prescription with BH4.

2.5. Data Extraction

Data were collected by 2 independent authors (F.I. and C.M.) using a standardized data extraction form and were checked by a third author (A.P.). Information extracted was (1) study characteristics (authors, publication year, country, and design of the study), (2) description of population (sample size and number of BH4 responders, methodology for assessment of BH4 responsiveness, gender, age, type of HPA/PKU, and ethnicity), (3) description of BH4 treatment (time of initiation, dose, drug type, duration, and adherence), (4) primary outcomes (intakes, before BH4 treatment and at follow-up, of protein substitute, natural and total protein, special low-protein foods, any additional supplements, and Phe tolerance), and (5) secondary outcomes (authors' protocols for natural protein and protein substitute prescriptions, nutritional status, growth, and blood Phe control). Authors of papers where relevant information was missing or ambiguous were contacted to obtain further information/clarification.

2.6. Quality Appraisal and Risk of Bias Assessment

Two reviewers (F.I. and C.M.) independently assessed the quality of the evidence and the risk of bias of the included studies using the "Quality Assessment Tool for Before-After (Pre-Post) Studies with No Control Group" [33]. This tool was developed jointly by the U.S. National Heart, Lung and Blood Institute (NHLBI, National Institutes of Health) and Research Triangle Institute (RTI) International. It includes 12 items to evaluate potential flaws in study methods or implementation, including sources of bias (e.g., patient selection, performance, attrition, and detection), confounding, study power, the strength of causality in the association between interventions and outcomes, as well as other factors. Each item was rated as "yes", "no", "cannot determine", "not reported", or "not applicable". Based on the ratings, we made an overall judgement regarding the quality of each study: (1) "good quality" if the study had minimal risk of bias, (2) "fair quality" if the study was susceptible to some bias but not deemed sufficient to invalidate its results, and (3) "poor quality" if the study raised substantial concerns. Differing ratings between reviewers were discussed until consensus was reached.

2.7. Data Analysis

The analyses considered long-term BH4 responders (as defined by the authors, and, if individual data were available, by considering both the long-term increase in Phe intake and the long-term decrease in blood Phe levels), who had need for a protein substitute before BH4 therapy, and who had been treated with BH4 for at least 3 months to ensure any changes in outcomes were reliable.

For the main outcomes, meta-analyses were performed to compare means before and after start of BH4 therapy, if a minimum of 2 studies were available. Heterogeneity between studies was estimated using the I^2 statistic, with values of 25%, 50%, and 75% considered to indicate low, medium, and high heterogeneity, respectively. Given the heterogeneity level between studies, we used a random-effects model to calculate pooled estimates with the "metafor 2.4-0" package of R software version 4.0.3 (R foundation for statistical computer, Vienna, Austria) [34]. Because of the relatively small number of studies, we preferred to calculate the 95% confidence intervals using a t-distribution (with degrees of freedom = number of studies-1). As studies reported dietary Phe and protein intakes using different scales (e.g., mg/kg/day or mg/day for Phe intake), the standardized mean difference (SMD) was used to standardize the results to the same scale (SMD = mean change/standard deviation of change). Mean change was obtained by subtracting the mean at follow-up from the baseline mean. However, this method cannot be used to determine the standard deviation of changes because it is not known whether the changes were consistent or variable across individuals. Hence, the standard deviations for the changes were calculated by using 1 of the 2 following methods: (1) the original baseline and final follow-up measurements if individual data were available, or (2) statistical analyses comparing the changes (e.g., confidence intervals, t-values, or *p*-values) if they were presented in the original articles [35]. For the meta-analyses, a *p*-value less than 0.05 was considered statistically significant. Sensitivity analyses considering only studies reporting outcomes in the same unit were also performed. Furthermore, when heterogeneity was particularly high (e.g., $I^2 > 95\%$), additional sensitivity analyses considered the exclusion of studies that were suspected to contribute most to the heterogeneity. Studies that could not be included in the meta-analyses (i.e., insufficient data or results not reported as means and SDs but only as medians and interquartile range) were analyzed qualitatively. Secondary outcomes were analyzed qualitatively (no meta-analysis).

All data analyzed (both quantitatively and qualitatively) are discussed and used to derive recommendations for which data should be reported at a minimum when investigating responsiveness of PKU patients to BH4; the being aim to improve future data reporting.

3. Results

3.1. Study Selection

Of 2349 unique published articles and conference abstracts identified, 19 eligible articles [16,17,29,36–51] and 3 conference abstracts [52–54], describing a total of 18 studies, were included in the systematic review (Figure 1). Three articles [39,42,43] reported data for the same study first published by Singh et al. (2011) [48], and 1 conference abstract [52] reported additional data for the same study published by Ünal et al. (2015) [50]. All 4 were included in the systematic review. Out of a total of 18 studies, 15 were included in the meta-analyses (pre-/post-BH4 data were lacking for two studies [51,53], and only medians were provided in Aldámiz et al. [36]).

3.2. Study Characteristics

The characteristics of the 18 studies included are summarized in Table 1. These studies described a total of 306 PKU patients with long-term use of BH4. Most studies were longitudinal (retrospective or prospective) and conducted in Europe (Europe, *n* = 14; USA, *n* = 3; and Turkey, *n* = 1). Sample size varied from 6 to 51, after excluding some patients from the original dataset who did not meet our inclusion criteria (i.e., long-term responders treated with BH4 for ≥3 months and who had been on a Phe-restricted diet and protein substitute(s) before BH4). Different protocols were used to evaluate BH4 responsiveness (Table 1 and Table S1). BH4 loading tests were conducted from 8 to 48 h in most studies (but for 1 week to 4 months in 4 studies [16,29,48,49]), and the dose of BH4 prescribed ranged from 5 to 24 mg/kg/day. BH4 therapy was started at a mean age between 5 months and 18 years. Mean duration of follow-up ranged from 3 months to 5.7 years, with some patients on BH4 treatment for up to 8.8 years [17] (Table 1).

Table 1. Main characteristics of included studies.

Reference	Country	Study Design	No. of Patients Tested/No. of Long-Term Responders [a]		Gender of Long-Term Responders (M/F)	Duration of BH4 Loading Test	BH4 Dose (Mean or Range; mg/kg/day)	Age at Initiation of BH4 (Mean or Range; Years)	Duration of Follow-up (Mean or Range; Years)
Bélanger-Quintana 2005 [38]	Spain	Retrospective longitudinal single-center study	Total: mHPA: mPKU: mo/cPKU:	50/7 [b] 7/- 22/7 21/-	n/a	24 h	5–20 [†]	7.8 (range: 0.7–18)	0.9 (range: 0.4–1.5)
Lambruschini 2005 [45]	Spain	Prospective longitudinal single-center study	Total: mHPA: mPKU: moPKU: cPKU:	73/11 [c] - -/9 -/2 -	4/7	24 h [d]	5–10 [†]	5.0 (range: 0.2–12.2)	1.0
Burlina 2009 [40]	Italy	Retrospective longitudinal single-center study	Total:	30/12 [e]	n/a	24 h	10 [†]	5.5 (range: 2.0–16.0)	3.5 (range: 0.5–7.0)
Singh 2010 [49]	USA	Prospective longitudinal single-center study	Total:	10/6 [f]	6/0	1 week	20 [‡]	8.7 (range: 5–12)	2.0
Vilaseca 2010 [51]	Spain	Cross-sectional single-center study	Total: mHPA: mPKU: moPKU: cPKU:	61/10 [g] - 5/3 21/7 35/-	n/a	21 h	5–15 [†]	7.4 (range: 1.0–16.0)	5.7 (range: 5.3–6.0)
Singh 2011 [48] Douglas 2013a [42] Douglas 2013b [43] Brantley 2018 [39]	USA	Prospective longitudinal single-center study	Total:	57/17 [h]	10/7	4 months	20 [‡]	16.6 (range: 6.1–36.8)	1.0
Hennermann 2012 [17]	Germany	Prospective longitudinal single-center study	Total:	84/18 [i]	n/a	24 h ($n = 56$) 8 h ($n = 26$)	8–19 [§]	n/a	4.0 (range: 0.7–8.8)
Leuret 2012 [46]	France	Retrospective longitudinal multicenter study	Total: mHPA: mPKU: moPKU: cPKU:	-/8 [j] - -/8 - -	n/a	24 h	8–24 [§]	1.1 (range: 0.4–2.9)	1.9 [j] (range: 0.6–6.7)

Table 1. *Cont.*

Reference	Country	Study Design	No. of Patients Tested/No. of Long-Term Responders [a]	Gender of Long-Term Responders (M/F)	Duration of BH4 Loading Test	BH4 Dose (Mean or Range; mg/kg/day)	Age at Initiation of BH4 (Mean or Range; Years)	Duration of Follow-up (Mean or Range; Years)
Aldámiz-Echevarría 2013 [36]	Spain	Retrospective longitudinal multicenter study			*Cohort 1: Patients with 2 y follow-up* [k]			
			Total: -/36 mHPA: - mPKU: -/7 moPKU: -/24 cPKU: -/5	18/18	24 h (24 h or 1 week at one hospital after 2005)	5–20 [§]	5.0	2.0
					Cohort 2: Patients with 5 y follow-up [k]			
			Total: -/10 mHPA: - mPKU: -/1 moPKU: -/9 cPKU: -	6/4	24 h (24 h or 1 week at one hospital after 2005)	5–20 [§]	5.2	5.0
Demirdas 2013 [41]	The Netherlands	Prospective multicenter cohort study	Total: 45/8 [l]	n/a	48 h	n/a [‡]	n/a	range: 1.4–2.0
Aldámiz-Echevarría 2015 [37]	Spain	Retrospective longitudinal multicenter study	Total: -/22 mHPA: - mPKU: -/5 moPKU: -/14 cPKU: -/3	12/10	8 h or 12 h; (24 h or 1 week at one hospital after 2005)	5–20 [§]	1.4 (neonatal in $n = 4$)	1.0
Scala 2015 [47]	Italy	Prospective longitudinal multicenter study	Total: 43/17 [m] mHPA: -/3 mPKU: -/8 moPKU: -/4 cPKU: -/2	11/6	48 h	10 [§]	15.1 (range: 7.0–22.0)	5.7 (range: 1.0–7.0)

Table 1. *Cont.*

Reference	Country	Study Design	No. of Patients Tested/No. of Long-Term Responders [a]		Gender of Long-Term Responders (M/F)	Duration of BH4 Loading Test	BH4 Dose (Mean or Range; mg/kg/day)	Age at Initiation of BH4 (Mean or Range; Years)	Duration of Follow-up (Mean or Range; Years)
Thiele 2015 [29]	Germany	Retrospective longitudinal single-center study	Total: mHPA: mPKU: moPKU: cPKU:	-/8 -/3 -/3 -/1 -/1	5/3	6 weeks	10–19 ‡	8.8 (range: 5.0–15.0)	2.0
Ünal 2015 [50] Gökmen Özel 2014 [52]	Turkey	Cross-sectional single-center study	Total: mHPA: mPKU: moPKU: cPKU:	-/51 [n] -/18 -/23 -/6 -/3	27/24	48 h	20 ‡	5.4 (range: 0.5–14.0)	2.5 (range: 0.5–4.0)
Feldmann 2017 [16]	Germany	Prospective longitudinal single-center study	Total:	112/30 [o]	n/a	2 weeks	20 ‡	n/a	0.5
Rocha 2017 [54]	Portugal	Retrospective single-center cohort study	Total: mHPA: mPKU: moPKU: cPKU:	-/9 [p] - - -/8 -/1	3/6	48 h	n/a ‡	16.6 (range: 9.0–28.0)	1.0 (range: 0.3–1.4)
Evers 2018 [44]	The Netherlands	Retrospective multicenter cohort study	Total:	-/18 [q]	5/13	48 h	10–20 ‡	12.0 (range: 4.0–19.0)	5.0 (range: 4.5–5.5)

Table 1. *Cont.*

Reference	Country	Study Design	No. of Patients Tested/No. of Long-Term Responders [a]	Gender of Long-Term Responders (M/F)	Duration of BH4 Loading Test	BH4 Dose (Mean or Range; mg/kg/day)	Age at Initiation of BH4 (Mean or Range; Years)	Duration of Follow-up (Mean or Range; Years)
Paras 2018 [53]	USA	Retrospective longitudinal single-center study	Total: -/8 [r]	n/a	n/a	20 [‡]	5.8 (range: 0.4–18.0)	≥0.3

Abbreviations: (6R-)BH4: tetrahydrobiopterin; M/F: male/female; mHPA: mild hyperphenylalaninemia; cPKU: classic phenylketonuria; mPKU: mild phenylketonuria; moPKU: moderate phenylketonuria; No: number; Phe: phenylalanine; n/a: not available. [†] BH4 given as 6R-BH4 (Bélanger-Quintana 2005; Lambruschini 2005; Burlina 2009; Vilaseca 2010). [‡] BH4 given as sapropterin dihydrochloride (Singh 2010; Singh 2011; Demirdas 2013; Thiele 2015; Ünal 2015; Feldmann 2017; Rocha 2017; Evers 2018; Paras 2018). [§] BH4 given as 6R-BH4 before 2009 and as sapropterin dihydrochloride after 2009 (Hennermann 2012; Leuret 2012; Aldámiz-Echevarría 2013; Aldámiz-Echevarría 2015; Scala 2015). [a] Only long-term responders (follow-up ≥3 months) who were on a Phe-restricted diet and protein substitute before BH4 were included in the analyses. Long-term responsiveness as judged by the original authors. [b] Bélanger-Quintana 2005: Long-term BH4 treatment was initiated only in 7 responders with mild PKU who were able to liberalize their diet. [c] Lambruschini 2005: Only 11 out of 14 responders were included in the analyses: BH4 therapy was stopped in 3 patients (1 cPKU and 2 moPKU) who were not able to increase their Phe tolerance and continued to take medical formula. [d] Lambruschini 2005: BH4 loading test was performed after neonatal screening before starting the Phe-restricted diet in 7 patients. A combined 24 h-long Phe/BH4 loading test was used in the remaining 66 patients. [e] Burlina 2009: Long-term BH4 treatment was initiated only in 12 responders who had a baseline Phe level >450 µmol/L. [f] Singh 2010: From a total of 7 responders, 6 male patients were included in the analyses (the female patient dropped out of the study). Age reported here is for total sample of 10 patients. [g] Vilaseca 2010: Only 10 out of 13 patients were included in the analyses: 3 patients (2 mPKU and 1 moPKU) were excluded since the BH4 loading test was performed just after neonatal screening before starting the Phe-restricted diet and protein substitutes. Age reported here is for the 13 patients. [h] Singh 2011: Thirty-two patients who experienced at least a 15% decrease in plasma Phe at 1 month were described as "preliminary responders". Of these, 20 patients who could increase Phe tolerance by at least 300 mg/d, and decrease prescribed medical food needs by at least 25% with good metabolic control were defined as "definitive/true responders" (long-term responders). Nine patients were considered "provisional responders" (long-term non-responders: 6 males and 3 females aged between 4.6 to 17.8 years) and excluded from the analyses. Two long-term responders had dropped out by 4 months of follow-up and a third dropped out between 4 months and 1 year; hence, 18 and 17 long-term responders were included in the analyses for the 4 months and 1 year follow-ups, respectively. n = 17 is shown here as it was the number of responders at last follow-up. Age reported here is for the responders including a dropout and 1 patient never on protein substitute. [i] Hennermann 2012: Neonatal BH4 loading test was performed in 84 patients. Long-term responsiveness was described on the basis of the increase in Phe tolerance after 3 months of BH4 initiation. Only 18 out of 23 patients (11 males, 12 females) who met the criteria were included in the analyses. The other 5 patients were considered long-term non-responders. [j] Leuret 2012: From a total of 15 responders (7 males, 8 females), only 8 were treated by conventional diet therapy (i.e., Phe-restricted diet supplemented with protein substitutes) before initiation of BH4 and were included in the analyses. The other 7 patients who started BH4 therapy during the neonatal period were excluded. However, duration of BH4 treatment was only available for the total sample of 15 patients. [k] Aldámiz-Echevarría 2013: Unclear if patients with a 5 y follow-up were also described in the group of patients with a 2 y follow-up. It was assumed that the 2 cohorts comprised different patients. [l] Demirdas 2013: Only 8 out of 10 responders (mean age 13.8 years) with complete data on dietary intakes were included in the analyses. [m] Scala 2015: From a total of 19 responders, 17 were included: 2 mPKU patients who did not agree to participate in the long-term treatment were excluded from the analyses. One of the 17 patients turned out to be a pseudo-responder and discontinued therapy at 12 months; however, it was not possible to exclude this patient from the analyses. [n] Ünal 2015: Type of PKU was unknown in 1 patient. Only 51 out of 75 responders were included: 21 patients who were not treated with protein substitute before BH4 were excluded, as well as 3 patients for whom BH4 treatment was stopped due to unsatisfactory metabolic control with little improvement in Phe tolerance (long-term non-responders). [o] Feldmann 2017: Out of 46 responders, 30 were included in the analyses: 35 patients completed the study but 5 patients who were not able to increase Phe tolerance after BH4 were excluded (long-term non-responders). [p] Rocha 2017: From a total of 13 responders, 9 were included: 4 patients either not taking any protein substitute before BH4 (n = 1 due to non-compliance, n = 1 not required), or with a follow-up duration less than 3 months (n = 1), or with unsatisfactory treatment results (n = 1 long-term non-responder) were excluded. [q] Evers 2018: From a total of 21 responders, 18 were included in the analyses: 2 patients with missing data on protein substitute intakes and 1 patient who was not treated with protein substitute before BH4 treatment were excluded. [r] Paras 2018: In this conference poster, the authors chose to only report on those patients who could be treated solely with BH4. From a total of 22 responders, only 8 were included: 13 patients who were not treated with protein substitutes before BH4 and 1 patient with maternal PKU were excluded.

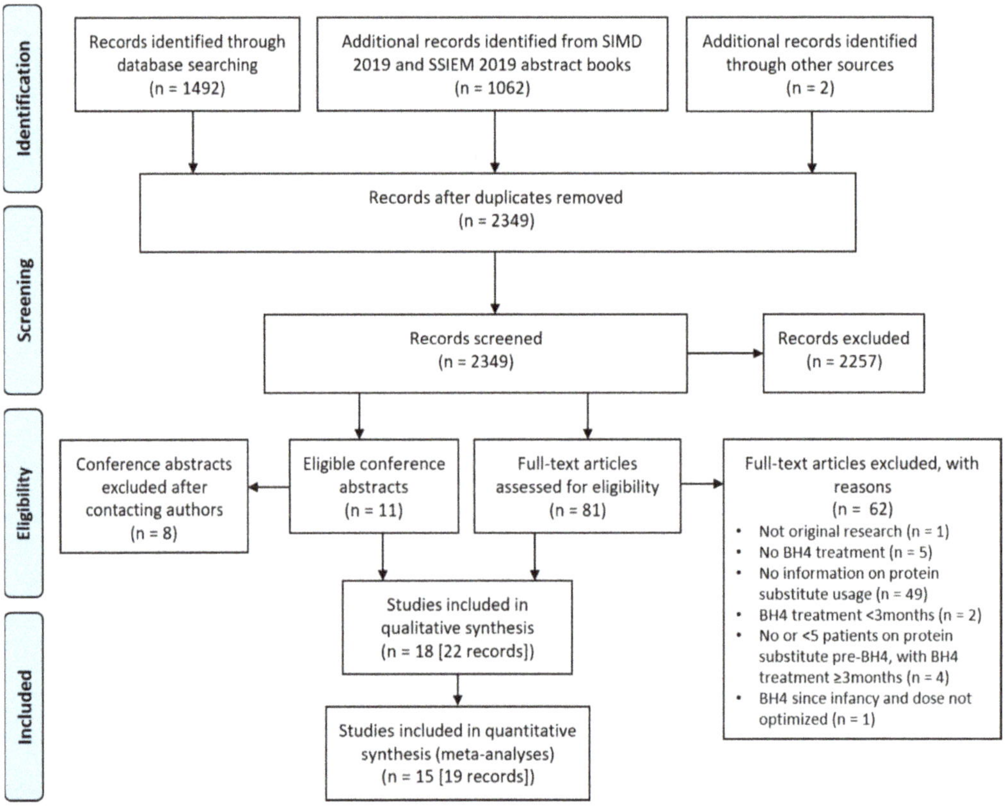

Figure 1. Study selection process according to the Preferred Reporting Items for Systematic Reviews and Meta-Analysis (PRISMA) flow chart.

3.3. *Systematic Review of Key Findings and Meta-Analyses*

Table 2 and Table S2 summarize the main outcomes of the studies included in the systematic review, i.e., the changes in Phe and protein intakes with long-term ($\geq$3 months) BH4 treatment. Meta-analyses of the data were performed, and the overall effect estimate is presented (SMD with confidence intervals (CI)) and illustrated in forest plots.

3.3.1. Change in Phe Intake with BH4 Treatment

Long-term changes in Phe intake were evaluated in 13/18 studies. Phe intakes were self-reported in most studies (self-reported data, n = 10; both self-reported and prescribed data, n = 2; not specified, n = 1; Table 2). Meta-analysis of 12/13 studies showed that Phe intake increased significantly with BH4 treatment (SMD [95% CI] = 1.66 [1.20, 2.12]; $p < 0.0001$; $I^2 = 65.9\%$; n = 186 subjects; Figure 2). The effect was consistent across studies (Figure 2 and Table 2). Although only a small increase in Phe intake ($\approx$1.5-fold) was reported in 2/12 studies [37,54], improvement was seen in 90% of long-term responders, and Phe intake increased >2-fold (range: 2.2 to 4.3-fold) in the other 10/12 studies (increase observed in 100% of long-term responders). The study that could not be included in the meta-analysis (no means and SDs) [36] showed only small increases in median Phe intake, and no change in Phe intake was observed for 22% and 40% of long-term responders after 2 and 5 y of BH4 treatment, respectively (Table 2).

Table 2. Overview of study results: changes in phenylalanine and protein intakes (total protein, natural protein, and protein equivalent from protein substitute) of long-term responders on tetrahydrobiopterin (BH4) treatment [1].

Reference	Duration on BH4 (Mean or Range; Years)	Change in Phe Intake		Relative Change in Natural Protein Intake from Baseline [2]	Change in Protein Equivalent Intake from Protein Substitute		Relative Change in Total Protein Intake from Baseline [2]
		Relative Change from Baseline [2]	No. of Responders with Increased Intake (%)		Relative Change from Baseline [2]	No. of Responders with Change in Dose (%) [3]	
Bélanger-Quintana 2005 [38]	0.9 (range: 0.4–1.5)	**3.5-fold ↑** *(mean; mg/kg/day)* **2.7-fold ↑** *(median; mg/kg/day)*	**7/7 (100)**	n/a	**90% ↓** *(mean; g/kg/day)* **100% ↓** *(median; g/kg/day)*	Decreased: **2/7 (29)** Stopped: **5/7 (71)** No change: -	n/a
Lambruschini 2005 [45]	1.0	**4.3-fold ↑** * *(mean SR intake; mg/day)*	**11/11 (100)**	n/a	**100% ↓** *(mean and median; g/day)*	Decreased: - Stopped: **11/11 (100)** No change: -	n/a
Burlina 2009 [40]	3.5 (range: 0.5–7.0)	**3.2-fold ↑** *(mean SR intake; mg/day)*	**12/12 (100)**	n/a	**100% ↓** *(mean and median; g/day)*	Decreased: - Stopped: **12/12 (100)** No change: -	n/a
Singh 2010 [49]	2.0	*3mo FU:* **2.2-fold ↑** * *1y FU:* **2.3-fold ↑** * *2y FU:* **2.2-fold ↑** * *(mean SR intake; mg/kg/day)* *3mo FU:* **3.4-fold ↑** * *1y FU:* **3.4-fold ↑** * *2y FU:* **3.1-fold ↑** * *(mean prescription; mg/kg/day)*	**6/6 (100)**	*3mo FU:* **114% ↑** * *1y FU:* **119% ↑** * *2y FU:* **125% ↑** * *(mean SR intake; g/kg/day)*	*3mo FU:* **77% ↓** * *1y FU:* **70% ↓** * *2y FU:* **84% ↓** * *(mean SR intake; g/kg/day)*	*3mo FU:* Decreased: **3/6 (50)** Stopped: **3/6 (50)** No change: - *2y FU:* Decreased: **4/6 (67)** Stopped: **2/6 (33)** No change: -	*3mo FU:* **25% ↓** ns *1y FU:* **18% ↓** ns *2y FU:* **27% ↓** ns *(mean SR intake; g/kg/day)*
Vilaseca 2010 [51]	5.7 (range: 5.3–6.0)	n/a	n/a	n/a	**100% ↓** *(mean and median; g/day)*	Decreased: - Stopped: **10/10 (100)** No change: -	n/a

Table 2. *Cont.*

Reference	Duration on BH4 (Mean or Range; Years)	Change in Phe Intake		Relative Change in Natural Protein Intake from Baseline [2]	Change in Protein Equivalent Intake from Protein Substitute		Relative Change in Total Protein Intake from Baseline [2]
		Relative Change from Baseline [2]	No. of Responders with Increased Intake (%)		Relative Change from Baseline [2]	No. of Responders with Change in Dose (%) [3]	
Singh 2011 [48], Douglas 2013a [42], Douglas 2013b [43], Brantley 2018 [39]	**1.0**	*4mo FU*: **2.7-fold ↑** * † *(mean prescription; mg/day)* *1y FU*:**2.9-fold ↑** *(mean prescription; mg/day)* *1y FU*:**1.5-fold ↑** ns *(median SR intake; mg/day)*	*4mo FU*: **18/18 (100)** *1y FU*:**17/17 (100)**	n/a	*4mo FU*: **83% ↓** * *(mean prescription; g/day)* *1y FU*:**77% ↓** *(mean prescription; g/day)* **75 to 100% ↓ (*n* = 6/17)** **50 to 75% ↓ (*n* = 8/17)** **20 to 25% ↓ (*n* = 1/17)** **<20% ↓ (*n* = 2/17)** *(prescription; g/day)*	*4mo FU*: Decreased: **7/18 (39)** Stopped: **9/18 (50)** No change: **2/18 (11)** *1y FU*: Decreased: **10/17 (59)** Stopped: **5/17 (29)** No change: **2/17 (12)**	n/a
Hennermann 2012 [17]	**4.0** (range: 0.7–8.8)	**3.8-fold ↑** *(mean; mg/day)* **3.1-fold ↑** *(median; mg/day)*	**18/18 (100)**	n/a	n/a	Decreased/ No change: **10/18 (56)** Stopped: **8/18 (44)**	n/a
Leuret 2012 [46]	**1.9** § (range: 0.6–6.7)	**3.2-fold ↑** * *(mean SR intake; mg/day)*	**8/8 (100)**	n/a	n/a	Decreased: - Stopped: **7/8 (87)** No change: **1/8 (13)**	n/a
Aldámiz-Echevarría 2013 [36]	**2.0** (cohort 1) #	*2y FU*:**1.4-fold ↑** *(median SR intake; mg/kg/day)*	*2y FU*:**28/36 (78)**	*2y FU*:**14% ↑** *(median SR intake; g/kg/day)*	*2y FU*:**44% ↓** *(median SR intake; g/kg/day)*	*2y FU*: Decreased/ No change: **25/36 (69)** Stopped: **11/36 (31)**	*2y FU*:**17% ↓** *(median SR intake; g/kg/day)*
	5.0 (cohort 2) #	*5y FU*:**1.2-fold ↑** *(median SR intake; mg/kg/day)*	*5y FU*:**6/10 (60)**	*5y FU*:**13% ↑** *(median SR intake; g/kg/day)*	*5y FU*:**57% ↓** *(median SR intake; g/kg/day)*	*5y FU*: Decreased/ No change: **8/10 (80)** Stopped: **2/10 (20)**	*5y FU*:**29% ↓** *(median SR intake; g/kg/day)*

Table 2. *Cont.*

Reference	Duration on BH4 (Mean or Range; Years)	Change in Phe Intake		Relative Change in Natural Protein Intake from Baseline [2]	Change in Protein Equivalent Intake from Protein Substitute		Relative Change in Total Protein Intake from Baseline [2]
		Relative Change from Baseline [2]	No. of Responders with Increased Intake (%)		Relative Change from Baseline [2]	No. of Responders with Change in Dose (%) [3]	
Demirdas 2013 [41]	range: **1.4–2.0**	n/a	**8/8 (100)**	**311% ↑ *** *(mean SR intake; g/day)*	**100% ↓** (n = 3/8) **>60% ↓** (n = 3/8) **<20% ↓** (n = 2/8) *(SR intake; g/day)*	Decreased: **5/8 (63)** Stopped: **3/8 (37)** No change: -	n/a
Aldámiz-Echevarría 2015 [37]	**1.0**	**1.4-fold ↑ *** *(mean SR intake; mg/kg/day)*	**20/22 (90)**	**14% ↓ ns** *(mean SR intake; g/kg/day)*	**22% ↓ ns** *(mean SR intake; g/kg/day)*	Decreased/ No change: **20/22 (91)** Stopped: **2/22 (9)**	**14% ↓ ns** *(mean SR intake; g/kg/day)*
Scala 2015 [47]	**5.7** (range: 1.0–7.0)	**2.5-fold ↑ *** *(mean SR intake; mg/day)* **2.7-fold ↑ *** *(median SR intake; mg/day)*	**17/17 (100)**	n/a	n/a	Decreased: **2/17 (12)** Stopped: **9/17 (53)** No change: **6/17 (35)**	n/a
Thiele 2015 [29]	**2.0**	*3mo FU:***4.5-fold ↑ *** *2y FU:***4.1-fold ↑ *** *(mean SR intake; mg/day)*	**8/8 (100)**	*3mo FU:* **307% ↑ *** *(g/day)* **244% ↑ *** *(g/kg/day)* *(median SR intake)* *2y FU:* **244% ↑ *** *(g/day)* **157% ↑ *** *(g/kg/day)* *(median SR intake)*	*3mo FU:* **100% ↓ *** *(g/day)* **100% ↓ *** *(g/kg/day)* *(median SR intake)* *2y FU:* **84% ↓ *** *(g/day)* **88% ↓ *** *(g/kg/day)* *(median SR intake)*	Decreased: - Stopped: **4/8 (50)** No change: **4/8 (50)**	*3mo FU:* **12% ↑ ns** *(g/day)* **4% ↓ ns** *(g/kg/day)* *(median SR intake)* *2y FU:* **27% ↑ ns** *(g/day)* **2% ↑ ns** *(g/kg/day)* *(median SR intake)*
Ünal 2015 [50] Gökmen Özel 2014 [52]	**2.5** (range: 0.5–4.0)	**3.8-fold ↑** *(mg/day)* **2.9-fold ↑** *(mg/kg/day)* *(mean SR intake)* **3.7-fold ↑** *(mg/day)* **2.8-fold ↑** *(mg/kg/day)* *(median SR intake)*	**51/51 (100)**	n/a	**87% ↓** *(g/day)* **89% ↓** *(g/kg/day)* *(mean SR intake)* **100% ↓** *(g/day)* **100% ↓** *(g/kg/day)* *(median SR intake)*	Decreased: **5/51 (10)** Stopped: **43/51 (84)** No change: **3/51 (6)**	**79% ↑** *(g/day)* **35% ↑** *(g/kg/day)* *(mean SR intake)* **78% ↑** *(g/day)* **33% ↑** *(g/kg/day)* *(median SR intake)*
Feldmann 2017 [16]	**0.5**	n/a	n/a	n/a	**49% ↓** *(mean; g/kg/day)*	Decreased/ No change: **23/30 (77)** Stopped: **7/30 (23)**	**92% ↑** *(mean; g/day)*

Table 2. *Cont.*

Reference	Duration on BH4 (Mean or Range; Years)	Change in Phe Intake		Relative Change in Natural Protein Intake from Baseline [2]	Change in Protein Equivalent Intake from Protein Substitute		Relative Change in Total Protein Intake from Baseline [2]
		Relative Change from Baseline [2]	No. of Responders with Increased Intake (%)		Relative Change from Baseline [2]	No. of Responders with Change in Dose (%) [3]	
Rocha 2017 [54]	**1.0** (range: 0.3–1.4)	**1.8-fold** ↑ *(mg/day)* **1.5-fold** ↑ *(mg/kg/day)* *(median SR intake)*	**8/9 (89)**	**79%** ↑ *(g/day)* **51%** ↑ *(g/kg/day)* *(median SR intake)*	**16%** ↓ *(g/day)* **23%** ↓ *(g/kg/day)* *(median SR intake)*	Decreased: **4/9 (44)** Stopped: - No change: **5/9 (56)**	**19%** ↑ *(g/day)* **8%** ↑ *(g/kg/day)* *(median SR intake)*
Evers 2018 [44]	**5.0** (range: 4.5–5.5)	n/a	n/a	**59%** ↑ *(mean prescription; g/kg/day)* **100%** ↑ *(median prescription; g/kg/day)*	**69%** ↓ *(mean prescription; g/kg/day)* **61%** ↓ *(median prescription; g/kg/day)*	Decreased: **10/18 (56)** Stopped: **8/18 (44)** No change: -	**33%** ↓ *(mean prescription; g/kg/day)*
Paras 2018 [53]	**≥0.3** (range: ≥0.3–≥3.5)	n/a	**8/8 (100)**	n/a	**100%** ↓ *(mean and median; g/day)*	Decreased: - Stopped: **8/8 (100)** No change: -	n/a

Abbreviations: FU: follow-up; No: number; ns: not statistically significant; Phe: phenylalanine; SR: self-reported; y: year; mo: month; n/a: not available. ↑: increase; ↓: decrease. [1] Only long-term responders (follow-up ≥3 months) who were on a Phe-restricted diet and protein substitute before BH4 were included in the analyses. Long-term responsiveness as reported by the original authors, except for Rocha 2017 where 1 patient was considered long-term non-responder after discussing with the authors (lack of changes in Phe tolerance and natural protein intake, while Phe levels only decreased by 10%). [2] Superscripts indicate that a statistical analysis was performed by the original authors. *: statistically significant change; [ns]: change not statistically significant. Otherwise, no statistical analysis was performed with the exception of Ünal 2015, Rocha 2017, and Evers 2018, who performed statistical analyses with their original samples. However, statistical significance is not reported here because some patients included in the original analyses did not meet our inclusion criteria (i.e., long-term responders followed up ≥3 months who were on a Phe-restricted diet and protein substitute before BH4). [3] Change as reported by the original authors. If individual data were available (i.e., reported or provided upon request), change in protein substitute intake was considered a "decrease" only if the reduction was ≥25% compared with baseline, as this was deemed clinically meaningful. Reductions <25% of baseline were counted as "no change". [†] Singh 2011:n = Change in Phe tolerance at 4mo FU included 1 patient never taking any protein substitute but who could not be removed from this analysis, and thus n = 19 instead of 18. One other patient was lost to follow-up between 4mo and 1y FU. [§] Leuret 2012: Median duration of BH4 treatment, not mean. Only 8/15 patients were on a Phe-restricted diet before BH4 and were therefore included in our analyses; however, duration of BH4 treatment was only available for the total sample of 15 patients. [#] Aldámiz-Echevarría 2013: Unclear if patients with a 5y follow-up were also described in the group of patients with a 2y follow-up. It was assumed that the 2 cohorts comprised different patients.

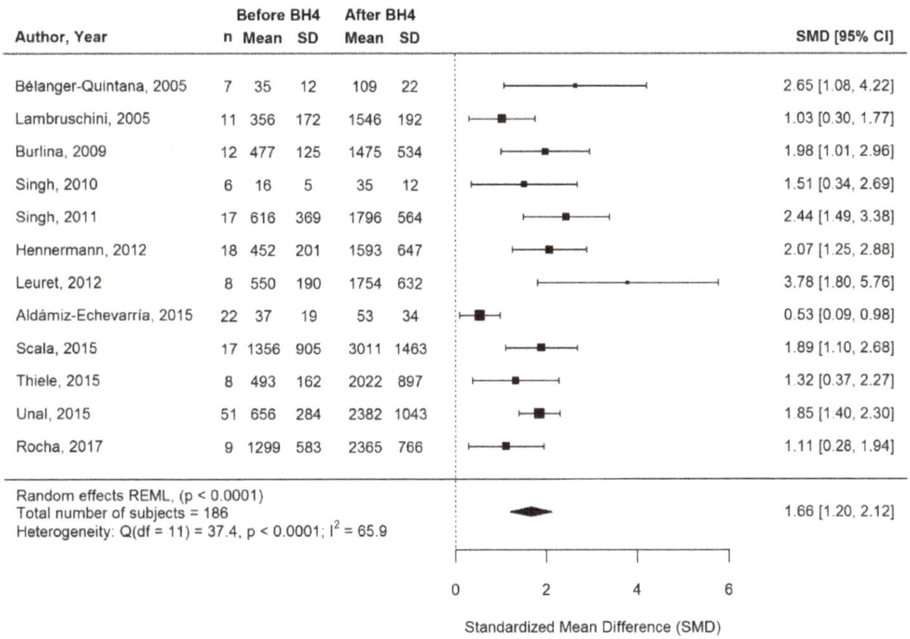

Figure 2. Standardized change in phenylalanine intake of long-term responders on BH4 treatment. Means and SDs before/after BH4 are milligram phenylalanine per kilogram bodyweight per day for Belanger-Quintana (2005), Singh (2010), and Aldámiz-Echevarría (2015), and milligram per day for all other studies. Abbreviations: BH4, tetrahydrobiopterin; CI: confidence interval; n: sample size; SD: standard deviation; SMD, standardized mean difference.

3.3.2. Change in Natural Protein Intake with BH4 Treatment

Only 7/18 studies assessed long-term changes in natural protein intake (self-reported data, n = 6; prescribed data, n = 1; Table 2). Meta-analysis of 6/7 studies demonstrated a significant increase with BH4 treatment (SMD [95% CI] = 1.17 [0.17, 2.16]; p = 0.0298; I^2 = 81.4%; n = 71 subjects; Figure 3). The effect was consistent across 5/6 studies, although heterogeneity was high and effect sizes varied widely (range: 51 to 157% when considering the increase from baseline in g natural protein/kg/day and 79 to 311% in g/day). The remaining two studies (one not included in the meta-analysis [36]), from the same Spanish metabolic centers, showed little to no change in natural protein intake after 1 to 5 y of BH4 treatment [36,37] (Table 2).

3.3.3. Change in Protein Equivalent Intake from Protein Substitute with BH4 Treatment

Protein equivalent intake from protein substitute was self-reported in most studies (self-reported data, n = 13; prescribed data, n = 1; both, n = 1; not specified, n = 3; Table 2). Meta-analysis of 10/18 studies showed a significant, consistent reduction in protein equivalent intake from protein substitute (SMD [95% CI] = −1.44 [−1.96, −0.92]; p = 0.0001; I^2 = 74.3%; n = 179 subjects; Figure 4). The result did not change when Belanger-Quintana et al. [38] and Singh et al. [49] were excluded in a sensitivity analysis (data not shown). This result was also broadly consistent with the findings in the remaining studies not included in the meta-analysis (Table 2). Overall, long-term BH4 treatment led to a mean decrease in protein equivalent intake from protein substitute (both when expressed as mg/day and mg/kg/day) of at least 80% compared with baseline in 9/18 studies, and at least 40% in 5/18 studies. However, the decrease in protein equivalent intake from protein substitute was <25% in 2/18 studies, and almost all patients continued to require a substantial amount of protein substitutes in both studies, despite BH4 treatment [37,54] (Table 2). For 2/18 studies, the reduction in protein equivalent intake from protein substitute could not be estimated [17,47] (Table 2).

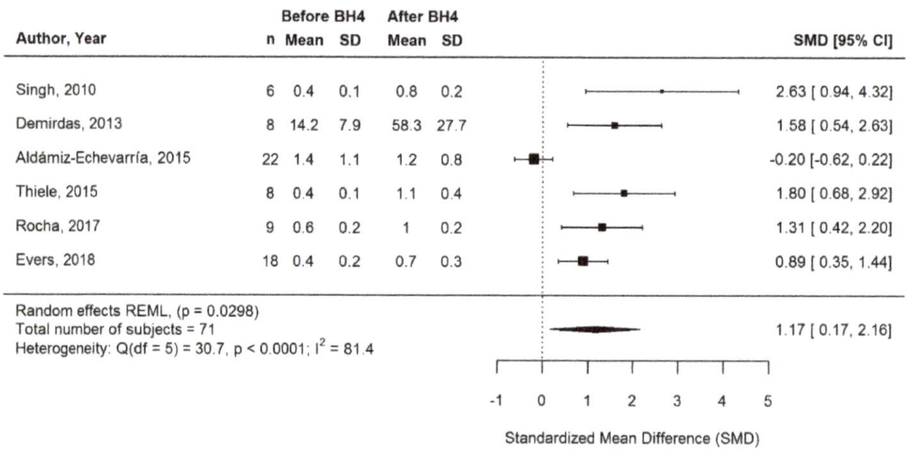

Figure 3. Standardized change in natural protein intake of long-term responders on BH4 treatment. Means and SDs before/after BH4 are gram natural protein per day for Demirdas (2013), and gram per kilogram bodyweight per day for all other studies. Abbreviations: BH4, tetrahydrobiopterin; CI: confidence interval; n: sample size; SD: standard deviation; SMD, standardized mean difference.

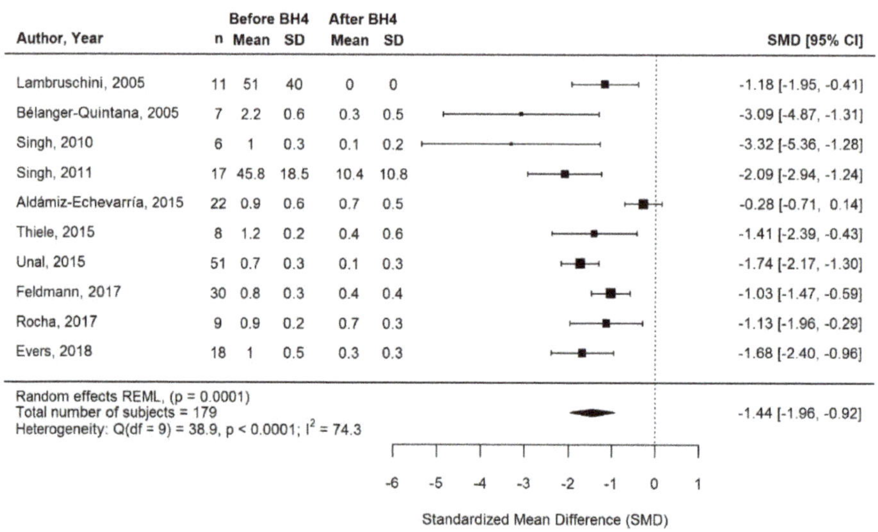

Figure 4. Standardized change in protein equivalent intake from protein substitute of long-term responders on BH4 treatment. Means and SDs before/after BH4 are gram protein equivalent per day for Lambruschini (2005) and Singh (2011), and gram per kilogram bodyweight per day for all other studies. Abbreviations: BH4, tetrahydrobiopterin; CI: confidence interval; n: sample size; SD: standard deviation; SMD, standardized mean difference.

Thereby, approximately half of all long-term responders (149/306) continued to require protein substitutes with BH4 treatment, and half (157/306) stopped protein substitute usage (Table 2). For 63/149, the dose of protein substitute was reduced in 67% (*n* = 42) but remained unchanged in 33% (*n* = 21) on long-term BH4 treatment (Table 2). In the other 86/149 patients still requiring protein substitutes, it was unreported if the amount could be decreased or remained unchanged [16,17,36,37] (Table 2).

3.3.4. Change in Total Protein Intake after BH4 Treatment

Only 8/18 studies evaluated long-term changes in total protein intake (Table 2), and meta-analysis of 7/8 studies showed no significant change with BH4 treatment (SMD [95% CI] = 0.02 [−0.94, 0.99]; $p = 0.9516$; $I^2 = 92.9\%$; $n = 144$ subjects; Figure 5). However, there was a considerable amount of heterogeneity within the data. Although results across studies were inconsistent, the mean/median total protein intakes (per kg of body weight) met dietary reference values for protein intake throughout the evaluation periods [55].

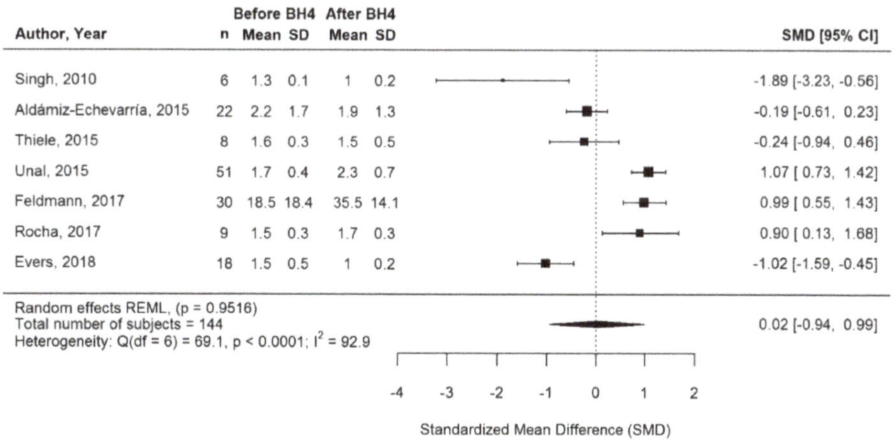

Figure 5. Standardized change in total protein intake of long-term responders on BH4 treatment. Means and SDs before/after BH4 are gram total protein per day for Feldmann (2017), and gram per kilogram per day for all other studies. Abbreviations: BH4, tetrahydrobiopterin; CI: confidence interval; n: sample size; SD: standard deviation; SMD, standardized mean difference.

3.3.5. Supplementary Sensitivity Meta-Analyses

Some authors reported dietary/nutritional outcomes in gram per day (mg/day for Phe intake), whereas others expressed their results per kilogram bodyweight (g/kg/day or mg/kg/day), and thus SMDs were used in the main meta-analyses in order to compare data in different units. However, for each dietary outcome, two sets of meta-analyses were also performed by pooling only studies expressing data in the same unit (Figures S1–S8). Despite the generally high heterogeneity within the data, results were similar irrespective of the units used and in line with the main meta-analyses reported above. One exception was total daily protein intake, where, although no significant change was observed per kilogram bodyweight, total protein intake significantly increased by 16.71 g/day with BH4 treatment (95% CI = [6.91, 26.50]; $p = 0.0123$; $I^2 = 73.9\%$; 4 studies; $n = 98$ subjects; Figure S8). Finally, because of the particularly high heterogeneity in the meta-analyses of the changes in milligram Phe intake per kilogram bodyweight per day (Figure S1; $I^2 = 96.4$) and gram protein equivalent intake from protein substitute per kilogram bodyweight per day (Figure S5; $I^2 = 97.5$), sensitivity analyses excluding Belanger-Quintana et al. [38] were performed; however, results remained similar (data not shown).

3.4. Systematic Review of Findings Related to Secondary Outcomes

3.4.1. Change in Micronutrient Intakes and Serum Concentrations with BH4

Only 8/18 studies investigated the change in micronutrient intakes [17,29,39,45,49] and/or markers of nutritional status [17,39,44,45,47,49,54] with long-term BH4 treatment (data not shown). Thiele et al. reported significant decreases in vitamin (OH)D_3, vitamin B_{12}, folic acid, iron, and calcium intakes, and in one patient, protein substitute had to be re-introduced because of severe atopic skin lesions, lowering of serum zinc concentration below normal range, and decreased protein intake below 80% of the recommended amount [29]. Similar changes in intakes of these micronutrients were reported by Brantley et al., along with significant decreases in serum iron, folate, and vitamin B_{12} concentrations compared to baseline [39]. Diet was not fully liberalized in all patients, but protein substitute intake was reduced by at least 50% in both studies. Lower intakes of calcium, iron, and vitamin B_{12} were also observed by Hennermann et al. [17], but only in patients who could liberalize their diet without protein substitute, and serum levels remained within the normal range. In contrast, other authors found no significant change in dietary intakes or serum concentrations of several micronutrients [44,45,47,49], except for a decrease in zinc concentrations in 5 patients in one study [54].

3.4.2. Change in Growth with BH4

Of the 18 studies, 9 investigated changes in weight and height z-scores during long-term BH4 treatment (data not shown). In general, weight- and height-for-age z-scores remained within the normal range [17,29,38,44,45]. Improvement in linear growth was observed in two studies after diet liberalization with BH4 treatment, which may be attributable to a marked increase in Phe/natural protein intake [49,52]. In two other studies, weight and height z-scores were below average at baseline (z-scores < 0) and did not improve after 1 to 5 years of BH4 treatment. In both studies, the increase in Phe intake was limited (<1.5-fold), while protein equivalent intake from protein substitute intake was reduced by 22–57%, resulting in slight decrease in total protein intake [36,37].

3.4.3. Change in Metabolic Control with BH4

Of the 18 studies, 15 evaluated metabolic control after BH4 treatment (data not shown). Overall, blood Phe concentrations did not change compared to baseline in 8/18 [17,36–38,44,45,49,50], significantly increased in 2/18 [29,47], and decreased in the remaining 5/18 studies [40,46,48,53,54]. Mean/median blood Phe levels remained in age-specific therapeutic ranges in most subjects. In one study [40], long-term BH4 treatment was only started in initial responders who were non-adherent with the low-Phe diet and had a baseline blood Phe level higher than the recommended range. At last follow-up (range: 6 months to 7 years), blood Phe levels had lowered into the therapeutic range in all subjects, and their diet was liberalized.

3.5. Quality Appraisal and Risk of Bias Assessment

Overall, the quality was rated as "fair" for most studies (13/18) (Table 3). The main concerns were small sample sizes and likely selection bias, making it unclear if the study samples were representative of PKU patients who would benefit from long-term BH4 treatment. A statistical analysis for pre–post treatment comparisons was also lacking in most cases. Three studies with low risks of bias were rated as "good quality" [44,48,50]. The remaining two studies were judged "poor" due to unreliability or inadequacy of outcome measurements, serious selection bias, small sample size, and lack of information on the intervention (i.e., BH4 treatment) [41,53].

Table 3. Quality appraisal and risk of bias.

Study (Author, Year)	Items of "Quality Assessment Tool for Before-After (Pre-Post) Studies with No Control Group"												Overall
	1	2	3	4	5	6	7	8	9	10	11	12	
Bélanger-Quintana 2005 [38]	x	+	?	?	+	+	+	NA	?	x	+	NA	Fair
Lambruschini 2005 [45]	+	+	?	+	+	+	+	NA	+	+	+	NA	Fair
Burlina 2009 [40]	+	+	?	?	+	+	+	NA	?	x	+	NA	Fair
Singh 2010 [49]	+	+	?	x	+	+	+	NA	+	+	+	NA	Fair
Vilaseca 2010 [51]	+	+	?	?	+	+	+	NA	?	x	+	NA	Fair
Singh 2011 [48] Douglas 2013a [42] Douglas 2013b [43] Brantley 2018 [39]	+	+	?	+	+	+	+	NA	+	+	+	NA	Good
Hennermann 2012 [17]	+	+	?	x	?	+	+	NA	+	x	+	NA	Fair
Leuret 2012 [46]	x	+	?	?	+	?	+	NA	+	+	?	NA	Fair
Aldámiz-Echevarría 2013 [36]	+	+	?	?	?	+	+	NA	?	x	+	NA	Fair
Demirdas 2013 [41]	+	+	?	x	?	?	x	NA	?	+	?	NA	Poor
Aldámiz-Echevarría 2015 [37]	+	+	?	?	?	+	+	NA	?	+	+	NA	Fair
Scala 2015 [47]	x	+	?	x	?	+	+	NA	+	+	+	NA	Fair
Thiele 2015 [29]	+	+	?	?	+	+	+	NA	?	+	+	NA	Fair
Ünal 2015 [50] Gökmen Özel 2014 [52]	+	+	+	?	+	+	+	NA	?	+	+	NA	Good
Feldmann 2017 [16]	+	+	?	x	?	+	+	NA	x	x	+	NA	Fair
Rocha 2017 [54]	+	x	?	?	+	?	+	NA	?	+	?	NA	Fair
Evers 2018 [44]	+	+	?	+	+	+	+	NA	?	+	+	NA	Good
Paras 2018 [53]	+	x	x	?	?	x	+	NA	?	x	?	NA	Poor

Each item was rated as low risk ("yes" = +), unclear ("cannot determine/not reported" = ?), or high risk ("no" = x) for the following type of bias: objective study question (1); description of eligibility/selection criteria for the study population (2); representativeness of study population of general/clinical population of interest (3); selection bias (4); sample size, power, effect estimate (5); description of intervention, adherence, and deviations from intended interventions (6); measurement of outcomes (defined, valid, and reliable) (7); blinding of outcome assessors (8); loss to follow-up < 20% (9); statistical comparison for pre-to-post changes (10); frequency of repeated measurements (11); group-level interventions (12). NA, not applicable.

4. Discussion

This is the first time that changes in protein equivalent intake from protein substitute with BH4 treatment have been assessed systematically, although other systematic reviews or meta-analyses have investigated the effects of BH4 treatment on blood Phe control and dietary Phe tolerance [56–58]. We have demonstrated that PKU patients with long-term BH4 responsiveness had a significant increase in dietary Phe and natural protein intake when on BH4 treatment. This enabled the majority of responsive patients to reduce the dose of protein substitute, and 51% (157/306) were able to stop protein substitute. However, almost half (149/306) of long-term responders continued to require some protein substitute, even though Phe and natural protein tolerance substantially improved. In this group, the protein substitute dose could be reduced in 28% (42/149) but remained unchanged in 14% of patients (21/149). In 58% (86/149) of patients on BH4 with protein substitute, the authors did not report if the dose was adjusted. Overall, the extent of reduction of protein equivalent intake from protein substitute, the time needed for change, as well as approaches to adjusting the PKU diet varied widely between studies. These findings highlight the need for guidance on when and how to decrease or stop protein substitute intake with BH4 treatment Pooled analysis of 10 studies showed that protein equivalent intake from protein substitute significantly decreased after a median BH4 treatment of one year (range: 0.5–5 years). Where half or more of the responsive patients were able to reduce or stop the use of protein substitutes, dietary Phe tolerance (as either expressed in mg/kg/day or mg/day) had increased by 2.5- to 4.3-fold [29,38,40,45–50]. In contrast, three studies reported a Phe tolerance increase <1.5-fold [36,37,54], and two of them failed to show a meaningful reduction (i.e., ≥25% from baseline) in median [54] or mean [37] protein equivalent intake from protein substitute after 1 year of BH4 treatment. Aldámiz et al. [37] attributed these findings to the inability of the BH4 loading test "cut off" of 30% decrease in blood Phe concentrations to identify true (i.e., long-term) responders correctly. When a 50% decrease in blood Phe as cut-off was used in a new loading test protocol [59], all responders were able to consume normal diets without protein substitute in the long term [37]. Most studies included in this systematic review used ≥30% decrease in blood Phe levels as a criterion to define BH4 responsiveness and showed successful long-term outcomes. However, BH4 therapy was discontinued in some patients (n = 27) mainly due to unsatisfactory blood Phe control when additional Phe/natural protein was added longer term [16,17,45,47,48,50,60].

Meeting nutritional requirements while maintaining blood Phe concentrations within therapeutic range is a central consideration when prescribing pharmaceutical therapies for PKU. Daily protein and micronutrient requirements increase throughout childhood and in women during pregnancy and lactation. With BH4 treatment, it is important to use a stepwise approach to increasing natural protein whilst in parallel reducing protein equivalent intake from protein substitute by similar amounts. Attention should be paid to the quantity as well as quality of natural protein. It is critical to ensure a good mix of animal and plant protein so that natural foods can supply all the nutrients in the amounts that meet requirements. Ongoing evaluation about the need for protein substitute supplementation as well as education about appropriate food choices is essential. We identified only a few studies [17,45,48] that have described in detail how natural protein is increased with BH4 therapy (see Table S1). Of these, the protocol by Singh et al. (2011) was the most thorough [48]. All responsive patients were instructed to add 20g of non-fat dry milk powder (≈350 mg Phe or 6.8 g protein) to their diet each week until new Phe tolerance was established [48], although this may be considered a rapid increase in natural protein intake by some. In practice, it may take several months to determine the final Phe tolerance and establish the ongoing need for a source of protein equivalent from protein substitute. Paras et al. reported a range of 3 months to 3.5 years until full diet liberalization occurred [53]. Caution is necessary in the case of illness episodes, injury, or trauma, as these may all adversely affect metabolic control, and it is established that BH4 is less effective in illness [38]. Protein substitutes offer a protective role by counteracting protein catabolism. It may be considered that, in young children, a small dose of protein substitute should be maintained as it is difficult to re-establish intake specifically for illness episodes or to meet the increased age-appropriate protein requirements during growth phase [61,62]. For others, it will be necessary to evaluate the need for protein substitute re-introduction or an increase in dose might be required. Some studies have described patients who could initially stop using protein substitute, but for whom it had to be re-introduced [29,48].

Most protein substitutes provide a major supply of vitamins and minerals, and one of the concerns associated with long-term BH4 treatment is the nutritional adequacy of a relaxed diet when protein substitute is stopped or reduced [29]. We found inconsistent results about the impact on micronutrient status. Overall, the reduction in usage of protein substitutes or change in dietary habits with BH4 led to a decreased intake of several essential micronutrients in some [17,29,39,54] but not all studies [44,45,47,49]. Nutritional inadequacies were generally observed when diet was not fully liberalized, particularly when the dose of protein substitute was reduced by at least half of the baseline prescription [29,39], but it was also reported in a subgroup of patients who could relax their diet and stop protein substitute intake [17]. Another concern has been the establishment of healthy eating habits in BH4-treated patients who were well established in their dietary patterns before initiation of BH4 therapy. One of the two studies that investigated change in eating habits after diet relaxation demonstrated poorer eating habits in patients treated with BH4, despite training and education [29]. Although there was some recovery (e.g., re-increase of fruit intake) after 2 years of treatment, consumption of fish and dairy products remained markedly lower than healthy peers and was replaced by a higher intake of potatoes and pasta [29]. Similar findings were also reported by Hennermann et al. [17] who observed that normal bread, normal pasta, eggs, sausages, and meat were well accepted when dietary treatment was relaxed, while milk and dairy products were poorly accepted, and fish was completely refused by all patients. Growth impairment was found only in 2/9 studies [36,37]. This was evident at baseline and it did not improve with BH4 therapy, possibly due to the limited increase in dietary Phe tolerance coupled with a slight decrease in protein equivalent from protein substitute and thus total protein intake. Overall, our results indicate that long-term BH4 therapy does not seem to have a negative impact on total protein intake, and hence on growth. Nonetheless, there is still a risk of inadequate protein quality and of micronutrient deficiencies, which may be attributable to an embedded high-carbohydrate, low-protein disordered eating pattern that may take many months and years of education and counselling to improve. Further investigations in larger prospective studies including patients from different age groups and with all forms of PKU are needed to confirm the effects of BH4 treatment on dietary adequacy and growth.

Strengths and Limitations

The main strength of this systematic review and meta-analysis is that we only included patients who demonstrated long-term BH4 responsiveness. Some patients who appeared BH4-responsive immediately following a loading test in the long-term were unable to increase their Phe tolerance/natural protein intake without a detrimental impact on metabolic control [16,17,45,47,48,50,60]. In this patient category, protein substitute prescription usually remained unchanged, and if dose was decreased, a later increase was necessary. We decided to exclude these patients (i.e., long-term non-responders) in order to evaluate the impact of BH4 supplementation on change in protein equivalent intake from protein substitute in patients for whom the drug was "justly" efficacious. Furthermore, we believe that the duration of follow-up strengthens the reliability of these findings. We elected to include only studies where patients had been on BH4 for at least 3 months. In fact, the majority (55%) of studies included had a mean BH4 treatment duration of $\geq$2 years, with some patients on cofactor therapy for almost 9 years [17].

Our work also had several limitations. Many articles were excluded during the screening process due to inadequate information about protein substitute intake (47/62). It is crucial in any study investigating new treatments for PKU to measure and report any changes in protein intake (including both natural and protein equivalent from protein substitute). Furthermore, one of the inclusion criteria was that prior to BH4 treatment, a Phe-restricted diet supplemented with protein substitute was necessary, which led to the exclusion of a limited number of patients on a normal diet at baseline from the analyses. The meta-analyses showed a medium-to-high level of heterogeneity between study results for the main outcomes of interest. This may be explained by the wide differences in age and phenotypes of patients, as well as the variation in the definition of BH4 responsiveness, duration of follow-up, target blood Phe levels, or the protocols followed by centers for dietary changes with BH4. Authors usually described self-reported intakes rather than prescribed amounts of protein. Non-adherence to the prescribed amount of protein substitute is common in PKU, and hence the change in self-reported intakes may not reflect the true effect of BH4. Finally, the quality of most included studies was rated as fair only for several reasons, e.g., small sample size, lack of power analysis, or absence of statistical comparison, even though some of these limitations are due to the rarity of the disorder.

5. Recommendations

This work, as well as our clinical experience, call for several recommendations, which will help guide healthcare professionals when adjusting dietary prescriptions of patients with PKU on BH4 treatment. Some of these recommendations will also be valid for other new therapies such as pegvaliase.

5.1. BH4 Treatment Trial and Adjusting Phe Intake

- BH4 responsiveness requires careful assessment—the aim is to maintain blood Phe within target therapeutic range while maintaining normal growth but also (1) establish an increase in Phe tolerance, (2) reduce protein equivalent intake from protein substitute in alignment with any increase in natural protein intake, and (3) establish the maintenance dose of BH4.

- Once BH4 is administered, if three consecutive blood Phe levels are maintained within target therapeutic range, then Phe intake should be increased by at least 20%, and then this process should be repeated until natural protein tolerance is established. If the mean blood Phe level exceeds target therapeutic range, then the Phe intake should be reduced by approximately 10 to 30%, depending on the degree of elevation of the blood Phe levels (adapted from Muntau et al. [63]).
- With BH4 treatment, it is expected that the final Phe tolerance should be increased by $\geq$100% of baseline, provided natural protein intake is below safe levels of protein intake. If natural protein intake already exceeds safe levels of protein intake at baseline, an improvement in blood Phe control may be an appropriate alternative goal. Maintenance of blood Phe levels within target therapeutic range and an increase in Phe tolerance should be observed for at least 3 months to ascertain BH4 responsiveness.

5.2. *Quality of Natural Protein Intake*

- Natural protein intake should be sourced from different proteins, e.g., dairy and eggs, cereals, lentils, and protein-rich vegetables if tolerated. Food choices should be made according to national and international recommendations. Natural protein sources should provide micronutrients to minimize the need for extra micronutrient supplements. Continuous patient education and support about the need for a healthy diet with appropriate food choices will be necessary with BH4 treatment.

5.3. *Adapting Protein Substitute Dose*

- Protein equivalent from substitute intake should be reduced in parallel with any increase in natural protein intake. The more natural protein that is tolerated, the lower the requirement should be for protein substitute. For every increase in natural protein, the protein equivalent from protein substitute should be reduced accordingly.
- It is possible that the natural protein intake meets or exceeds safe levels of protein intake so that a protein substitute is not needed to meet protein requirements. However, some protein substitute might be necessary for micronutrient requirements to be met. Micronutrient supply should be monitored carefully, especially if patients cannot be allowed an unlimited Phe intake. Moreover, it may be better for patients to remain familiar with and accepting of the taste of protein substitute in case it needs to be reintroduced in illness, pre-conception, pregnancy, or lactation, or if BH4 therapy is discontinued. It is also good practice to give a small dose of protein substitute each day to infants who may appear fully responsive to BH4 and without immediate need for a protein restriction. It is possible protein restriction may be necessary at a later age when daily protein requirements increase.

5.4. *Monitoring*

- Once patients are established on BH4 therapy and the diet is stabilized, clinic visits and blood monitoring should occur at the same frequency as for other patients with PKU who are not on BH4 treatment. If there are any concerns about adherence with BH4 or diet, more frequent monitoring may be required.
- Continue to assess that at least 75% of blood Phe levels remain within target therapeutic range and that more than 100% of original prescription of Phe intake is maintained (unless patients are already meeting safe levels of protein intake). If more than 25% of blood Phe levels are outside target therapeutic range, consider adjusting BH4 dosage or reduce Phe intake. BH4 treatment continuation should be evaluated.
- Evaluate if protein substitute should be re-introduced, or prescription increased, in any event of increased protein requirements (rapid growth, illness, injury/trauma, pregnancy, lactation).

- Patient's nutritional status including height/length, weight, and body mass index (BMI) should be conducted at least 6-monthly. It is important that patients are encouraged to maintain a healthy BMI.
- Assessment of patient's nutritional biochemical markers such as plasma amino acids, homocysteine/or methyl malonic acid, hemoglobin, mean corpuscular volume, ferritin, zinc, calcium, selenium, vitamin D, vitamin B12, and folic acid should be completed annually for patients on BH4 therapy.
- Monitor nutritional intake adequacy by 3-day dietary assessments regularly, at least every 3 months in the first year of BH4 therapy. Vitamin and mineral supplements may be required if dietary assessment or patient's nutritional biomarkers indicate they are necessary. Patients may be more vulnerable to nutritional deficiency if they have stopped or reduced protein substitute intake.
- The ongoing prescription for BH4 should be reassessed and adjusted as appropriate at each clinic visit.

5.5. Clinical Trials of (New) Treatments

- Any future studies investigating treatment strategies for PKU should evaluate long-term (at least 6 months) changes in nutrient intake, in particular natural protein, the need for protein substitute, and micronutrient supplementation. Data about prescribed as well as self-reported protein/Phe intakes should be collected and reported (both gram (or milligram) per day and gram (or milligram) per kilogram bodyweight per day). In published studies, individual data should be provided rather than only summary statistics such as means or medians.

6. Conclusions

In BH4-responsive patients with PKU, protein equivalent intake from protein substitute significantly decreased with long-term BH4 treatment, with half of the patients able to stop protein substitute and follow a liberalized diet. However, the other half of BH4 responders still required at least some protein substitute to meet their protein requirements and to achieve good metabolic control, even though Phe tolerance substantially improved. It is important to follow a systematic protocol to increase natural protein intake while reducing the dose of protein substitutes in order to ensure protein and micronutrient requirements are met and sustained. Normal growth was maintained with BH4 treatment, but micronutrient deficiency associated with a decreased intake of protein substitute is a potential risk. Special attention is required in any situations where protein requirements are increased (e.g., rapid growth, illness, or pregnancy), and increase in prescription or re-introduction of protein substitute should be evaluated.

Supplementary Materials: The following are available online at https://www.mdpi.com/2072-6643/13/3/1040/s1: Figure S1: Change in phenylalanine intake (mg/kg/day) of long-term responders on BH4 treatment. Figure S2: Change in phenylalanine intake (mg/day) of long-term responders on BH4 treatment. Figure S3: Change in natural protein intake (g/kg/day) of long-term responders on BH4 treatment. Figure S4: Change in natural protein intake (g/day) of long-term responders on BH4 treatment. Figure S5: Change in protein equivalent intake from protein substitute (g/kg/day) of long-term responders on BH4 treatment. Figure S6: Change in protein equivalent intake from protein substitute (g/day) of long-term responders on BH4 treatment. Figure S7: Change in total protein intake (g/kg/day) of long-term responders on BH4 treatment. Figure S8: Change in total protein intake (g/day) of long-term responders on BH4 treatment. Table S1: Assessment and definition of BH4 responsiveness, long-term BH4 treatment, and protocol for adjusting dietary management. Table S2: Phenylalanine and protein intakes (total protein, natural protein, and protein equivalent from protein substitute) before and on BH4 treatment.

Author Contributions: Conceptualization, M.K., C.M., and A.M.; methodology, C.M., F.I., and A.M.; study selection, data extraction, formal analysis, and risk of bias assessment and quality appraisal, F.I. and C.M.; data curation, A.P.; statistical analysis, E.K.; writing—original draft preparation, F.I. and C.M.; writing—review and editing, F.I., C.M., A.P., R.S., C.R., E.K., H.G.-Ö., M.K., A.M.; supervision,

A.M. and M.K.; project administration, C.M. All authors have read and agreed to the published version of the manuscript.

Funding: This research received no external funding. Publication costs were covered by Danone Nutricia Research.

Institutional Review Board Statement: Not applicable.

Informed Consent Statement: Not applicable.

Data Availability Statement: Data is contained within the article or supplementary material.

Acknowledgments: We are greatly indebted to Julio Rocha, Annemiek van Wegberg, Roeland Evers, Reinhold Feldmann, and Alena Thiele for providing important clarifications and/or additional data. We also wish to thank Daan Snoeks for carrying out the search in ProQuest.

Conflicts of Interest: F.I. and E.K. declare no conflict of interest. A.P. has received an educational grant from Cambrooke Therapeutics and grants from Vitaflo, Danone Nutricia, Merck Serono, Biomarin, and Mevalia to attend scientific meetings. C.R. has received honoraria from Danone Nutricia, MetaX, Schär, Biomarin, and Vitaflo. H.G.-Ö. is a member of the European nutrition expert panel (Biomarin) and she is on the scientific advisory board of Danone Nutricia Metabolics Turkey. R.S. has served on the global medical advisory board for Danone Nutricia and Biomarin. A.M. has received research funding and honoraria from Danone Nutricia, Vitaflo International, Biomarin, MetaX, Applied Pharma Research, and Merck Serono; she is a member of the advisory board for Danone Nutricia, Arla, and Applied Pharma Research. C.M. and M.K. are Danone Nutricia Research employees.

References

1. Blau, N.; van Spronsen, F.J.; Levy, H.L. Phenylketonuria. *Lancet* **2010**, *376*, 1417–1427. [CrossRef]
2. Hillert, A.; Anikster, Y.; Belanger-Quintana, A.; Burlina, A.; Burton, B.K.; Carducci, C.; Chiesa, A.E.; Christodoulou, J.; Đorđević, M.; Desviat, L.R.; et al. The Genetic Landscape and Epidemiology of Phenylketonuria. *Am. J. Hum. Genet.* **2020**, *107*, 234–250. [CrossRef] [PubMed]
3. Van Wegberg, A.M.J.; Macdonald, A.; Ahring, K.; BãLanger-Quintana, A.; Blau, N.; Bosch, A.M.; Burlina, A.; Campistol, J.; Feillet, F.; Giżewska, M.; et al. The complete European guidelines on phenylketonuria: Diagnosis and treatment. *Orphanet J. Rare Dis.* **2017**, *12*, 1–56. [CrossRef]
4. Mitchell, J.J.; Trakadis, Y.J.; Scriver, C.R. Phenylalanine hydroxylase deficiency. *Genet. Med.* **2011**, *13*, 697–707. [CrossRef]
5. Macdonald, A.; Van Wegberg, A.M.J.; Ahring, K.; Beblo, S.; Bélanger-Quintana, A.; Burlina, A.; Campistol, J.; Coşkun, T.; Feillet, F.; Giżewska, M.; et al. PKU dietary handbook to accompany PKU guidelines. *Orphanet J. Rare Dis.* **2020**, *15*, 1–21. [CrossRef]
6. MacDonald, A.; White, F. Amino Acid Disorders. In *Clinical Paediatric Dietetics*; John Wiley & Sons: Chichester, UK, 2015; pp. 391–455. [CrossRef]
7. Ford, S.; O'Driscoll, M.; MacDonald, A. Living with Phenylketonuria: Lessons from the PKU community. *Mol. Genet. Metab. Rep.* **2018**, *17*, 57–63. [CrossRef] [PubMed]
8. Acosta, P.B.; Yannicelli, S. Protein intake affects phenylalanine requirements and growth of infants with phenylketonuria. *Acta Paediatr.* **1994**, *83*, 66–67. [CrossRef]
9. Kindt, E.; Motzfeldt, K.; Halvorsen, S.; Lie, O.S. Protein requirements in infants and children: A longitudinal study of children treated for phenylketonuria. *Am. J. Clin. Nutr.* **1983**, *37*, 778–785. [CrossRef] [PubMed]
10. Macdonald, A.; Rylance, G.; Davies, P.; Asplin, D.; Hall, S.K.; Booth, I.W. Administration of protein substitute and quality of control in phenylketonuria: A randomized study. *J. Inherit. Metab. Dis.* **2003**, *26*, 319–326. [CrossRef]
11. Brown, C.S.; Lichter-Konecki, U. Phenylketonuria (PKU): A problem solved? *Mol. Genet. Metab. Rep.* **2016**, *6*, 8–12. [CrossRef]
12. Burnett, J.R. Sapropterin dihydrochloride (Kuvan/phenoptin), an orally active synthetic form of BH4 for the treatment of phenylketonuria. *IDrugs Investig. Drugs J.* **2007**, *10*, 805–813.
13. Hydery, T.; Coppenrath, V.A. A Comprehensive Review of Pegvaliase, an Enzyme Substitution Therapy for the Treatment of Phenylketonuria. *Drug Target Insights* **2019**, *13*, 1177392819857089. [CrossRef]
14. Kure, S.; Hou, D.-C.; Ohura, T.; Iwamoto, H.; Suzuki, S.; Sugiyama, N.; Sakamoto, O.; Fujii, K.; Matsubara, Y.; Narisawa, K. Tetrahydrobiopterin-responsive phenylalanine hydroxylase deficiency. *J. Pediatr.* **1999**, *135*, 375–378. [CrossRef]
15. Burton, B.K.; Grange, D.K.; Milanowski, A.; Vockley, G.; Feillet, F.; Crombez, E.A.; Abadie, V.; Harding, C.O.; Cederbaum, S.; Dobbelaere, D.; et al. The response of patients with phenylketonuria and elevated serum phenylalanine to treatment with oral sapropterin dihydrochloride (6R-tetrahydrobiopterin): A phase II, multicentre, open-label, screening study. *J. Inherit. Metab. Dis.* **2007**, *30*, 700–707. [CrossRef]
16. Feldmann, R.; Wolfgart, E.; Weglage, J.; Rutsch, F. Sapropterin treatment does not enhance the health-related quality of life of patients with phenylketonuria and their parents. *Acta Paediatr.* **2017**, *106*, 953–959. [CrossRef]

17. Hennermann, J.B.; Roloff, S.; Gebauer, C.; Vetter, B.; Von Arnim-Baas, A.; Mönch, E. Long-term treatment with tetrahydrobiopterin in phenylketonuria: Treatment strategies and prediction of long-term responders. *Mol. Genet. Metab.* **2012**, *107*, 294–301. [CrossRef]
18. Trefz, F.K.; Burton, B.K.; Longo, N.; Casanova, M.M.-P.; Gruskin, D.J.; Dorenbaum, A.; Kakkis, E.D.; Crombez, E.A.; Grange, D.K.; Harmatz, P.; et al. Efficacy of Sapropterin Dihydrochloride in Increasing Phenylalanine Tolerance in Children with Phenylketonuria: A Phase III, Randomized, Double-Blind, Placebo-Controlled Study. *J. Pediatr.* **2009**, *154*, 700–707.e1. [CrossRef] [PubMed]
19. Utz, J.R.J.; Lorentz, C.P.; Markowitz, D.; Rudser, K.D.; Diethelm-Okita, B.; Erickson, D.; Whitley, C.B. START, a double blind, placebo-controlled pharmacogenetic test of responsiveness to sapropterin dihydrochloride in phenylketonuria patients. *Mol. Genet. Metab.* **2012**, *105*, 193–197. [CrossRef]
20. Blau, N.; Erlandsen, H. The metabolic and molecular bases of tetrahydrobiopterin-responsive phenylalanine hydroxylase deficiency. *Mol. Genet. Metab.* **2004**, *82*, 101–111. [CrossRef] [PubMed]
21. Kure, S.; Sato, K.; Fujii, K.; Aoki, Y.; Suzuki, Y.; Kato, S.; Matsubara, Y. Wild-type phenylalanine hydroxylase activity is enhanced by tetrahydrobiopterin supplementation in vivo: An implication for therapeutic basis of tetrahydrobiopterin-responsive phenylalanine hydroxylase deficiency. *Mol. Genet. Metab.* **2004**, *83*, 150–156. [CrossRef] [PubMed]
22. Erlandsen, H.; Pey, A.L.; Gámez, A.; Pérez, B.; Desviat, L.R.; Aguado, C.; Koch, R.; Surendran, S.; Tyring, S.; Matalon, R.; et al. From the Cover: Correction of kinetic and stability defects by tetrahydrobiopterin in phenylketonuria patients with certain phenylalanine hydroxylase mutations. *Proc. Natl. Acad. Sci. USA* **2004**, *101*, 16903–16908. [CrossRef] [PubMed]
23. Dobrowolski, S.F.; Pey, A.L.; Koch, R.; Levy, H.; Ellingson, C.C.; Naylor, E.W.; Martinez, A. Biochemical characterization of mutant phenylalanine hydroxylase enzymes and correlation with clinical presentation in hyperphenylalaninaemic patients. *J. Inherit. Metab. Dis.* **2008**, *32*, 10–21. [CrossRef] [PubMed]
24. Anjema, K.; Van Rijn, M.; Hofstede, F.C.; Bosch, A.M.; Hollak, C.E.; Rubio-Gozalbo, E.; De Vries, M.C.; Janssen, M.C.; Boelen, C.C.; Burgerhof, J.G.; et al. Tetrahydrobiopterin responsiveness in phenylketonuria: Prediction with the 48-hour loading test and genotype. *Orphanet J. Rare Dis.* **2013**, *8*, 103. [CrossRef]
25. Blau, N. Genetics of Phenylketonuria: Then and Now. *Hum. Mutat.* **2016**, *37*, 508–515. [CrossRef] [PubMed]
26. Karačić, I.; Meili, D.; Sarnavka, V.; Heintz, C.; Thöny, B.; Ramadža, D.P.; Fumić, K.; Mardešic, D.; Baric, I.; Blau, N. Genotype-predicted tetrahydrobiopterin (BH4)-responsiveness and molecular genetics in Croatian patients with phenylalanine hydroxylase (PAH) deficiency. *Mol. Genet. Metab.* **2009**, *97*, 165–171. [CrossRef] [PubMed]
27. Macdonald, A.; Rocha, J.C.; Van Rijn, M.; Feillet, F. Nutrition in phenylketonuria. *Mol. Genet. Metab.* **2011**, *104*, S10–S18. [CrossRef]
28. Pinto, A.; Almeida, M.F.; Macdonald, A.; Ramos, P.C.; Rocha Guimas, A.; Ribeiro, R.; Martins, E.; Bandeira, A.; Jackson, R.; van Spronsen, F.; et al. Over Restriction of Dietary Protein Allowance: The Importance of Ongoing Reassessment of Natural Protein Tolerance in Phenylketonuria. *Nutrients* **2019**, *11*, 995. [CrossRef]
29. Thiele, A.G.; Rohde, C.; Mütze, U.; Arelin, M.; Ceglarek, U.; Thiery, J.; Baerwald, C.; Kiess, W.; Beblo, S. The challenge of long-term tetrahydrobiopterin (BH4) therapy in phenylketonuria: Effects on metabolic control, nutritional habits and nutrient supply. *Mol. Genet. Metab. Rep.* **2015**, *4*, 62–67. [CrossRef]
30. Thiele, A.G.; Weigel, J.F.; Ziesch, B.; Rohde, C.; Mütze, U.; Ceglarek, U.; Thiery, J.; Müller, A.S.; Kiess, W.; Beblo, S. Nutritional Changes and Micronutrient Supply in Patients with Phenylketonuria Under Therapy with Tetrahydrobiopterin (BH4). *JIMD Rep.* **2012**, *9*, 31–40. [CrossRef]
31. Macdonald, A.; Ahring, K.; Dokoupil, K.; Gokmen-Ozel, H.; Lammardo, A.M.; Motzfeldt, K.; Robert, M.; Rocha, J.C.; Van Rijn, M.; Bélanger-Quintana, A. Adjusting diet with sapropterin in phenylketonuria: What factors should be considered? *Br. J. Nutr.* **2011**, *106*, 175–182. [CrossRef]
32. Moher, D.; Liberati, A.; Tetzlaff, J.; Altman, D.G.; The PRISMA Group. Preferred reporting items for systematic reviews and meta-analyses: The PRISMA statement. *PLoS Med.* **2009**, *6*, e1000097. [CrossRef]
33. National Heart, Lung and Blood Institute (NHLBI). Quality Assessment Tool for Before-After (Pre-Post) Studies with No Control Group. Available online: https://www.nhlbi.nih.gov/health-topics/study-quality-assessment-tools (accessed on 30 September 2020).
34. R Development Core Team. *R: A Language and Environment for Statistical Computing*; Foundation for Statistical Computing: Vienna, Austria, 2020.
35. Higgins, J.P.T.; Li, T.; Deeks, J.J. Chapter 6: Choosing effect measures and computing estimates of effect. In *Cochrane Handbook for Systematic Reviews of Interventions*, 2nd ed.; Higgins, J.P.T., Thomas, J., Chandler, J., Cumpston, M., Li, T., Page, M.J., Welch, V.A., Eds.; John Wiley & Sons: Chichester, UK, 2019.
36. Aldámiz-Echevarría, L.; Bueno, M.A.; Couce, M.L.; Lage, S.; Dalmau, J.; Vitoria, I.; Andrade, F.; Llarena, M.; Blasco-Alonso, J.; Alcalde, C.; et al. Tetrahydrobiopterin therapy vs phenylalanine-restricted diet: Impact on growth in PKU. *Mol. Genet. Metab.* **2013**, *109*, 331–338. [CrossRef]
37. Aldámiz-Echevarría, L.; Bueno, M.A.; Couce, M.L.; Lage, S.; Dalmau, J.; Vitoria, I.; Llarena, M.; Andrade, F.; Blasco-Alonso, J.; Alcalde, C.; et al. 6R-tetrahydrobiopterin treated PKU patients below 4years of age: Physical outcomes, nutrition and genotype. *Mol. Genet. Metab.* **2015**, *115*, 10–16. [CrossRef] [PubMed]

38. Bélanger-Quintana, A.; García, M.J.; Castro, M.; Desviat, L.R.; Pérez, B.; Mejía, B.; Ugarte, M.; Martínez-Pardo, M. Spanish BH4-responsive phenylalanine hydroxylase-deficient patients: Evolution of seven patients on long-term treatment with tetrahydrobiopterin. *Mol. Genet. Metab.* **2005**, *86*, 61–66. [CrossRef] [PubMed]
39. Brantley, K.D.; Douglas, T.D.; Singh, R.H. One-year follow-up of B vitamin and Iron status in patients with phenylketonuria provided tetrahydrobiopterin (BH4). *Orphanet J. Rare Dis.* **2018**, *13*, 192. [CrossRef] [PubMed]
40. Burlina, A.; Blau, N. Effect of BH4 supplementation on phenylalanine tolerance. *J. Inherit. Metab. Dis.* **2008**, *32*, 40–45. [CrossRef]
41. Demirdas, S.; Maurice-Stam, H.; Boelen, C.C.; Hofstede, F.C.; Janssen, M.C.; Langendonk, J.G.; Mulder, M.F.; Rubio-Gozalbo, M.E.; Van Spronsen, F.J.; De Vries, M.; et al. Evaluation of quality of life in PKU before and after introducing tetrahydrobiopterin (BH4); a prospective multi-center cohort study. *Mol. Genet. Metab.* **2013**, *110*, S49–S56. [CrossRef]
42. Douglas, T.D.; Jinnah, H.A.; Bernhard, D.; Singh, R.H. The effects of sapropterin on urinary monoamine metabolites in phenylketonuria. *Mol. Genet. Metab.* **2013**, *109*, 243–250. [CrossRef]
43. Douglas, T.D.; Ramakrishnan, U.; Kable, A.J.; Singh, R.H. Longitudinal quality of life analysis in a phenylketonuria cohort provided sapropterin dihydrochloride. *Health Qual. Life Outcomes* **2013**, *11*, 218. [CrossRef]
44. Evers, R.A.; Van Wegberg, A.M.; Van Dam, E.; De Vries, M.C.; Janssen, M.C.; Van Spronsen, F.J. Anthropomorphic measurements and nutritional biomarkers after 5 years of BH 4 treatment in phenylketonuria patients. *Mol. Genet. Metab.* **2018**, *124*, 238–242. [CrossRef]
45. Lambruschini, N.; Pérez-Dueñas, B.; Vilaseca, M.A.; Mas, A.; Artuch, R.; Gassió, R.; Gómez, L.; Gutiérrez, A.; Campistol, J. Clinical and nutritional evaluation of phenylketonuric patients on tetrahydrobiopterin monotherapy. *Mol. Genet. Metab.* **2005**, *86*, 54–60. [CrossRef] [PubMed]
46. Leuret, O.; Barth, M.; Kuster, A.; Eyer, D.; De Parscau, L.; Odent, S.; Gilbert-Dussardier, B.; Feillet, F.; Labarthe, F. Efficacy and safety of BH4 before the age of 4 years in patients with mild phenylketonuria. *J. Inherit. Metab. Dis.* **2012**, *35*, 975–981. [CrossRef]
47. Scala, I.; Concolino, D.; Della Casa, R.; Nastasi, A.; Ungaro, C.; Paladino, S.; Capaldo, B.; Ruoppolo, M.; Daniele, A.; Bonapace, G.; et al. Long-term follow-up of patients with phenylketonuria treated with tetrahydrobiopterin: A seven years experience. *Orphanet J. Rare Dis.* **2015**, *10*, 12–14. [CrossRef] [PubMed]
48. Singh, R.H.; Quirk, M.E. Using change in plasma phenylalanine concentrations and ability to liberalize diet to classify responsiveness to tetrahydrobiopterin therapy in patients with phenylketonuria. *Mol. Genet. Metab.* **2011**, *104*, 485–491. [CrossRef]
49. Singh, R.H.; Quirk, M.E.; Douglas, T.D.; Brauchla, M.C. BH4 therapy impacts the nutrition status and intake in children with phenylketonuria: 2-year follow-up. *J. Inherit. Metab. Dis.* **2010**, *33*, 689–695. [CrossRef]
50. Ünal, Ö.; Gökmen-Özel, H.; Coşkun, T.; Özgül, R.K.; Yücel, D.; Hişmi, B.; Tokatlı, A.; Dursun, A.; Sivri, H.S. Sapropterin dihydrochloride treatment in Turkish hyperphenylalaninemic patients under age four. *Turk. J. Pediatr.* **2015**, *57*, 213–218. [PubMed]
51. Vilaseca, M.A.; Lambruschini, N.; Gomez-Lopez, L.; Gutiérrez, A.; Moreno, J.; Tondo, M.; Artuch, R.; Campistol, J. Long-chain polyunsaturated fatty acid status in phenylketonuric patients treated with tetrahydrobiopterin. *Clin. Biochem.* **2010**, *43*, 411–415. [CrossRef] [PubMed]
52. Gökmen Özel, H.; Ünal, C.; Dursun, A.; Tokatli, A.; Çoşkun, T.; Sivri, H.S. Does diet liberalization impact anthropometric measures in PKU patients on relatively long-term BH4 treatment. *J. Inherit. Metab. Dis.* **2014**, *37*, 27–185. [CrossRef]
53. Paras, A.; Bausell, H.; Arduini, K.; Johnson, A.; Kalb, F.; Widera, S.; Burton, B. Sapropterin Dihydrochloride As Sole Treatment In Subset of Patients With Non-Classical Phenylketonuria. In Proceedings of the American College of Medical Genetics Annual Clinical Genetics Meeting 2018, Charlotte, NC, USA, 10 April 2018.
54. Rocha, J.C.; Almeida, M.; Rocha, S.; Guimas, A.; Ribeiro, R.; Martins, E.; Bandeira, A.; Borges, N.; Macdonald, A.; Van Spronsen, F. Nutritional status in BH4 treated patients with phenylketonuria: Preliminary data from TNSPKU project. *J. Inborn Errors Metab. Screen.* **2017**, *5*, 90–91. [CrossRef]
55. Energy and Protein Requirements. In *Report of a joint FAO/WHO/UNU Expert Consultation*; World Health Organization Technical Report Series; World Health Organization: Geneva, Switzerland, 1985; Volume 724, pp. 1–206.
56. Lindegren, M.L.; Krishnaswami, S.; Reimschisel, T.; Fonnesbeck, C.; Sathe, N.A.; McPheeters, M.L. A Systematic Review of BH4 (Sapropterin) for the Adjuvant Treatment of Phenylketonuria. *JIMD Rep.* **2012**, *8*, 109–119. [CrossRef]
57. Somaraju, U.R.; Merrin, M. Sapropterin dihydrochloride for phenylketonuria. *Cochrane Database Syst. Rev.* **2015**, *2015*, CD008005. [CrossRef] [PubMed]
58. Qu, J.; Yang, T.; Wang, E.; Li, M.; Chen, C.; Ma, L.; Zhou, Y.; Cui, Y. Efficacy and safety of sapropterin dihydrochloride in patients with phenylketonuria: A meta-analysis of randomized controlled trials. *Br. J. Clin. Pharmacol.* **2019**, *85*, 893–899. [CrossRef]
59. Bueno, M.A.; Lage, S.; Delgado, C.; Andrade, F.; Couce, M.L.; González-Lamuño, M.; Pérez, M.; Aldámiz-Echevarría, L. New evidence for assessing tetrahydrobiopterin (BH4) responsiveness. *Metabolism* **2012**, *61*, 1809–1816. [CrossRef] [PubMed]
60. Van Wegberg, A.M.; Evers, R.A.; Van Dam, E.; De Vries, M.C.; Janssen, M.C.; Heiner-Fokkema, M.R.; Van Spronsen, F.J. Does the 48-hour BH4 loading test miss responsive PKU patients? *Mol. Genet. Metab.* **2020**, *129*, 186–192. [CrossRef]
61. Ozel, H.G.; Lammardo, A.; Motzfeldt, K.; Robert, M.; Rocha, J.C.; Van Rijn, M.; Ahring, K.; Belanger-Quintana, A.; Macdonald, A.; Dokoupil, K. Use of sapropterin in the management of phenylketonuria: Seven case reports. *Mol. Genet. Metab.* **2013**, *108*, 109–111. [CrossRef] [PubMed]
62. Trefz, F.K.; Scheible, D.; Frauendienst-Egger, G.; Korall, H.; Blau, N. Long-term treatment of patients with mild and classical phenylketonuria by tetrahydrobiopterin. *Mol. Genet. Metab.* **2005**, *86*, 75–80. [CrossRef]
63. Muntau, A.C.; Du Moulin, M.; Feillet, F. Diagnostic and therapeutic recommendations for the treatment of hyperphenylalaninemia in patients 0–4 years of age. *Orphanet J. Rare Dis.* **2018**, *13*, 1–9. [CrossRef]

nutrients

MDPI

Article

Growth and Body Composition in PKU Children—A Three-Year Prospective Study Comparing the Effects of L-Amino Acid to Glycomacropeptide Protein Substitutes

Anne Daly [1,*], Wolfgang Högler [2], Nicola Crabtree [1], Nick Shaw [1], Sharon Evans [1], Alex Pinto [1], Richard Jackson [3], Boyd J. Strauss [4,5], Gisela Wilcox [4,6], Júlio C. Rocha [7,8], Catherine Ashmore [1] and Anita MacDonald [1]

1 Birmingham Women's and Children's Hospital, NHS Foundation Trust, Birmingham B4 6NH, UK; nicola.crabtree@nhs.net (N.C.); nick.shaw@nhs.net (N.S.); evanss21@me.com (S.E.); alex.pinto@nhs.net (A.P.); catherine.ashmore@nhs.net (C.A.); anita.macdonald@nhs.net (A.M.)
2 Department of Paediatrics and Adolescent Medicine, Johannes Kepler University, Kepler University Hospital, Krankenhausstraße 26-30, 4020 Linz, Austria; wolfgang.hoegler@kepleruniklinikum.at
3 Liverpool Clinical Trials Centre, University of Liverpool, Brownlow Hill, Liverpool L69 3GL, UK; r.j.jackson@liverpool.ac.uk
4 School of Medical Sciences, Faculty of Biology, Medicine and Health Sciences, University of Manchester, Manchester M13 9PL, UK; boyd.strauss@manchester.ac.uk (B.J.S.); gisela.wilcox@srft.nhs.uk (G.W.)
5 School of Clinical Sciences, Faculty of Medicine, Nursing and Health Sciences, Monash University, Clayton, VIC 3800, Australia
6 The Mark Holland Metabolic Unit, Salford Royal Foundation NHS Trust, Ladywell NW2, Salford, Manchester M6 8HD, UK
7 Nutrition and Metabolism, NOVA Medical School, Faculdade de Ciências Médicas, Universidade Nova de Lisboa, 1169-056 Lisboa, Portugal; rochajc@nms.unl.pt
8 Centre for Health and Technology and Services Research (CINTESIS), 4200-450 Porto, Portugal
* Correspondence: a.daly3@nhs.net

Citation: Daly, A.; Högler, W.; Crabtree, N.; Shaw, N.; Evans, S.; Pinto, A.; Jackson, R.; Strauss, B.J.; Wilcox, G.; Rocha, J.C.; et al. Growth and Body Composition in PKU Children—A Three-Year Prospective Study Comparing the Effects of L-Amino Acid to Glycomacropeptide Protein Substitutes. *Nutrients* **2021**, *13*, 1323. https://doi.org/10.3390/nu13041323

Academic Editor: Shanon L. Casperson

Received: 1 April 2021
Accepted: 10 April 2021
Published: 16 April 2021

Publisher's Note: MDPI stays neutral with regard to jurisdictional claims in published maps and institutional affiliations.

Copyright: © 2021 by the authors. Licensee MDPI, Basel, Switzerland. This article is an open access article distributed under the terms and conditions of the Creative Commons Attribution (CC BY) license (https://creativecommons.org/licenses/by/4.0/).

Abstract: Protein quality and quantity are important factors in determining lean body (muscle) mass (LBM). In phenylketonuria (PKU), protein substitutes provide most of the nitrogen, either as amino acids (AA) or glycomacropeptide with supplementary amino acids (CGMP-AA). Body composition and growth are important indicators of long-term health. In a 3-year prospective study comparing the impact of AA and CGMP-AA on body composition and growth in PKU, 48 children were recruited. $N = 19$ (median age 11.1 years, range 5–15 years) took AA only, $n = 16$ (median age 7.3 years, range 5–15 years) took a combination of CGMP-AA and AA, (CGMP50) and 13 children (median age 9.2 years, range 5–16 years) took CGMP-AA only (CGMP100). A dual energy X-ray absorptiometry (DXA) scan at enrolment and 36 months measured LBM, % body fat (%BF) and fat mass (FM). Height was measured at enrolment, 12, 24 and 36 months. No correlation or statistically significant differences (after adjusting for age, gender, puberty and phenylalanine blood concentrations) were found between the three groups for LBM, %BF, FM and height. The change in height z scores, (AA 0, CGMP50 +0.4 and CGMP100 +0.7) showed a trend that children in the CGMP100 group were taller, had improved LBM with decreased FM and % BF but this was not statistically significant. There appeared to be no advantage of CGMP-AA compared to AA on body composition after 3-years of follow-up. Although statistically significant differences were not reached, a trend towards improved body composition was observed with CGMP-AA when it provided the entire protein substitute requirement.

Keywords: phenylketonuria; body composition; glycomacropeptide; protein substitute

1. Introduction

There are concerns about increasing obesity and consequential long-term comorbidities in both the general and phenylketonuria (PKU) populations [1–3]. A reliance on an "artificial" diet may alter normal physiological processes such as the distribution of fat and lean mass, adversely affecting long-term health outcomes [4]. Body composition is a key component of health, and it typically refers to the quantification of body fat and muscle mass—changes that cannot be adequately assessed by body weight or body mass index (BMI) [5,6]. In PKU, reports of body composition are few and there are no long-term prospective studies or systematic/meta-analyses reviewing body composition. Therefore, it is difficult to extrapolate any association between body composition and other factors such as muscular fitness, adiposity and longer-term health outcomes.

In classical PKU, a low-phenylalanine diet requires substantial modification of usual dietary patterns. For most patients with PKU, high-biological-value proteins are excluded (e.g., meat, fish, eggs and dairy products), with low-phenylalanine/phenylalanine-free protein substitutes providing the principle source of obligatory nitrogen, which is essential to maintain metabolic control and enable optimal growth and lean body mass [7,8]. Muscle contributes up to 40–45% of body weight, and skeletal muscle is the largest store of peptides and free amino acids [9]. Reliance on "synthetic" nitrogen sources may compromise body composition; lean mass is dependent on amino acid availability and, compared to natural protein, delivery and utilisation of amino acids from protein substitutes is sub-optimal [10–12].

A consistent finding from a systematic meta-analysis in healthy subjects [13] is that lean mass is a strong predictor of bone mass. Lean mass, and therefore bone mass, particularly in teenagers with PKU, may be compromised as it coincides with a time when adherence with diet and protein substitutes is being challenged [14,15]. Additionally, the physiological increase in lean body mass and fat mass differs in adolescent boys and girls, further exacerbating the difficulties in body composition interpretation [16].

The synthetic protein given in PKU is usually derived from amino acids without phenylalanine (AA). More recently, casein glycomacropeptide (CGMP), a bioactive phosphoglycopeptide, has been used as an alternative low-phenylalanine protein substitute (CGMP-AA). It is associated with better palatability, so adherence is improved [17], but it is unknown if this bioactive macropeptide will alter body composition in PKU. The type of protein, its absorption and amino acid composition alters insulin and glucagon responses. Insulin stimulates protein synthesis [18,19], while glucagon increases amino acid catabolism. Some non-PKU studies have shown that nitrogen retention is improved when protein is in the form of oligopeptides compared to whole protein or amino acids [20–22]. It is possible that CGMP, a whey-derived macropeptide, may promote nitrogen retention, improving lean body mass synthesis [23] and growth potential in children with PKU.

This 3-year, prospective, longitudinal study in children with PKU aimed to compare the impact of two different sources of protein substitute, AA and low-phenylalanine CGMP-AA, on growth and body composition by comparing height, lean body mass and fat mass.

2. Materials and Methods

2.1. Methods

Children were included in the study if they were diagnosed with PKU by newborn screening, aged 5–16 years of age, on dietary treatment only and adherent with protein substitute, with 70% of routine blood phenylalanine concentrations within phenylalanine target range for 6 months before study enrolment. Target blood phenylalanine range for children aged 5–12 years was 120 to $\leq$ 360 µmol/L and for 13 years and older was 120 to < 600 µmol/L, as recommended by the European PKU guidelines [24].

CGMP-AA and AA Protein Substitutes

Two types of protein substitute were studied: AA and CGMP-AA. AA were either powders made up with water to a semi-solid consistency or ready-to-drink liquids providing 10, 15 or 20 g of protein equivalent, tailored to a child's protein requirements. The CGMP-AA powdered protein substitute (a test product via Vitaflo International Ltd. Liverpool, UK) contained 36 mg of phenylalanine for each 20 g protein equivalent and was reconstituted by adding 120 mL of water. Both products had a similar energy profile per 20 g protein equivalent; CGMP-AA, 120 Kcal, 6.5 g carbohydrate and 1.5 g fat; AA, 124 Kcal, 9.4 g carbohydrate and 0.7 g fat. Threonine and leucine were higher in the CGMP-AA product.

2.2. *Study Design*

In this prospective, longitudinal, 3-year study, home visits were conducted 3 monthly collecting dietary information, weight and height. Dual energy X-ray absorptiometry scans (DXA) measured body composition at enrolment and at 36 months. At enrolment, all the children were on AA protein substitute and had a Tanner pubertal assessment. Following the DXA scan, the patients were divided into 3 subgroups:

(1) AA: protein substitute given as AA only;
(2) CGMP50: patients tolerating a combination of CGMP-AA and AA;
(3) CGMP100: patients tolerating all their protein substitute as CGMP-AA.

Due to the negative impact on blood phenylalanine control, only some children were able to meet their protein requirements using only CGMP-AA [25]. Therefore, in addition to the AA group, a third group (CGMP50) was introduced where a combination of CGMP-AA and AA provided approximately 50% of the protein equivalent intake.

2.2.1. Selection into CGMP-AA or AA Group

The children chose CGMP-AA or AA, depending on their taste preference. Those in the CGMP-AA group entered CGMP50 or CGMP100 groups depending on their phenylalanine blood concentrations.

2.2.2. Dual X-ray Absorptiometry (DXA)

A DXA scan of the total body to assess body composition (fat and lean body mass) was carried out by two trained operators, using a GE Lunar iDXA and Encore ™ software version 13.1 (GE Healthcare, Wisconsin, MD, USA). Trunk thickness and body weight were utilised to ensure that each child was scanned in the most appropriate acquisition mode. Children lay supine on a bed, while the DXA scan was completed. At baseline and 36 months, the following parameters were measured: lean body mass (LBM) g, fat mass (FM) g, % body fat (%BF), weight (kg), height (cm) and body mass index (BMI) (kg/m^2). Daily quality assurance tests were performed according to the manufacturer's instructions. The precision of the instrument was calculated as 1.0% for fat and 0.5% for lean in normal-weight subjects.

2.2.3. Anthropometric Measurements

Weight and height were measured by one of two metabolic dietitians. Height was measured using a Harpenden stadiometer (Holtain Ltd., Crymych, UK) and weight on calibrated digital scales (Seca, Medical Measuring Systems and Scales, Birmingham, UK. Model 875); weight was measured to the nearest 0.1 g and height to the nearest 0.1 cm. Weight, height and BMI were analysed over four time points; baseline, 12, 24 and 36 months.

2.2.4. Blood Phenylalanine Levels

Throughout the study, trained caregivers collected weekly early morning fasted blood spots on filter cards, Perkin Elmer 226 (UK Standard NBS) at home. Blood specimens were sent via first class post to the laboratory at Birmingham Children's Hospital. All the cards had a standard thickness, and the blood phenylalanine concentrations were calculated on a 3.2 mm punch by MS/MS tandem mass spectrometry.

2.2.5. Pubertal Status

A general medical examination was conducted and pubertal status was measured at enrolment using the Tanner picture index [26]. Stage 1 and 2 were classified as pre-pubertal and stage 3, 4 and 5 as pubertal.

2.3. Statistical Analysis of Anthropometry and Body Composition

Continuous data are presented as medians with associated inter-quartile ranges (IQR); categorical data are presented as frequencies of counts with associated percentages. Outcome data were divided into anthropometric data, weight (kg), height (cm) and body mass index (BMI) (kg/m^2), which were measured as standardised scores, and body composition was measured as lean body mass g, % body fat (%BF) and fat mass g. Anthropometric and body composition data were compared with blood phenylalanine concentrations. Standardised height was represented as the change in height and height z scores at each time point relative to baseline. Given the number of patients and the difference in ages between the groups, analysis was performed using longitudinal regression, which adjusts for patient age. Although standardised measures implicitly account for patient age, it was retained as a covariate in the analysis to avoid any confounding due to age when comparing treatment groups.

Correlations between the outcome data were calculated using Pearson's correlation coefficient. Anthropometric data were analysed using longitudinal modelling techniques, including main effects for time and age and evaluating the effect of treatment within each time-point. Models were further adjusted for patient age, accounting for the differences in the enrolment age between the treatment groups. Body composition outcomes were measured at two time points; data were analysed using analysis of covariance (ANCOVA) techniques, analysing the 36-month data as the outcome and adjusting for the enrolment data, including patient age at enrolment as well as their pre-pubescent status and gender.

Power Calculation

This prospective intervention study took as a primary outcome measure a conservative difference between CGMP and AA groups using day-to-day blood phenylalanine concentrations from a previous study. Twenty children maintaining blood phenylalanine concentrations between 100 and 400 µmol/L would detect a 5% reduction in blood phenylalanine concentrations outside the expected target range, at a power of detection of 88% and at a significance level of $p = 0.05$. A minimum of 45 children was the target recruitment aim.

2.4. Ethical Permission

The South Birmingham Research Ethics committee granted a favourable ethical opinion, reference 13/WM/0435 and IRAS (integrated research application system) number 129497. Written informed consent was obtained for all subjects from at least one caregiver with parental responsibility and written assent obtained from the subject if appropriate for their age and level of understanding.

3. Results

3.1. Subjects

Fifty children (28 boys, 22 girls) with PKU were recruited. Forty-seven children were European and three were of Asian origin. Forty-eight completed the study: 29 in the CGMP-AA group and 19 in the AA group. A significant difference in age was noted between the AA and CGMP50 groups ($p = 0.005$) and between the CGMP50 and CGMP100 groups ($p = 0.04$) (Table 1).

Table 1. Subject characteristics at recruitment.

	AA	CGMP50	CGMP100
Number recruited	$n = 19$	$n = 16$	$n = 13$
Girls	$n = 8$	$n = 8$	$n = 5$
Boys	$n = 11$	$n = 8$	$n = 8$
Median age y (range)	11.1 (5–15)	7.3 (5–15)	9.2 (5–16)
% of children prepubertal (stage 1 and 2)	32%	69%	62%
Girls	$n = 2$	$n = 6$	$n = 5$
Boys	$n = 4$	$n = 5$	$n = 3$
% of children pubertal (stage 3 to 5)	68%	31%	38%
Girls	$n = 6$	$n = 2$	$n = 0$
Boys	$n = 7$	$n = 3$	$n = 5$

AA, amino acid; CGMP, glycomacropeptide; CGMP50, protein substitute based on combination of CGMP and AA; CGMP100, protein substitute based on CGMP only.

3.2. Pubertal Status

Pre-pubertal status (stage 1 and 2): 32% ($n = 6/19$) were pre-pubertal in the AA group, 69% ($n = 11/16$) in the CGMP50 group and 62% ($n = 8/13$) in the CGMP100 group.

Late puberty (stage 3 to 5): 68% ($n = 13/19$) in the AA group, 31% ($n = 5/16$) in the CGMP50 group and 38% ($n = 5/13$) in the CGMP100 group.

All had classical PKU, except two with mild PKU based on untreated blood phenylalanine levels at diagnosis and dietary phenylalanine tolerance.

3.3. Subject Withdrawal

One boy and one girl (aged 12 and 11 years, respectively) in the CGMP-AA group were excluded from the study, as both were unable to adhere to the study protocol. One failed to return blood phenylalanine samples and both had poor adherence to their phenylalanine-restricted diet.

3.4. Protein Substitutes and Phenylalanine Concentrations

We have previously reported the types/manufacturers of protein substitutes taken by the AA, CGMP50 and CGMP100 groups. Similarly, the median phenylalanine concentrations have been reported at baseline and year 3 [27]. Median phenylalanine concentrations were within recommended target reference ranges for children aged ≤ 11 and ≥ 12 years old [24].

The median daily dose of protein equivalent from protein substitute was 60 g/day (range 40–80 g), and the median amount of prescribed natural protein was 5.5 g protein/day (range 3–30 g) or 275 mg/day of phenylalanine (range 150–1500 mg) in all groups.

3.5. Body Composition Lean Mass, Fat Mass and % Body Fat

Body composition was analysed using ANCOVA, adjusting for patient age, gender, phenylalanine concentration and pre-pubescent status (Table 2). No statistically significant differences were found between the three treatment groups for lean body mass, %BF or fat mass. All parameters increased over the 3-year study period.

3.6. Lean Body Mass, % Body Fat and Fat Mass

ANCOVA showed no significant differences in lean body mass, fat mass or % body fat between the treatment groups, although a trend for improved lean body mass, fat mass and % body fat was observed in the CGMP100 group.

Table 2. Median (range) lean mass, fat mass and % body fat in the AA, CGMP50 and CGMP 100 groups at enrolment and 36 months.

Body Composition	Time of Assessment	AA (Range) $n = 19$	GMP50 (Range) $n = 13$	GMP100 (Range) $n = 16$
Lean mass (g)	Enrolment	26,702 (16,920–34,209)	16,334 (14,280–17,686)	20,060 (16,451–21,947)
	36 m	32,560 (25,893–40,511)	23,921 (22,725–26,477)	31,268 (25,561–35,875)
Delta		5858 (8973–6302)	7587 (8445–8791)	11,208 (9110–13,928)
Fat mass (g)	Enrolment	9528 (6961–15,018)	5764 (4504–6758)	6688 (5057–8811)
	36 m	17,216 (10,930–20,687)	12,945 (10,678–16,519)	12,220 (8347–13,101)
Delta		7688 (3969–5669)	7181 (6174–9761)	5532 (3290–4290)
% body fat	Enrolment	29 (23–36)	24 (22–28)	25 (19–30)
	36 m	35 (25–39)	33 (30–36)	28 (20–33)
Delta		6	9	3

AA, amino acid; CGMP, glycomacropeptide; CGMP50, protein substitute based on combination of CGMP and AA; CGMP100, protein substitute based on CGMP only; g, grams; kg, kilograms.

3.7. Changes in Height Z Scores

Accounting for the age and gender differences, there were no statistically significant differences for height within or between the groups. At the end of the 3-year study, all groups had a positive height z score. We have previously reported weight and BMI z scores over the 3 year period [27], showing no statistical differences between the groups (Tables 3 and 4, Figure 1).

Table 3. Median z scores (range) for height in AA, CGMP50 and CGMP100 groups measured annually from enrolment to 36 months in PKU children.

Time (Months)	AA Height z Score $n = 19$	CGMP50 Height z Score $n = 16$	CGMP100 Height z Score $n = 13$
Enrolment (range)	0.2 (−0.2 to 0.8)	−0.1 (−0.6 to 0.6)	−0.1 (−0.4 to 0.3)
12 months (range)	0.2 (−0.2 to 0.6)	0.1 (−0.4 to 0.5)	0.1 (−0.1 to 0.3)
24 months (range)	0.2 (−0.1 to 0.5)	0.2 (−0.2 to 0.5)	0.4 (0.0 to 0.7)
36 months (range)	0.2 (0.0 to 0.5)	0.3 (−0.1 to 0.7)	0.6 (0.1 to 0.7)
Delta height z score	0	+0.4	+0.7

AA, amino acid; CGMP, glycomacropeptide; CGMP50, protein substitute based on combined CGMP and AA; CGMP100, protein substitute based on CGMP only.

Statistical modelling showed a trend in the CGMP100 group towards improved growth and a reduction in total body fat percentage and improved lean body mass. Analysis of the delta change in height was divided by age for those ≤ 10 and ≥ 11 years.

Table 4. ANOVA regression model showing values for age, treatment (AA, CGMP50, CGMP100) and treatment/time.

	Df	Sum Sq	Mean Sq	*F* Value	Pr (>F)
Age	1	0.015	0.015	0.020	0.887
Treatment	2	1.225	0.613	0.843	0.432
Treatment/time	3	2.081	0.694	0.954	0.415
Residuals	185	134.418	0.727		

1 = AA, 2 = CGMP50, 3 = CGMP100.

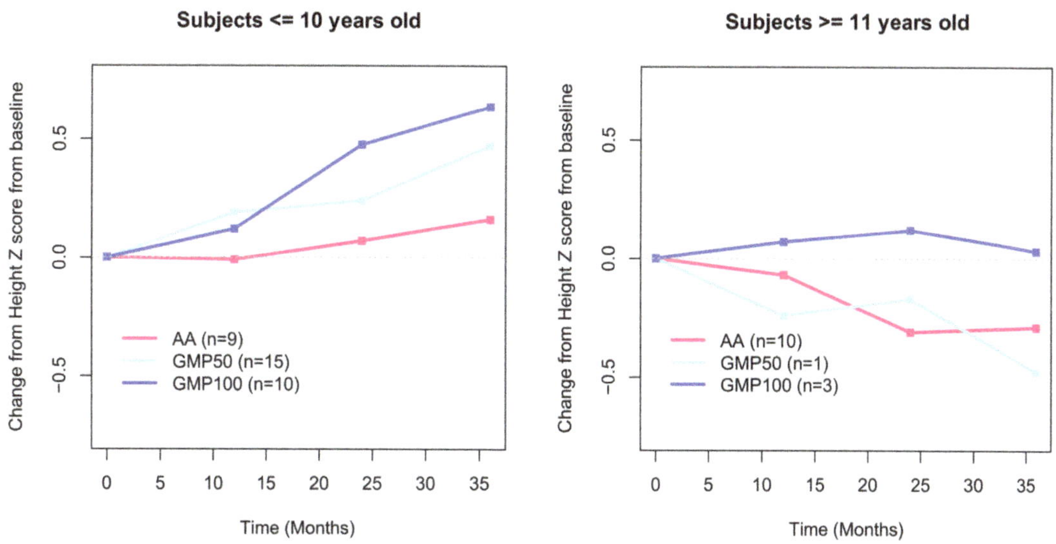

Figure 1. Change in median height z score from baseline to 36 months in the amino acids (AA), GCMP100 and CGMP 50 groups. Legend: AA, amino acid; CGMP, glycomacropeptide; CGMP50, protein substitutes based on combined CGMP and AA; CGMP100, protein substitute based on CGMP only.

4. Discussion

Body composition and height over 36 months in a group of children with PKU taking CGMP or AA protein substitutes showed no statistically significant changes in any of the measured parameters. However, in the CGMP100 group, statistical modelling indicated a trend ($p = 0.42$) towards improved longitudinal growth, a reduction in fat mass and % body fat and improved lean body mass. When growth was represented as a median change from baseline over time, it showed that the CGMP100 group had the greatest change in height. However, age modified this trend and, although the CGMP100 group continued to show improved height growth, it did not reach significance.

We can only speculate about a suggested trend in improved body composition when taking CGMP as the complete source of protein substitute. One possible explanation is the bioactivity of CGMP; it is rich in the branched chain amino acids isoleucine and leucine, which are potent modulators of protein turnover and have been shown to have a significant effect on insulin and glucose metabolism [28,29]. If CGMP, by the action of these amino acids, improves insulin sensitivity, it is possible that growth may be improved. However, we did not collect any information on insulin resistance in this group of subjects, so it is difficult to draw any firm conclusions.

Huemer et al. [30] measured growth and body composition over 12 months in 34 children with classical PKU. Total protein intake was 124% of the German recommended daily allowance. A significant correlation was found between lean body mass and intake of natural protein, suggesting that improved natural protein intake was beneficial. Evans et al. [31] also reported a similar significant relationship between a lower % fat mass and a higher total, natural and protein substitute intake, with natural protein > 0.5 g/kg/day associated with an improved body composition; no relationship was found between natural protein intake and improved height z scores. Evans, similarly, to Hoeksma et al. [32], observed that neither natural protein nor energy intake correlated with linear growth, as reported by Aldamiz Echevarria et al. [33]. The effect of a low-protein diet on energy balance and postprandial fat oxidation has received little attention in subjects with PKU. A study by Alfheeaid et al. [34] reported a lower thermal effect of feeding and fat oxidation after healthy subjects had taken a meal containing special low-protein foods and protein substitutes, possibly leading to a higher fat mass and altered body composition. Patients with milder PKU, responsive to sapropterin dihydrochloride (BH4), have a higher natural protein tolerance but there appeared to be no advantage in height, weight, body mass index or growth velocity when BH4 was compared to conventional PKU therapy [35]. There are no studies reporting body composition in BH4-responsive patients who use fewer special low-protein food products and have a wider range of natural protein sources compared to the classically treated patients with PKU.

The importance of adequate protein intake from protein substitutes (both quality and quantity) in patients with PKU has been documented by many authors [36–41]. No studies have identified the protein digestibility score or absorption kinetics of CGMP protein substitutes; this is important to ascertain protein efficiency. In healthy adults, protein-containing meals taken at regular intervals improve skeletal muscle protein by 25%, reinforcing the need to consume protein substitutes in divided doses [42–44]. The optimal amount of protein substitute based on free amino acids or CGMP remains undefined, but any factor leading to protein inefficiency may compromise body composition and optimal height and increase the incidence of overweight. Another confounding factor that may affect body composition is the effect of a long-term low-phenylalanine diet higher in carbohydrates, which may be associated with a higher risk of adiposity and insulin resistance [45,46]. All of these factors may lead to under achievement of optimal growth potential in children with PKU [47].

Comparison of body composition by gender, regardless of group, showed that lean body mass was statistically significantly higher in males than females, consistent with reports in the literature ($p = 0.013$) [48]. Fat mass and lean body mass vary with age, gender and pubertal status. Various authors have reported an age-related increase in lean body mass index being more rapid in males compared with females, particularly between the ages of 11 and 16 years, which is in line with the rapid accrual of lean body mass during male puberty [49]. Children gain lean mass disproportionately to height and this is more pronounced in boys compared to girls [48].

A multitude of methods exist for assessing body composition, including DXA, bioelectrical impedance (BIA) and whole-body air displacement plethysmograph (Bodpod), each having their own assumptions, advantages and inadequacies [50]. Unfortunately, there is a lack of standardised reference data, making interpretation and comparison of results challenging. Sensitive and accurate measurements are needed to detect differences in visceral, compared to central, fat accumulation, as ponderal and body mass index alone are unable to detect subtle differences. The gold standard for measuring body composition is a four-compartment (4C) model [51]. DXA has been evaluated against 4C models in children, and although it overestimates body fat by 1–4% depending on age, sex and body size, the correlation compared to a 4C model is good despite the small error [52,53]. Reference data for comparison of body composition parameters are limited; a recent publication by Ofenheimer [54] has produced age and gender specific reference percentiles of body composition parameters for European children and adolescents. Comparison of our data would indicate appropriate body composition for fat and lean body mass when calculated as a median between baseline and 36 months.

There are limitations to this study: we did not have a healthy reference control group or UK-based reference data to compare body composition parameters, an inherent problem when using the DXA for body composition analysis. Endocrine parameters such as bone age or growth hormone were not measured, but they may have explained differences in linear growth. Until kinetic studies are conducted, it is unknown if a peptide compared to amino acids alters the delivery and assimilation of amino acids, leading to improved lean body mass and growth. We did not collect parental height data, which may have been useful as a comparison within the groups. Not all the children were able to completely replace their full amino acid requirement with CGMP-AA, which may have reduced the strength of our findings. There were small numbers of children in each study group. More older children chose to stay on their AA supplement compared to the younger age group, who were more willing to try an alternative protein substitute, which may have led to some bias.

5. Conclusions

In this 3-year longitudinal study, we found no noticeable differences in body composition between the groups taking CGMP-AA and AA. However, there was a trend towards improved body composition in the group taking all of their protein substitute as CGMP-AA. This may suggest that CGMP does confer some biological benefit. Proof of concept will only be possible via larger controlled studies and over a longer duration throughout childhood.

Author Contributions: Conceptualization, A.D., C.A. and A.M.; Formal analysis, A.D. and R.J.; Investigation, A.D.; Methodology, A.D.; Supervision, C.A. and A.M.; Validation, C.A. and A.M.; Writing—original draft, A.D.; Writing—review & editing, A.D., W.H., N.C., N.S., S.E., A.P., R.J., B.J.S., G.W., J.C.R., C.A. and A.M. All authors have read and agreed to the published version of the manuscript.

Funding: This research was funded by Vitaflo International as part of a Ph.D. grant.

Institutional Review Board Statement: The study was conducted according to the guidelines of the Declaration of Helsinki, and approved by the Ethics Committee. The South Birmingham Research Ethics committee granted a favourable ethical opinion, reference 13/WM/0435 and IRAS (integrated research application system) number 129497.

Informed Consent Statement: Informed consent was obtained for all subjects involved in the study.

Data Availability Statement: Data available at the request of the corresponding author.

Conflicts of Interest: A.D., research funding from Vitaflo International, financial support from Nutricia, Mevalia & Vitaflo International to attend study days and conferences. S.E., research funding from Nutricia, financial support from Nutricia & Vitaflo International to attend study days and conferences. A.P. has received an educational grant from Cambrooke Therapeutics and grants from Vitaflo International, Nutricia, Merck Serono, Biomarin and Mevalia to attend scientific meetings. C.A. received honoraria from Nutricia and Vitaflo International to attend study days and conferences. J.C.R. is a member of the European Nutritionist Expert Panel (Biomarin), the Advisory Board for Applied Pharma Research and Nutricia, and received honoraria as a speaker from APR, Merck Serono, Biomarin, Nutricia, Vitaflo, Cambrooke, PIAM and Lifediet. A.M. received research funding and honoraria from Nutricia, Vitaflo International & Merck Serono, and is a member of the European Nutrition Expert Panel (Merck Serono International), the Sapropterin Advisory Board (Merck Serono international) and the Advisory Board Element (Danone-Nutricia). The funders had no role in the design of the study; in the collection, analyses, or interpretation of data; in the writing of the manuscript, or in the decision to publish the results.

References

1. Abdullah, A.; Peeters, A.; de Courten, M.; Stoelwinder, J. The magnitude of association between overweight and obesity and the risk of diabetes: A meta-analysis of prospective cohort studies. *Diabetes Res. Clin. Pract.* **2010**, *89*, 309–319. [CrossRef] [PubMed]
2. Moreno, L.A.; Pineda, I.; Rodriguez, G.; Fleta, J.; Giner, A.; Juste, M.G.; Sarria, A.; Bueno, M. Leptin and metabolic syndrome in obese and non-obese children. *Horm. Metab. Res.* **2002**, *34*, 394–399. [CrossRef]
3. Sarria, A.; Moreno, L.A.; Garcí-LIop, L.A.; Fleta, J.; Morellon, M.P.; Bueno, M. Body mass index, triceps skinfold and waist circumference in screening for adiposity in male children and adolescents. *Acta Paediatr.* **2001**, *90*, 387–392. [CrossRef]
4. Kelly, D.A. Nutritional factors affecting growth before and after liver transplantation. *Pediatr. Transpl.* **1997**, *1*, 80–84.
5. Cole, T.J.; Freeman, J.V.; Preece, M.A. Body mass index reference curves for the UK, 1990. *Arch. Dis. Child.* **1995**, *73*, 25–29. [CrossRef]
6. Javed, A.; Jumean, M.; Murad, M.H.; Okorodudu, D.; Kumar, S.; Somers, V.K.; Sochor, O.; Lopez-Jimenez, F. Diagnostic performance of body mass index to identify obesity as defined by body adiposity in children and adolescents: A systematic review and meta-analysis. *Pediatr. Obes.* **2015**, *10*, 234–244. [CrossRef]
7. Dokoupil, K.; Gokmen-Ozel, H.; Lammardo, A.M.; Motzfeldt, K.; Robert, M.; Rocha, J.C.; van Rijn, M.; Ahring, K.; Bélanger-Quintana, A.; MacDonald, A.; et al. Optimising growth in phenylketonuria: Current state of the clinical evidence base. *Clin. Nutr.* **2012**, *31*, 16–21. [CrossRef]
8. MacDonald, A.; Chakrapani, A.; Hendriksz, C.; Daly, A.; Davies, P.; Asplin, D.; Hall, K.; Booth, I.W. Protein substitute dosage in PKU: How much do young patients need? *Arch. Dis. Child.* **2006**, *91*, 588–593. [CrossRef]
9. Wu, G. Dietary protein intake and human health. *Food Funct.* **2016**, *7*, 1251–1265. [CrossRef]
10. Gropper, S.S.; Acosta, P.B. Effect of simultaneous ingestion of L-amino acids and whole protein on plasma amino acid and urea nitrogen concentrations in humans. *J. Parenter. Enteral Nutr.* **1991**, *15*, 48–53. [CrossRef]
11. Gropper, S.S.; Gropper, D.M.; Acosta, P.B. Plasma amino acid response to ingestion of L-amino acids and whole protein. *J. Pediatr. Gastroenterol. Nutr.* **1993**, *16*, 143–150. [CrossRef]
12. Monch, E.; Herrmann, M.E.; Brösicke, H.; Schöffer, A.; Keller, M. Utilisation of amino acid mixtures in adolescents with phenylketonuria. *Eur. J. Pediatr.* **1996**, *155*, S115–S120. [CrossRef]
13. Smith, J.J.; Eather, N.; Morgan, P.J.; Plotnikoff, R.C.; Faigenbaum, A.D.; Lubans, D.R. The health benefits of muscular fitness for children and adolescents: A systematic review and meta-analysis. *Sports Med.* **2014**, *44*, 1209–1223. [CrossRef] [PubMed]
14. MacDonald, A.; Van Rijn, M.; Feillet, F.; Lund, A.M.; Bernstein, L.; Bosch, A.M.; Gizewska, M.; Van Spronsen, F.J. Adherence issues in inherited metabolic disorders treated by low natural protein diets. *Ann. Nutr. Metab.* **2012**, *61*, 289–295. [CrossRef] [PubMed]
15. Walter, J.H.; White, F.J.; Hall, S.K.; MacDonald, A.; Rylance, G.; Boneh, A.; Francis, D.E.; Shortland, G.J.; Schmidt, M.; Vail, A.; et al. How practical are recommendations for dietary control in phenylketonuria? *Lancet* **2002**, *360*, 55–57. [CrossRef]
16. Kleinman, R.E.; Greer, F.R. *Pediatric Nutrition*; American Academy of Pediatrics: Elk Grove Village, IL, USA, 2019; p. 1731.
17. Lim, K.; van Calcar, S.C.; Nelson, K.L.; Gleason, S.T.; Ney, D.M. Acceptable low-phenylalanine foods and beverages can be made with glycomacropeptide from cheese whey for individuals with PKU. *Mol. Genet. Metab.* **2007**, *92*, 176–178. [CrossRef] [PubMed]
18. Boucher, J.; Softic, S.; El Ouaamari, A.; Krumpoch, M.T.; Kleinridders, A.; Kulkarni, R.N.; O'Neill, B.T.; Kahn, C.R. Differential Roles of Insulin and IGF-1 Receptors in Adipose Tissue Development and Function. *Diabetes.* **2016**, *65*, 2201–2213. [CrossRef]
19. James, H.A.; O'Neill, B.T.; Nair, K.S. Insulin Regulation of Proteostasis and Clinical Implications. *Cell Metab.* **2017**, *26*, 310–323. [CrossRef] [PubMed]
20. Adibi, S.A. Intestinal transport of dipeptides in man: Relative importance of hydrolysis and intact absorption. *J. Clin. Investig.* **1971**, *50*, 2266–2275. [CrossRef]
21. Akhavan, T.; Luhovyy, B.L.; Brown, P.H.; Cho, C.E.; Anderson, G.H. Effect of premeal consumption of whey protein and its hydrolysate on food intake and postmeal glycemia and insulin responses in young adults. *Am. J. Clin. Nutr.* **2010**, *91*, 966–975. [CrossRef]
22. Silk, D.B.; Fairclough, P.D.; Clark, M.L.; Hegarty, J.E.; Addison, J.M.; Burston, D.; Clegg, K.M.; Matthews, D.M. Use of a peptide rather than free amino acid nitrogen source in chemically defined "elemental" diets. *JPEN J. Parenter Enteral Nutr.* **1980**, *4*, 548–553. [CrossRef]
23. Van Calcar, S.C.; MacLeod, E.L.; Gleason, S.T.; Etzel, M.R.; Clayton, M.K.; Wolff, J.A.; Ney, D.M. Improved nutritional management of phenylketonuria by using a diet containing glycomacropeptide compared with amino acids. *Am. J. Clin. Nutr.* **2009**, *89*, 1068–1077. [CrossRef] [PubMed]
24. Van Wegberg, A.M.J.; MacDonald, A.; Ahring, K.; Bélanger-Quintana, A.; Blau, N.; Bosch, A.M.; Burlina, A.; Campistol, J.; Feillet, F.; Giżewska, M.; et al. The complete European guidelines on phenylketonuria: Diagnosis and treatment. *Orphanet J. Rare Dis.* **2017**, *12*, 1–56. [CrossRef] [PubMed]
25. Daly, A.; Evans, S.; Chahal, S.; Santra, S.; Macdonald, A. Glycomacropeptide in children with phenylketonuria: Does its phenylalanine content affect blood phenylalanine control? *J. Hum. Nutr. Diet.* **2017**, *30*, 515–523. [CrossRef] [PubMed]
26. Tanner, J.M.; Whitehouse, R.H. Clinical longitudinal standards for height, weight, height velocity, weight velocity, and stages of puberty. *Arch. Dis. Child.* **1976**, *51*, 170–179. [CrossRef] [PubMed]
27. Daly, A.; Evans, S.; Pinto, A.; Jackson, R.; Ashmore, C.; Rocha, J.C.; Macdonald, A. The Impact of the Use of Glycomacropeptide on Satiety and Dietary Intake in Phenylketonuria. *Nutrients* **2020**, *12*, 2704. [CrossRef] [PubMed]

28. Bifari, F.; Nisoli, E. Branched-chain amino acids differently modulate catabolic and anabolic states in mammals: A pharmacological point of view. *Br. J. Pharmacol.* **2017**, *174*, 1366–1377. [CrossRef] [PubMed]
29. Lynch, C.J.; Adams, S.H. Branched-chain amino acids in metabolic signalling and insulin resistance. *Nat. Rev. Endocrinol.* **2014**, *10*, 723–736. [CrossRef]
30. Huemer, M.; Huemer, C.; Möslinger, D.; Huter, D.; Stöckler-Ipsiroglu, S. Growth and body composition in children with classical phenylketonuria: Results in 34 patients and review of the literature. *J. Inherit. Metab. Dis.* **2007**, *30*, 694–699. [CrossRef]
31. Evans, M.; Truby, H.; Boneh, A. The relationship between dietary intake, growth and body composition in Phenylketonuria. *Mol. Genet. Metab.* **2017**, *122*, 36–42. [CrossRef] [PubMed]
32. Hoeksma, M.; Van Rijn, M.; Verkerk, P.H.; Bosch, A.M.; Mulder, M.F.; De Klerk, J.B.C.; De Koning, T.J.; Rubio-Gozalbo, E.; De Vries, M.; Sauer, P.J.J.; et al. The intake of total protein, natural protein and protein substitute and growth of height and head circumference in Dutch infants with phenylketonuria. *J. Inherit. Metab. Dis.* **2005**, *28*, 845–854. [CrossRef]
33. Aldámiz-Echevarría, L.; Bueno, M.A.; Couce, M.L.; Lage, S.; Dalmau, J.; Vitoria, I.; Andrade, F.; Blasco-Alonso, J.; Alcalde, C.; Gil, D.; et al. Anthropometric characteristics and nutrition in a cohort of PAH-deficient patients. *Clin. Nutr.* **2014**, *33*, 702–717. [CrossRef]
34. Alfheeaid, H.; Gerasimidis, K.; Năstase, A.-M.; Elhauge, M.; Cochrane, B.; Malkova, D.; Gerasimidis, K. Impact of phenylketonuria type meal on appetite, thermic effect of feeding and postprandial fat oxidation. *Clin. Nutr.* **2018**, *37*, 851–857. [CrossRef] [PubMed]
35. Aldámiz-Echevarría, L.; Bueno, M.A.; Couce, M.L.; Lage, S.; Dalmau, J.; Vitoria, I.; Andrade, F.; Llarena, M.; Blasco-Alonso, J.; Alcalde, C.; et al. Tetrahydrobiopterin therapy vs phenylalanine-restricted diet: Impact on growth in PKU. *Mol. Genet. Metab.* **2013**, *109*, 331–338. [CrossRef]
36. Acosta, P.B.; Yannicelli, S. Protein intake affects phenylalanine requirements and growth of infants with phenylketonuria. *Acta Paediatr.* **1994**, *407*, 66–67. [CrossRef] [PubMed]
37. Acosta, P.B.; Yannicelli, S.; Singh, R.; Mofidi, S.; Steiner, R.; DeVincentis, E.; Jurecki, E.; Bernstein, L.; Gleason, S.; Chetty, M.; et al. Nutrient intakes and physical growth of children with phenylketonuria undergoing nutrition therapy. *J. Am. Diet. Assoc.* **2003**, *103*, 1167–1173. [CrossRef]
38. Macdonald, A.; Rylance, G.; Davies, P.; Asplin, D.; Hall, S.K.; Booth, I.W. Administration of protein substitute and quality of control in phenylketonuria: A randomized study. *J. Inherit. Metab. Dis.* **2003**, *26*, 319–326. [CrossRef] [PubMed]
39. Rocha, J.C.; Van Rijn, M.; Van Dam, E.; Ahring, K.; Bélanger-Quintana, A.; Dokoupil, K.; Ozel, H.G.; Lammardo, A.M.; Robert, M.; Heidenborg, C.; et al. Weight Management in Phenylketonuria: What Should Be Monitored? *Ann. Nutr. Metab.* **2015**, *68*, 60–65. [CrossRef] [PubMed]
40. Weglage, J.; Brämswig, J.H.; Koch, H.G.; Karassalidou, S.; Ullrich, K. Growth in patients with phenylketonuria. *Eur. J. Nucl. Med. Mol. Imaging* **1994**, *153*, 537–538. [CrossRef]
41. Rocha, J.C.; Macdonald, A. Dietary intervention in the management of phenylketonuria: Current perspectives. *Pediatr. Heal. Med. Ther.* **2016**, *ume 7*, 155–163. [CrossRef]
42. Herrmann, M.E.; Brosicke, H.G.; Keller, M.; Monch, E.; Helge, H. Dependence of the utilization of a phenylalanine-free amino acid mixture on different amounts of single dose ingested. A case report. *Eur. J. Pediatr.* **1994**, *153*, 501–503. [CrossRef]
43. Macdonald, A.; Rylance, G.; Hall, S.K.; Asplin, D.; Booth, I.W. Factors affecting the variation in plasma phenylalanine in patients with phenylketonuria on diet. *Arch. Dis. Child.* **1996**, *74*, 412–417. [CrossRef]
44. Macdonald, A.; Rylance, G.W.; Asplin, D.; Hall, S.K.; Booth, I.W. Does a single plasma phenylalanine predict quality of control in phenylketonuria? *Arch. Dis. Child.* **1998**, *78*, 122–126. [CrossRef] [PubMed]
45. Couce, M.L.; Guler, I.; Anca-Couce, A.; Lojo, M.; Mirás, A.; Leis, R.; Pérez-Muñuzuri, A.; Fraga, J.M.; Gude, F. New insights in growth of phenylketonuric patients. *Eur. J. Nucl. Med. Mol. Imaging* **2014**, *174*, 651–659. [CrossRef]
46. Moretti, F.; Pellegrini, N.; Salvatici, E.; Rovelli, V.; Banderali, G.; Radaelli, G.; Scazzina, F.; Giovannini, M.; Verduci, E. Dietary glycemic index, glycemic load and metabolic profile in children with phenylketonuria. *Nutr. Metab. Cardiovasc. Dis.* **2017**, *27*, 176–182. [CrossRef]
47. Ilgaz, F.; Pinto, A.; Gökmen-Özel, H.; Rocha, J.C.; Van Dam, E.; Ahring, K.; Bélanger-Quintana, A.; Dokoupil, K.; Karabulut, E.; Macdonald, A. Long-Term Growth in Phenylketonuria: A Systematic Review and Meta-Analysis. *Nutrients* **2019**, *11*, 2070. [CrossRef] [PubMed]
48. Wells, J.C.K.; E Williams, J.; Chomtho, S.; Darch, T.; Grijalva-Eternod, C.; Kennedy, K.; Haroun, D.; Wilson, C.; Cole, T.J.; Fewtrell, M.S. Body-composition reference data for simple and reference techniques and a 4-component model: A new UK reference child. *Am. J. Clin. Nutr.* **2012**, *96*, 1316–1326. [CrossRef] [PubMed]
49. Weber, D.R.; Moore, R.H.; Leonard, M.B.; Zemel, B.S. Fat and lean BMI reference curves in children and adolescents and their utility in identifying excess adiposity compared with BMI and percentage body fat. *Am. J. Clin. Nutr.* **2013**, *98*, 49–56. [CrossRef] [PubMed]
50. Wells, J.C.K.; Fewtrell, M.S.; E Williams, J.; Haroun, D.; Lawson, M.S.; Cole, T.J. Body composition in normal weight, overweight and obese children: Matched case–control analyses of total and regional tissue masses, and body composition trends in relation to relative weight. *Int. J. Obes.* **2006**, *30*, 1506–1513. [CrossRef]
51. Roemmich, J.N.; Clark, P.A.; Weltman, A.; Rogol, A.D. Alterations in growth and body composition during puberty. I. Comparing multicompartment body composition models. *J. Appl. Physiol.* **1997**, *83*, 927–935. [CrossRef]

52. Sopher, A.B.; Thornton, J.C.; Wang, J.; Pierson, R.N.; Heymsfield, S.B.; Horlick, M. Measurement of Percentage of Body Fat in 411 Children and Adolescents: A Comparison of Dual-Energy X-Ray Absorptiometry With a Four-Compartment Model. *Pediatrics* **2004**, *113*, 1285–1290. [CrossRef] [PubMed]
53. Williams, J.E.; Wells, J.C.K.; Wilson, C.M.; Haroun, D.; Lucas, A.; Fewtrell, M.S. Evaluation of Lunar Prodigy dual-energy X-ray absorptiometry for assessing body composition in healthy persons and patients by comparison with the criterion 4-component model. *Am. J. Clin. Nutr.* **2006**, *83*, 1047–1054. [CrossRef] [PubMed]
54. Ofenheimer, A.; Breyer-Kohansal, R.; Hartl, S.; Burghuber, O.C.; Krach, F.; Schrott, A.; Franssen, F.M.E.; Wouters, E.F.M.; Breyer, M. Reference charts for body composition parameters by dual-energy X-ray absorptiometry in European children and adolescents aged 6 to 18 years—Results from the Austrian LEAD (Lung, hEart, sociAl, boDy) cohort. *Pediatr. Obes.* **2021**, *16*, e12695. [CrossRef] [PubMed]

nutrients

Article

A Three-Year Longitudinal Study Comparing Bone Mass, Density, and Geometry Measured by DXA, pQCT, and Bone Turnover Markers in Children with PKU Taking L-Amino Acid or Glycomacropeptide Protein Substitutes

Anne Daly [1,*], Wolfgang Högler [2], Nicola Crabtree [1], Nick Shaw [1], Sharon Evans [1], Alex Pinto [1], Richard Jackson [3], Catherine Ashmore [1], Júlio C. Rocha [4,5], Boyd J. Strauss [6,7], Gisela Wilcox [6,8], William D. Fraser [9], Jonathan C. Y. Tang [9,10] and Anita MacDonald [1]

1 Birmingham Women's and Children's Hospital, NHS Foundation Trust, Birmingham B4 6NH, UK; nicola.crabtree@nhs.net (N.C.); nick.shaw@nhs.net (N.S.); evanss.21@me.com (S.E.); alex.pinto@nhs.net (A.P.); catherine.ashmore@nhs.net (C.A.); anita.macdonald@nhs.net (A.M.)
2 Department of Paediatrics and Adolescent Medicine, Johannes Kepler University, Kepler University Hospital, Krankenhausstraße 26-30, 4020 Linz, Austria; wolfgang.hoegler@keplerunikliniku m.at
3 Liverpool Clinical Trials Centre, University of Liverpool, Brownlow Hill, Liverpool L69 3GL, UK; r.j.jackson@liverpool.ac.uk
4 Nutrition and Metabolism, NOVA Medical School, Faculdade de Ciências Médicas, Universidade Nova de Lisboa, 1169-056 Lisboa, Portugal; rochajc@nms.unl.pt
5 Centre for Health and Technology and Services Research (CINTESIS), 4200-450 Porto, Portugal
6 School of Medical Sciences, Faculty of Biology, Medicine and Health Sciences, University of Manchester, Manchester M13 9PL, UK; boyd.strauss@manchester.ac.uk (B.J.S.); gisela.wilcox@manchester.ac.uk (G.W.)
7 School of Clinical Sciences, Faculty of Medicine, Nursing and Health Sciences, Monash University, Melbourne 3800, Australia
8 The Mark Holland Metabolic Unit, Salford Royal Foundation NHS Trust, Ladywell NW2, Salford, Manchester M6 8HD, UK
9 BioAnalytical Facility, BCRE Builiding University or East Anglia, Norwich NR4 7TJ, UK; Fraser@uea.ac.uk (W.D.F.); Jonathan.Tang@uea.ac.uk (J.C.Y.T.)
10 Departments of Clinical Biochemistry and Endocrinology, Norfolk and Norwich University Hospitals Trust, Norwich NR4 7UY, UK
* Correspondence: a.daly3@nhs.net

Citation: Daly, A.; Högler, W.; Crabtree, N.; Shaw, N.; Evans, S.; Pinto, A.; Jackson, R.; Ashmore, C.; Rocha, J.C.; Strauss, B.J.; et al. A Three-Year Longitudinal Study Comparing Bone Mass, Density, and Geometry Measured by DXA, pQCT, and Bone Turnover Markers in Children with PKU Taking L-Amino Acid or Glycomacropeptide Protein Substitutes. *Nutrients* **2021**, *13*, 2075. https://doi.org/10.3390/nu13062075

Academic Editor: Roberto Iacone

Received: 30 May 2021
Accepted: 9 June 2021
Published: 17 June 2021

Publisher's Note: MDPI stays neutral with regard to jurisdictional claims in published maps and institutional affiliations.

Copyright: © 2021 by the authors. Licensee MDPI, Basel, Switzerland. This article is an open access article distributed under the terms and conditions of the Creative Commons Attribution (CC BY) license (https://creativecommons.org/licenses/by/4.0/).

Abstract: In patients with phenylketonuria (PKU), treated by diet therapy only, evidence suggests that areal bone mineral density (BMDa) is within the normal clinical reference range but is below the population norm. Aims: To study longitudinal bone density, mass, and geometry over 36 months in children with PKU taking either amino acid (L-AA) or casein glycomacropeptide substitutes (CGMP-AA) as their main protein source. Methodology: A total of 48 subjects completed the study, 19 subjects in the L-AA group (median age 11.1, range 5–16 years) and 29 subjects in the CGMP-AA group (median age 8.3, range 5–16 years). The CGMP-AA was further divided into two groups, CGMP100 (median age 9.2, range 5–16 years) (n = 13), children taking CGMP-AA only and CGMP50 (median age 7.3, range 5–15 years) (n = 16), children taking a combination of CGMP-AA and L-AA. Dual X-ray absorptiometry (DXA) was measured at enrolment and 36 months, peripheral quantitative computer tomography (pQCT) at 36 months only, and serum blood and urine bone turnover markers (BTM) and blood bone biochemistry at enrolment, 6, 12, and 36 months. Results: No statistically significant differences were found between the three groups for DXA outcome parameters, i.e., BMDa (L2–L4 BMDa g/cm^2), bone mineral apparent density (L2–L4 BMAD g/cm^3) and total body less head BMDa (TBLH g/cm^2). All blood biochemistry markers were within the reference ranges, and BTM showed active bone turnover with a trend for BTM to decrease with increasing age. Conclusions: Bone density was clinically normal, although the median z scores were below the population mean. BTM showed active bone turnover and blood biochemistry was within the reference ranges. There appeared to be no advantage to bone density, mass, or geometry from taking a macropeptide-based protein substitute as compared with L-AAs.

Keywords: PKU; bone mineral density; bone turnover markers; osteoporosis; blood biochemistry; casein glycomacropeptide; amino acid protein substitute

1. Introduction

Optimal bone mass is key to preventing the risk of fractures later in life, and many factors influence peak bone mass accretion including genetics, physical activity, body composition, and quality of diet. Severe dietary restriction may be problematic in conditions such as phenylketonuria (PKU) which require rigorous exclusion of many natural foods [1]. In children with classical PKU, the majority of protein is provided by a low phenylalanine, semisynthetic protein (protein substitute), with some limited dietary phenylalanine given from natural foods according to individual metabolic tolerance and disorder severity. Dependency on a synthetic protein may compromise both peak bone mass attainment and bone geometry [2,3].

Protein substitutes, are traditionally derived from essential and non-essential amino acids and are usually supplemented with added vitamins, minerals, and trace minerals aimed at achieving optimal growth, bone mass, and body composition. Protein substitutes are necessary lifelong, but long-term adherence is difficult to sustain particularly during adolescence [4,5], which is a vulnerable time for maximising bone mass, density, mineralization, and growth potential. Amino acids (AAs) contribute to the structural components of bone in addition to those of growth and tissue maintenance [2,6,7].

Protein has a positive effect on bone [6,7], and protein intake promotes peripubertal bone growth and delays bone loss [8,9]. Several long-term prospective observational studies [10,11] have shown significant positive associations between protein intake and bone mineral content, periosteal circumference, cortical area, and an index of strength strain. These studies reinforce that a moderate to high protein diet promotes bone accretion. The acid ash theory suggests that a high protein intake including protein substitutes based on amino acids are detrimental to bone accretion [8,12]. Protein substitutes are acidic, producing sulphuric acid from sulphur containing amino acids. The hypothesis suggests that calcium stored primarily in bones is slowly excreted to buffer the acidic pH, and this process leads to a decreased bone mineral density [13–16]. However, systematic and meta-analysis studies have dismissed this theory [17,18]. Although the urine pH is lower when taking a protein rich diet, the pH of the extracellular fluid is undisturbed due to regulatory control by the kidneys [8].

The use of casein glycomacropeptide supplemented with amino acids (CGMP-AA) has been associated with improved bone mass in PKU animal models [19], but CGMP (a bioactive peptide) compared with AAs and their influence on bone mass, density, and geometry has not been studied in children with PKU.

In this longitudinal prospective controlled study over 36 months, we investigated the efficacy of CGMP-AA as compared with L-AA protein substitutes on bone mass, density, geometry, and turnover markers in children with PKU.

2. Materials and Methods

2.1. Methods

The inclusion criteria included the following: children with PKU diagnosed by newborn screening, children aged 5–16 years and not treated with sapropterin dihydrochloride, known adherence with protein substitutes, and maintenance of 70% of blood phenylalanine concentrations within the European PKU target therapeutic range for 6 months prior to study enrolment [20]. Target blood phenylalanine ranges for children aged 5–12 years were from 120 to ≤360 µmol/L, and for children 12 years and older they were from 120 to ≤600 µmol/L.

2.1.1. Ethical Approval

This study was registered by the Health Research Authority and was given a favourable ethical opinion by the South Birmingham Research Ethical Committee (referenced 13/WM/0435 and IRAS (integrated research application system) number 129497). Written informed consent was given by at least one caregiver with parental responsibility and written consent was obtained from the subjects if appropriate for their age and level of understanding.

2.1.2. CGMP-AA and L-AA Protein Substitutes

The CGMP-AA (a test product by Vitaflo International Ltd., Liverpool, UK) was a flavoured powder. Each 35 g sachet contained 20 g protein equivalent, and 36 mg phenylalanine, mixed with 120 mL of water. The flavoured L-AA was either a powder mixed with water or a ready-prepared liquid that provided 10, 15, or 20 g of protein equivalent. The CGMP-AA and L-AA products both had a similar nutritional and AA profile, except CGMP-AA contained residual phenylalanine and higher amounts of threonine and leucine.

2.1.3. Selection into the CGMP Group or L-AA Group

The children chose the product they preferred, depending on their taste preference, i.e., the CGMP-AA group or L-AA group. They remained on this formula for the duration of the study.

2.2. Study Design

The primary aim of this 3-year longitudinal study was to compare bone mass, density and geometry of children with PKU taking CGMP-AA or L-AA as their primary protein source. The following examinations were conducted: dual-energy X-ray absorptiometry (DXA), together with blood bone biochemistry and blood and urine bone turnover markers. Peripheral quantitative computed tomography (pQCT) of the forearm was performed at 36 months only (Figure 1 and Table 1).

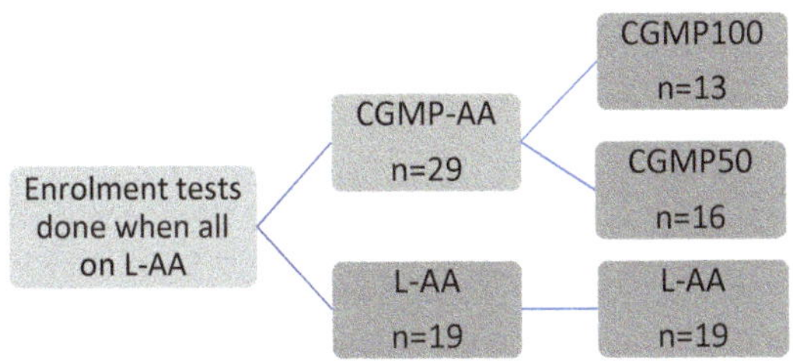

Figure 1. Diagram of the scheme for study methodology, from enrolment to 36 months. Legend: CGMP, casein glycomacropeptide; CGMP100, children taking all their protein substitute as casein glycomacropeptide; CGMP50, children taking a combination of casein glycomacropeptide and amino acids; L-AA, amino acids.

A previous pilot study [21] demonstrated that the residual phenylalanine in the CGMP-AA group led to compromised phenylalanine control in some children. Therefore, the CGMP-AA group was subdivided into: (1) CGMP100 group, in which the children took the entire protein substitute as CGMP-AA and (2) CGMP50 group, in which children took a combination of L-AA and CGMP-AA. There was also a third group of children who remained on their usual L-AA only (L-AA group).

Table 1. Frequency of nutritional blood biochemistry, bone blood and urine markers, DXA and pQCT scans, over study duration from enrolment to 36 months.

Enrolment	6 Months	12 Months	36 Months
• Fasting blood biochemistry • Serum bone markers • 2nd void urine bone markers • DXA	• Fasting blood biochemistry • Serum bone markers • 2nd void urine bone markers	• Fasting blood biochemistry • Serum bone markers • 2nd void urine bone markers	• Fasting blood biochemistry • Serum bone markers • 2nd void urine bone markers • DXA • pQCT
Anthropometry: 3/month			
Blood phenylalanine: weekly			

Legend: CGMP, casein glycomacropeptide; CGMP100, children taking all their protein substitute as casein glycomacropeptide; CGMP50, children taking a combination of casein glycomacropeptide and amino acids; L-AA, amino acids; DXA, dual-energy x-ray absorptiometry; pQCT, peripheral quantitative computerised tomography.

2.2.1. Dual-Energy X-ray Absorptiometry (DXA) and Peripheral Quantitative Computed Tomography (pQCT)

A GE Lunar iDXA and Encore™ software version 13.1 g (GE Healthcare, Madison, WI, USA) was used to measure bone density at enrolment and at the end of 36 months. Trunk thickness and body weight were used as a guide for scanning each child in the most appropriate acquisition mode. Children lay supine on a bed, while the DXA scan was completed. The following measurements were performed: lumbar spine (L2–L4) areal bone mineral density (L2–L4 BMDa) in g/cm^2, lumbar spine (L2–L4) bone mineral content (L2–L4 BMC) in g, total body mineral content (BMC) in g, total body less head BMDa (TBLH) in g/cm^2, and size corrected outcome measures included lumbar spine bone mineral apparent density (L2–L4 BMAD) in g/cm^3. At 36 months, in addition to the DXA assessment, pQCT was also performed.

2.2.2. pQCT

The pQCT (Stratec XCT 2000 L, Pfozheim, Germany) measurements were taken at the 4% and 66% region of the non-dominant forearm, evaluating volumetric bone mineral density, together with muscle and bone geometry, size, and strength. At the 4% site, trabecular and total cross-sectional area were measured, while at the 66% site, cortical density, as well as muscle, bone, and fat area were measured. The pQCT also measured the strength strain index as a surrogate marker of bone strength.

2.2.3. Serum Blood and Urine Bone Turnover Markers

Fasting, early morning, venous blood samples were collected at enrolment, 6, 12, and 36 months for the following serum bone markers: procollagen type 1 N-terminal propeptide (P1NP), type 1 collagen β crosslinked C-telopeptide (β-CTX), and bone alkaline phosphatase (bone ALP). A urine sample, the second sample of the day, was collected at enrolment, 6, 12, and 36 months for urine creatinine adjusted free urine pyridinoline (fPYD/Ur Cr) and urine free deoxypyridinoline crosslinks (fDPD/Ur Cr), and urinary calcium/creatinine ratio (Ur Ca/Cr). Urine samples were collected in containers, which were wrapped in tin foil and put into an envelope to shield them from any light. All urine samples were taken immediately to the laboratory for processing and stored at −80 °C. β-CTX and P1NP were analysed using an electro-chemiluminescence immunoassay (ECLIA) on a COBAS e601 analyser (Roche Diagnostics, Mannheim, Germany). The inter-assay coefficient of variation (CV) for β-CTX was <3% across the analytical range, between 0.01 and 6.0 μg/L, with a sensitivity of 0.01 μg/L. The inter-assay CV for P1NP was <3%, between 5 and 1200 μg/L, with a sensitivity of 5 μg/L. The serum bone ALP was determined by MicroVue™ enzyme-linked immunosorbent assay ELISA kit (Quidel Corporation, San Diego, USA). The inter-assay CV for bone ALP was <5.8%, between 0.5 and 150 U/L, with a detection limit of 0.7 U/L.

The analyses for urinary fPYD and fDPD were performed using the liquid chromatography tandem mass spectrometry (LC-MS/MS) method, as described by Tang et al [22]. In brief, 0.5 mL of urine sample/calibration/quality control materials pretreated with 0.5 mL hydrochloric acid (40% concentrate) was extracted using a solid phase extraction (SPE) column packed with cellulose slurry. Pyridinium crosslinks were eluted from the SPE columns and analysed by LC–MS/MS coupled with an electrospray ionisation (ESI) source operated in positive mode. The inter-assay CVs were $\leq$10.3% for PYD in the concentration range of 5–2000 nmol/L and $\leq$13.1% for DPD between 2 and 1000 nmol/L. The lower limit of quantification was 6 nmol/L for fPYD and 2.5 nmol/L for fDPD.

Urine creatinine was measured to obtain the fPYD/ and fDPD/urine creatinine ratios and the urine calcium/creatine ratio. Samples were analysed using Roche kinetic colorimetric assays performed on a COBAS® C501 analyser (Roche, Burgess Hill, UK), according to the manufacturer's instructions. The inter-assay CV ranged from 1.3 to 2.1% across the assay working range for Ur Ca of 0.20–7.5 mmol/L and Ur creatinine of 0.355 mmol/L.

2.2.4. Blood Biochemistry Markers

Overnight fasting blood samples for serum calcium, magnesium, phosphate, vitamin D, and parathyroid hormone were collected at enrolment, 6, 12, and 36 months.

2.2.5. Blood Phenylalanine/Tyrosine Monitoring

Throughout the 36-month study, trained caregivers collected weekly overnight fasting morning blood spots at home for phenylalanine and tyrosine. Blood specimens were sent via the post to the Birmingham Women's and Children's Hospital Laboratory. The blood spot filter cards used were Perkin Elmer 226 UK standard NBS (Perkin Elmer, Waltham, MA, USA). All the cards had a standard thickness, and the blood phenylalanine and tyrosine concentrations were calculated on a 3.2 mm punch by tandem mass spectrometry.

2.2.6. Pubertal Status

A general medical examination and pubertal status was measured at enrolment using the Tanner picture index. Stages 1 and 2 are classified as pre-pubertal, and Stages 3, 4, and 5 are classified as pubertal.

2.2.7. Anthropometric Measurements

Weight and height were measured once every 3 months by one of two metabolic dietitians. Height was measured using a Harpenden stadiometer (Holtain Ltd., Crymych, Wales, UK).

2.3. Statistical Methods

Continuous data are presented as median and interquartile ranges and categorical data are presented as frequencies of counts with associated percentages. Longitudinal data are presented graphically using profile plots to show the average change over time. Correlations between continuous covariates were evaluated using Pearson's correlation coefficient. Comparisons between treatment groups were performed using analysis of covariance (ANCOVA) techniques, to analyse the follow-up data, while including baseline measures as adjusting covariates. Models also included covariates for patients' gender, age, and puberty status (supplementary data are provided for these parameters). A p-value of 0.05 was used throughout to determine statistical significance. All analyses were performed using R (Version 3).

3. Results

3.1. Subjects

Fifty children (28 boys and 22 girls) with PKU were recruited. Forty-seven children were of European origin and three children were of Asian origin. Forty-eight children completed the study, 29 children in the CGMP-AA group and 19 children in the L-AA group. At enrolment, the median age (range) in the CGMP100 group was 9.2 years (5–16 years) ($n = 13$); in the CGMP50 group, the median age was 7.3 years (5–15 years) ($n = 16$), and in the L-AA group, the median age was 11.1 years (5–16 years) ($n = 19$). Only six children were able to tolerate >10 g/day of natural protein (CGMP100 $n = 2$, CGMP50 $n = 1$, and L-AA $n = 3$), all the others received <10 g/day of natural protein.

3.1.1. Subject Drop Out

One boy and one girl (both aged 12 years) in the CGMP-AA group were excluded from the study as both failed to comply with the study protocol. One failed to return blood phenylalanine samples and both had poor adherence to the low phenylalanine diet.

3.1.2. Pubertal Status

The number of children prepubertal (Stages 1 and 2) at enrolment were: CGMP100 group, 62% ($n = 8/13$); CGMP50 group, 69% ($n = 11/16$); and L-AA group, 32% ($n = 6/19$).

The number of children in puberty (Stages 3 to 5) were: CGMP100 group, 38% ($n = 5/13$); CGMP50 group, 31% ($n = 5/16$); and L-AA group, 68% ($n = 13/19$).

3.1.3. Median DXA Z Score Measurements for CGMP100, CGMP50, and L-AA Groups

Overall, there were no significant differences among the groups for any of the measured DXA parameters. Bone density was on the lower side of normal but within a normal reference range (Table 2).

Table 2. Median z scores (range) for L2–L4 bone mineral density (BMDa), lumbar spine bone mineral apparent density (L2–L4 BMAD), and total body less head BMDa (TBLH). Other parameters measured include median (range) L2–L4 bone mineral content and total bone mineral content for CGMP100, CGMP50, and L-AA groups, at enrolment and 36 months.

Group	Enrolment z Score (Range)	36 Months z Score (Range)
	L2–L4 BMDa (g/ cm²)	
CGMP100	−0.2 (−0.9 to 0.8)	−0.6 (−0.9 to 0.6)
CGMP50	−0.1 (−0.5 to 0.5)	−0.1 (−0.6 to 0.4)
L-AA	−0.1 (−0.7 to 0.4)	−0.5 (−0.8 to 0.0)
	L2–L4 BMAD (g/cm³)	
CGMP100	0.2 (−0.9 to 0.6)	0.2 (−0.4 to 0.5)
CGMP50	−0.2 (−0.5 to 0.9)	−0.2 (−0.4 to 0.3)
L-AA	−0.3 (−0.8 to 0.4)	−0.6 (−1.2 to −0.1)

Table 2. *Cont.*

Group	Enrolment z Score (Range)	36 Months z Score (Range)
	TBLH BMDa (g/cm^2)	
CGMP100	−0.6 (−1 to −0.5)	−0.5 (−0.6 to −0.2)
CGMP50	−0.8 (−1.3 to −0.1)	−0.6 (−0.9 to −0.3)
L-AA	−0.2 (−0.5 to 0.1)	−0.2 (−0.4 to −0.1)
	Median values (range) for Total and L2–L4 BMC g	
	Total body BMC g	
CGMP100	832.8 (672.9 to 1543.5)	1258.4 (1082.8 to 1816.9)
CGMP50	604.9 (532.9 to 680.3)	1019.1 (963.4 to 1134.8)
L-AA	1183.8 (672.9 to 1543.5)	1650.2 (1082.8 to 1816.9)
	L2–L4 BMC g	
CGMP100	18.9 (14.1 to 22.9)	28.1 (24.1 to 38.3)
CGMP50	14.2 (13.0 to 16.6)	22.1 (20.4 to 25.1)
L-AA	25.6 (15.9 to 34.9)	40.2 (25.0 to 45.4)

Legend: CGMP, casein glycomacropeptide; CGMP100, children taking all their protein substitute as casein glycomacropeptide; CGMP50, children taking a combination of casein glycomacropeptide and amino acids; L-AA, amino acids; L2–L4 BMD, bone mineral density lumbar vertebrae 2 to 4; BMAD, bone mineral apparent density; TBLH, total body less head; L2–L4 BMC, bone mineral content lumbar vertebrae 2 to 4; TBMC, total bone mineral content.

3.1.4. Median pQCT Z Score Measurements at 36 Months for CGMP100, CGMP50, and L-AA Groups

Similar to the DXA z score measurements, overall, there were no significant differences among the groups, but cortical density at the 66% site was statistically significantly different between the CGMP100 and L-AA groups (Table 3).

Table 3. Results from the pQCT scan measuring median z scores (range) for trabecular, cortical, and total densities at the 4% site; bone, muscle, and fat areas; strength strain index; and bone area/muscle area at 36 months in the CGMP100, CGMP50, and L-AA groups.

Group		36 Months Z Score (Range)
	Trabecular density: 4%	
CGMP100		−1.0 (−1.3 to −0.5)
CGMP50		−1.0 (−1.2 to −0.7)
L-AA		−0.5 (−1.2 to −0.1)
	Total density: 4%	
CGMP100		−0.7 (−1.1 to −0.6)
CGMP50		−0.7 (−0.9 to −0.3)
L-AA		−0.4 (−0.9 to 0.5)
	Cortical density: 66%	
CGMP100		0.1 (−0.1 to 0.3) *
CGMP50		−0.5 (−1.4 to −0.1)
L-AA		−0.4 (−1.0 to 0.5)

Table 3. *Cont.*

Group	36 Months Z Score (Range)
Bone area: 66%	
CGMP100	1.9 (1.4 to 4.0)
CGMP50	0.9 (0.2 to 1.8)
L-AA	2.0 (1.5 to 3.7)
Muscle area: 66%	
CGMP100	−1.1 (−1.8 to −0.5)
CGMP50	−1.2 (−1.4 to −0.6)
L-AA	−1.0 (−1.8 to −0.5)
Fat area: 66%	
CGMP100	0.5 (−0.3 to 0.9)
CGMP50	1.0 (0.4 to 1.8)
L-AA	1.2 (0.1 to 2.3)
Bone area/muscle area: 66% area	
CGMP100	0.5 (0.2 to 1.1)
CGMP50	−0.4 (−1.2 to 0.5)
L-AA	0.5 (0.2 to 1.6)
Strength strain index (SSI): 66%	
CGMP100	−0.7 (−1.0 to 1.3)
CGMP50	−0.1 (−0.6 to 0.5)
L-AA	0.4 (−0.3 to 0.6)

* CGMP100 as compared with L-AA ($p = 0.05$). Legend: CGMP, casein glycomacropeptide; CGMP100, children taking all their protein substitute as casein glycomacropeptide; CGMP50, children taking a combination of casein glycomacropeptide and amino acids; L-AA, amino acids.

3.2. Nutritional Bone Biochemistry Markers

Median concentrations for all the biochemistry markers (calcium, phosphate, magnesium, vitamin D, and parathyroid hormone) were within normal reference ranges for all the groups over the 36-month study period (Table 4). There were no statistically significant differences within or among the groups.

Table 4. Median (range) biochemical bone markers at enrolment and 36 months for CGMP100, CGMP50, and L-AA groups.

	Calcium mmol/L (Range)		Phosphate mmol/L (Range)		Magnesium mmol/L (Range)		25 (OH) Vit D nmol/L (Range)		PTH ng/L (Range)	
	Enrolment	36 m	Enrolment	36 m	Enrolment	36 m	Enrolment	36 m	Enrolment	36 m
CGMP100	2.5 (2.3, 2.6)	2.4 (2.3, 2.5)	1.4 (1.0, 1.5)	1.3 (1.0, 1.5)	0.9 (0.7, 1.0)	0.8 (0.8, 0.9)	112 (81, 162)	79 (43.7, 113)	17 (11, 42)	32 (22, 57)
CGGMP50	2.5 (2.3, 2.6)	2.4 (2.3, 2.5)	1.4 (1.1, 1.6)	1.3 (1.1, 1.5)	0.8 (0.8, 1.0)	0.8 (0.8, 0.9)	94.6 (61.8, 135)	95.2 (56.3, 137)	15.5 (6, 37)	31 (19, 46)
L-AA	2.5 (2.3, 2.6)	2.4 (2.3, 2.5)	1.3 (1.0, 1.5)	1.2 (0.8, 1.7)	0.8 (0.8, 0.9)	0.8 (0.7, 0.9)	93.9 (38.8, 182)	91.8 (60.3, 161)	21 (6, 44)	31 (19, 46)

Normal reference ranges (references from Birmingham Children's Hospital Clinical Chemistry Laboratory): Calcium 2.2–2.7 mmol/L, phosphate 0.8–1.9 mmol/L, magnesium 0.7–1.0 mmol/L, 25 (OH) vitamin D ≥50 nmol/L; parathyroid hormone (PTH) 15–60 ng/. Legend: CGMP, casein glycomacropeptide; CGMP100, children taking all their protein substitute as casein glycomacropeptide; CGMP50, children taking a combination of casein glycomacropeptide and amino acids; L-AA, amino acids.

Measurement for Bone Formation Markers and Urine Calcium

The urine calcium/creatinine ratio (Ur Ca/Cr) a measure of renal acid excretion was normal with no indication of excess calcium excretion (Table 5). Similarly, serum and urine BTM showed a physiological decrease with age, and no evidence of a disturbance between formation and resorption.

Table 5. Median (range) serum bone and urine turnover markers calculated from enrolment, 12, 24, and 36 months for CGMP100, CGMP50 and L-AA groups in girls and boys.

	CGMP100 Boys	CGMP100 Girls	CGMP50 Boys	CGMP50 Girls	L-AA Boys	L-AA Girls
β-CTX µg/L	1.2 (1.2, 1.6)	1.2 (1, 1.5)	1.2 (1.1, 1.4)	1.2 (1.2, 1.3)	1.4 (1.3, 1.4)	1.2 (0.9, 1.3)
Bone ALP U/L	86 (76, 95)	103 (92, 106)	125 (114, 131)	108 (76, 116)	85 (75, 95)	83 (46, 97)
P1NP µg/L	503 (488, 509)	476 (387, 663)	470 (434, 543)	507 (487, 649)	522 (418, 556)	445 (175, 553)
fDPD nmol/L	178 (68, 307)	114 (71, 338)	207 (91, 227)	147 (98, 265)	157 (96, 247)	107 (93, 114)
fDPD/Ur Cr nmol/mmol	22 (9, 27)	24 (12, 28)	26 (10, 30)	23 (13, 28)	25 (8, 27)	14 (8, 26)
fPYD nmol/L	735 (276, 1514)	429 (275, 700)	825 (310, 951)	624 (347, 1134)	615 (331, 876)	413 (290, 436)
fPYD/Ur Cr nmol/mmol	96 (33, 118)	90 (40, 121)	96 (37, 111)	105 (49, 110)	94 (27, 109)	58 (24, 100)
Ur Ca/Cr mmol/L	1 (0.4, 1.2)	1.1 (0.8, 1.4)	1.3 (0.7, 1.5)	0.8 (0.4, 1.3)	1.6 (1.3, 2.4)	1.9 (1.3, 2.5)
Ur Cr mmol/L	12 (1, 15)	6 (5, 11)	8 (7, 9)	8 (8, 10)	10 (8, 16)	7 (6, 8)

Legend: M, males; F, females; CGMP, casein glycomacropeptide; CGMP100, children taking all their protein substitute as casein glycomacropeptide; CGMP50, children taking a combination of casein glycomacropeptide and amino acids; L-AA, amino acids; β-CTX, type 1 collagen β crosslinked C-telopeptide; bone ALP, bone alkaline phosphatase; P1NP, procollagen type 1 N-terminal propeptide; fDPD, urine free deoxypyridinoline; fDPD/Ur Cr, deoxypyridinoline (free)/creatinine ratio; fPYD, urine free pyridinoline; fPYD/Ur Cr, pyridinoline (free)/creatinine ratio; Ur Ca/Cr, urine calcium/creatinine ratio; Ur Cr, urine creatinine. Standard references for children are not available.

A strong positive correlation was observed between PN1P and β-CTX at 36 months ($r = 0.82$) (Figure 2). The ANCOVA analysis performed on PN1P indicated that the level of PN1P was somewhat dependent on age, with older subjects having a lower PN1P level. Furthermore, there was evidence of an increase in PN1P at 36 months associated with CGMP100 as compared with L-AA ($p = 0.041$) (Figure 3). There was no difference between the CGMP50 and L-AA groups ($p = 0.80$).

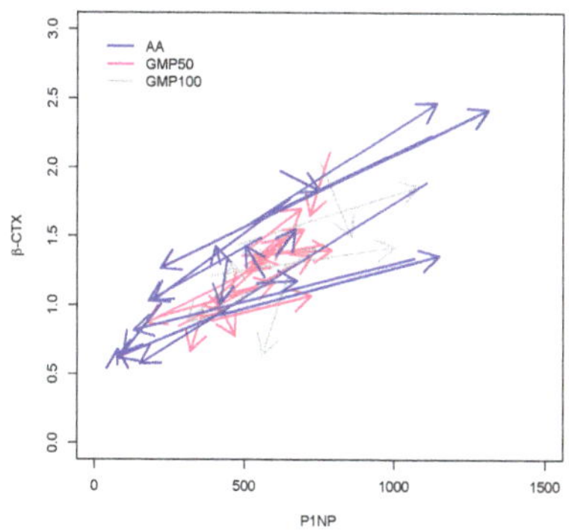

Figure 2. Correlation of β-CTX with PINP for CGMP100, CGMP50, and L-AA, at 36 months (↑ CGMP100, glycomacropeptide only; ↑ CGMP50, combination of CGMP and L-AA; and ↑ L-AA only).

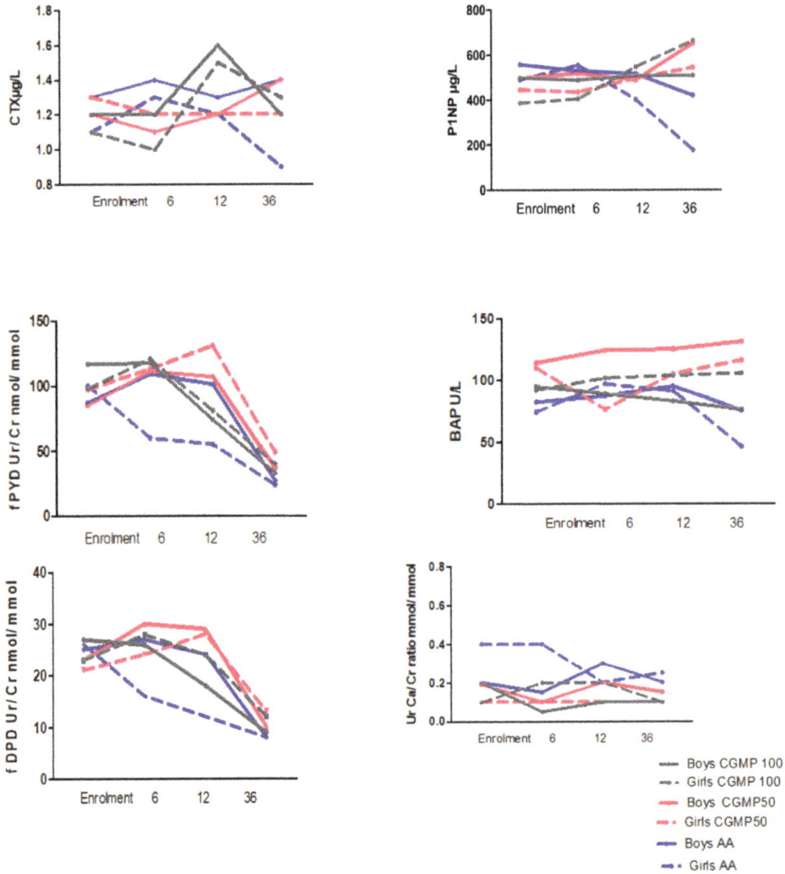

Figure 3. Graphs showing serum and urine bone turnover markers at enrolment, 6, 12, and 36 months separated by gender for CGMP100, CGMP50, and L-AA groups.

3.3. Anthropometry

We have previously reported height, weight, and body mass index in this group of children [23]. At 36 months, all groups had a median positive height z score: L-AA, 0.2 (range 0 to 0.5); for CGMP50, 0.3 (range −0.1 to 0.7); and for CGMP100, 0.6 (range 0.1 to 0.7). Median weight for height z scores and BMI z scores were above the ideal reference mean, indicating an overweight group of children (Table 6).

Table 6. Median z scores (range) for height, weight, and BMI in the L-AA, CGMP100, and CGMP50 groups, measured annually from enrolment to 36 months in PKU children taking either L-AA, CGMP50, or CGMP100.

Time (Months)	**L-AA Height Z Score *n* = 19**	**CGMP50 Height Z Score *n* = 16**	**CGMP100 Height Z Score *n* = 13**
Enrolment	0.2	−0.1	−0.1
(range)	(−0.2 to 0.8)	(−0.6 to 0.6)	(−0.4 to 0.3)
36 Months	0.2	0.3	0.6
(range)	(0.0 to 0.5)	(−0.1 to 0.7)	(0.1 to 0.7)
	L-AA Weight Z score *n* = 19	**CGMP50 Weight Z score *n* = 16**	**CGMP100 Weight Z score *n* = 13**
Enrolment	0.9	0.6	0.4
(range)	(−1.1 to 3.1)	(−1.9 to 1.8)	(−0.6 to 2.3)
36 Months	1.0	1.2	0.9
(range)	(−1.3 to 2.6)	(−2.4 to2.1)	(−0.4 to 1.8)
	L-AA BMI Z score *n* = 19	**CGMP50 BMI Z score *n* = 16**	**CGMP100 BMI Z score *n* = 13**
Enrolment	1.2	0.8	0.4
(range)	(−2.5 to 2.0)	(−0.2 to 2.0)	(−0.6 to 2.8)
36 Months	1.0	1.3	0.9
(range)	(−0.8 to 2.8)	(−1.2 to 2.4)	(−0.9 to 1.8)

3.4. Blood Phenylalanine Concentrations

The median phenylalanine concentrations for this study have been previously reported. Median phenylalanine concentrations were within recommended target reference ranges for children aged $\leq$11 and $\geq$12 years old [23].

The median daily dose of protein equivalent from protein substitute was 60 g/day (range 40–80 g), and the median amount of prescribed natural protein was 5.5 g protein/day (range 3–30 g) or 275 mg/day of phenylalanine (range 150–1500 mg), in all three groups. Eighty-eight percent (n = 42) of the children tolerated $\leq$10 g/day natural protein and 12% (n = 6) >10 g/day (CGMP100, n = 2; CGMP50, n = 1; and L-AA, n = 3).

4. Discussion

In this 36-month longitudinal study in children with PKU, bone mass, density, and geometry were comprehensively examined by DXA and pQCT, in addition to serum BTM and blood biochemistry. With the exception of cortical density at the 66% site, none of the other bone measurements showed any benefit of CGMP100 over L-AA or CGMP50, suggesting that CGMP-AA had no advantage over L-AA for bone development. Similarly, there was no evidence to suggest any differences in bone mass, density, or geometry by gender, age, or puberty (Supplementary Tables S1 and S2).

A strong positive correlation between β-CTX and P1NP was observed in all three study groups, with P1NP being lower in the older age subjects, and an increased P1NP being evident in the CGMP100 group. This synergy between bone formation and resorption shows active bone turnover and reflects appropriate bone growth, since these markers derive from physiological processes. Our results contrast with those reported by Casto et al. [24], which suggested a trend towards increased bone resorption in subjects with PKU. This controlled study, was the first to monitor bone mass and density using two separate imaging technologies (DXA and pQCT), and holistically assesses serum bone, urine, and blood biochemistry parameters in PKU. Similar to findings from two systematic reviews [24,25], the overall bone density values for the groups were below the population mean but within the normal reference values. Imaging results met the International Society for Clinical Densitometry (ISCD) recommendations (ISCD 2013) [26]. There were no differences in biochemical or BTM among the groups, suggesting no changes in bone metabolism attributed to the type of protein substitute. Naturally, BTM concentrations decreased in older adolescents towards those of lower adult levels, as a physiological phenomenon expected in a healthy population [27].

Unlike the findings of Schwahn et al., Mc Murry et al., and Fernandez et al. [28–30], we found no evidence to suggest that mineralization defects began in childhood, and then became more evident in adolescents. In this study, the groups of children were overweight. The relationship between overweight, obesity and bone is contentious.

Evidence [31] suggests that in early childhood obesity confers a structural advantage on the developing skeleton, but with age this relationship is reversed and becomes detrimental to skeletal development. Clarke et al [32] reported a positive relationship between adiposity and bone mass accrual in 3082 healthy children, while others [33,34] have reported opposite findings. Lean body mass has been shown to be the strongest predictor of bone mineral content [35,36] and relates to bone mass and skeletal development in children. Our previous study [37] indicated a trend towards improved lean body mass in the CGMP100 group; however, there was no evidence to suggest a similar beneficial effect on bone density in this group.

In PKU mouse models, CGMP as compared with L-AA has been shown to increase bone strength measured by biochemical mechanisms. Solverson [19] gave PKU and wild type mice different dietary regimens, i.e., a normal diet or a low phenylalanine diet supplemented with L-AA or CGMP protein substitutes. The PKU mice, regardless of protein substitute type, had lower bone density as compared with wild type mice, and those taking L-AA had inferior bone strength as compared with the CGMP protein substitute group. The authors proposed that the peptide structure of CGMP could possibly account for the positive influence on bone radial size improving biochemical performance. Alternatively, the high acid load due to L-AA could decrease bone strength via excreting higher amounts of calcium. However, both these suggestions were conjecture, as they did not measure net acid excretion, bone collagen, and markers of bone biomechanical performance. The results from our study in our cohort of children would suggest that neither of these mechanisms are active. BTM monitoring collagen were physiologically normal and there was no evidence of net acid excretion with a normal calcium/creatinine ratio.

Although many studies have identified lower BMD in PKU [38–41], not all of these studies included a size correlation for DXA output and there has been little agreement about lower BMD pathophysiology. Dobrowolski et al. [42] studied bone mineralization in PKU mice and showed phenylalanine toxicity inhibited bone mineralization. However, in human studies, there is a discord on the link between hyperphenylalaninemia and bone mass, with some studies showing a correlation and others not [38,40,43,44].

Within the three groups (CGMP100, CGMP50, and L-AA) there were expected physiological changes in the concentrations of BTM. In adults, BTM mainly represent bone remodelling; in children, BTM are released during bone remodelling, modelling, and perpendicular growth. Millet et al. [44] measured urine DPD and bone ALP in patients with PKU and compared these with a healthy paediatric group; bone remodelling was active in children with PKU aged 7–14 years, and bone ALP, as expected, was found to be significantly lower in the oldest group of patients (aged >18 years), although significantly higher DPD concentrations independent of age were reported. In our study, bone resorption and formation markers were consistently lowest in the L-AA group, particularly noticeable in the L-AA girls who had reached late puberty with a median age of 17 (8–18 years) at 36 months [27,43,45,46]. In contrast, the youngest group of CGMP50 boys showed an increase in BTM over the 36 months.

The interpretation of BTM is difficult and their concentrations vary widely in children, affected by a multitude of factors including age, gender, puberty, growth velocity, the rate of mineral accrual, hormonal regulation, nutritional status, circadian, and even day-to-day changes [47]. Paediatric reference data are available for some BTM [48–51], although UK specific data are lacking, which hampers appropriate interpretation. Specificity for bone tissue as well as sensitivity and specificity of the measurement assays lead to variations, rendering comparisons among study groups difficult [50,52]. Despite these challenges, in our study in which children were followed for 36 months, BTM followed the expected variations for age with no differences between the groups. These children had an active bone turnover profile, supportive of a normal bone mineral density. The reason why their bone mineral densities were below the population median was unclear, but these groups were not at any increased clinical risk of fractures.

There are limitations to this study. Patient numbers in each group were small which reduced the power of this study. An extended follow-up period of >3 years may be needed for any differences to emerge between protein substitute sources, as noted, P1NP was increased in the CGMP100 group. We also did not have a healthy control group, which would have been beneficial to compare differences with the children with PKU. The ages of the children were significantly different in all three groups, and CGMP was given at two different concentrations making any absolute differences difficult to recognize, although statistical modelling was used to account for this variable. Age influences bone changes and children entered puberty over the study period. In children, no bone marker is specific for any of the three different biological processes of modelling, remodelling, and changes in endochondral ossification. However, our findings were consistent, i.e., all measurements were taken via DXA or pQCT and showed a below average bone density, with no significant differences among the groups taking CGMP-AA or L-AA. Bone markers appeared to follow a similar pattern to that in healthy children. We did not measure exercise activity in these groups of children, but a high proportion (60%) participated in regular activities such as football, dancing, and gymnastics.

5. Conclusions

In this detailed and comprehensive study measuring global bone development, using both two- and three-dimensional imaging in addition to serum BTM and blood biochemistry, a complete assessment of bone mass, density, geometry, and bone turnover was conducted. There were no statistical differences in the groups of children, who had good metabolic control when taking either L-AA or CGMP-AA protein substitutes. Bone density was normal and similar to the findings from systematic reviews, which suggests it was lower than the population norm but carried no increased osteoporotic risk. Bone remodelling processes appear to be active in children with PKU, with both L-AA and CGMP-AA protein substitutes supporting normal bone growth.

Supplementary Materials: The following are available online at https://www.mdpi.com/article/10.3390/nu13062075/s1. Table S1: Differentiation of median DXA z scores (range): lumbar vertebrae L2–L4 areal bone mineral density (L2–L4 aBMD); lumbar vertebrae L2–L4 bone mineral apparent density (L2–L4 BMAD) and total body less head areal bone mineral density (TBLH BMDa) by gender. Median value (range) for total body bone mineral content (BMC) by gender. Table S2: Differentiation of peripheral quantitative computerised tomography(pQCT) z scores (range) by gender at 36 months.

Author Contributions: Conceptualization, N.S., A.M. and A.D.; methodology, A.M. and A.D.; software, R.J. and A.D.; validation, A.M., A.D., W.H., N.C., S.E., A.P., C.A., J.C.R., B.J.S., G.W., W.D.F. and J.C.Y.T.; formal analysis, A.D. and R.J.; investigation, A.D. and A.M.; writing—original draft preparation, A.D. and A.M.; writing—review and editing, A.M., A.D., W.H., N.C., S.E., A.P., C.A., J.C.R., B.J.S., G.W., W.D.F. and J.C.Y.T.; supervision, A.M. Funding was part of Ph.D. project from Vitaflo International. All authors have read and agreed to the published version of the manuscript.

Funding: This research was funded by Vitaflo International as part of Ph.D. grant.

Institutional Review Board Statement: The study was conducted according to the guidelines of the Declaration of Helsinki, and approved by the Ethics Committee.

Informed Consent Statement: Informed consent was obtained from all subjects involved in the study.

Data Availability Statement: Please contact the corresponding author.

Conflicts of Interest: A.D. research funding from Vitaflo International, financial support from Nutricia, Mevalia & Vitaflo International to attend study days & conferences. S.E. research funding from Nutricia, financial support from Nutricia & Vitaflo International to attend study days & conferences. A.P. has received an educational grant from Cambrooke Therapeutics and grants from Vitaflo International, Nutricia, Merck Serono, Biomarin and Mevalia to attend scientific meetings. C.A. received honoraria from Nutricia and Vitaflo International to attend study days and conferences. J.C.R. member of the European Nutritionist Expert Panel (Biomarin), the Advisory Board for Applied Pharma Research and Nutricia, and received honoraria as a speaker from APR, Merck Serono, Biomarin, Nutricia, Vitaflo, Cambrooke, PIAM and Lifediet. A.M. research funding & honoraria from Nutricia, Vitaflo International & Merck Serono, Member of European Nutrition Expert Panel (Merck Serono international), member of Sapropterin Advisory Board (Merck Serono international), member of the Advisory Board Element (Danone-Nutricia). The funders had no role in the design of the study; in the collection, analyses, or interpretation of data; in the writing of the manuscript, or in the decision to publish the results.

References

1. MacDonald, A.; van Wegberg, A.M.J.; Ahring, K.; Beblo, S.; Bélanger-Quintana, A.; Burlina, A.; Campistol, J.; Coşkun, T.; Feillet, F.; Giżewska, M.; et al. PKU dietary handbook to accompany PKU guidelines. *Orphanet J. Rare Dis.* **2020**, *15*, 171. [CrossRef] [PubMed]
2. Budek, A.Z.; Hoppe, C.; Ingstrup, H.; Michaelsen, K.F.; Bügel, S.; Mølgaard, C. Dietary protein intake and bone mineral content in adolescents—The Copenhagen Cohort Study. *Osteoporos. Int.* **2007**, *18*, 1661–1667. [CrossRef]
3. Ferrari, S.L.; Chevalley, T.; Bonjour, J.P.; Rizzoli, R. Childhood fractures are associated with decreased bone mass gain during puberty: An early marker of persistent bone fragility? *J. Bone Miner. Res.* **2006**, *21*, 501–507. [CrossRef] [PubMed]
4. Macdonald, A.; Van Rijn, M.; Feillet, F.; Lund, A.M.; Bernstein, L.; Bosch, A.M.; Gizewska, M.; Van Spronsen, F.J. Adherence Issues in Inherited Metabolic Disorders Treated by Low Natural Protein Diets. *Ann. Nutr. Metab.* **2012**, *61*, 289–295. [CrossRef]
5. Walter, J.; White, F.; Hall, S.; MacDonald, A.; Rylance, G.; Boneh, A.; Francis, D.; Shortland, G.; Schmidt, M.; Vail, A. How practical are recommendations for dietary control in phenylketonuria? *Lancet* **2002**, *360*, 55–57. [CrossRef]
6. Rizzoli, R.; Biver, E.; Bonjour, J.-P.; Coxam, V.; Goltzman, D.; Kanis, J.A.; Lappe, J.; Rejnmark, L.; Sahni, S.; Weaver, C.; et al. Benefits and safety of dietary protein for bone health—An expert consensus paper endorsed by the European Society for Clinical and Economical Aspects of Osteopororosis, Osteoarthritis, and Musculoskeletal Diseases and by the International Osteoporosis Foundation. *Osteoporos. Int.* **2018**, *29*, 1933–1948. [CrossRef]
7. Thorpe, M.P.; Evans, E.M. Dietary protein and bone health: Harmonizing conflicting theories. *Nutr. Rev.* **2011**, *69*, 215–230. [CrossRef] [PubMed]
8. Conigrave, A.D.; Brown, E.M.; Rizzoli, R. Dietary Protein and Bone Health: Roles of Amino Acid–Sensing Receptors in the Control of Calcium Metabolism and Bone Homeostasis. *Annu. Rev. Nutr.* **2008**, *28*, 131–155. [CrossRef]
9. Rizzoli, R.; Bianchi, M.L.; Garabédian, M.; McKay, H.A.; Moreno, L.A. Maximizing bone mineral mass gain during growth for the prevention of fractures in the adolescents and the elderly. *Bone* **2010**, *46*, 294–305. [CrossRef]
10. Alexy, U.; Remer, T.; Manz, F.; Neu, C.M.; Schoenau, E. Long-term protein intake and dietary potential renal acid load are associated with bone modeling and remodeling at the proximal radius in healthy children. *Am. J. Clin. Nutr.* **2005**, *82*, 1107–1114. [CrossRef]
11. Bonjour, J.-P.; Ammann, P.; Chevalley, T.; Rizzoli, R. Protein Intake and Bone Growth. *Can. J. Appl. Physiol.* **2001**, *26*, S153–S166. [CrossRef] [PubMed]
12. Patience, J.F. A review of the role of acid-base balance in amino acid nutrition. *J. Anim. Sci.* **1990**, *68*, 398–408. [CrossRef] [PubMed]
13. Kerstetter, J.E.; O'Brien, K.O.; Caseria, D.M.; Wall, D.E.; Insogna, K.L. The Impact of Dietary Protein on Calcium Absorption and Kinetic Measures of Bone Turnover in Women. *J. Clin. Endocrinol. Metab.* **2005**, *90*, 26–31. [CrossRef]
14. New, S.A. Nutrition Society Medal Lecture: The role of the skeleton in acid—Base homeostasis. *Proc. Nutr. Soc.* **2002**, *61*, 151–164. [CrossRef] [PubMed]
15. Remer, T.; Manz, F. Potential Renal Acid Load of Foods and its Influence on Urine pH. *J. Am. Diet. Assoc.* **1995**, *95*, 791–797. [CrossRef]
16. Sebastian, A.; Frassetto, L.A.; Sellmeyer, D.E.; Merriam, R.L.; Morris, R.C. Estimation of the net acid load of the diet of ancestral preagricultural Homo sapiens and their hominid ancestors. *Am. J. Clin. Nutr.* **2002**, *76*, 1308–1316. [CrossRef]
17. Fenton, T.R.; Lyon, A.W.; Eliasziw, M.; Tough, S.C.; Hanley, D.A. Meta-Analysis of the Effect of the Acid-Ash Hypothesis of Osteoporosis on Calcium Balance. *J. Bone Miner. Res.* **2009**, *24*, 1835–1840. [CrossRef]
18. Fenton, T.R.; Lyon, A.W.; Eliasziw, M.; Tough, S.C.; Hanley, D.A. Phosphate decreases urine calcium and increases calcium balance: A meta-analysis of the osteoporosis acid-ash diet hypothesis. *Nutr. J.* **2009**, *8*, 41. [CrossRef]
19. Solverson, P.; Murali, S.G.; Litscher, S.J.; Blank, R.D.; Ney, D.M. Low bone strength is a manifestation of phenylketonuria in mice and is attenuated by a glycomacropeptide diet. *PLoS ONE* **2012**, *7*, e45165. [CrossRef]

20. Wegberg, A.M.J.; MacDonald, A.; Ahring, K.; Belanger-Quintana, A.; Blau, N.; Bosch, A.M.; Burlina, A.; Campistol, J.; Feillet, F.; Giżewska, M.; et al. The complete European guidelines on phenylketonuria: Diagnosis and treatment. *Orphanet J. Rare Dis.* **2017**, *12*, 162. [CrossRef]
21. Daly, A.; Evans, S.; Chahal, S.; Santra, S.; Macdonald, A. Glycomacropeptide in children with phenylketonuria: Does its phenylalanine content affect blood phenylalanine control? *J. Hum. Nutr. Diet.* **2017**, *30*, 515–523. [CrossRef] [PubMed]
22. Tang, J.C.C.; Dutton, J.J.; Piec, L.; Green, D.; Fisher, E.; Washbourne, C.J.; Fraser, W.D. LC-MS/MS application for urine free pyridinoline and free deoxypyridionine: Urine markers of collagen and bone degradation. *Clin. Mass Spectrom.* **2016**, *1*, 11–18. [CrossRef]
23. Daly, A.; Evans, S.; Pinto, A.; Jackson, R.; Ashmore, C.; Rocha, J.C.; Macdonald, A. The Impact of the Use of Glycomacropeptide on Satiety and Dietary Intake in Phenylketonuria. *Nutrients* **2020**, *12*, 2704. [CrossRef]
24. de Castro, M.J.; de Lamas, C.; Sanchez-Pintos, P.; Gonzalez-Lamuno, D.; Couce, M.L. Bone Status in Patients with Phenylketonuria: A Systematic Review. *Nutrients* **2020**, *12*, 2154. [CrossRef] [PubMed]
25. Demirdas, S.; Coakley, E.K.; Bisschop, P.H.; Hollak, C.E.M.; Bosch, A.M.; Singh, R.H. Bone health in phenylketonuria: A systematic review and meta-analysis. *Orphanet J. Rare Dis.* **2015**, *10*, 17. [CrossRef] [PubMed]
26. Schousboe, J.T.; Shepherd, J.A.; Bilezikian, J.P.; Baim, S. Executive Summary of the 2013 International Society for Clinical Densitometry Position Development Conference on Bone Densitometry. *J. Clin. Densitom.* **2013**, *16*, 455–466. [CrossRef]
27. Mora, S.; Pitukcheewanont, P.; Kaufman, F.R.; Nelson, J.C.; Gilsanz, V. Biochemical Markers of Bone Turnover and the Volume and the Density of Bone in Children at Different Stages of Sexual Development. *J. Bone Miner. Res.* **1999**, *14*, 1664–1671. [CrossRef]
28. Fernandez Espuelas, C.; Manjon Llorente, G.; Gonzalez Lopez, J.M.; Ruiz-Echarri, M.P.; Baldellou Vazquez, A. Bone mineral turnover and bone densitometry in patients with a high-risk diet: Hyperphenylalaninemia and galactosemia. *An. Pediatr.* **2005**, *63*, 224–229.
29. McMurry, M.P.; Chan, G.M.; Leonard, C.O.; Ernst, S.L. Bone mineral status in children with phenylketonuria—Relationship to nutritional intake and phenylalanine control. *Am. J. Clin. Nutr.* **1992**, *55*, 997–1004. [CrossRef]
30. Schwahn, B.; Mokov, E.; Scheidhauer, K.; Lettgen, B.; Schönau, E. Decreased trabecular bone mineral density in patients with phenylketonuria measured by peripheral quantitative computed tomography. *Acta Paediatr.* **1998**, *87*, 61–63. [CrossRef]
31. Dimitri, P. Fat and bone in children—Where are we now? *Ann. Pediatr. Endocrinol. Metab.* **2018**, *23*, 62–69. [CrossRef]
32. Clark, E.M.; Ness, A.R.; Bishop, N.J.; Tobias, J.H. Association Between Bone Mass and Fractures in Children: A Prospective Cohort Study. *J. Bone Miner. Res.* **2006**, *21*, 1489–1495. [CrossRef]
33. Rauch, F.; Bailey, D.A.; Baxter-Jones, A.; Mirwald, R.; Faulkner, R. The 'muscle-bone unit' during the pubertal growth spurt. *Bone* **2004**, *3*, 771–775. [CrossRef] [PubMed]
34. Russell, M.; Mendes, N.; Miller, K.K.; Rosen, C.J.; Lee, H.; Klibanski, A.; Misra, M. Visceral Fat Is a Negative Predictor of Bone Density Measures in Obese Adolescent Girls. *J. Clin. Endocrinol. Metab.* **2010**, *95*, 1247–1255. [CrossRef] [PubMed]
35. Carbone, J.W.; Pasiakos, S.M. Dietary Protein and Muscle Mass: Translating Science to Application and Health Benefit. *Nutrients* **2019**, *11*, 1136. [CrossRef]
36. Crabtree, N.; Kibirige, M.; Fordham, J.; Banks, L.; Muntoni, F.; Chinn, D.; Boivin, C.; Shaw, N. The relationship between lean body mass and bone mineral content in paediatric health and disease. *Bone* **2004**, *35*, 965–972. [CrossRef] [PubMed]
37. Daly, A.; Högler, W.; Crabtree, N.; Shaw, N.; Evans, S.; Pinto, A.; Jackson, R.; Strauss, B.; Wilcox, G.; Rocha, J.; et al. Growth and Body Composition in PKU Children—A Three-Year Prospective Study Comparing the Effects of L-Amino Acid to Glycomacropeptide Protein Substitutes. *Nutrients* **2021**, *13*, 1323. [CrossRef]
38. Adamczyk, P.; Morawiec-Knysak, A.; Płudowski, P.; Banaszak, B.; Karpe, J.; Pluskiewicz, W. Bone metabolism and the muscle–bone relationship in children, adolescents and young adults with phenylketonuria. *J. Bone Miner. Metab.* **2010**, *29*, 236–244. [CrossRef]
39. Al-Qadreh, A.; Schulpis, K.H.; Athanasopoulou, H.; Mengreli, C.; Skarpalezou, A.; Voskaki, I. Bone mineral status in children with phenylketonuria under treatment. *Acta Paediatr.* **1998**, *87*, 1162–1166. [CrossRef] [PubMed]
40. Hillman, L.; Schlotzhauer, C.; Lee, D.; Grasela, J.; Witter, S.; Allen, S.; Hillman, R. Decreased bone mineralization in children with phenylketonuria under treatment. *Eur. J. Pediatr.* **1996**, *155* (Suppl. 1), S148–S152. [CrossRef]
41. Roato, I.; Porta, F.; Mussa, A.; D'Amico, L.; Fiore, L.; Garelli, D.; Spada, M.; Ferracini, R. Bone Impairment in Phenylketonuria Is Characterized by Circulating Osteoclast Precursors and Activated T Cell Increase. *PLoS ONE* **2010**, *5*, e14167. [CrossRef] [PubMed]
42. Dobrowolski, S.F.; Tourkova, I.L.; Robinson, L.J.; Secunda, C.; Spridik, K.; Blair, H.C. A bone mineralization defect in the Pah(enu2) model of classical phenylketonuria involves compromised mesenchymal stem cell differentiation. *Mol. Genet. Metab.* **2018**, *125*, 193–199. [CrossRef] [PubMed]
43. Blumsohn, A.; Hannon, R.A.; Wrate, R.; Barton, J.; Ai-Dehaimi, A.W.; Colwell, A.; Eastell, R. Biochemical markers of bone turnover in girls during puberty. *Clin. Endocrinol.* **1994**, *40*, 663–670. [CrossRef] [PubMed]
44. Millet, P.; Vilaseca, M.A.; Valls, C.; Pérez-Dueñas, B.; Artuch, R.; Gómez, L.; Lambruschini, N.; Campistol, J. Is deoxypyridinoline a good resorption marker to detect osteopenia in phenylketonuria? *Clin. Biochem.* **2005**, *38*, 1127–1132. [CrossRef] [PubMed]
45. Bollen, A.-M.; Eyre, D. Bone resorption rates in children monitored by the urinary assay of collagen type I cross-linked peptides. *Bone* **1994**, *15*, 31–34. [CrossRef]

46. Shaw, N.J.; Bishop, N.J. Mineral accretion in growing bones—A framework for the future? *Arch. Dis. Child.* **1995**, *72*, 177–179. [CrossRef]
47. Gennai, I.; Di Iorgi, N.; Reggiardo, G.; Gatti, C.; Bertelli, E.; Allegri, A.E.M.; Barco, S.; Maghnie, M.; Tripodi, G.; Cangemi, G. Age- and sex-matched reference curves for serum collagen type I C-telopeptides and bone alkaline phosphatase in children and adolescents: An alternative multivariate statistical analysis approach. *Clin. Biochem.* **2016**, *49*, 802–807. [CrossRef]
48. Huang, Y.; Eapen, E.; Steele, S.; Grey, V. Establishment of reference intervals for bone markers in children and adolescents. *Clin. Biochem.* **2011**, *44*, 771–778. [CrossRef]
49. Rauchenzauner, M.; Schmid, A.; Heinz-Erian, P.; Kapelari, K.; Falkensammer, G.; Griesmacher, A.; Finkenstedt, G.; Högler, W. Sex- and Age-Specific Reference Curves for Serum Markers of Bone Turnover in Healthy Children from 2 Months to 18 Years. *J. Clin. Endocrinol. Metab.* **2006**, *92*, 443–449. [CrossRef] [PubMed]
50. Szulc, P.; Seeman, E.; Delmas, P.D. Biochemical Measurements of Bone Turnover in Children and Adolescents. *Osteoporos. Int.* **2000**, *11*, 281–294. [CrossRef]
51. Tisè, M.; Ferrante, L.; Mora, S.; Tagliabracci, A. A biochemical approach for assessing cutoffs at the age thresholds of 14 and 18 years: A pilot study on the applicability of bone specific alkaline phosphatase on an Italian sample. *Int. J. Leg. Med.* **2016**, *130*, 1149–1158. [CrossRef] [PubMed]
52. Delmas, P.D. Biochemical markers of bone turnover. *Acta Orthop. Scand. Suppl.* **1995**, *266*, 176–182. [CrossRef] [PubMed]

nutrients

MDPI

Article

Current Practices and Challenges in the Diagnosis and Management of PKU in Latin America: A Multicenter Survey

Soraia Poloni [1,*], Bruna Bento dos Santos [1,2], Ana Chiesa [3], Norma Specola [4], Marcela Pereyra [5], Manuel Saborío-Rocafort [6], María Florencia Salazar [7], María Jesús Leal-Witt [7], Gabriela Castro [7], Felipe Peñaloza [7], Sunling Palma Wong [8], Ramsés Badilla Porras [9], Lourdes Ortiz Paranza [10], Marta Cristina Sanabria [11], Marcela Vela Amieva [12], Marco Morales [13], Amanda Rocío Caro Naranjo [14], Antonieta Mahfoud [15], Ana Rosa Colmenares [16], Aida Lemes [17], José Fernando Sotillo-Lindo [18], Ceila Perez [19], Laritza Martínez Rey [20], Georgina María Zayas Torriente [21], Lilia Farret Refosco [1], Ida Vanessa Doederlein Schwartz [1,2] and Veronica Cornejo [7]

1 Hospital de Clínicas de Porto Alegre, Porto Alegre 90035-903, Brazil; brunabentods@gmail.com (B.B.d.S.); lrefosco@hcpa.edu.br (L.F.R.); idadschwartz@gmail.com (I.V.D.S.)
2 Genetics and Molecular Biology, Universidade Federal do Rio Grande do Sul, Porto Alegre 91501-970, Brazil
3 Centro de Investigaciones Endocrinologicas DR Cesar Bergadá, CEDIE-CONICET-Fundación de Endocrinologia Infantil-Division de Endocrinologia Hospital de Niños R Gutierrez, Gallo 1330, Buenos Aires C1425EFD, Argentina; achiesa@cedie.org.ar
4 Unidad de Metabolismo, Hospital de Niños de La Plata, La Plata B1904, Argentina; normaspecola@gmail.com
5 Servicio de Crecimiento y Desarrollo del Hospital Pediátrico Dr. H. Notti, 2603, Mendoza M5519, Argentina; marcela_pereyra@hotmail.com
6 Hospital Nacional de Niños, Caja Costarricense de Seguro Social & Sistema de Estudios de Posgrado, Universidad de Costa Rica, San José 11501, Costa Rica; MANUEL.SABORIO@ucr.ac.cr
7 Instituto de Nutrición y Tecnología de los Alimentos (INTA), Universidad de Chile, Santiago de Chile 1058, Chile; mfsalazar@inta.uchile.cl (M.F.S.); mj.leal@inta.uchile.cl (M.J.L.-W.); gcastro@inta.uchile.cl (G.C.); felipe.penaloza@inta.uchile.cl (F.P.); vcornejo@inta.uchile.cl (V.C.)
8 Programa Nacional de Tamizaje, Hospital Nacional de Niños, San José 267-1005, Costa Rica; spalmaw@tamizajecr.com
9 FCCMG Servicio de Genética Médica y Metabolismo, Hospital Nacional de Niños, San José 267-1005, Costa Rica; rbadillap@tamizajecr.com
10 Ministerio de Salud Pùblica y Bienestar Social, Asunciòn 1735, Paraguay; lrortizp@yahoo.es
11 Pediatric Department and Department of the Hospital de Clínicas, Universidad Nacional de Asunciòn, Asunción 1102, Paraguay; marta.sanabria@gmail.com
12 Laboratorio de Errores Innatos del Metabolismo y Tamiz-Instituto Nacional de Pediatría, Ciudad de México 04530, Mexico; dravelaamieva@yahoo.com
13 Hospital Rebagliati, Lima 15072, Peru; moralesmarco2004@yahoo.com
14 Instituto de Errores Innatos del Metabolismo de la, Pontificia Universidad Javeriana, Bogota 110231, Colombia; amanda.caro@javeriana.edu.co
15 Instituto de Estudios Avanzados, Caracas 17606, Venezuela; amahfoud@gmail.com
16 Hospital Clinica Caracas-Materno Infantil de Caricuao, Caracas 1000, Venezuela; anarosacol@hotmail.com
17 Instituto de la Seguridad Social, Montevideo 11000, Uruguay; lemesa@adinet.com.uy
18 Hospital de especialidades Pediátricas "Omar Torrijos Herrera", Ciudad de Panamá 07136, Panama; jfsotillotuto_2000@hotmail.com
19 Robert Reid Cabral Children's Hospital, Santo Domingo 10101, Dominican Republic; cepef@hotmail.com
20 Centro Nacional de Genética Médica, La Habana 11300, Cuba; laritzam@infomed.sld.cu
21 Centro de Nutrición e Higiene de los Alimentos del Instituto Nacional de Higiene, Epidemiología y Microbiología (INHEM), La Habana 10300, Cuba; georgina.zayas@infomed.sld.cu
* Correspondence: spoloni@hcpa.edu.br

Citation: Poloni, S.; dos Santos, B.B.; Chiesa, A.; Specola, N.; Pereyra, M.; Saborío-Rocafort, M.; Salazar, M.F.; Leal-Witt, M.J.; Castro, G.; Peñaloza, F.; et al. Current Practices and Challenges in the Diagnosis and Management of PKU in Latin America: A Multicenter Survey. *Nutrients* **2021**, *13*, 2566. https://doi.org/10.3390/nu13082566

Academic Editor: Shanon L. Casperson

Received: 30 May 2021
Accepted: 8 July 2021
Published: 27 July 2021

Publisher's Note: MDPI stays neutral with regard to jurisdictional claims in published maps and institutional affiliations.

Copyright: © 2021 by the authors. Licensee MDPI, Basel, Switzerland. This article is an open access article distributed under the terms and conditions of the Creative Commons Attribution (CC BY) license (https://creativecommons.org/licenses/by/4.0/).

Abstract: This study aimed to describe the current practices in the diagnosis and dietary management of phenylketonuria (PKU) in Latin America, as well as the main barriers to treatment. We developed a 44-item online survey aimed at health professionals. After a pilot test, the final version was sent to 25 practitioners working with inborn errors of metabolism (IEM) in 14 countries. Our results include 22 centers in 13 countries. Most countries (12/13) screened newborns for PKU. Phenylalanine (Phe) targets at different ages were very heterogeneous among centers, with greater consistency at the 0–1 year age group (14/22 sought 120–240 μmol/L) and the lowest at >12 years (10 targets reported).

Most countries had only unflavored powdered amino acid substitutes (10/13) and did not have low-protein foods (8/13). Only 3/13 countries had regional databases of the Phe content of foods, and only 4/22 centers had nutrient analysis software. The perceived obstacles to treatment were: low purchasing power (62%), limited/insufficient availability of low-protein foods (60%), poor adherence, and lack of technical resources to manage the diet (50% each). We observed a heterogeneous scenario in the dietary management of PKU, and most countries experienced a lack of dietary resources for both patients and health professionals.

Keywords: phenylketonuria; PKU; low-protein diet; newborn screening

1. Introduction

Latin America comprises 20 countries and has an ethnically diverse population of over 650 million people. With a complex political and economic background, these countries face many challenges in the diagnosis and care of patients with inborn errors of metabolism (IEM), such as phenylketonuria (PKU, OMIM #261600). The success of early diagnosis and dietary treatment of PKU has been well described since the 1960s. Since then, and until the mid-1970s, most developed countries have initiated national newborn screening (NBS) programs for PKU [1]. In Latin America, the first organized NBS programs were only started in 1986 in Cuba, followed by Costa Rica (1990), and Chile (1992) [2]. At present, 16 countries have national or regional NBS programs, but only 6 have coverage ≥90% [3].

Similarly to NBS programs, PKU management faces many challenges in Latin America. Despite significant health system reforms in the 1980s, inequality and impaired access to health care remains a major problem in the region [4]. A recent report by the Organization for Economic Co-operation and Development (OECD) showed that government and compulsory health insurance represented only 54.3% of the current expenditure on health in Latin America, with 34% of all health spending being paid out-of-pocket. Nearly 8% of the Latin American population spends more than 10% of their household consumption or income on health care services. Latin American countries also have a much lower availability of medical technologies and health professionals when compared to other countries [5].

Treatment of PKU inflicts a substantial time and cost burden on patients and their families [6], and this can be a significant obstacle to encouraging patients to remain on a restricted diet [7]. Moreover, a trained health care team is needed to manage the extremely restrictive diet and to educate patients and families, and frequent laboratory tests are required to guide the treatment. The current situation of PKU diagnosis and management in Latin America is unknown, since only sparse and country-based reports have been published [8–12]. Therefore, the aim of this study was to map the current practices in the diagnosis and dietary management of PKU in Latin America, as well as the main barriers to treatment perceived by health care providers.

2. Materials and Methods

2.1. Study Design

A questionnaire containing 44 questions on the diagnosis and management of PKU was developed by a team of experts from Brazil (S.P., B.B.S., I.V.D.S., and L.F.R.) and Chile (M.J.L., F.S., G.C., and V.C.). These were experienced metabolic dietitians and geneticists, all co-authors of this paper. The survey had multiple choice and short answer questions and was aimed at health care professionals following patients with PKU. Five main issues were addressed: features and professional training of the health care team, newborn screening, treatment goals and dietary practices, availability of alternative treatments, and perceived barriers to treatment.

After the first Portuguese and Spanish versions of the questionnaire were finished, a pilot study was performed with 6 PKU experts (3 Portuguese and 3 Spanish speakers)

to identify possible flaws or misinterpretations of the questions. Only minor adaptations were made, and the final version was then shared on an online platform.

To disseminate the survey, practitioners of IEM were searched for (through public archives of the Sociedad Latinoamericana de Errores Innatos del Metabolismo y Pesquisa Neonatal in all Latin American countries, and were found in 14 of them. The coordinator team designated 1 responsible person in each country to distribute the survey to other centers nationally. The aim was to distribute the survey to as many centers as possible in each country. The only exception was Brazil, the largest country with the most PKU treatment centers (>20); to avoid overrepresentation, we chose 1 center from each region of the country. The invitation and distribution of the survey was performed by e-mail from July to November 2020. The final version of the questionnaire is available by request.

2.2. *Ethical Aspects*

The study was approved by the Research Ethics Committee of Hospital de Clínicas de Porto Alegre, Brazil (CAAE 36401120.6.0000.5327), and the survey was initiated only after the participants agreed with the online informed consent form.

3. Results

3.1. *Participants*

Out of the 14 contacted country representatives, 13 were following patients with PKU, and all of them agreed to participate in the study. In total, 22 treatment centers were enrolled from the following countries: Brazil (n = 5), Argentina (n = 4), Colombia (n = 2), Venezuela (n = 2), Costa Rica (n = 1), Chile (n = 1), Mexico (n = 1), Paraguay (n = 1), Peru (n = 1), Dominican Republic (n = 1), Panama (n = 1), Uruguay (n = 1), and Cuba (n = 1).

The respondents were mostly female (91%), were aged ≥45 years (61%), and had worked with PKU for over 10 years (70%). Physicians represented 59% of the respondents, with the remaining respondents being dietitians. Regarding professional training, 45% (n = 10/22) stated that they had a specialization course in the field, 41% (n = 9/22) had only short-term courses, and 9% (n = 2) had no formal training. The number of patients with PKU who were followed up by the professionals varied considerably: 18.2% (n = 4/22) had <10 patients, 18.2% (n = 4/22) had 10–25 patients, 18.2% (n = 4/22) had 26–50 patients, 13.6% (n = 3/22) had 51–75 patients, and 32% (n = 7/22) had >75 patients.

3.2. *Newborn Screening and Phenylalanine (Phe) Monitoring*

Regarding NBS, all countries but one (Dominican Republic) had a national NBS program, the most recent one being in Colombia (2019). When inquired on the Phe cutoff level used to start dietary treatment, 13/22 centers (59%) responded ≥360 µmol/L (≥6 mg/dL), 5/22 (23%) responded ≥600 µmol/L (≥10 mg/dL), 2/22 (9%) responded <360 µmol/L (<6 mg/dL), and 1 (4.5%) responded ≥480 µmol/L (≥8 mg/dL). In most centers (19/22), blood Phe was measured in dried blood spots. The most used method to analyze blood Phe was the fluorometric assay (12/22), followed by tandem mass spectrometry (5/22). Figure 1 shows the recommended frequency of Phe and tyrosine (Tyr) monitoring in the studied centers.

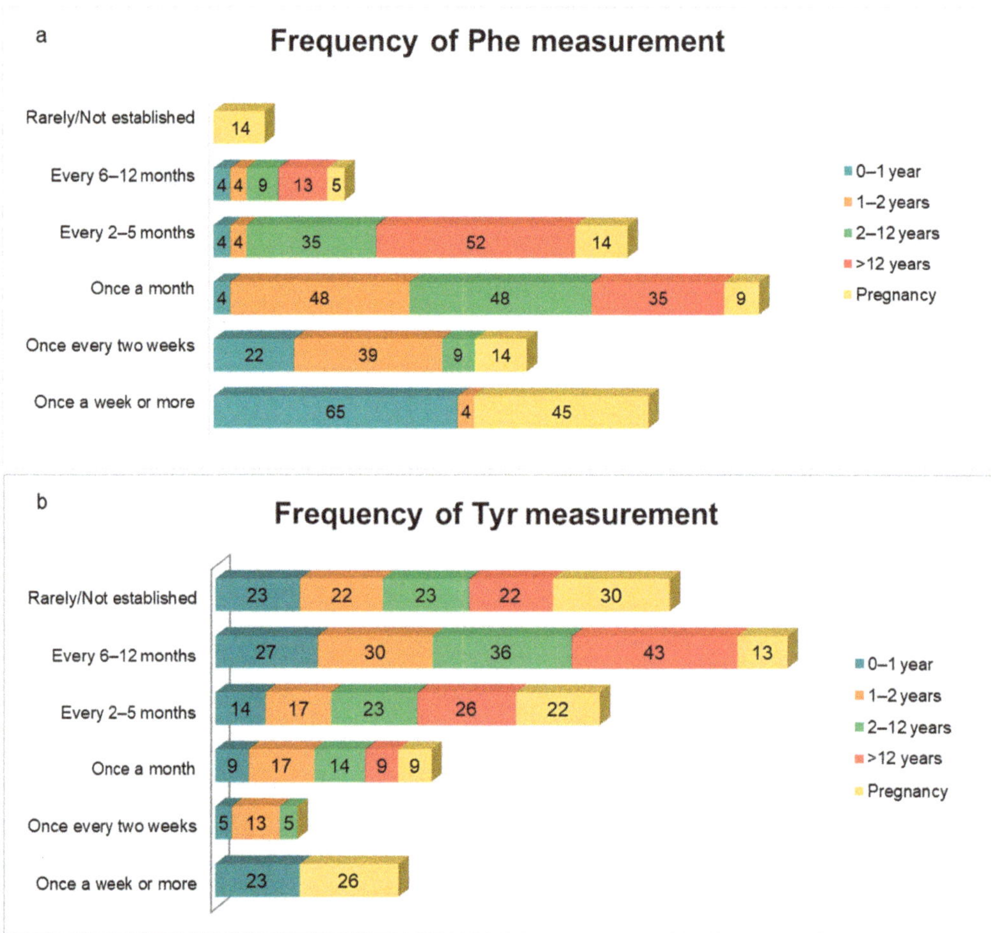

Figure 1. Frequencies of blood phenylalanine (Phe, **a**) and tyrosine (Tyr, **b**) monitoring for each age group as adopted by the Latin American centers included in the study (n = 22). Numbers within columns represent relative percentages.

3.3. Treatment Targets and Dietary Practices

Figure 2 shows Phe target levels at different ages in the studied centers. Dietary guidance was most frequently performed through the simplified method of high/medium/low Phe content of foods (10/22), followed by individualized meal plans (8/22), and protein counting (3/22). A 24 h dietary recall (or similar) was performed at every appointment in most centers (17/22). Total protein prescriptions are described in Figure 3.

All but two respondents reported that the maintenance of partial breastfeeding was encouraged in classical PKU patients. Most respondents (80%) said that they instructed mothers to offer the protein substitute right before breastfeeding to control Phe intake.

Regarding nutritional monitoring, all centers reported weight and height measurements at every appointment, and 19/22 always assessed head circumference. The evaluation of body composition was less frequent; 12/22 (54%) did not assess skinfolds and none performed bioelectrical impedance analyses on a regular basis. The blood tests that were performed at least once a year were: a complete blood count (95%), fasting glucose (91%), total protein (91%), creatinine (91%), urea (77%), lipoproteins and triglycerides (77%), albumin (73%), vitamins B12 and D (60%), and ferritin (60%). A complete amino acid profile was requested in 10/22 centers (45%), and only 1 center evaluated essential fatty acids on a regular basis. Bone densitometry was routinely performed in 11/22 centers (50%).

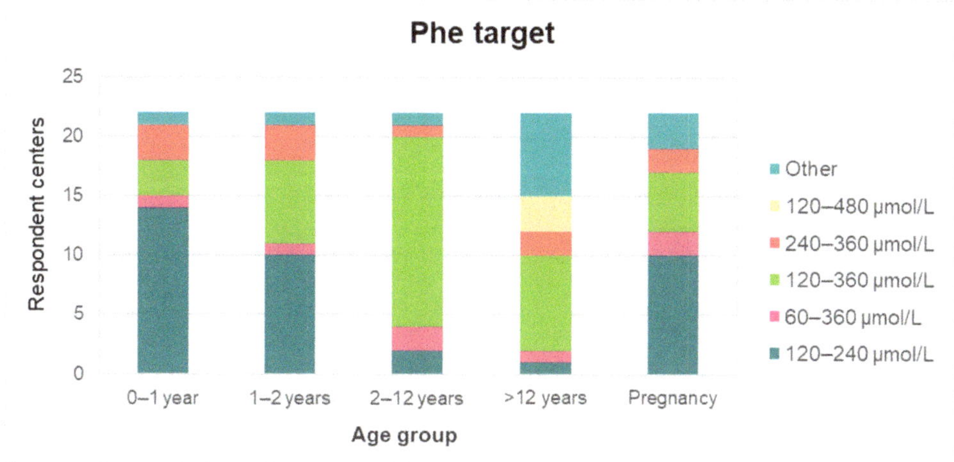

Figure 2. Target Phe levels during treatment in different age groups, as adopted by the studied centers (n = 22). Phe: phenylalanine.

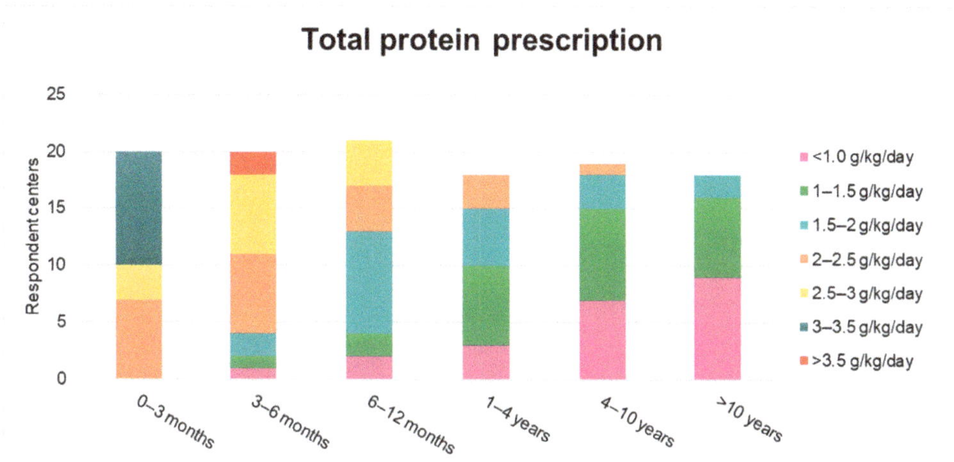

Figure 3. Total protein (natural + protein substitute) prescriptions, in different age groups, in the studied centers (n = 22) Some centers did not fully answer this question; therefore, the sample size varies in different age groups.

3.4. Nutritional Resources

Out of the 13 included countries, 9 reported having national guidelines for PKU management, and 12/22 (54%) centers had local management protocols. The theoretical background most commonly used by the respondents was: international guidelines (61%), scientific papers (56%), and national guidelines (48%). Regarding dietary resources, only 18% of the centers (4/22) reported having an adequate regional database of the Phe content of foods, and 33% stated that only an incomplete database was available. The remaining (49%) centers utilized a variety of international databases or considered only the protein content of foods for guiding the diet. Food recalls were usually calculated manually (48%)

or through a customized spreadsheet developed by the center (3%). Only 3/22 centers reported having specialized nutrition software.

Except for one country, none of the participant countries had the Phe content available on food labels. Regarding protein substitutes, 11/13 countries had only unflavored powdered amino acid formulas; only 1 country (Argentina) had several options, such as gels and tablets, available. In 10/13 countries, the protein substitute was fully subsidized by the government. Specific low-protein foods for PKU were not available in 8/13 countries; even when these were available, 58% of the centers stated that they were not affordable. These products were subsidized by the government in only 2/13 countries.

3.5. Alternative Treatments and Challenges

Six countries had no alternative treatments available. Among those that had them, sapropterin (BH4) was the most frequent (six countries—Argentina, Brazil, Costa Rica, Dominican Republic and Mexico; approximately 60 patients in total); large neutral amino acids (LNAA) were available in two countries (Argentina and Peru), and glycomacropeptide (GMP) was available in one country (Argentina). Argentina was the only country that had all three options available, also with the most patients using them (>20 patients on BH4 and GMP and nearly 10 patients on LNAA).

Participants were asked to provide a score from 0 to 100 on how much they believed each category had contributed to hampering therapeutic success. Median scores are depicted in Figure 4. In addition to the aspects shown in Figure 4, other cited barriers to treatment were: low accessibility due to geographic location, limited access to alternative treatments, high cost of treatments, and long periods of time for samples to arrive at the laboratory.

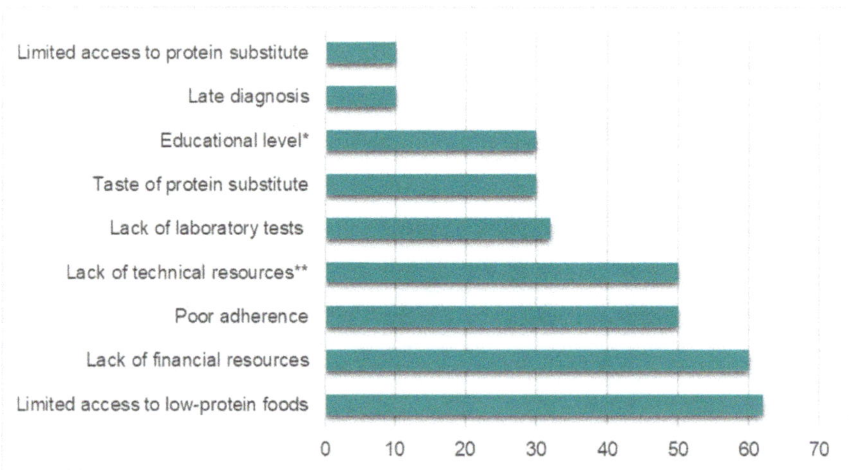

Figure 4. Barriers to treatment most commonly perceived by the respondents (n = 22). Values represent the median scores assigned by the respondents. * Educational level of patients and caregivers; ** Technical resources required or desirable to manage the diet, such as a local database of the Phe content of foods and specialized nutrition software.

4. Discussion

This study reports a broad and unprecedented characterization of the current state of diagnosis and management of PKU across Latin America. Data on NBS, laboratory tests, professional training, treatment targets, dietary practices, and resources, among other aspects, were compiled from 13 different countries and 22 treatment centers. These countries represent 87% of the Latin American population. Respondents were physicians and dietitians, most of whom were experienced in PKU treatments and were following a variable amount of PKU patients of all ages.

NBS for PKU began mostly after the 1990s in Latin America, nearly 30 years after the USA and some European countries had initiated their screening programs [1,2]. Nevertheless, most Latin American countries currently have wide-coverage national NBS programs for PKU and multidisciplinary reference centers for the follow-up of these patients, as shown in our study. Although the need for NBS and early treatment of PKU was generally agreed upon, other practices were not. Whereas both American and European guidelines [13,14] recommend that treatment should be started when Phe levels are ≥360 µmol/L, 35% (n = 8) of the centers in our sample employed different cutoffs, with most of them (n = 6) using higher levels. Higher cutoffs could miss mild PKU patients and raise concern due to the detrimental effect of high Phe levels in early life. A meta-analysis showed that each 100 µmol/L increase in Phe in early life predicted a 1.3- to 3.9-point decrease in intelligence quotient (IQ) over a Phe range of 394 to 750 µmol/L [15]. However, the exact cutoff at which treatment should begin is still debatable. There is a consensus that individuals with Phe levels >600 µmol/L should be treated, but the evidence regarding the initiation of treatment with blood Phe concentrations between 360 and 600 µmol/L is inconsistent. Given the risk of neurocognitive consequences, most guidelines recommend initiating treatment when blood Phe concentrations are >360 µmol/L [8,13,14].

The frequency of Phe and Tyr monitoring was highly heterogeneous among centers. The highest agreement (65%) found was in respect to measuring Phe once a week or more in infants younger than 1 year of age (Figure 1), which is in line with both American and European recommendations [13,14,16]. An even greater disagreement was observed for Tyr measurements, in all age groups. This probably reflects the lower availability of Tyr analyses in several centers: more than 20% of them rarely or never measured Tyr, regardless of the patient's age group or condition. Tyr monitoring is critical in PKU, since this amino acid cannot be synthetized properly due to the metabolic blockage, and a decreased availability of Tyr in the brain likely contributes to the cognitive impairment found in untreated patients [17]. American Genetic Metabolic Dietitians International (GMDI) guidelines recommend that Tyr measurements be performed as frequently as Phe measurements [16].

A similar heterogeneity was observed for Phe target values throughout life (Figure 2). The highest agreement (69%) was for children aged 2–12 years, where the Phe target was 120–360 µmol/L; this was in agreement with both international guidelines [13,14]. For infants younger than 1 year of age, most (61%) centers aimed for Phe levels to be between 120 and 240 µmol/L, a goal that differed from the American and European guidelines, which recommend a Phe target of 120–360 µmol/L [13,14]. Chilean guidelines support the 120–240 µmol/L target at this age, since in this period many factors interfere with the Phe level, such as growth, teething, infections, and frequent vaccinations [8]. The age group with the highest heterogeneity was >12 years, with 11 different targets reported. Eight (35%) centers agreed with the American target (120–360 µmol/L), and two (9%) agreed with the European values (120–600 µmol/L). A trend towards more restrictive targets was observed in all age groups and in pregnant patients. However, the theoretical basis for some of the reported targets was not clear; in some cases they were unique and diverged within the same country, even when national guidelines were available.

Greater consistency was found in dietary practices. The simplified method was the most frequently used approach to manage dietary intake (in 48% of the centers). The simplified diet approach has been shown to be easier to follow, encourages healthy food choices, and can improve the quality of life and adherence of patients with PKU [18,19]. Breastfeeding was encouraged in most centers, reflecting its clear evidence-based benefits in PKU [20,21]. Regarding nutritional monitoring, basic measurements such as weight, height, head circumference, and food recalls were performed in most centers at all appointments, meeting international recommendations [13,14]. Blood tests for nutritional monitoring were usually performed on a regular basis. It is noteworthy, however, that pivotal examinations such as amino acid profiles and bone densitometry were not regularly performed in most centers (56%). These assessments are required in the follow-up of patients with PKU who are being treated, since they are at risk of amino acid deficiencies and osteopenia [13,14,17,22]. A likely explanation for this is that these two technologies are less available due to their high costs and need for specialized facilities. There are substantial differences in the availability of technologies across Latin American countries [5]. The total protein prescriptions showed some heterogeneity among centers (Figure 3). However, the well-established recommendation that a higher protein intake is necessary for patients with PKU [23] has been mostly followed.

Nutritional resources to support patients, families, and professionals were scarce. Although most countries had national guidelines, most respondents reached for international guidelines (61%) as theoretical background. This might be due to outdated or incomplete local guidelines. While most professionals used Phe intake for managing the diet, most of them (78%) did not have a suitable regional database of the Phe content of foods and had to rely on international databases. However, nutrient contents of foods can vary due to environmental factors, production, and processing, and might differ between countries [24]. Health professionals also face difficulties calculating the diets: only four (17%) centers had specialized nutrition software.

For patients, unflavored powdered amino acid formulas were the only protein substitute available in most countries. Specific low-protein foods for PKU were unavailable in 61% of the countries. Even when available, they were usually not affordable, since these were rarely subsidized by the government. Specially designed low-protein products are important for satiety and diet variety [25], and were also proven valuable in improving metabolic control and growth in patients with PKU [26]. However, they inflict a significant financial burden to the PKU diet: in an American study, low-protein foods represented the highest annual out-of-pocket costs (child = US$1651.00; adult = US$967.00) when compared to other categories of care [6]. Considering the gross national income per capita in 2019 [27], this would be equivalent to 20% of the income of a Latin American citizen. The average expenditure on food of a Brazilian citizen, for instance, is USD 866.00 per year. Therefore, it is completely unreasonable to expect that Latin American patients with PKU would be able to afford low-protein foods without subsidy.

Alternative therapies are also a reality for a few in Latin America. BH4 was the most common therapy, available in 7/13 countries. However, even when approved, this therapy was only used in a few patients. LNAA and GMP were even rarer, despite several products being available in Europe and the USA for years [28]. Alternative treatments are highly relevant in PKU since most patients struggle to follow the restrictive diet and to take the protein substitute [29,30]. As a consequence of suboptimal metabolic control and restrictive dietary management, psychiatric illness is common in adult PKU patients. The advent of new treatments that do not require such a restricted diet might improve metabolic control, mental health, and cognitive functioning in these patients [31].

Finally, we asked the respondents to score the topics they considered the greatest barriers to the adequate treatment of PKU in their realities. The answers largely reflected the major gaps found throughout the study: lack of nutritional resources for patients and professionals and the high cost of therapies. One of the highest assigned scores was for "poor adherence", which may also be an outcome of the difficulties mentioned above. Another barrier cited by the respondents was low accessibility due to geographic location. In Latin America, most of the sophisticated technologies that are required for the follow-up of patients with IEM are geographically concentrated in larger and wealthier urban areas, contributing to health inequalities in this population [5].

5. Conclusions

In conclusion, here we have reported the first compilation of the status of PKU care in Latin America. Despite most countries having national NBS programs and guidelines, we found a highly heterogeneous scenario considering practices across countries and even within the same country. The struggles, however, were similar. Most countries experienced a lack of resources for both patients and health care professionals, which may be impairing treatment outcomes. Together, these results indicate an urgent need for a comprehensive Latin American guideline that must be able to integrate the latest evidence-based recommendations with the challenges and possibilities faced by Latin American countries.

Author Contributions: Conceptualization, data collection and analysis, writing—review and editing: S.P., B.B.d.S., I.V.D.S., V.C. The review of the project, data collection and proofreading of the manuscript: M.F.S., M.J.L.-W. and G.C. Data collection and proofreading of the manuscript: A.C., N.S., M.P., M.S.-R., F.P., S.P.W., R.B.P., L.O.P., M.C.S., M.V.A., M.M., A.R.C.N., A.M., A.R.C., A.L., J.F.S.-L., C.P., L.M.R., G.M.Z.T. and L.F.R. All authors have read and agreed to the published version of the manuscript.

Funding: This research was funded by FIPE-HCPA, grant number 2020-0439, Coordenação de Aperfeiçoamento de Pessoal de Nível Superior – Brasil (CAPES) and Fundação de Amparo à Pesquisa do Estado do Rio Grande do Sul (FAPERGS), grant number: 88887.192047/2018-00.

Institutional Review Board Statement: The study was approved by the Research Ethics Committee of Hospital de Clínicas de Porto Alegre, Brazil (CAAE 36401120.6.0000.5327).

Informed Consent Statement: Informed consent was obtained from all participants.

Data Availability Statement: Data available on request due to restrictions eg privacy or ethical. The data presented in this study are available on request from the corresponding author. The data are not publicly available due to containing information that could compromise the privacy of research participants.

Acknowledgments: The authors thank Dévora Random, Laura Fernández and Mariana Castro for their assistance with data collection. The authors also thank all Latin American PKU patients and their families.

Conflicts of Interest: The authors declare no conflict of interest.

References

1. Therrell, B.L.; Padilla, C.D.; Loeber, J.G.; Kneisser, I.; Saadallah, A.; Borrajo, G.J.C.; Adams, J. Current status of newborn screening worldwide: 2015. *Semin. Perinatol.* **2015**, *39*, 171–187. [CrossRef] [PubMed]
2. Borrajo, G.J.C. Newborn Screening for Phenylketonuria: Latin American Consensus Guidelines. *J. Inborn Errors Metab. Screen.* **2016**, *4*. [CrossRef]
3. Borrajo, G.J.C. Newborn screening in Latin America: A brief overview of the state of the art. *Am. J. Med. Genet. C Semin. Med. Genet.* **2021**. [CrossRef] [PubMed]
4. Atun, R.; De Andrade, L.O.M.; Almeida, G.; Cotlear, D.; Dmytraczenko, T.; Frenz, P.; Garcia, P.; Gómez-Dantés, O.; Knaul, F.M.; Muntaner, C.; et al. Health-system reform and universal health coverage in Latin America. *Lancet* **2015**, *28*, 1230–1247. [CrossRef]
5. Organisation for Economic Co-Operation and Development/The World Bank. *Health at a Glance: Latin America and the Caribbean 2020*; OECD: Paris, France, 2020.
6. Rose, A.M.; Grosse, S.D.; Garcia, S.P.; Bach, J.; Kleyn, M.; Simon, N.J.E.; Prosser, L.A. The financial and time burden associated with phenylketonuria treatment in the United States. *Mol. Genet. Metab. Rep.* **2019**, *21*, 1–9. [CrossRef]
7. Guest, J.F.; Bai, J.J.; Taylor, R.R.; Sladkevicius, E.; Lee, P.J.; Lachmann, R.H. Costs and outcomes over 36 years of patients with phenylketonuria who do and do not remain on a phenylalanine-restricted diet. *J. Intellect. Disabil. Res.* **2013**, *57*, 567–579. [CrossRef]
8. Castro, G.; Hamilton, V.; Cornejo, V. Chilean Nutrition Management Protocol for Patients with Phenylketonuria. *J. Inborn Errors Metab. Screen.* **2017**, *5*. [CrossRef]
9. Chiesa, A.; Fraga, C.; Prieto, L.; Pardo, M.L. Modelo de atención de pacientes con fenilcetonuria (PKU) en Argentina. *Acta Pediatr. Mex.* **2012**, *33*, 308–310.
10. García-Restrepo, N.; Hernández, G.J.; Londoño, M.L.; Muriel-Ramírez, R. Deficiencia de fenilalanina hidroxilasa: Espectro clínico y estado actual del diagnóstico en Colombia. *Biosalud* **2018**, *17*, 49–64.
11. Vieira Neto, E.; Maia Filho, H.S.; Monteiro, C.B.; Carvalho, L.M.; Tonon, T.; Vanz, A.P.; Schwartz, I.V.D.; Ribeiro, M.G. Quality of life and adherence to treatment in early-treated Brazilian phenylketonuria pediatric patients. *Braz. J. Med. Biol. Res.* **2018**, *51*. [CrossRef]
12. Lemes, A.; Queijo, C.; Zabala, C.; Queiruga, G. Modelo de atención de la Fenilcetonuria (PKU) en Uruguay. *Acta Pediatr. Mex.* **2012**, *33*, 311–314.

13. Van Wegberg, A.M.J.; MacDonald, A.; Ahring, K.; Bélanger-Quintana, A.; Blau, N.; Bosch, A.M.; Burlina, A.; Campistol, J.; Feillet, F.; Giżewska, M.; et al. The complete European guidelines on phenylketonuria: Diagnosis and treatment. *Orphanet J. Rare Dis.* **2017**, *12*, 1–56. [CrossRef] [PubMed]
14. Vockley, J.; Andersson, H.C.; Antshel, K.M.; Braverman, N.E.; Burton, B.K.; Frazier, D.M.; Mitchell, J.; Smith, W.E.; Thompson, B.H.; Berry, S.A. Phenylalanine hydroxylase deficiency: Diagnosis and management guideline. *Genet. Med.* **2014**, *16*, 188–200. [CrossRef] [PubMed]
15. Waisbren, S.E.; Noel, K.; Fahrbach, K.; Cella, C.; Frame, D.; Dorenbaum, A.; Levy, H. Phenylalanine blood levels and clinical outcomes in phenylketonuria: A systematic literature review and meta-analysis. *Mol. Genet. Metab.* **2007**, *92*, 63–70. [CrossRef]
16. Singh, R.H.; Cunningham, A.C.; Mofidi, S.; Douglas, T.D.; Frazier, D.M.; Hook, D.G.; Jeffers, L.; McCune, H.; Moseley, K.D.; Ogata, B.; et al. Updated, web-based nutrition management guideline for PKU: An evidence and consensus based approach. *Mol. Genet. Metab.* **2016**, *118*, 72–83. [CrossRef] [PubMed]
17. Van Spronsen, F.J.; Van Rijn, M.; Bekhof, J.; Koch, R.; Smit, P.G.A. Phenylketonuria: Tyrosine supplementation in phenylalanine-restricted diets. *Am. J. Clin. Nutr.* **2001**, *73*, 153–157. [CrossRef] [PubMed]
18. Bernstein, L.; Burns, C.; Sailer-Hammons, M.; Kurtz, A.; Rohr, F. Multiclinic Observations on the Simplified Diet in PKU. *J. Nutr. Metab.* **2017**, *2017*. [CrossRef]
19. Zimmermann, M.; Jacobs, P.; Fingerhut, R.; Torresani, T.; Thöny, B.; Blau, N.; Baumgartner, M.R.; Rohrbach, M. Positive effect of a simplified diet on blood phenylalanine control in different phenylketonuria variants, characterized by newborn BH 4 loading test and PAH analysis. *Mol. Genet. Metab.* **2012**, *106*, 264–268. [CrossRef]
20. Banta-Wright, S.A.; Shelton, K.C.; Lowe, N.D.; Knafl, K.A.; Houck, G.M. Breast-feeding Success Among Infants with Phenylketonuria. *J. Pediatr. Nurs.* **2012**, *27*, 319–327. [CrossRef]
21. Kanufre, V.C.; Starling, A.L.P.; Leão, E.; Aguiar, M.J.B.; Santos, J.S.; Soares, R.D.L.; Silveira, A.M. Breastfeeding in the treatment of children with phenylketonuria. *J. Pediatr.* **2007**, *83*. [CrossRef]
22. de Castro, M.J.; de Lamas, C.; Sánchez-Pintos, P.; González-Lamuño, D.; Couce, M.L. Bone status in patients with phenylketonuria: A systematic review. *Nutrients* **2020**, *12*, 2154. [CrossRef]
23. Acosta, P.B. Recommendations for protein and energy intakes by patients with phenylketonuria. *Eur. J. Pediatr.* **1996**, *155*, S121–S124. [CrossRef]
24. Greenfield, H.; Southgate, D.A.T. *Food Composition Data*; FAO: Rome, Italy, 1992.
25. Cochrane, B.; Schwahn, B.; Galloway, P.; Robinson, P.; Gerasimidis, K. A questionnaire survey on the usage of low protein staple foods by people with phenylketonuria in Scotland. *J. Hum. Nutr. Diet.* **2014**, *27*, 533–541. [CrossRef] [PubMed]
26. Handoom, B.; Megdad, E.; Al-Qasabi, D.; Al Mesned, M.; Hawary, R.; Al-Nufiee, S.; Al-Hassnan, Z.; Alsayed, M.D.; Eldali, A. The effects of low protein products availability on growth parameters and metabolic control in selected amino acid metabolism disorders patients. *Int. J. Pediatr. Adolesc. Med.* **2018**, *5*, 60–68. [CrossRef] [PubMed]
27. The World Bank GNI Per Capita, Atlas method (Current US$)—Latin America & Caribbean (Excluding High Income). Available online: https://data.worldbank.org/indicator/NY.GNP.PCAP.CD?locations=XJ (accessed on 18 January 2021).
28. Pena, M.J.; De Almeida, M.F.; Van Dam, E.; Ahring, K.; Bélanger-Quintana, A.; Dokoupil, K.; Gokmen-Ozel, H.; Lammardo, A.M.; MacDonald, A.; Robert, M.; et al. Protein substitutes for phenylketonuria in Europe: Access and nutritional composition. *Eur. J. Clin. Nutr.* **2016**, *70*, 785–789. [CrossRef] [PubMed]
29. Ford, S.; O'Driscoll, M.; MacDonald, A. Living with Phenylketonuria: Lessons from the PKU community. *Mol. Genet. Metab. Rep.* **2018**, *17*, 57–63. [CrossRef] [PubMed]
30. MacDonald, A.; Van Rijn, M.; Gokmen-Ozel, H.; Burgard, P. The reality of dietary compliance in the management of phenylketonuria. *J. Inherit. Metab. Dis.* **2010**, *33*, 665–670. [CrossRef] [PubMed]
31. Ashe, K.; Kelso, W.; Farrand, S.; Panetta, J.; Fazio, T.; De Jong, G.; Walterfang, M. Psychiatric and Cognitive Aspects of Phenylketonuria: The Limitations of Diet and Promise of New Treatments. *Front. Psychiatry* **2019**, *10*, 1–20. [CrossRef]

nutrients

MDPI

Article

Metabolic Control of Patients with Phenylketonuria in a Portuguese Metabolic Centre Comparing Three Different Recommendations

Viviane Kanufre [1,2,3,4], Manuela Ferreira Almeida [1,2,5], Catarina Sousa Barbosa [1,2], Carla Carmona [1,2,5], Anabela Bandeira [2], Esmeralda Martins [2,5], Sara Rocha [2], Arlindo Guimas [2], Rosa Ribeiro [2], Anita MacDonald [6], Alex Pinto [6] and Júlio César Rocha [1,2,7,8,*]

1 Centro de Genética Médica, Centro Hospitalar Universitário do Porto (CHUPorto), 4099-028 Porto, Portugal; vikanufre@gmail.com (V.K.); manuela.almeida@chporto.min-saude.pt (M.F.A.); catarina-s-@hotmail.com (C.S.B.); carla.carmona@chporto.min-saude.pt (C.C.)
2 Centro de Referência na Área de Doenças Hereditárias do Metabolismo, Centro Hospitalar Universitário do Porto-CHUPorto, 4099-001 Porto, Portugal; anabela.ol.bandeira@sapo.pt (A.B.); esmeralda.g.martins@gmail.com (E.M.); saraisabelrocha@gmail.com (S.R.); arlguimas@gmail.com (A.G.); rocrff@gmail.com (R.R.)
3 Núcleo de Ações e Pesquisa em Apoio Diagnóstico (NUPAD), School of Medicine, Federal University of Minas Gerais (UFMG), Avenida Professor Alfredo Balena, 190, Belo Horizonte 30130-100, Brazil
4 Hospital das Clínicas, UFMG, Avenida Professor Alfredo Balena, 110, Santa Efigênia, Belo Horizonte 30130-100, Brazil
5 Unit for Multidisciplinary Research in Biomedicine, Abel Salazar Institute of Biomedical Sciences, University of Porto-UMIB/ICBAS/UP, 4050-313 Porto, Portugal
6 Birmingham Women's and Children's Hospital, Birmingham B4 6NH, UK; anita.macdonald@nhs.net (A.M.); alex.pinto@nhs.net (A.P.)
7 Nutrition & Metabolism, NOVA Medical School, Faculdade de Ciências Médicas, Universidade Nova de Lisboa, Campo Mártires da Pátria, 130, 1169-056 Lisbon, Portugal
8 Centre for Health Technology and Services Research (CINTESIS), 4200-450 Porto, Portugal
* Correspondence: rochajc@nms.unl.pt; Tel.: +351-21-880-3000

Citation: Kanufre, V.; Almeida, M.F.; Barbosa, C.S.; Carmona, C.; Bandeira, A.; Martins, E.; Rocha, S.; Guimas, A.; Ribeiro, R.; MacDonald, A.; et al. Metabolic Control of Patients with Phenylketonuria in a Portuguese Metabolic Centre Comparing Three Different Recommendations. *Nutrients* **2021**, *13*, 3118. https://doi.org/10.3390/nu13093118

Academic Editor: Gary David Lopaschuk

Received: 26 July 2021
Accepted: 1 September 2021
Published: 6 September 2021

Publisher's Note: MDPI stays neutral with regard to jurisdictional claims in published maps and institutional affiliations.

Copyright: © 2021 by the authors. Licensee MDPI, Basel, Switzerland. This article is an open access article distributed under the terms and conditions of the Creative Commons Attribution (CC BY) license (https://creativecommons.org/licenses/by/4.0/).

Abstract: Blood phenylalanine (Phe) is used as the primary marker to evaluate metabolic control. Our study aimed to describe the metabolic control of patients with phenylketonuria (PKU) comparing three different treatment recommendations (European guidelines/US guidelines/Portuguese consensus). This was a retrospective, observational, single centre study in patients with PKU collecting data on blood Phe levels from 2017. Nutritional intake data and sapropterin (BH4) prescription were collected at the last appointment of 2017. The final sample studied included 87 patients (48% females) [13 hyperphenylalaninemia; 47 mild PKU; 27 classical PKU] with a median age of 18 y (range: 1–36 y). The median number of blood Phe measurements for patients was 21 (range: 6–89). In patients aged < 12 y, the median blood Phe level was 300 μmol/L (range 168–480) and 474 μmol/L (range 156–1194) for patients ≥ 12 y. Overall, a median of 83% of blood Phe levels were within the European PKU guidelines target range. In patients aged ≥ 12 years, there was a higher median % of blood Phe levels within the European PKU guidelines target range (≥12 y: 84% vs. <12 y: 56%). In children < 12 y with classical PKU (n = 2), only 34% of blood Phe levels were within target range for all 3 guidelines and 49% with mild PKU (n = 11). Girls had better control than boys (89% vs. 66% median Phe levels within European Guidelines). Although it is clear that 50% or more patients were unable to achieve acceptable metabolic control on current treatment options, a globally agreed upper Phe target associated with optimal outcomes for age groups is necessary. More studies need to examine how clinics with dissimilar resources, different therapeutic Phe targets and frequency of monitoring relate to metabolic control.

Keywords: phenylketonuria; phenylalanine; metabolic control; guidelines

1. Introduction

Phenylketonuria (PKU) is an inborn error of amino acid metabolism characterized by persistent hyperphenylalaninemia due to a deficiency of the hepatic phenylalanine hydroxylase enzyme. This prevents the hydroxylation of the essential amino acid phenylalanine (Phe) into tyrosine (Tyr) leading to increased blood and brain Phe concentrations. Immediate and sustained treatment following newborn screening is crucial to enable normal development, health and well-being throughout life [1]. While high blood Phe concentrations during childhood are known to primarily affect intellectual functioning, increased levels during adulthood are associated with neurological, mental health, executive functioning and behavioural problems, as well as deficits in social skills [2].

The main treatment objective is to maintain blood Phe levels within a safe target therapeutic range, while providing necessary macro and micronutrients to enable proper growth and development [3]. The core treatment is a Phe restricted diet (by restriction of natural protein), supplemented with a low/free Phe protein substitute (PS) and special low protein foods (SLPFs) [3–6]. The availability of special medical foods for PKU varies widely across countries [5,6], potentially influencing blood Phe control. The dose (g/kg) and protein source of protein substitutes also impacts metabolic control [3,7–9]. It is well recognized that adhering to a Phe restricted diet is particularly challenging [10,11].

A subgroup of patients with PKU, usually with a milder PKU phenotype may benefit from a drug therapy, tetrahydrobiopterin (BH4) (sapropterin dihydrochloride). BH4 acts as a pharmaceutical chaperone increasing the residual activity of the phenylalanine hydroxylase enzyme; BH4 responsive patients are generally able to increase their natural protein intake by 2 to 4-fold and/or reduce blood Phe levels [12]. Pegvaliase, the newest treatment option, is an enzyme substitution therapy that converts Phe to trans-cinnamic acid and ammonia, and was approved in the United States (US) in 2018 and Europe in 2019 [13]. It is administered by subcutaneous injection, and it is effective in lowering blood Phe levels and increasing Phe tolerance. However, pegvaliase may be associated with immunologic reactions, requiring careful management [13]. This treatment is directed at adults with blood Phe levels $\geq$ 600 µmol/L and is not recommended in pregnancy. Further studies are required looking at its long-term efficacy and safety.

Blood Phe is used as the primary marker to evaluate metabolic control. Routine blood Phe is the only practical marker even though the aim of treatment is to prevent high Phe levels in the brain due to Phe transport across the blood brain barrier [1]. This is particularly important in early life when high Phe levels have a severe impact in brain and neurological development [14–16]. However, chronic long term high blood Phe levels in adulthood may also impact cognition and there are an increasing case studies of people with PKU developing neurological issues with increasing age [17–19]. International PKU guidelines give recommendations for target blood Phe levels which are essential to guide and monitor treatment, assess patient's outcomes, and compare effectiveness of treatments [1,20].

International guidelines representing different nations have recommended different target therapeutic ranges for blood Phe levels and there is conflict amongst professionals regarding the optimal target range, particularly in adult patients with PKU [21]. In 2017, the European PKU Guidelines gave scientific evidence to support an upper blood Phe target level of 360 µmol/L for children aged < 12 y (years) and for patients aged $\geq$ 12 y, 600 µmol/L [1]. The United States (US) 2014 guidelines recommended a target blood Phe level of 360 µmol/L for all age groups [20]. In 2007, in Portugal, a PKU working group from the Portuguese Society for Metabolic Disorders (SPDM) suggested an upper Phe target in the Portuguese Consensus of 360 µmol/L up to 12 y and 480 µmol/L onwards [22]. A recent publication from Portugal showed that professionals did not agree with all the key statements from the European PKU Guidelines [23]. Although 100% of professionals agreed with target blood Phe levels for patients < 12 y, this decreased to only 32% for the recommendations $\geq$ 12 y [23]. In practice, Portuguese centres already aimed for a lower 'upper' blood Phe target level (480 µmol/L) and were uncomfortable relaxing this to a maximum upper limit of 600 µmol/L without conclusive evidence.

This study aims to describe the metabolic control of patients with PKU in a single Portuguese centre comparing three different recommendations (European guidelines, US guidelines and Portuguese consensus) [1,20,22].

2. Materials and Methods

2.1. Participants

All patients with PKU being treated at Centro Hospitalar Universitário do Porto were considered for this study ($n = 136$).

The severity of the disorder was classified according to neonatal blood Phe levels at newborn screening, as defined by the Portuguese Consensus: hyperphenylalaninemia (blood Phe < 360 µmol/L), mild PKU (blood Phe ≥ 360 and ≤1200 µmol/L), and classical PKU (blood Phe > 1200 µmol/L) [22].

Exclusion criteria included: failure to attend clinic appointments, <6 blood Phe measurements in the year of study, late diagnosed with PKU and pre-conception diet/pregnancy during the 12-month study period.

2.2. Study Design

This was a retrospective, observational, single centre study about blood Phe control of patients with PKU. All blood Phe and Tyr levels were taken during 2017. Nutritional intake data and BH4 prescription were also collected at the last appointment of 2017. All data were collected from electronic patient clinical records.

2.3. Data Collection

2.3.1. Nutritional Intake

Dietary intake data were collected for natural protein (NP; g/kg/day), protein equivalent from protein substitute (PS; g/kg/day) and total protein (TP; g/kg/day). Twenty-four hour dietary recalls were performed by experienced nutritionists (M.F.A. and J.C.R.) at each clinic to assess dietary intake. These data were transferred to an Excel sheet (Microsoft, Washington, DC, USA) which calculated nutritional intake. This excel is formatted with the nutritional composition from the Portuguese Food Composition Tables for normal foods and composition of the SLPFs and PS available in Portugal.

The three main treatment types used were defined in our analysis:

- **PKU diet only**: Phe restricted diet supplemented with PS and SLPFs
- **BH4 + diet:** BH4 treated patients with Phe restriction and ±PS
- **Non-restricted diet:** without PS or BH4 prescription

2.3.2. BH4

In patients taking BH4, data were collected on dose prescribed in mg/kg. Kuvan® from Biomarin was the BH4 molecule prescribed.

2.3.3. Metabolic Control

Blood Phe levels were measured from fasting dried blood spots taken by patients/caregivers and analyzed using a tandem mass spectrometry. Patients/caregivers were instructed by a nurse about the dried blood spot taking technique.

Data, stored on the patient database, were collected by a dietetic researcher (V.K.). Median blood Phe and Tyr levels were calculated and % of annual blood Phe measurements within target range from the year of data collection. Frequency of recommended monitoring was once weekly until 1 y, once every 2 weeks until 12 y and once monthly ≥ 12 y [20].

Blood Phe levels were compared with the European PKU Guidelines (recommended blood Phe levels 120–360 µmol/L up to 12 y and 120–600 µmol/L onwards) [1], US PKU guidelines (120–360 µmol/L throughout life) [20] and Portuguese Consensus (blood Phe levels 120–360 µmol/L up to 12 y and 120–480 µmol/L onwards) [22].

2.4. Ethical Statement

The study protocol was approved by the ethical committee of Centro Hospitalar Universitário do Porto on the 19 December 2018 (Reference 2018.199). Written informed consent was obtained from either each patient or caregiver (according to age). Participants were identified by a code to maintain patient anonymity.

2.5. Statistical Analysis

Descriptive statistics were used to present the results. Categorical variables were presented as absolute values or percentages, while continuous variables were presented as medians.

3. Results

3.1. Study Cohort

From 136 patients followed up in clinic, 49 patients were excluded due to: no attendance to scheduled appointments, either no blood Phe measurements during 2017 (n = 18) or <6 blood Phe measurements during the study period (n = 16); late diagnosed patients (n = 12); and pre-conception diet/pregnancy (n = 3).

The final sample studied included 87 patients (48% females) with a median age of 18 y (range from age 1 to 36 y). Nineteen patients were <12 y (median age of 8 y; range 1–11 y) and 68 patients ≥ 12 y (median age 22 y; range 12–36 y). Of the 68 patients ≥ 12 y, 36 patients (53%) were >20 y.

There were 13 patients with hyperphenylalaninemia (15%), 47 with mild PKU (54%), and 27 patients with classical PKU (31%).

Table 1 presents patients characteristics by age, gender, disorder severity and type of treatment prescribed.

Table 1. Patient's characteristics by type of treatment, gender, age and severity of PKU.

Variable		<12 Years n (%)	≥12 Years n (%)	Total n (%)
N		19 (22)	68 (78)	87 (100)
Gender	Female	8 (42)	34 (50)	42 (48)
	Male	11 (58)	34 (50)	45 (52)
PKU severity	Classical PKU	2 (11)	25 (37)	27 (31)
	Mild PKU	11 (58)	36 (53)	47 (54)
	HPA	6 (32)	7 (10)	13 (15)
Type of treatment	PKU diet only	12 (63)	38 (56)	50 (57)
	BH4 + diet	3 (16)	19 (28)	22 (26)
	Non-restricted diet	4 (21)	11 (16)	15 (17)

Abbreviations: HPA: hyperphenylalaninemia; **PKU**: Phenylketonuria; **BH4**: sapropterin; **PKU diet only**: Phenylalanine restricted diet supplemented with protein substitute and special low protein foods; **BH4 + diet**: BH4 treatment with Phe restriction and ±protein substitute; **Non-restricted diet:** without protein substitute or BH4.

3.2. Nutritional Intake

Patients were prescribed three main types of treatment; (1) PKU diet only, which is a Phe restricted diet supplemented with PS and SLPFs, n = 50; (2) BH4 (BH4 treated patients with Phe restriction and ±PS), n = 22; and (3) a non-restricted diet (without PS or BH4 prescription), n = 15. Table 2 presents nutritional protein intake regarding age, disorder severity and type of treatment prescribed.

Table 2. Median intake of natural protein, protein equivalent and total protein.

	Variable	Median Natural Protein (P25–P75) g/kg/Day	Median Protein Equivalent (P25–P75) g/kg/Day	Median Total Protein (P25–P75) g/kg/Day
	Total	0.69 (0.12–4.09)	0.74 (0.00–1.55)	1.54 (0.68–4.09)
Age	**<12 y (*n* = 19)**	0.69 (0.28–4.09)	0.89 (0.00–1.55)	1.84 (1.23–4.09)
	≥12 y (*n* = 68)	0.69 (0.12–2.55)	0.72 (0.00–1.32)	1.46 (0.68–2.55)
PKU Severity	**Classical PKU (*n* = 27)**	0.40 (0.17–1.80)	0.85 (0.00–1.32)	1.37 (0.95–1.80)
	Mild PKU (*n* = 47)	0.69 (0.12–2.4)	0.74 (0.00–1.55)	1.54 (0.68–2.40)
	HPA (*n* = 13)	1.97 (1.39–4.09)	0.00 (0.00–0.51)	2.10 (1.60–4.09)
Type of treatment	**PKU diet only (*n* = 50)**	0.48 (0.17–1.80)	0.87 (0.08–1.55)	1.47 (0.95–3.60)
	BH4 + diet (*n* = 22)	0.99 (0.24–1.84)	0.63 (0.00–1.07)	1.53 (0.68–2.19)
	Non-restricted diet (*n* = 15)	1.97 (1.26–4.09)	0.00 (0.00–0.00)	1.97 (1.26–4.09)

Abbreviations: HPA: hyperphenylalaninemia; **PKU**: Phenylketonuria; **BH4**: sapropterin; **PKU diet only**: Phenylalanine restricted diet supplemented with protein substitute and special low protein foods; **BH4 + diet**: BH4 treatment with Phe restriction and ±protein substitute; **Non-restricted diet:** without protein substitute or BH4.

Of 50 patients prescribed a Phe restricted diet, n = 41 were given Phe-free amino acids, n = 4 Phe-free amino acid together with glycomacropeptide (CGMP-AA); and 5 patients CGMP-AA supplement only. For the patients on long term BH4 treatment (n = 22), (duration of BH4 treatment = median 2 years), 18 required supplementation with a PS. The median daily dose of BH4 was 15.5 mg/kg/day (range 11.6–20.6 mg/kg).

Patients in the non-restricted diet group (n = 15) all met safe levels of protein intake [21] without the use of PS.

3.3. Metabolic Control—Portuguese Consensus

The median number of blood Phe measurements for each patient recorded in 2017 was 21 (range 6–89). In patients aged < 12 y, the median blood Phe level was 300 μmol/L (range 168–480); blood Tyr was 71 μmol/L (range 43–96). In patients aged ≥ 12 years, the median blood Phe level was 474 μmol/L (range 156–1194) and Tyr was 67 μmol/L (range 40–94). Median results were within the Portuguese targets for both age groups [22]. However, in children < 12 years with mild PKU, only 49% of levels were within target range. Girls had overall better control than boys (median % of blood Phe levels within target range was females: 66% (aged ≥ 12 y) and 74% (aged < 12 y); males: 41% (aged ≥ 12 y) and 45% (aged < 12 y)). When assessing the younger and older age groups, the percentage of blood Phe within target range improved with age in the mild PKU group and remained unchanged in the classical and HPA group.

Table 3 presents the median % of blood Phe levels within target range, recommended by the Portuguese consensus [22] stratified by age, sex, disease severity and type of treatment.

Annual median blood Phe levels are presented for each patient by age and type of treatment in HPA patients (Figure 1), Mild PKU (Figure 2) and Classical PKU (Figure 3).

Table 3. Median percentage of blood Phe measurements within target range recommended by the Portuguese consensus.

		<12 Years	≥12 Years
		Median % of Blood Phe Levels within Target Range *	**Median % of Blood Phe Levels within Target Range ***
Sex	**Female (*n* = 42)**	74	66
	Male (*n* = 45)	45	41
PKU severity	**Classical PKU (*n* = 27)**	34	28
	Mild PKU (*n* = 47)	49	77
	HPA (*n* = 13)	91	100
Type of treatment	**PKU diet only (*n* = 50)**	54	27
	BH4 + diet (*n* = 22)	49	84
	Non-restricted diet (*n* = 15)	73	100

Abbreviations: HPA: hyperphenylalaninemia; **PKU**: Phenylketonuria; **BH4**: sapropterin; **PKU diet only**: Phenylalanine restricted diet supplemented with protein substitute and special low protein foods; **BH4 + diet**: BH4 treatment with Phe restriction and ±protein substitute; **Non-restricted diet**: without protein substitute or BH4. * Portuguese consensus.

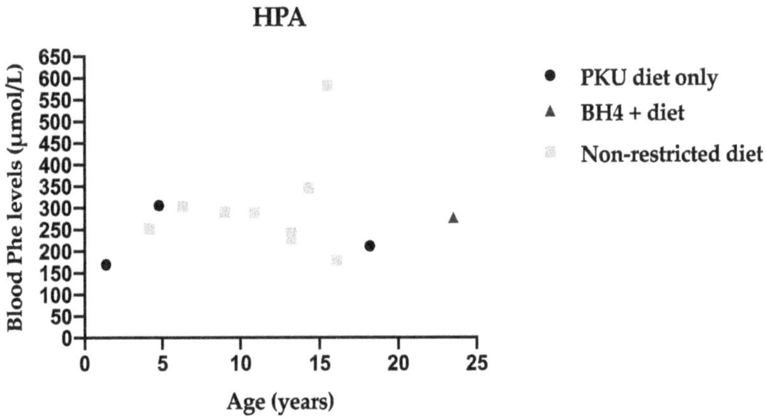

Figure 1. Median annual blood Phe levels for each individual patient with HPA by age and type of treatment.

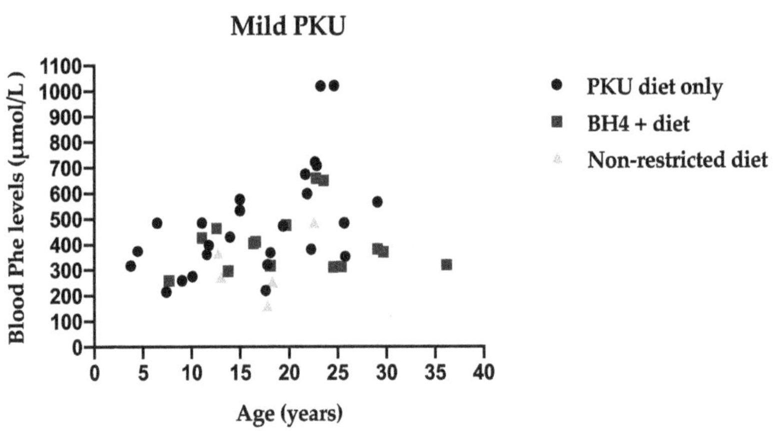

Figure 2. Median annual blood Phe levels for each individual patient with mild PKU by age and type of treatment.

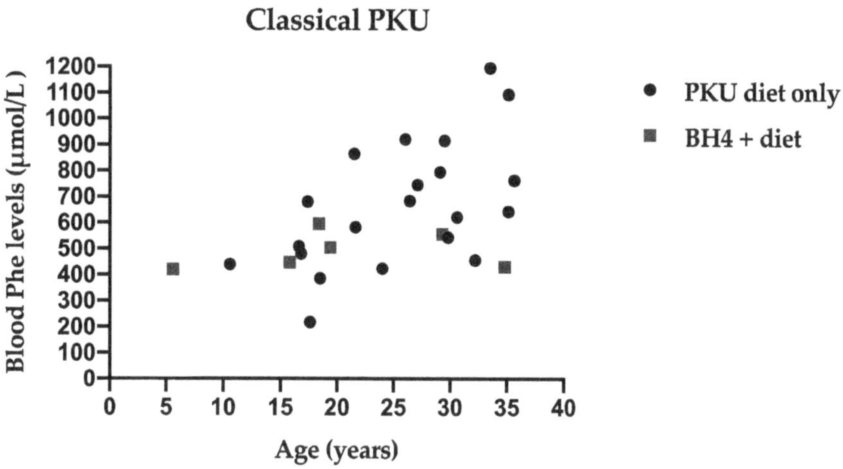

Figure 3. Median annual blood Phe levels for each individual patient with Classical PKU by age and type of treatment.

3.4. Metabolic Control Comparing Three Different Recommendations

Annual median blood Phe levels increased with age. Figure 4 shows the annual median of blood Phe levels for each patient studied, comparing to the upper target levels of the Portuguese consensus, European and US guidelines.

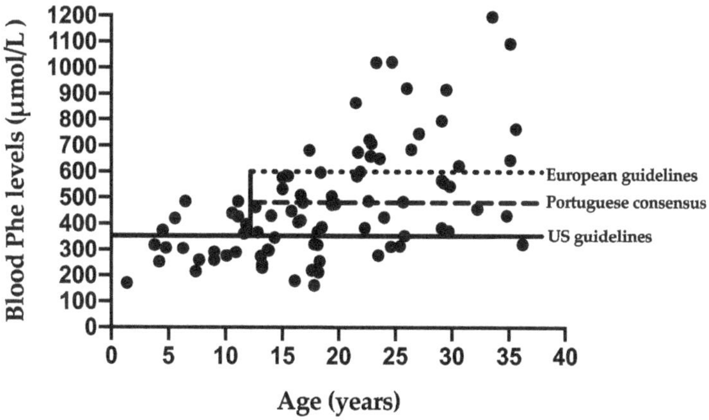

Figure 4. Median annual blood Phe levels presented for each patient compared to three different recommendations.

The median percentage of blood Phe levels within target range according to the two International guidelines and the Portuguese consensus are given in Table 4.

Table 4. Median percentage of blood Phe measurements within target range according to the Portuguese consensus, European and US guidelines.

		Median % of Blood Phe Levels within Target Range		
	Variable	**Portuguese Consensus % (p25–p75)**	**European Guidelines % (p25–p75)**	**US Guidelines % (p25–p75)**
	Total	56 (19–94)	83 (36–100)	26 (0–78)
Age	**<12 years (*n* = 19)**	56 (23–87)	56 (23–87)	56 (23–87)
	≥12 years (*n* = 68)	54 (13–96)	84 (38–100)	17 (0–60)
Gender	**Female (*n* = 42)**	70 (23–98)	89 (42–100)	29 (0–81)
	Male (*n* = 45)	41 (13–89)	66 (22–99)	22 (0–59)
PKU severity	**HPA (*n* = 13)**	100 (84–100)	100 (85–100)	96 (7–100)
	Mild PKU (*n* = 47)	63 (19–94)	84 (39–84)	41 (0–76)
	Classical PKU (*n* = 27)	27 (0–44)	47 (4–83)	6 (0–26)
Type of treatment	**PKU diet only (*n* = 50)**	32 (8–74)	59 (14–91)	15 (0–51)
	BH4 diet (*n* = 22)	76 (44–92)	91 (66–100)	32 (17–76)
	Non-restricted diet (*n* = 15)	97 (33–100)	100 (83–100)	83 (13–100)

Abbreviations: US: United States; **HPA**: hyperphenylalaninemia; **PKU**: Phenylketonuria; **BH4**: sapropterin; **PKU diet only**: Phenylalanine restricted diet supplemented with protein substitute and special low protein foods; **BH4 + diet**: BH4 treatment with Phe restriction and ±protein substitute; **Non-restricted diet:** without protein substitute or BH4.

4. Discussion

This study demonstrated that patients with classical PKU struggled to achieve an acceptable level of blood Phe control on dietary treatment only, irrespective of age or the upper target blood Phe level guideline. Their blood Phe control was suboptimal compared with mild PKU and HPA. These figures most likely underestimate poor control, as they do not consider the excluded patients from this study who either did not attend clinic appointments or failed to return blood Phe spots. Overall, this was a group of patients who were well supported by their clinical multidisciplinary team (nutritionist, psychologist, clinician). Patients with classical PKU only tolerated 0.4 g/kg (range: 0.17–1.80) of natural protein and a total protein intake of 1.37 g/kg (range: 0.95–1.80). It is possible that a higher dose of PS may have improved blood Phe control as their total protein intake was lower than the other 2 groups, although safe levels of protein intake were met [24].

Patients with classical PKU may have been unable to maintain their severe and onerous dietary restriction. Not only is the Phe restricted diet very limited, but it also involves maintaining strict dietary routines, preparation of SLPFs, planning daily Phe consumption, preparing low-Phe meals and meticulously planning every activity that involves food. It incurs a time management burden of 19 h per week [22]. Moreover, maintaining a Phe restrictive diet seems even more challenging than in the past. Some protein substitutes such as CGMP-AA contain Phe which may complicate gaining acceptable blood Phe control in children [7]. Persistent consumer pressure from the food industry and busy working adult lives has led to increased dependence on processed foods which are commonly not low in Phe or may have unreliable protein labelling information [25]. In addition, societal efforts to reduce the sugar content of foods has also led to sugar replacement by artificial sweeteners such as aspartame, another unquantified source of Phe [26]. Pre-existing social disadvantages such as parental poor literacy, health literacy and poverty may render some children particularly vulnerable.

The evidence to support an upper blood Phe level of 360 µmol/L in children aged under <12 y is convincing and is supported by all three PKU Guidelines/Consensus [1,20,22]. Therefore, the low percentage of blood Phe levels within target range for children aged < 12 y was a concern. Even children with mild PKU achieved <50% of blood Phe levels < 360 µmol/L. There is much evidence to suggest that the inability to sustain good metabolic control in childhood is associated with a decline in IQ (intelligence quotient) score and executive function and will have a negative influence in adulthood [14,17,18]. A meta-analysis estimated that an increase of 100 µmol/L in lifetime Phe levels predicts an average 1.9 to 4.1 point reduction in IQ over a range of Phe from 394 to 666 µmol/L [15]. Jaha et al. in 2017 showed that high blood Phe levels in childhood, affect adult cognitive flexibility, executive motor control, executive function in daily life and adult mental health [17]. Weglage et al. (2013) also showed that high blood Phe levels in childhood and adolescence were related to poorer IQ, information processing and attention in adulthood [27]. It is evident that alternative treatment choices are necessary to help improve the control of this group of patients with PKU. Much attention is directed at identifying effective non dietary treatments for adults but is essential that the paediatric population is not neglected. However, even when children were on BH4, only 49% had blood Phe levels within the target therapeutic range.

Deterioration of blood Phe control is well described with age. However, for our patient cohort, the overall % of blood Phe levels within the target therapeutic range (Portuguese consensus) did not deteriorate in patients aged $\geq$ 12 y. In fact, over 80% of Phe levels were within the European PKU target range but only 17% below the US guideline upper target range. Older patients with BH4 treatment for 2 y duration benefited from a relaxed dietary treatment without loss of metabolic control. Overall, the differences in upper blood Phe target ranges between local recommendations, Europe and USA are confusing and unsatisfactory for both patients and health professionals, particularly when the same evidence-based approach has been used to develop the two different guidelines. Both upper target levels aim to maintain safety of adult patients and prevent neurological and mental health complications, but neither recommendation is supported by robust clinical studies.

Even so, evidence is accumulating that significant sub-optimal outcomes exist in early treated adult patients. Pilotto et al. (2020) [28] provided evidence from 19 adult patients (median blood Phe level 873 µmol/L) showing that blood Phe levels were highly correlated with the number of failed neuropsychological tests, neuropsychiatric symptoms, motor evoked potential latency and parietal lobe atrophy high and there was direct association between brain function and metabolic control in adulthood. Historically diet was discontinued in many children and teenagers, and there is evidence of neurological symptoms in some patients [29]. Early treated patients with PKU have only reached 50 y, and little is known about their aging process in later adulthood. Due to this uncertainty of outcome and historical errors and missteps that have been made over treatment duration and degree of metabolic control, it is unsurprising that some international guidelines suggest stricter metabolic control for their adult populations.

We consider that the focus should be on seeking alternative treatments and home monitoring tools to help self-care and alleviation of strict dietary treatment. Treatments associated with minimal side effects, that are easy to administer and associated with optimal neurocognitive and mental health outcomes are essential. More resources/tools are needed to allow patients to achieve the lowest blood Phe level within target therapeutic range with no negative impact on quality of life. This is especially needed for classical patients with PKU who particularly struggle to meet the defined targets. These patients face bigger challenges with much lower Phe tolerances compared to other patient subgroups.

There are several limitations in this project. This is a retrospective uncontrolled study, only reflecting results of routine clinical practice for 1 year only in a single clinic. Protein intake in patients on a non-restrictive diet was not controlled as in other patients. Blood Phe levels are associated with error, reflective of blood specimen quality and concentration [30]. Blood Phe levels may also not directly reflect neurotransmitter metabolism. Exclusion of patients who did not attend clinic or perform blood samples may have altered the median percentage of blood Phe results observed. Also, the number of blood spots returned varied between patients. Twenty-four hours dietary recalls used for assessing dietary intake are associated with error and inaccuracy.

5. Conclusions

In general, blood Phe levels were around 56% within therapeutic target according to the Portuguese consensus although there is a tendency for increasing median blood Phe levels with age. The number of blood Phe levels within target range according to the European guidelines and US guidelines blood Phe levels were around 83% and 26%, respectively. In consideration of the different Phe upper limits recommended, we must strive for safe levels that are associated with the best patient outcomes. There should be focus on improving alternative treatment options and clinical resources to enable patients to achieve lower blood Phe levels. More studies are needed comparing outcome of centres using different blood Phe targets, their frequency of monitoring and the resources that they have available to them to determine optimal blood Phe control.

Author Contributions: V.K. was involved in the project design, collection of data, data analysis and writing of the manuscript. J.C.R. was involved in the project design, data analysis and writing of the manuscript. A.M. was involved in data analysis, interpretation of data and writing of the manuscript. A.P. was involved in data analysis and writing of the manuscript. C.S.B., M.F.A., C.C., S.R., A.G., R.R., E.M. and A.B. were involved in critical revisions of the paper for important intellectual content and final approval of the version submitted for publication. All authors have read and agreed to the published version of the manuscript.

Funding: CINTESIS—UIDB/4255/2020 a program financially supported by Fundação para a Ciência e Tecnologia/Ministério da Educação e Ciência, through national funds is acknowledged.

Institutional Review Board Statement: The study was conducted according to the guidelines of the Declaration of Helsinki, and approved by the ethical committee of Centro Hospitalar Universitário do Porto on the 19 December 2018 (Reference 2018.199).

Informed Consent Statement: Informed consent was obtained from all subjects involved in the study.

Data Availability Statement: The data will be made available from the authors upon reasonable request.

Conflicts of Interest: M.F.A. received grants from Glutamine, Nutricia, Merck Serono, Biomarin, Orphan and Lifediet to attend scientific meetings. A.M. has received research funding and honoraria from Nutricia, Vitaflo International and Merck Serono. She is a member of the European Nutritionist Expert Panel (Biomarin), a member of Sapropterin Advisory Board (Biomarin), a member of the advisory board entitled ELEMENT (Danone-Nutricia), and a member of an advisory board for Arla and Applied Pharma Research. A.P. received an educational grant from Cambrooke Therapeutics and grants from Vitaflo, Nutricia, Merck Serono, Biomarin, and Mevalia to attend scientific meetings. J.C.R. is a member of the European Nutritionist Expert Panel (Biomarin), the Advisory Board for Applied Pharma Research and Nutricia, and received honoraria as a speaker from APR, Merck Serono, Biomarin, Nutricia, Vitaflo, Cambrooke, PIAM and Lifediet.

References

1. van Wegberg, A.M.J.; MacDonald, A.; Ahring, K.; Bélanger-Quintana, A.; Blau, N.; Bosch, A.M.; Burlina, A.; Campistol, J.; Feillet, F.; Giżewska, M.; et al. The complete European guidelines on phenylketonuria: Diagnosis and treatment. *Orphanet J. Rare Dis.* **2017**, *12*, 162. [CrossRef]
2. Didycz, B.; Nitecka, M.; Bik-Multanowski, M. The Use of d2 and Benton Tests for Assessment of Attention Deficits and Visual Memory in Teenagers with Phenylketonuria. *JIMD Rep.* **2018**, *40*, 23–29. [CrossRef]
3. MacDonald, A.; van Wegberg, A.M.J.; Ahring, K.; Beblo, S.; Bélanger-Quintana, A.; Burlina, A.; Campistol, J.; Coşkun, T.; Feillet, F.; Giżewska, M.; et al. PKU dietary handbook to accompany PKU guidelines. *Orphanet J. Rare Dis.* **2020**, *15*, 171. [CrossRef]
4. MacDonald, A.; Rocha, J.C.; van Rijn, M.; Feillet, F. Nutrition in phenylketonuria. *Mol. Genet. Metab.* **2011**, *104*, S10–S18. [CrossRef]
5. Pena, M.J.; Almeida, M.F.; van Dam, E.; Ahring, K.; Bélanger-Quintana, A.; Dokoupil, K.; Gokmen-Ozel, H.; Lammardo, A.M.; MacDonald, A.; Robert, M.; et al. Special low protein foods for phenylketonuria: Availability in Europe and an examination of their nutritional profile. *Orphanet J. Rare Dis.* **2015**, *10*, 162. [CrossRef] [PubMed]
6. Pena, M.J.; de Almeida, M.F.; van Dam, E.; Ahring, K.; Bélanger-Quintana, A.; Dokoupil, K.; Gokmen-Ozel, H.; Lammardo, A.M.; MacDonald, A.; Robert, M.; et al. Protein substitutes for phenylketonuria in Europe: Access and nutritional composition. *Eur. J. Clin. Nutr.* **2016**, *70*, 785–789. [CrossRef]

7. Daly, A.; Evans, S.; Chahal, S.; Santra, S.; Pinto, A.; Jackson, R.; Gingell, C.; Rocha, J.C.; Van Spronsen, F.J.; MacDonald, A. Glycomacropeptide: Long-term use and impact on blood phenylalanine, growth and nutritional status in children with PKU. *Orphanet J. Rare Dis.* **2019**, *14*, 44. [CrossRef] [PubMed]
8. Pena, M.J.; Pinto, A.; de Almeida, M.F.; de Sousa Barbosa, C.; Ramos, P.C.; Rocha, S.; Guimas, A.; Ribeiro, R.; Martins, E.; Bandeira, A.; et al. Continuous use of glycomacropeptide in the nutritional management of patients with phenylketonuria: A clinical perspective. *Orphanet J. Rare Dis.* **2021**, *16*, 84. [CrossRef] [PubMed]
9. Pinto, A.; Almeida, M.F.; Ramos, P.C.; Rocha, S.; Guimas, A.; Ribeiro, R.; Martins, E.; Bandeira, A.; MacDonald, A.; Rocha, J.C. Nutritional status in patients with phenylketonuria using glycomacropeptide as their major protein source. *Eur. J. Clin. Nutr.* **2017**, *71*, 1230–1234. [CrossRef]
10. Jurecki, E.R.; Cederbaum, S.; Kopesky, J.; Perry, K.; Rohr, F.; Sanchez-Valle, A.; Viau, K.S.; Sheinin, M.Y.; Cohen-Pfeffer, J.L. Adherence to clinic recommendations among patients with phenylketonuria in the United States. *Mol. Genet. Metab.* **2017**, *120*, 190–197. [CrossRef]
11. Walter, J.H.; White, F.J.; Hall, S.K.; MacDonald, A.; Rylance, G.; Boneh, A.; Francis, D.E.; Shortland, G.J.; Schmidt, M.; Vail, A. How practical are recommendations for dietary control in phenylketonuria? *Lancet* **2002**, *360*, 55–57. [CrossRef]
12. Muntau, A.C.; Adams, D.J.; Bélanger-Quintana, A.; Bushueva, T.V.; Cerone, R.; Chien, Y.H.; Chiesa, A.; Coşkun, T.; de Las Heras, J.; Feillet, F.; et al. International best practice for the evaluation of responsiveness to sapropterin dihydrochloride in patients with phenylketonuria. *Mol. Genet. Metab.* **2019**, *127*, 1–11. [CrossRef]
13. Longo, N.; Dimmock, D.; Levy, H.; Viau, K.; Bausell, H.; Bilder, D.A.; Burton, B.; Gross, C.; Northrup, H.; Rohr, F.; et al. Evidence- and consensus-based recommendations for the use of pegvaliase in adults with phenylketonuria. *Genet. Med.* **2019**, *21*, 1851–1867. [CrossRef]
14. Waisbren, S.E.; Schnell, R.R.; Levy, H.L. Diet termination in children with phenylketonuria: A review of psychological assessments used to determine outcome. *J. Inherit. Metab. Dis.* **1980**, *3*, 149–153. [CrossRef]
15. Waisbren, S.E.; Noel, K.; Fahrbach, K.; Cella, C.; Frame, D.; Dorenbaum, A.; Levy, H. Phenylalanine blood levels and clinical outcomes in phenylketonuria: A systematic literature review and meta-analysis. *Mol. Genet. Metab.* **2007**, *92*, 63–70. [CrossRef] [PubMed]
16. Palermo, L.; Geberhiwot, T.; MacDonald, A.; Limback, E.; Hall, S.K.; Romani, C. Cognitive outcomes in early-treated adults with phenylketonuria (PKU): A comprehensive picture across domains. *Neuropsychology* **2017**, *31*, 255–267. [CrossRef] [PubMed]
17. Jahja, R.; Huijbregts, S.C.J.; de Sonneville, L.M.J.; van der Meere, J.J.; Legemaat, A.M.; Bosch, A.M.; Hollak, C.E.M.; Rubio-Gozalbo, M.E.; Brouwers, M.; Hofstede, F.C.; et al. Cognitive profile and mental health in adult phenylketonuria: A PKU-COBESO study. *Neuropsychology* **2017**, *31*, 437–447. [CrossRef]
18. Koch, R.; Burton, B.; Hoganson, G.; Peterson, R.; Rhead, W.; Rouse, B.; Scott, R.; Wolff, J.; Stern, A.M.; Guttler, F.; et al. Phenylketonuria in adulthood: A collaborative study. *J. Inherit. Metab. Dis.* **2002**, *25*, 333–346. [CrossRef]
19. Romani, C.; Palermo, L.; MacDonald, A.; Limback, E.; Hall, S.K.; Geberhiwot, T. The impact of phenylalanine levels on cognitive outcomes in adults with phenylketonuria: Effects across tasks and developmental stages. *Neuropsychology* **2017**, *31*, 242–254. [CrossRef] [PubMed]
20. Vockley, J.; Andersson, H.C.; Antshel, K.M.; Braverman, N.E.; Burton, B.K.; Frazier, D.M.; Mitchell, J.; Smith, W.E.; Thompson, B.H.; Berry, S.A. Phenylalanine hydroxylase deficiency: Diagnosis and management guideline. *Genet. Med.* **2014**, *16*, 188–200. [CrossRef] [PubMed]
21. Burgard, P.; Ullrich, K.; Ballhausen, D.; Hennermann, J.B.; Hollak, C.E.M.; Langeveld, M.; Karall, D.; Konstantopoulou, V.; Maier, E.M.; Lang, F.; et al. Issues with European guidelines for phenylketonuria. *Lancet Diabetes Endocrinol.* **2017**, *5*, 681–683. [CrossRef]
22. Rocha, J.C.; Vilarinho, L.; Cabral, A.; Osório, R.V.; Almeida, M.F. Consenso para o tratamento nutricional de fenilcetonúria (Article in Portuguese). *Acta Pediatr. Port.* **2007**, *38*, 44–54.
23. Sousa, C.; Almeida, M.F.; Sousa Barbosa, C.; Martins, E.; Janeiro, P.; Tavares de Almeida, I.; MacDonald, A.; Rocha, J.C. The European Phenylketonuria Guidelines and the challenges on management practices in Portugal. *J. Pediatr. Endocrinol. Metab.* **2019**, *32*, 623–629. [CrossRef] [PubMed]
24. Joint WHO/FAO/UNU Expert Consultation. *Protein and Amino Acid Requirements in Human Nutrition*; World Health Organization Technical Report Series; World Health Organization: Geneva, Switzerland, 2007.
25. Kraleva, D.; Evans, S.; Pinto, A.; Daly, A.; Ashmore, C.; Pointon-Bell, K.; Rocha, J.C.; MacDonald, A. Protein Labelling Accuracy for UK Patients with PKU Following a Low Protein Diet. *Nutrients* **2020**, *12*, 3440. [CrossRef]
26. Newbould, E.; Pinto, A.; Evans, S.; Ford, S.; O'Driscoll, M.; Ashmore, C.; Daly, A.; MacDonald, A. Accidental Consumption of Aspartame in Phenylketonuria: Patient Experiences. *Nutrients* **2021**, *13*, 707. [CrossRef] [PubMed]
27. Weglage, J.; Fromm, J.; van Teeffelen-Heithoff, A.; Möller, H.E.; Koletzko, B.; Marquardt, T.; Rutsch, F.; Feldmann, R. Neurocognitive functioning in adults with phenylketonuria: Results of a long term study. *Mol. Genet. Metab.* **2013**, *110*, S44–S48. [CrossRef]

28. Pilotto, A.; Zipser, C.M.; Leks, E.; Haas, D.; Gramer, G.; Freisinger, P.; Schaeffer, E.; Liepelt-Scarfone, I.; Brockmann, K.; Maetzler, W.; et al. Phenylalanine Effects on Brain Function in Adult Phenylketonuria. *Neurology* **2021**, *96*, e399–e411. [CrossRef]
29. Jaulent, P.; Charriere, S.; Feillet, F.; Douillard, C.; Fouilhoux, A.; Thobois, S. Neurological manifestations in adults with phenylketonuria: New cases and review of the literature. *J. Neurol.* **2020**, *267*, 531–542. [CrossRef]
30. van Vliet, K.; van Ginkel, W.G.; van Dam, E.; de Blaauw, P.; Koehorst, M.; Kingma, H.A.; van Spronsen, F.J.; Heiner-Fokkema, M.R. Dried blood spot versus venous blood sampling for phenylalanine and tyrosine. *Orphanet J. Rare Dis.* **2020**, *15*, 82. [CrossRef]

nutrients

MDPI

Article

Nitrogen Balance after the Administration of a Prolonged-Release Protein Substitute for Phenylketonuria as a Single Dose in Healthy Volunteers

Mika Scheinin [1,2,*], Jouni Junnila [3,4], Giorgio Reiner [5], Anita MacDonald [6] and Ania C. Muntau [7]

1 CRST Oy, Itäinen Pitkäkatu 4B, FI-20520 Turku, Finland
2 Institute of Biomedicine, University of Turku and TYKSLAB, Turku University Hospital, Kiinamyllynkatu 4-8, FI-20520 Turku, Finland
3 Oy 4Pharma Ltd., Arkadiankatu 7, FI-00100 Helsinki, Finland; jouni.junnila@estimates.fi
4 EstiMates Oy, Kunnallissairaalantie 56 as 1, FI-20810 Turku, Finland
5 APR Applied Pharma Research SA, Via Corti 5, CH-6828 Balerna, Switzerland; giorgio.reiner@apr.ch
6 Dietetic Department, Birmingham Women's and Children's Hospital NHS Foundation Trust, Birmingham B4 6NH, UK; anita.macdonald@nhs.net
7 University Medical Center Hamburg Eppendorf, University Children's Hospital, Martinistrasse 52, D-20246 Hamburg, Germany; muntau@uke.de
* Correspondence: mika.scheinin@crst.fi; Tel.: +358-40-5014762; Fax: +358-2-2510046

Abstract: Nitrogen balance is the difference between nitrogen excreted as urea and nitrogen ingested, mainly in proteins. Increased circulating concentrations of amino acids (AA) in the bloodstream are usually associated with proportional increases in the production and excretion of urea. Previously, we reported results from a randomized, controlled, single-dose, crossover trial in healthy adult volunteers (n = 30) (Trial Registration: ISRCTN11016729), in which a Test product (prolonged-release AA mixture formulated with Physiomimic Technology™ (PT™)) significantly slowed down the release and reduced the peak plasma concentrations of essential AAs compared with a free AA mixture (Reference product) while maintaining essential AA bioavailability. Here, we report an assessment of the nitrogen balance from the same study. The amount of nitrogen contained in plasma AAs, levels of blood urea nitrogen (BUN) ($p < 0.0001$) and changes in BUN ($p < 0.0001$) were smaller after the Test product compared with the Reference product. These findings suggest that the production of urea in proportion to systemic AA availability was significantly smaller after the administration of the Test product compared with the Reference product and that the test product conferred the increased utilization of AAs for protein synthesis and reduced their oxidation and conversion to urea. In the clinical setting, it is possible that the effects of PT™ observed on the disposition of free AAs in this study may translate to health benefits in terms of physiological body composition and growth if used for the treatment of subjects with phenylketonuria (PKU). Further investigation in patients with PKU is warranted.

Keywords: phenylketonuria; nitrogen balance; amino acid catabolism; blood urea nitrogen; prolonged release

Citation: Scheinin, M.; Junnila, J.; Reiner, G.; MacDonald, A.; Muntau, A.C. Nitrogen Balance after the Administration of a Prolonged-Release Protein Substitute for Phenylketonuria as a Single Dose in Healthy Volunteers. *Nutrients* **2021**, *13*, 3189. https://doi.org/10.3390/nu13093189

Academic Editor: Christophe Moinard

Received: 15 July 2021
Accepted: 10 September 2021
Published: 14 September 2021

Publisher's Note: MDPI stays neutral with regard to jurisdictional claims in published maps and institutional affiliations.

Copyright: © 2021 by the authors. Licensee MDPI, Basel, Switzerland. This article is an open access article distributed under the terms and conditions of the Creative Commons Attribution (CC BY) license (https://creativecommons.org/licenses/by/4.0/).

1. Introduction

Phenylketonuria (PKU) is the most common inherited disease of amino acid (AA) metabolism, with a global prevalence of 0.3–38.1 per 100,000 newborns [1]. For more than half a century, patients with PKU have been treated with a phenylalanine-restricted diet combined with phenylalanine-free AA mixtures to compensate for the low intake of natural proteins. However, the administration of free AAs produces several metabolic imbalances not observed with equilibrated diets consisting of food containing intact natural proteins. A recent systematic review and meta-analysis reported that even with advances in dietary treatments, 'optimal' growth outcomes are not always attained in children with PKU on a

Phe-restricted diet. In contrast, growth is similar to reference populations in children with mild hyperphenylalaninemia not requiring dietary restriction [2]. The unsavoury taste of products containing free AAs often leads to a low adherence to dietary management and may significantly increase the burden of the disease [3–6]. Despite several improvements in recent years, a taste and odour-free protein substitute with the properties of natural protein still remains the main feature in order to guarantee an optimal physiological function and patient tolerance [4].

The more rapid absorption of free AAs compared with AAs from intact proteins is associated with a less efficient utilization, early oxidation and effects on insulin release, glycaemic control, and endocrine regulation [7,8]. The intake of free AAs results in higher plasma concentrations, earlier absorption peaks, and steeper blood concentration reductions compared with the intake of intact natural proteins [9] (Figure 1).

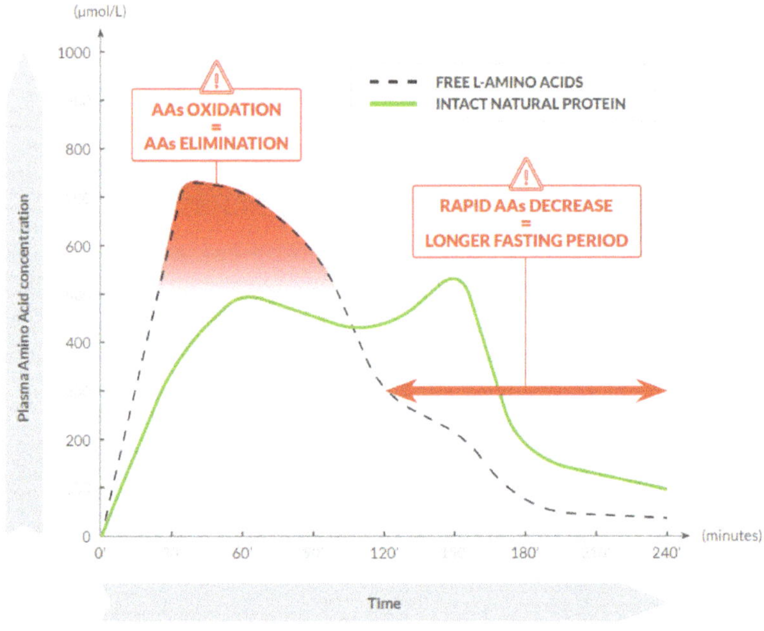

Figure 1. Consumption of free amino acids results in higher plasma concentrations, earlier absorption peak, and steeper blood concentration reductions compared with intact natural proteins.

AAs serve many functions in the body. Being the only source of nitrogen for mammals, AA-derived nitrogen is pivotal for synthetizing precursors of major energy molecules (i.e., ATP, ADP, IMP) and/or nucleic acids (i.e., DNA/RNA), and/or to produce compounds that can regulate major biochemical signalling pathways, such as nitric oxide [10]. Moreover, the deamination of AAs released from skeletal muscle and/or circulating dietary proteins generates a carbon skeleton rich in oxygen and hydrogen suitable for the subsequent biochemical transformation. This carbon skeleton can be used by the liver to produce glucose, through gluconeogenesis, and other important molecules, such as lipids. The AA-derived carbon skeleton is also relevant in producing intermediates fuelling the Krebs cycle that are, thereafter, transformed into energy and/or other metabolic intermediates. Therefore, AAs may be converted into energy, carbohydrates, lipids, and biochemical intermediates, dependent on the body's metabolic demands.

β oxidation, which is mostly mitochondrial, reduces the ratio of ATP/available oxygen, and obliges large amounts of essential AAs (EAAs) to be used as intermediates of the Krebs cycle. Such a metabolic shift is one of the main alterations leading to an imbalance between nitrogen demand and nitrogen intake observed in patients with chronic altered metabolic conditions, and is measured as the nitrogen balance [10].

Urea is the end product of protein catabolism in the liver, and the association between plasma AA concentrations and urea production is almost linear, i.e., increasing circulating concentrations of AAs result in proportional increases in the production and plasma concentration of urea [11]. Blood urea nitrogen (BUN) is a clinically employed indicator of nitrogen harboured by urea.

Mönch [12] reported that bolus administration of free AAs increases the amount of nitrogen excreted into urine, when the rapid increase in circulating AAs exceeds the capacity of anabolic processes to incorporate them into nascent proteins (protein synthesis). Similarly, when young healthy subjects were fed with 'slow' proteins (e.g., casein), protein retention was greater than in subjects fed with 'fast' proteins (e.g., whey); i.e., rapid AA uptake was associated with a rapid increase in blood AAs and higher oxidation rates [13,14].

The impact of free AAs on physiological and metabolic balance has prompted the search for nutritional strategies for patients with PKU that would closely match physiological circumstances [15,16]. Physiomimic Technology™ (PT™) is a pharmaceutical process that results in small granules coated with functional additives—ethylcellulose and sodium alginate—that allow the gradual release of their contents in the small intestine. PT™ modifies the release and absorption of AAs, while masking their taste and odour, with positive effects on the typically unpleasant aftertaste of traditional AA formulations. Preliminary evidence for use of this technology was obtained from a porcine model, where the application of PT™ to free AA mixtures reduced the peak blood concentration (C_{max}) by 18%, while maintaining a similar overall increase in plasma AAs [17].

As reported previously, we conducted a study in healthy adult volunteers to determine the effects on plasma AA profiles of a prolonged-release AA mixture formulated with PT™, comparing it with an immediate-release formulation of the same AA mixture, a commercially available free AA mixture and a natural intact protein, casein [18]. The study results showed that an AA mixture formulated with PT™ significantly prolonged the release of AAs, lowered peak EAA levels in plasma, and maintained an equivalent overall increase in plasma EAAs [18]. Here, we report more results from the same study, now comparing nitrogen balance after the administration of the PT™-formulated Test product and a Reference product containing free AAs.

2. Materials and Methods

2.1. Study Design

All study participants provided written informed consent. The study was conducted at CRST Oy in Turku, Finland, and was approved by the Ethics Committee of the Hospital District of Southwest Finland, Turku, Finland; ref: 78/1801/2017. Trial registration: ISRCTN11016729 [18]. Briefly, in this randomized, controlled, single-blind, crossover trial, the kinetic profiles of different AA preparations were assessed in 30 healthy volunteers (15 male, 15 female) aged between 18 and 45 years with body weight between 55 and 85 kg and body mass index ≤ 30 kg/m^2.

The Test product was a phenylalanine-free AA formulation, engineered with the PT™, containing 17 AAs, vitamins, minerals, other nutrients, ethylcellulose, and sodium alginate as food additives. The Reference product was a phenylalanine-free AA formulation with the same qualitative and quantitative composition as the Test product (in terms of AAs, vitamins, minerals, other nutrients, ethylcellulose, and sodium alginate). The only difference was that no coating layer was used.

The Test product and Reference product were administered in single doses (0.40 g AA/kg body weight) at time 0 (Figure 2). This single dose represented 1 of the 3 doses necessary to cover the daily AA requirements for adults with PKU. In this crossover study, the days on which the healthy volunteers received the study products were separated by a 9–14-day wash-out period. On each test day, venous blood samples and urine samples were obtained at regular intervals according to the analysis schedule.

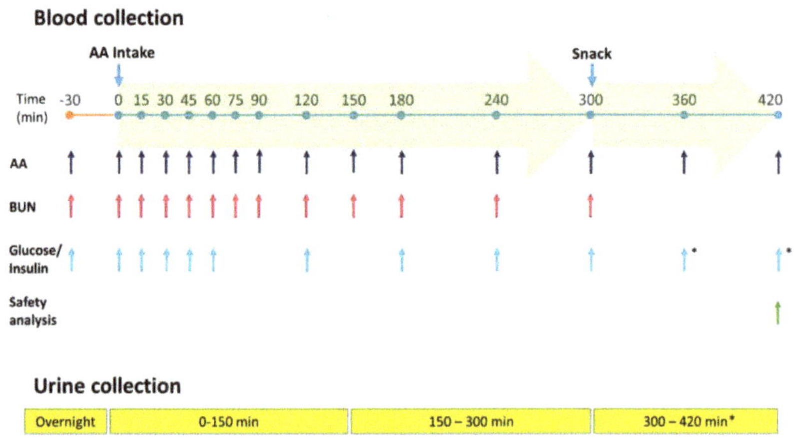

Figure 2. Administration, sample collection, and analysis schedule. AA, amino acids; BUN, blood urea nitrogen. * for safety assessment only.

2.2. Statistical Analysis

All results are described as mean and standard deviation. The nitrogen concentration data (separately for AA and BUN) were analysed with repeated measures analysis of variance (RMANOVA) models, where product sequence, product, timepoint, and the two-way interactions of sequence*timepoint and product*timepoint were used as fixed effects and subject within sequence and residual error term as random effects. The balance between nitrogen concentrations from AA administration in blood and BUN was analysed with a similar RMANOVA model, using the difference between blood and BUN nitrogen (within product) as the response variable.

The quantity of nitrogen (both in blood and BUN, calculated from the areas under the curve from 0 (baseline) to 300 min (AUC_{0-300}) contained in AAs and in BUN) was analysed with a linear mixed effect model on log-transformed data. The statistical model included sequence and product as fixed effects, and subject within sequence and residual error term as random effects. The 95% confidence interval (CI) for the nitrogen quantity was calculated from the model for equivalence evaluation. CI estimates were converted by anti-log transformation to obtain ratios of geometric least square means.

3. Results

Thirty subjects successfully completed the intervention with the Reference product, and twenty-eight subjects successfully completed the intervention with the Test product.

3.1. Blood AA and Nitrogen Concentrations

Total plasma AA concentrations for the Test and Reference products from baseline to 300 min after a dose intake and the differences in concentrations (mmol/L) of the Test product minus the Reference product (Delta AA) were determined (Table 1). For the time points from 15 min to 150 min, the concentrations of AAs were lower after the administration of the Test product compared with the Reference product. For the time points from 180 min to 300 min, the concentrations of AAs were higher after the administration of the Test product than the Reference product. Nitrogen concentrations and the nitrogen concentration differences (Delta) for the Test and Reference Products were calculated considering that some AAs contain two nitrogen atoms (such as glutamine, lysine, and tryptophan) or three nitrogen atoms (such as arginine and histidine). The Delta concentration of each AA and the Delta nitrogen concentrations were calculated for each time point, using a conversion factor that accounted for the different numbers of nitrogen atoms in each AA.

Table 1. Plasma amino acid and nitrogen concentrations after intake of the Test product and the Reference product.

Time	Test Product (AA, mmol/L) Mean (SD)	Reference Product (AA, mmol/L) Mean (SD)	Delta + AA Test vs. Reference (mmol/L) Mean (SD)	Delta ‡ Nitrogen Test vs. Reference (mmol/L) Mean (SD)
Baseline *	2.30 (0.30)	2.34 (0.26)	−0.05 (0.31)	−0.08 (0.43)
15 min	2.52 (0.33)	3.00 (0.46)	−0.43 (0.44)	−0.59 (0.60)
30 min	3.05 (0.44)	3.76 (0.59)	−0.65 (0.55)	−0.84 (0.72)
45 min	3.33 (0.48)	4.18 (0.61)	−0.80 (0.56)	−1.01 (0.76)
60 min	3.43 (0.48)	4.21 (0.54)	−0.80 (0.41)	−0.99 (0.58)
75 min	3.33 (0.48)	4.37 (0.64)	−1.06 (0.60)	−1.35 (0.82)
90 min	3.22 (0.44)	3.95 (0.55)	−0.77 (0.58)	−0.97 (0.81)
120 min	3.02 (0.43)	3.33 (0.37)	−0.35 (0.53)	−0.43 (0.73)
150 min	2.82 (0.38)	2.90 (0.31)	−0.10 (0.42)	−0.11 (0.58)
180 min	2.67 (0.38)	2.59 (0.26)	0.06 (0.35)	0.07 (0.49)
240 min	2.48 (0.30)	2.40 (0.24)	0.08 (0.33)	0.10 (0.48)
300 min	2.33 (0.29)	2.26 (0.23)	0.07 (0.29)	0.09 (0.40)

* Mean of concentration 30 min before and immediately before the intake of the products; + mmol/L AAs Test product minus mmol/L AAs Reference product; ‡ mmol/L total plasma nitrogen Test product minus mmol/L total plasma nitrogen Reference product. AA, amino acid; SD, standard deviation. AA concentrations in plasma and, consequently, amounts of nitrogen contained in plasma AAs were lower after the Test product than after the Reference product between 15 and 150 min after product intake and higher between 180 and 240 min, confirming the capacity of PT™ to slow down the absorption of free AAs.

3.2. BUN and Nitrogen Concentrations

Urea is a waste product that is formed in the liver when the body breaks down AAs; BUN reflects the nitrogen content in urea (molecular weight 28). The concentrations of BUN and urea are equal when expressed as mmol/L because both entities contain two nitrogen atoms.

Total plasma BUN concentrations for the Test and Reference products from baseline to 300 min and the differences in concentrations (mmol/L) of the Test product minus the Reference product (Delta BUN) were determined (Table 2). The concentration of BUN was lower at all time points following the administration of the Test product compared with the Reference product. To calculate the differences in the nitrogen concentration (mmol/L) of the Test product minus the Reference product (Delta nitrogen), the Delta BUN (mmol/L) had to be multiplied by two (since one molecule of urea contains two nitrogen atoms).

BUN concentrations and, consequently, the urea-bound nitrogen concentrations were consistently lower after the Test product than after the Reference product, indicating a lesser oxidation of AAs after the intake of the Test product. This overall difference between the products was statistically highly significant ($p < 0.0001$).

3.3. Balance between Nitrogen Concentrations from AA Administration in Blood and BUN

Based on the observations presented in Tables 1 and 2, it was possible to evaluate the balance between the Delta nitrogen concentrations contained in plasma AAs and in BUN after the administration of the Test and Reference products (Table 3). Delta nitrogen from AAs was negative over the first 120 min after the ingestion of the Test product, becoming positive from 180 min onwards, confirming the slower release and absorption of AAs from the Test product compared with the Reference product. Conversely, Delta nitrogen in BUN remained negative until 300 min, confirming a lower production of waste nitrogen (BUN) after the administration of the Test product compared with the Reference product.

Table 2. Blood urea nitrogen concentrations after the intake of the Test product and the Reference product and Delta nitrogen content.

	Concentration BUN (mmol/L)			
Time	**Test Product Mean (SD)**	**Reference Product Mean (SD)**	**Delta BUN * (mmol/L) Mean (SD)**	**Delta Nitrogen ** (mmol/L) Mean (SD)**
Baseline	3.89 (0.72)	4.08 (0.87)	−0.25 (0.56)	−0.49 (1.12)
15 min	3.82 (0.68)	4.16 (0.86)	−0.37 (0.56)	−0.74 (1.11)
30 min	3.94 (0.69)	4.30 (0.86)	−0.40 (0.86)	−0.79 (1.72)
45 min	4.04 (0.75)	4.57 (0.95)	−0.54 (0.58)	−1.07 (1.16)
60 min	4.15 (0.73)	4.77 (0.93)	−0.65 (0.57)	−1.29 (1.14)
75 min	4.28 (0.66)	5.08 (0.92)	−0.82 (0.58)	−1.64 (1.15)
90 min	4.37 (0.66)	5.26 (0.92)	−0.93 (0.65)	−1.85 (1.31)
120 min	4.50 (0.63)	5.47 (0.93)	−0.99 (0.62)	−1.99 (1.24)
150 min	4.58 (0.62)	5.45 (0.89)	−0.91 (0.65)	−1.83 (1.29)
180 min	4.55 (0.66)	5.45 (0.89)	−0.92 (0.76)	−1.84 (1.51)
240 min	4.58 (0.64)	5.26 (0.89)	−0.73 (0.64)	−1.45 (1.28)
300 min	4.53 (0.63)	5.10 (0.82)	−0.60 (0.63)	−1.21 (1.24)

* mmol/L BUN Test product minus mmol/L BUN Reference product; ** Delta mmol/L nitrogen = Delta mmol/L BUN × 2; BUN, blood urea nitrogen; SD, standard deviation.

Table 3. Comparison of Delta nitrogen contained in amino acids with Delta nitrogen as blood urea nitrogen *.

Time	Delta Nitrogen in AAs (mmol/L) Mean (SD)	Delta Nitrogen in BUN (mmol/L) Mean (SD)	Delta Nitrogen in AAs Minus Delta Nitrogen in BUN (mmol/L) Mean (SD)
Baseline	−0.08 (0.43)	−0.49 (1.12)	0.42 (1.07)
15 min	−0.59 (0.60)	−0.74 (1.11)	0.15 (0.93)
30 min	−0.84 (0.72)	−0.79 (1.72)	−0.05 (1.66)
45 min	−1.01 (0.76)	−1.07 (1.16)	0.11 (1.23)
60 min	−0.99 (0.58)	−1.29 (1.14)	0.30 (1.26)
75 min	−1.35 (0.82)	−1.64 (1.15)	0.29 (1.27)
90 min	−0.97 (0.81)	−1.85 (1.31)	0.88 (1.39)
120 min	−0.43 (0.73)	−1.99 (1.24)	1.56 (1.43)
150 min	−0.11 (0.58)	−1.83 (1.29)	1.72 (1.35)
180 min	0.07 (0.49)	−1.84 (1.51)	1.92 (1.55)
240 min	0.10 (0.48)	−1.45 (1.28)	1.55 (1.28)
300 min	0.09 (0.40)	−1.21 (1.25)	1.30 (1.24)

* Delta refers to total nitrogen in AAs or BUN after ingestion of the Test product minus the total nitrogen in AAs or BUN after ingestion of the Reference product. AA, amino acids; BUN, blood urea nitrogen; SD, standard deviation.

The Delta of nitrogen concentrations in BUN versus the Delta of nitrogen concentrations contained in plasma AAs increased over time, indicating a lesser oxidation of AAs after the Test product administration compared with the Reference product (Figures 3 and 4). The observed difference between the products was statistically highly significant ($p < 0.0001$).

Thus, the nitrogen balance was better after the Test product than after the Reference product, indicating that the PT™ coating employed in the Test product increased the utilization of AAs and reduced their oxidation (Figure 3).

3.4. Total Quantities of Nitrogen from the AUC_{0-300} Contained in AAs and in BUN

The total AUC_{0-300} of AAs and BUN after the intake of the Test and Reference products from baseline until 300 min after dose intake and the differences in mol/L of the Test product minus the Reference product (Delta) were calculated (Tables 4 and 5). To calculate the nitrogen content of circulating AAs, it was considered that individual AAs may contain one, two, or three nitrogen atoms. Starting from the AUC_{0-300} of each individual AA, the difference between the Test and Reference product was calculated. Starting from the contribution (in %) of each AA in the total AA $AUC_{0–300}$ difference (Delta), it was possible to calculate a conversion factor for each subject (within product). There was a difference of 0.098 mol of plasma nitrogen contained in free AAs per litre in the 300 min after dosing between the Test product and the Reference product (Table 4).

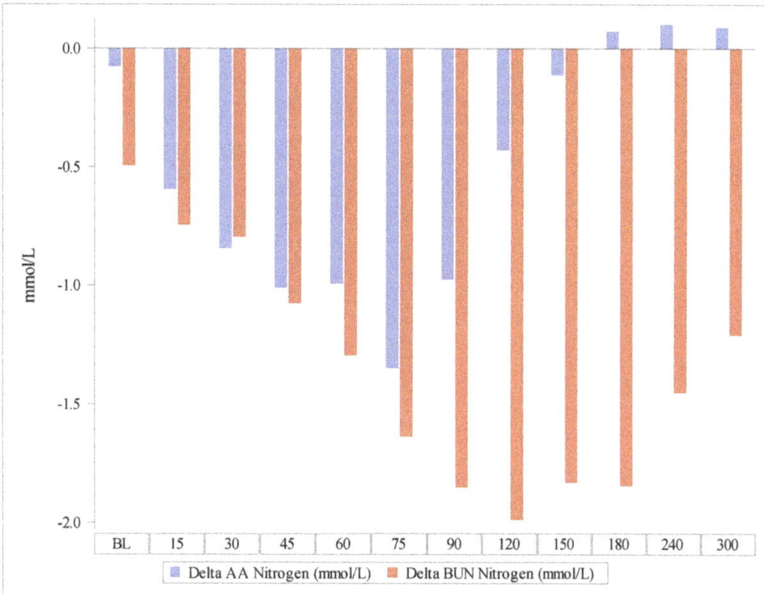

Figure 3. Delta of the nitrogen concentrations contained in plasma amino acids in comparison with Delta nitrogen concentrations as blood urea nitrogen. AAs, amino acids; BUN, blood urea nitrogen.

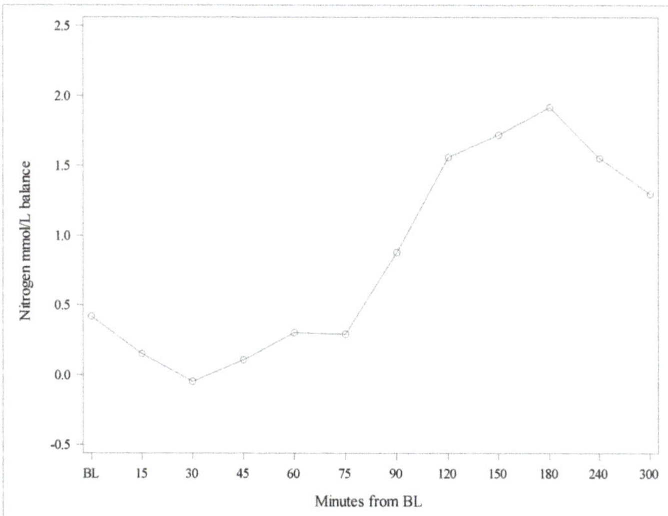

Figure 4. Differences between Deltas of nitrogen concentrations from amino acids, and nitrogen from blood urea nitrogen. BL, baseline.

Table 4. Comparison of total plasma amino acids $AUC_{0–300}$ and related quantity of nitrogen after administration of the Test and Reference product.

	Mean AAs $AUC_{0–300}$ (mol/L/300 min) Mean (SD)	Quantity of Nitrogen (mol/L/300 min) Mean (SD)
Reference product	0.9146 (0.075)	1.302 (0.100)
Test product	0.8391 (0.099)	1.209 (0.140)
Difference Test–Reference	−0.078 (0.096)	−0.098 (0.135)

AA, amino acid; $AUC_{0–300}$, area under the concentration–time curve from 0 to 300 min; SD, standard deviation.

Table 5. Comparison of total nitrogen present as BUN $AUC_{0–300}$ after the Test and Reference products.

	Mean BUN $AUC_{0–300min}$ (mol/L/300 min) Mean (SD)	Quantity of Nitrogen (mol/L/300 min) Mean (SD)
Reference product	1.5729 (0.266)	3.146 (0.532)
Test product	1.3574 (0.201)	2.715 (0.402)
Difference Test–Reference	−0.226 (0.187)	−0.452 (0.375)

$AUC_{0–300}$, area under the concentration–time curve from 0 to 300 min; BUN, blood urea nitrogen; SD, standard deviation.

To calculate the nitrogen differences of $AUC_{0–300}$ represented by BUN between the Test and Reference products, it was considered that one molecule of urea contains two nitrogen atoms. There was a difference of 0.452 mol of plasma nitrogen present as BUN per litre until 300 min between the Test product and the Reference product (Table 5).

The Delta nitrogen $AUC_{0–300}$ min (mol/L) contained in free AAs was −0.098 mol/L (geometric mean ratio 0.923, 90% CI 0.891–0.957) between the products, indicating that the nitrogen contents of AAs in plasma were rather similar after both products (Test and Reference). In contrast, the Delta nitrogen $AUC_{0–300}$ min (mol/L) present as BUN was −0.452 mol/L between the Test and Reference products with an associated *p*-value of <0.0001, indicating that the administration of the Reference product was associated with more AAs being metabolized to urea than the administration of the Test product.

4. Discussion

The Test product was previously reported to be bioequivalent with the Reference product for all subgroups of AAs [18]. However, peak concentrations (C_{max}) in plasma were significantly lower for all subgroups of AAs, indicating a delayed absorption of the Test product. In addition, BUN and the excretion of urea into the urine were significantly lower after the Test product compared with the Reference product. The present analyses indicate that the production of urea in proportion to systemic AA availability was significantly smaller after the administration of the Test product compared with the Reference product. This result supports the hypothesis that the Test product, manufactured using PT™, conferred the increased utilization of AAs for protein synthesis and reduced their oxidation and conversion to urea. The present analysis allowed to delineate how nitrogen balance is affected by the absorption kinetics of the ingested free AAs. The impact of the delayed absorption became most evident starting from 90 min after product ingestion and provided further support for the prolonged release obtained with PT™. Less wasted nitrogen means higher efficiency in the utilization of the AAs administered with the PT™.

Among experts caring for those with PKU, there is a consensus that the traditional phenylalanine-free AA formulations possess kinetic properties that lead to a suboptimal absorption. Furthermore, as children with PKU increase in age, adherence with dietary therapy commonly declines, particularly during adolescence and early adulthood [19–21]. Poor adherence with nutritionally supplemented AAs leads to both elevated blood phenylalanine concentrations and some nutritional deficiencies [22–24]. The distribution of the intake of AA mixtures evenly over the day and preferably together with or after meals [25] may have positive effects on blood phenylalanine levels as well as phenylalanine tolerance. However, from a behavioural point of view, this relentless routine may be a further challenge for adherence.

It is conceivable that the observed effects of PT™ on the disposition of free AAs, i.e., delayed and prolonged absorption, less oxidation, and thereby more efficient utilization compared with regular AA supplements, may be associated with clinically meaningful health benefits in terms of physiological body composition, better growth, and a more balanced supply of the ingredients of the AA mixture necessary for a successful treatment of subjects with PKU. The healthy subjects of the study were a limitation; a further investigation in patients with PKU is warranted.

Author Contributions: Conceptualization, G.R. and A.C.M.; methodology, G.R. and A.C.M.; validation, J.J.; formal analysis, G.R., J.J. and M.S.; investigation, G.R., J.J. and M.S.; data curation, J.J.; writing—original draft preparation, M.S., G.R., A.M., A.C.M. and J.J.; writing—review and editing, M.S., G.R., A.M., A.C.M. and J.J. All authors have read and agreed to the published version of the manuscript.

Funding: The clinical trial upon which this research is based and the article publication costs were funded by APR Applied Pharma Research S.A., Balerna, Switzerland.

Institutional Review Board Statement: The study was approved by the Ethics Committee of the Hospital District of Southwest Finland, Turku, Finland; ref: 78/1801/2017.

Informed Consent Statement: Informed consent was obtained from all subjects involved in the study.

Acknowledgments: We would like to extend a special thank you to Peter Burgard for his support with the data analyses and manuscript preparation. The authors would also like to thank Sam Coates for her assistance in preparing the manuscript for publication and Patrizia Marzorati at APR Applied Pharma Research S.A. for her review and comments.

Conflicts of Interest: The clinical trial upon which the current work is based was commissioned and funded by APR Applied Pharma Research S.A. (via Corti 5, CH-6828 Balerna, Switzerland). GR was employed by APR, and was involved in the interpretation of the study results and in the writing of the manuscript.

References

1. Shoraka, H.R.; Haghdoost, A.A.; Baneshi, M.R.; Bagherinezhad, Z.; Zolala, F. Global prevalence of classic phenylketonuria based on Neonatal Screening Program Data: Systematic review and meta-analysis. *Clin. Exp. Pediatr.* **2020**, *63*, 34–43. [CrossRef]
2. Ilgaz, F.; Pinto, A.; Gokmen-Ozel, H.; Rocha, J.C.; van Dam, E.; Ahring, K.; Belanger-Quintana, A.; Dokoupil, K.; Karabulut, E.; MacDonald, A. Long-Term Growth in Phenylketonuria: A Systematic Review and Meta-Analysis. *Nutrients* **2019**, *11*, 2070. [CrossRef] [PubMed]
3. Hood, A.; Grange, D.K.; Christ, S.E.; Steiner, R.; White, D.A. Variability in phenylalanine control predicts IQ and executive abilities in children with phenylketonuria. *Mol. Genet. Metab.* **2014**, *111*, 445–451. [CrossRef] [PubMed]
4. DeRoche, K.; Welsh, M. Twenty-five years of research on neurocognitive outcomes in early-treated phenylketonuria: Intelligence and executive function. *Dev. Neuropsychol.* **2008**, *33*, 474–504. [CrossRef] [PubMed]
5. Al Hafid, N.; Christodoulou, J. Phenylketonuria: A review of current and future treatments. *Transl. Pediatr.* **2015**, *4*, 304–317.
6. Simon, E.; Schwarz, M.; Roos, J.; Dragano, N.; Geraedts, M.; Siegrist, J.; Kamp, G.; Wendel, U. Evaluation of quality of life and description of the sociodemographic state in adolescent and young adult patients with phenylketonuria (PKU). *Health Qual. Life Outcomes* **2008**, *6*, 25. [CrossRef]
7. Daly, A.; Evans, S.; Chahal, S.; Santra, S.; Pinto, A.; Gingell, C.; Rocha, J.C.; van Spronsen, F.; Jackson, R.; MacDonald, A. The Effect of Glycomacropeptide versus Amino Acids on Phenylalanine and Tyrosine Variability over 24 Hours in Children with PKU: A Randomized Controlled Trial. *Nutrients* **2019**, *11*, 520. [CrossRef]
8. Evans, S.; Ford, S.; Adam, S.; Adams, S.; Ash, J.; Ashmore, C.; Caine, G.; Carruthers, R.; Cawtherley, S.; Chahal, S.; et al. Development of national consensus statements on food labelling interpretation and protein allocation in a low phenylalanine diet for PKU. *Orphanet J. Rare Dis.* **2019**, *14*, 2. [CrossRef]
9. Gropper, S.S.; Acosta, P.B. Effect of simultaneous ingestion of L-amino acids and whole protein on plasma amino acid and urea nitrogen concentrations in humans. *J. Parenter. Enter. Nutr.* **1991**, *15*, 48–53. [CrossRef]
10. Pasini, E.; Corsetti, G.; Aquilani, R.; Romano, C.; Picca, A.; Calvani, R.; Dioguardi, F.S. Protein-Amino Acid Metabolism Disarrangements: The Hidden Enemy of Chronic Age-Related Conditions. *Nutrients* **2018**, *10*, 391. [CrossRef]
11. Rafoth, R.J.; Onstad, G.R. Urea synthesis after oral protein ingestion in man. *J. Clin. Investig.* **1975**, *56*, 1170–1174. [CrossRef]

12. Mönch, E.; Herrmann, M.; Brösicke, H.; Schöffer, A.; Keller, M. Utilisation of amino acid mixtures in adolescents with phenylketonuria. *Eur. J. Pediatr.* **1996**, *155*, S115–S120. [CrossRef] [PubMed]
13. Boirie, Y.; Dangin, M.; Gachon, P.; Vasson, M.-P.; Maubois, J.-L.; Beaufrère, B. Slow and fast dietary proteins differently modulate postprandial protein accretion. *Proc. Natl. Acad. Sci. USA* **1997**, *94*, 14930–14935. [CrossRef]
14. Herrmann, M.-E.; Brösicke, H.; Keller, M.; Monch, E.; Helge, H. Dependence of the utilization of a phenylalanine-free amino acid mixture on different amounts of single dose ingested. A case report. *Eur. J. Pediatr.* **1994**, *153*, 501–503. [CrossRef]
15. Groschl, M.; Knerr, I.; Topf, H.; Schmid, P.; Rascher, W.; Rauh, M. Endocrine responses to the oral ingestion of a physiological dose of essential amino acids in humans. *J. Endocrinol.* **2003**, *179*, 237–244. [CrossRef]
16. Weigel, C.; Rauh, M.; Kiener, C.; Rascher, W.; Knerr, I. Effects of various dietary amino acid preparations for phenylketonuric patients on the metabolic profiles along with postprandial insulin and ghrelin responses. *Ann. Nutr. Metab.* **2007**, *51*, 352–358. [CrossRef]
17. Giarratana, N.; Gallina, G.; Panzeri, V.; Frangi, A.; Canobbio, A.; Reiner, G. A new Phe-free protein substitute engineered to allow a physiological absorption of free amino acids for phenylketonuria. *J. Inborn Errors Metab.* **2018**, *6*, 1–9. [CrossRef]
18. Scheinin, M.; Barassi, A.; Junnila, J.; Lovro, Z.; Reiner, G.; Sarkkinen, E.; MacDonald, A. Amino Acid Plasma Profiles from a Prolonged-Release Protein Substitute for Phenylketonuria: A Randomized, Single-Dose, Four-Way Crossover Trial in Healthy Volunteers. *Nutrients* **2020**, *12*, 1653. [CrossRef]
19. Brown, C.S.; Lichter-Konecki, U. Phenylketonuria (PKU): A problem solved? *Mol. Genet. Metab. Rep.* **2016**, *6*, 8–12. [CrossRef]
20. Walter, J.H.; White, F.J. Blood phenylalanine control in adolescents with phenylketonuria. *Int. J. Adolesc. Med. Health* **2004**, *16*, 41–45. [CrossRef] [PubMed]
21. Walter, J.H.; White, F.J.; Hall, S.K.; MacDonald, A.; Rylance, G.; Boneh, A.; Francis, D.E.; Shortland, G.J.; Schmidt, M.; Vail, A. How practical are recommendations for dietary control in phenylketonuria? *Lancet* **2002**, *360*, 55–57. [CrossRef]
22. Rohde, C.; von Teeffelen-Heithoff, A.; Thiele, A.G.; Arelin, M.; Mutze, U.; Kiener, C.; Gerloff, J.; Baerwald, C.; Schultz, S.; Heller, C.; et al. PKU patients on a relaxed diet may be at risk for micronutrient deficiencies. *Eur. J. Clin. Nutr.* **2014**, *68*, 119–124. [CrossRef] [PubMed]
23. Schulz, B.; Bremer, H.J. Nutrient intake and food consumption of adolescents and young adults with phenylketonuria. *Acta Paediatr.* **1995**, *84*, 743–748. [CrossRef]
24. Walter, J.H. Vitamin B12 deficiency and phenylketonuria. *Mol. Genet. Metab.* **2011**, *104*, S52–S54. [CrossRef]
25. MacDonald, A.; Rylance, G.; Davies, P.; Asplin, D.; Hall, S.K.; Booth, I.W. Administration of protein substitute and quality of control in phenylketonuria: A randomized study. *J. Inherit. Metab. Dis.* **2003**, *26*, 319–326. [CrossRef] [PubMed]

nutrients MDPI

Systematic Review

Is the Phenylalanine-Restricted Diet a Risk Factor for Overweight or Obesity in Patients with Phenylketonuria (PKU)? A Systematic Review and Meta-Analysis

Catarina Rodrigues [1,2], Alex Pinto [3], Ana Faria [1,2,4], Diana Teixeira [1,2,4], Annemiek M. J. van Wegberg [5], Kirsten Ahring [6], François Feillet [7], Conceição Calhau [1,4], Anita MacDonald [3], André Moreira-Rosário [1,4,*,†] and Júlio César Rocha [1,4,8,*,†]

[1] Nutrition and Metabolism, NOVA Medical School, Faculdade de Ciências Médicas, Universidade NOVA de Lisboa, 1169-056 Lisboa, Portugal; catarina.rodrigues@nms.unl.pt (C.R.); ana.faria@nms.unl.pt (A.F.); diana.teixeira@nms.unl.pt (D.T.); ccalhau@nms.unl.pt (C.C.)
[2] Comprehensive Health Research Centre, Universidade NOVA de Lisboa, 1169-056 Lisboa, Portugal
[3] Dietetic Department, Birmingham Children's Hospital, Steelhouse Lane, Birmingham B4 6NH, UK; alex.pinto@nhs.net (A.P.); anita.macdonald@nhs.net (A.M.)
[4] CINTESIS—Center for Health Technology and Services Research, NOVA Medical School, Campo dos Mártires da Pátria 130, 1169-056 Lisboa, Portugal
[5] Division of Metabolic Diseases, Beatrix Children's Hospital, University Medical Centre Groningen, University of Groningen, Hanzeplein 1, 9700 RB Groningen, The Netherlands; a.m.j.van.wegberg@umcg.nl
[6] Department of PKU, Copenhagen University Hospital, DK-2600 Glostrup, Denmark; kirsten.ahring@regionh.dk
[7] Department of Paediatrics, Reference Center for Inborn Errors of Metabolism, Hôpital d'Enfants Brabois, CHU Nancy, 54500 Vandoeuvre les Nancy, France; f.feillet@chru-nancy.fr
[8] Reference Centre of Inherited Metabolic Diseases, Centro Hospitalar Universitário de Lisboa Central, 1169-045 Lisboa, Portugal
* Correspondence: andre.rosario@nms.unl.pt (A.M.-R.); rochajc@nms.unl.pt (J.C.R.); Tel.: +351-21-880-3000 (A.M.-R. & J.C.R.)
† These authors contributed equally to this work.

Citation: Rodrigues, C.; Pinto, A.; Faria, A.; Teixeira, D.; van Wegberg, A.M.J.; Ahring, K.; Feillet, F.; Calhau, C.; MacDonald, A.; Moreira-Rosário, A.; et al. Is the Phenylalanine-Restricted Diet a Risk Factor for Overweight or Obesity in Patients with Phenylketonuria (PKU)? A Systematic Review and Meta-Analysis. *Nutrients* **2021**, *13*, 3443. https://doi.org/10.3390/nu13103443

Academic Editor: Arrigo Cicero

Received: 27 July 2021
Accepted: 22 September 2021
Published: 28 September 2021

Publisher's Note: MDPI stays neutral with regard to jurisdictional claims in published maps and institutional affiliations.

Copyright: © 2021 by the authors. Licensee MDPI, Basel, Switzerland. This article is an open access article distributed under the terms and conditions of the Creative Commons Attribution (CC BY) license (https://creativecommons.org/licenses/by/4.0/).

Abstract: Although there is a general assumption that a phenylalanine (Phe)-restricted diet promotes overweight in patients with phenylketonuria (PKU), it is unclear if this presumption is supported by scientific evidence. This systematic review aimed to determine if patients with PKU are at a higher risk of overweight compared to healthy individuals. A literature search was carried out on PubMed, Cochrane Library, and Embase databases. Risk of bias of individual studies was assessed using the Quality Assessment Tool for Observational Cohort and Cross-Sectional Studies, and the quality of the evidence for each outcome was assessed using the NutriGrade scoring system. From 829 articles identified, 15 were included in the systematic review and 12 in the meta-analysis. Body mass index (BMI) was similar between patients with PKU and healthy controls, providing no evidence to support the idea that a Phe-restricted diet is a risk factor for the development of overweight. However, a subgroup of patients with classical PKU had a significantly higher BMI than healthy controls. Given the increasing prevalence of overweight in the general population, patients with PKU require lifelong follow-up, receiving personalised nutritional counselling, with methodical nutritional status monitoring from a multidisciplinary team in inherited metabolic disorders.

Keywords: body mass index; obesity; overweight; phenylalanine restriction; phenylalanine-restricted diet; phenylketonuria

1. Introduction

In phenylketonuria (PKU), the prevalence and patient susceptibility to overweight and obesity has been widely discussed. Several retrospective studies have reported a higher body mass index (BMI) and a higher prevalence of overweight in patients with

PKU compared to the normal population [1–4], especially in females [1,5–9]. Generally, the prevalence of overweight worldwide has almost tripled since 1975 [10]. This multifactorial comorbidity is mainly associated with poor dietary habits and lack of physical activity, but other factors, such as social economic status and family history, may also influence outcome [11].

The World Health Organisation (WHO) defines overweight and obesity as abnormal or excessive fat accumulation. This has numerous negative health consequences including cardiovascular diseases, non-insulin-dependent diabetes mellitus, musculoskeletal disorders, pulmonary diseases, and cancer [12–14].

PKU is a rare autosomal recessive inborn error of phenylalanine (Phe) metabolism, and if untreated, can cause severe and irreversible neurological damage [15]. The main treatment is a Phe-restricted diet, composed of three parts: (1) strict control of natural protein intake according to individual Phe tolerance, (2) administration of a synthetic protein derived from Phe-free amino acids (L-AAs) or low-Phe glycomacropeptide supplemented with amino acids (GMP-AA), and (3) and low-Phe foods including the use of special low-protein foods (SLPFs). The primary aim is to prevent neurological sequelae by maintaining blood Phe levels within a therapeutic target range [14], whilst maintaining nutritional requirements to achieve normal growth and body composition.

Adequate dietary energy is essential to maintain blood Phe stability, particularly in patients with classical PKU, by promoting anabolism and counteracting catabolism, which increases blood Phe levels [15]. Energy is obtained from fruits and some vegetables, sugars, fats, and oils, as well as SLPFs such as bread, pasta, rice, cereals, and milk replacements, aiming to replace regular foods. Pena et al. [16] analysed the food labels of several SLPFs and found that, when compared to their regular foods, 75% had a higher energy content, 58% a higher fat content, and 92% a higher carbohydrate (CHO) content. Moreover, the quality of fat and fibre differs from regular foods [17]. Their consumption without moderation may lead to excessive energy intake, with a low supply of micronutrients, although these are usually supplied by protein substitutes (PS) [18,19]. Overall, a Phe-restricted diet is characterised by higher CHO intake compared with the general population [19,20].

Due to concerns over increasing obesity in PKU, industry has reformulated many of their PS, adding less CHO to their products [21]. Furthermore, a higher prevalence of overweight in patients with PKU is used to support the need for alternative treatments, even though a systematic analysis of published data is not available to verify this claim. In addition, some studies have found no differences in BMI and prevalence of overweight and obesity between patients with PKU and healthy individuals [22–26].

This lack of consensus highlights the need to assess the quality of evidence that reports the prevalence of overweight and obesity in PKU. This systematic review aims to (1) determine if patients with PKU are at a higher risk of overweight compared to healthy individuals, and to (2) understand the association between early exposure to Phe restriction and overweight in patients with PKU.

2. Materials and Methods

2.1. Protocol and Registration

This systematic review with meta-analysis was developed according to preferred reporting items for systematic reviews and meta-analyses (PRISMA) statement [27] and the Cochrane Handbook for Systematic Reviews of Interventions [28] guidelines. The protocol was registered (CRD42020214436) in the International Prospective Register of Systematic Reviews (PROSPERO).

2.2. Selection Criteria

Inclusion and exclusion criteria were defined according to the PECO (Population, Exposure, Comparator, Outcome) strategy. Inclusion criteria: (1) patients with PKU (Population) on a Phe-restricted diet (Exposure) and followed up at a PKU centre; (2) studies included healthy controls (Comparator); (3) reported anthropometric measures or prevalence of overweight (Outcome); (4) published as a full paper; and (5) included only randomised controlled trials (RCTs), non-randomised controlled trials (non-RCTs), or observational (case–control, cohort, and cross-sectional) studies.

Non-human studies, review articles, systematic reviews, meta-analysis, letters, conference abstracts, case reports, case series, position papers, and authors' replies were excluded. Only studies published in English were included.

2.3. Search Strategy

A literature search was carried out on PubMed, The Cochrane Library, and Embase databases on the 16 January 2020. Both medical subject headings (MeSH or Emtree) and text words related to overweight, obesity, and PKU were used. The PubMed search strategy was converted to search in other databases as described in detail in the Supplementary Materials, Section A.

2.4. Study Selection

All articles identified in the search were included in the screening process and duplicates excluded. Two independent reviewers (A.M. and J.C.R.) screened the titles and abstracts of the articles for relevance, and full-text articles were reviewed when title and abstract did not provide enough information. Once potentially relevant studies were identified, full-text articles were then assessed for eligibility according to previously established criteria. The reference lists of the included articles were screened to ensure that no relevant studies were missed.

2.5. Data Extraction

Data items were extracted by two authors (C.R. and A.P.) using a standard data extraction form. For each study, first author, year of publication, country of origin, study design, sample characteristics, methods, and outcomes were extracted. In cases where information was missing or incomplete, the correspondence authors were contacted requesting further information.

2.6. Assessment of Risk of Bias in Individual Studies

Risk of bias of individual studies was assessed by two independent reviewers (C.R. and A.P.) using the National Institutes of Health (NIH) Quality Assessment Tool for Observational Cohort and Cross-Sectional Studies [29]. The following domains were assessed: (1) research question; (2) study population; (3) eligibility criteria; (4) justification of the sample size; (5) exposure measures and assessment; (6) time frame between exposure and outcome assessment; (7) outcome measures; (8) blinding of outcome assessors; (9) follow-up rate; and (10) adjustment of confounders. Reviewers were blinded to each other's assessment, and disagreements were solved by reaching consensus.

2.7. Quantitative Synthesis

Standardised mean difference (SMD) was used as an effect measure for the continuous variable 'BMI'. Odds ratio (OR) was used as an effect measure for the dichotomous variable 'prevalence of overweight'. The SMD and OR were converted to a common metric and then combined across studies. A sensitivity analysis was performed to compare the meta-analysis results with and without the converted study [30]. Effect measures were reported along with the 95% confidence interval (CI).

The Cochran's Q (significance level of 0.1) and I^2 tests were used to assess heterogeneity. According to the Cochrane guidelines [28], the I^2 values were interpreted as follows: 0% to 40% might not be important; 30% to 60% may represent moderate heterogeneity; 50% to 90% may represent substantial heterogeneity; 75% to 100% represent considerable heterogeneity.

Mean BMI from Evans et al. [31] was calculated with values from the last evaluation (longest time-point of exposure). In the studies from Evans et al. [25] and Huemer et al. [26], only the mean BMI from the first evaluation (baseline) could be included. In the study from Schulpis et al. [32], consisting of patients both adhering to their diet and on a 'relaxed diet', only the BMI of the patients adhering to the diet was included in the meta-analysis.

Pooled estimates were computed and weighted using generic inverse-variance and random-effect modelling. A p-value < 0.05 was considered as statistically significant. Statistical analysis was performed using Review Manager (RevMan), version 5.4, The Cochrane Collaboration, 2020.

2.8. Grading the Evidence

Funnel plots were used to assess evidence of publication bias. Quality assessment of the evidence for each outcome was performed by two independent authors (C.R. and A.P.) using the NutriGrade scoring system [33]. The meta-analysis was scored with a maximum of 10 points, according to (1) risk of bias, (2) precision, (3) heterogeneity, (4) directness, (5) publication bias, (6) funding bias, (7) effect-size, and (8) dose–response. On the basis of the final score, we classified the quality of the evidence as high, moderate, low, or very low.

3. Results

3.1. Study Selection

A total of 829 articles were identified through database search (Figure 1). Titles and abstracts of 551 articles were screened for relevance, after removing duplicates. Once potentially relevant studies were identified, a total of 56 full-text articles were assessed for eligibility. Studies not fulfilling these criteria were excluded from the analysis ($n = 41$) (Supplementary Materials, Section B). Two studies by Rocha et al. [22,34] included two overlapping patient cohorts. To avoid duplicate publication bias, we included the study with more complete information [34]. From the included studies, only 12 provided data on BMI or the prevalence of overweight, qualifying them for quantitative analysis [7,18,25,26,30–32,34–38].

3.2. Study Characteristics

A summary of the main characteristics of included studies is given in Table 1. All studies were observational: 11 cross-sectional studies [7,18,30,32,34–40], 2 cross-sectional with nested longitudinal cohort studies [26,41], and 2 prospective studies [25,31]. Nine studies were conducted in Europe [7,26,30–32,34–37], three in Australia [25,39,41], two in Brazil [38,40], and one in the USA [18]. Studies were published between 1995 and 2020. In prospective studies, duration of follow-up ranged from 1 to 2 years. The total sample size of the 15 studies was 640 patients with PKU, and 503 were included in the meta-analysis (12 studies). All studies included patients with PKU from both genders (301 females and 299 males). Fisberg et al. [40] did not specify children's gender. The age range of the participants ranged from 2 months to 52 years. Most studies included children and adolescents, four included children, adolescents, and adults [30,34,37,38], and Azabdaftari et al. [36] included adults only.

The methods used to assess dietary intake varied between the included studies and are given in Table 2. No valid and reliable methods to assess exposure were used in five studies [7,35,37–39].

Patients with PKU were compared to 593 healthy controls, 455 of which were included in the meta-analysis. Healthy controls were from both genders, and the age range varied from 1 month to 50 years. The majority were matched for age and gender, and some studies included family relatives, friends, or healthy individuals with similar characteristics in the PKU group.

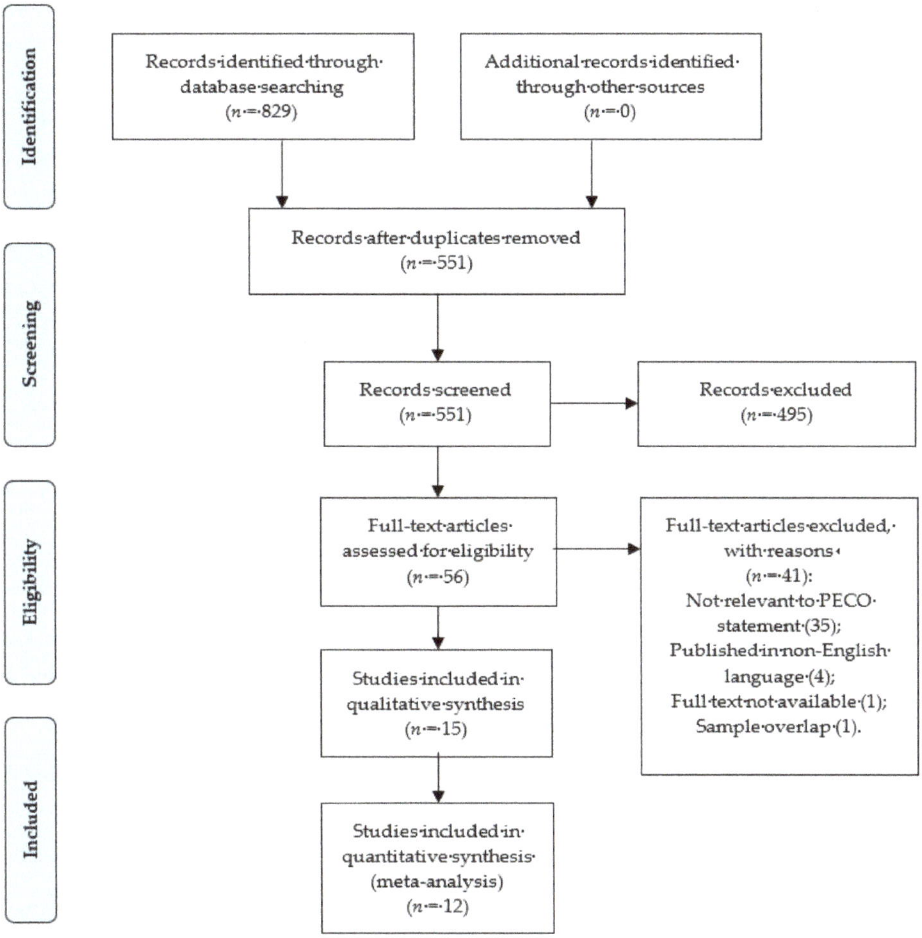

Figure 1. PRISMA study flow diagram describing the process of study selection. Abbreviation: PECO: Population, Exposure, Comparator, Outcome.

Most studies examined the association between a Phe-restricted diet and BMI [7,18,25,26,31,32,34–38]. Six studies examined the association between a Phe-restricted diet and overweight prevalence [18,30,31,34,37,38]. Eleven studies examined the association of different or additional parameters, such as weight-for-height and weight z-scores and body fat percentage [7,18,25,26,31,34,35,38–41].

From 15 studies included in the qualitative synthesis, 12 did not find significant differences in BMI and overweight prevalence between patients with PKU on a Phe-restricted diet, compared with healthy controls [7,18,25,26,31,32,34,37–41] (Table 1). Only 3 of 15 studies found a significantly higher BMI or higher prevalence of overweight in patients with PKU than controls [30,35,36].

Table 1. Characteristics of the studies included in the systematic review.

Reference	Country	Study Design (Duration of Follow-Up)	Sample Size (Phenotype)	Early and Continous Treatment	Age Range (Years)	Gender (F:M)	Annual Phe Levels (μmol/L)	Controls (Type)	Outcomes (Units)	Key Findings	Risk of Bias [1]
Allen et al. 1995 [39]	Australia	Cross-sectional	30 (NA)	NA	4.6–17.0	15:15	NA	76 family relatives, (age range: 4.3–18.4 y)	Body fat (%) Weight z-score	No significant differences between males with PKU and control subjects for weight scores, body fat, or fat free mass. Females with PKU had lower fat free mass and there was no difference in weight scores and body fat.	High
Allen et al. 1996 [41]	Australia	Cross-sectional with longitudinal cohort (1.1 y)	37 (37 classical)	Yes (NBS)	3.9–11.0 [2]	16:21	median at the time of the study: 652	27 unaffected siblings (PKU or cystic fibrosis), (age range: 4.0–11.5 y)	Body fat (%) Weight z-score	No significant differences between children with PKU and controls for body fat, lean body mass, or weight. Children with PKU were significantly shorter than controls.	High
Fisberg et al. 1999 [40]	Brazil	Cross-sectional	42 (NA)	NA	1.0–12.0	NA (both genders)	NA	31 with similar characteristics (age range: 1.0–12.0 y)	Weight for height z-score Weight z-score	No significant differences between patients with PKU and controls for weight for height and weight for age z-scores.	High
Schulpis et al. 2000 [32]	Greece	Cross-sectional	49 (49 classical–21 strict diet + 28 'relaxed' diet)	NA	strict diet: 5.2 ± 1.4 'relaxed' diet: 6.0 ± 1.5 (mean + SD)	23:26	mean ± SD: strict diet: 150 ± 40 'loose' diet: 800 ± 40	30 with similar age (mean age + SD: 7.9 ± 1.2 y)	BMI (kg/m^2)	No significant difference for BMI between patients with PKU adhering to their diet or on a 'relaxed diet' and controls. Patients with PKU on a 'relaxed diet' had significantly higher leptin concentrations compared to patients with PKU adhering to their diet and controls.	High
Huemer et al. 2007 [26]	Austria	Cross-sectional with longitudinal cohort (1 y)	34 (34 classical)	Yes (NBS)	0.2–15.0 [2]	12:22	mean ± SD at the time of the study: <10 y: 456 ± 432 10–15 y: 534 ± 324 >15 y: 444 ± 228	34 matched for age and gender (mean age difference: 0.5 y)	BMI (kg/m^2) BMI z-scores Weight z-score	No significant differences for BMI and body fat mass between patients with PKU and controls.	Moderate

Table 1. *Cont.*

Reference	Country	Study Design (Duration of Follow-Up)	Sample Size (Phenotype)	Early and Continous Treatment	Age Range (Years)	Gender (F:M)	Annual Phe Levels (µmol/L)	Controls (Type)	Outcomes (Units)	Key Findings	Risk of Bias [1]
Albersen et al. 2010 [7]	The Netherlands	Cross-sectional	20 (20 classical)	Yes (NBS)	6.0–16.0	13:7	mean ± SD: 375 ± 253? (F: 420 ± 303; M: 291 ± 77)	20 matched for age and gender (mean age difference: 0.5 y)	BMI (kg/m^2) Body fat (%)	No significant differences between children with PKU and controls for body weight and BMI. Body fat % was significantly higher in patients with PKU, especially in girls aged > 11 years.	High
Rocha et al. 2012 [34]	Portugal	Cross-sectional	89 (29 classical, 42 mild 18 HPA)	Yes (NBS)	3.0–30.0	41:48	mean ± SD: 393 ± 245	79 siblings, family or friends (mean age difference: 1.9 y)	BMI (kg/m^2) Body fat (%) Overweight prevalence (%)	No significant differences between patients with PKU and controls for overweight and obesity prevalence, BMI, waist circumference, and % body fat. Overweight prevalence was higher in patients with poor metabolic control and patients aged 10–16 years.	Moderate
Doulgeraki et al. 2014 [35]	Greece	Cross-sectional	80 (48 classical, 32 HPA)	Yes (NBS)	5.0–18.0	37:43	mean ± SD: PKU: 344 ± 178 HPA: 222 ± 51.6	57 matched for age and gender (mean age difference: 0.6 y)	BMI z-score Body fat (%) Weight z-score	Children with PKU had significantly higher BMI and weight when compared to healthy children. Fat mass increased significantly during puberty in PKU patients, especially in those with poor dietary adherence, and HPA.	High
Mazzola et al. 2016 [38]	Brazil	Cross-sectional	27 (13 classical, 14 mild)	Yes (11 early and 16 late diagnosed)	6.0–25.0	13:14	range at the time of the study: 102–1660	27 matched for age and gender (age NA)	BMI (kg/m^2) Body fat (%) Overweight prevalence (%)	No significant differences for anthropometric measures between patients with PKU and controls. PKU severity, time of diagnosis, or metabolic control had no effect on any body composition outcome measures.	High
Evans et al. 2017 [25]	Australia	Longitudinal prospective (2 y)	37 (NA)	Yes (NBS)	0.6–18.0 [2]	24:13	NA	21 matched for age and gender (mean age difference: 0.0 y)	BMI z-score Body fat (%) Weight z-score	No significant differences for BMI z-score and % body fat mass between patients treated with phe-restricted diet only, patients treated with BH4 + diet ($n = 5$), and controls.	High

Table 1. *Cont.*

Reference	Country	Study Design (Duration of Follow-Up)	Sample Size (Phenotype)	Early and Continous Treatment	Age Range (Years)	Gender (F:M)	Annual Phe Levels (µmol/L)	Controls (Type)	Outcomes (Units)	Key Findings	Risk of Bias [1]
Hermida-Ameijeiras et al. 2017 [37]	Spain	Cross-sectional	41 (22 classical, 19 mild–moderate)	Yes (early and late diagnosed)	6.0–50.0	30:11	NA	41 matched for age and gender (mean age difference: −2.9 y)	BMI (kg/m^2) Overweight prevalence (%)	No significant differences for BMI between patients with PKU and controls. Patients on BH4 therapy had lower BMI than those without BH4 therapy. Patients with lower Phe tolerance had higher body weight.	High
Couce et al. 2018 [30]	Spain	Cross-sectional	83 (37 classical, 20 mild-moderate 26 HPA)	Yes (70 early and 13 late diagnosed)	4.0–52.0	49:34	median: classical: 484 mild-moderate: 242 HPA: 296	68 matched for age and gender (age: NA)	Overweight prevalence (%)	Significantly higher % of overweight in patients with PKU than in patients with HPA and healthy controls, especially in those with good metabolic control.	Moderate
Evans et al. 2019 [31]	UK	Longitudinal Prospective (1.4–1.7 y; until 2 y of age)	20 (14 classical, 3 mild 3 moderate)	Yes (NBS)	0.2–0.6 [2]	6:14	mean ± SD: 249 ± 81	20 (18 matched for birth order and mother's educational level) (mean age difference: 0.0 y)	BMI z-score Overweight prevalence (%) Weight z-score	No significant differences between patients with PKU and controls for weight, head circumference, or BMI. Boys had lower mean BMI z-scores across both groups (PKU and controls).	Moderate
Azabdaftari et al. 2019 [36]	Germany	Cross-sectional	23 [3] (19 classical, 4 mild)	Yes (NBS)	18.0–47.0	10:13	mean ± SD: 1132 ± 321 (F: 1209 ± 316; M: 1068 ± 325)	28 healthy with similar age (mean age difference: −0.7 y)	BMI (kg/m^2)	Patients with PKU had significantly higher BMI than controls. Patients with poor metabolic control also had significantly higher BMI.	High
Sailer et al. 2020 [18]	USA	Cross-sectional	30 (30 classical)	Yes (NBS)	5.0–16.0	12:18	mean ± SD: 392 ± 184	30 matched for age and gender (mean age difference: −0.1 y)	BMI (kg/m^2) Body fat (%) Overweight prevalence (%)	No significant differences for BMI between patients with PKU and controls. Male subjects with PKU had significantly higher fat mass % and lower lean body mass % compared to male controls.	High

Abbreviations: BH4: sapropterin; BMI: body mass index; F: female; HPA: hyperphenylalaninaemia; M: male; NA: not available; NBS: newborn screening; Phe: phenylalanine; PKU: phenylketonuria; SD: standard deviation; UK: United Kingdom; USA: United States of America; y years. [1] Assessed using the National Institutes of Health (NIH) Quality Assessment Tool for Observational Cohort and Cross-Sectional Studies; [2] at baseline; [3] two patients refused physical examination.

Table 2. Exposure assessment method and nutritional intake of participants in the included studies.

Reference	Exposure Assessment Method	Natural Protein (g/kg/day)	PE from PS Supplements (g/kg/day)	Carbohydrate (%)	Lipids (%)	Energy (kcal)	BH4 Treatment	Additional Information
Allen et al. 1995 [39]	NA	NA	NA	NA	NA	NA	No	-
Allen et al. 1996 [41]	4 day dietary records	2.1		NA	NA	1.6xBMR	No	
Fisberg et al. 1999 [40]	3 day dietary records	<7 y: 105.0% RDA [1] ≥7 y: 109.4% RDA [1]		NA	NA	<7 y: 62.6% RDA ≥7 y: 60.5% RDA	No	-
Schulpis et al. 2000 [32]	1 week dietary record + 24 h dietary recall	strict diet: 7.5 ± 5.6 g/day; 'loose' diet: 15.8 ± 5.5 g/day	strict diet: 60.6 ± 7 g/day; 'loose' diet: 55.1 ± 14 g/day	strict diet: 49 'relaxed' diet: 43	strict diet: 21 'loose' diet: 38	strict diet: 2114 ± 463 'loose' diet: 2080 ± 487	No	-
Huemer et al. 2007 [26]	3 day dietary records	0.3	0.9	NA	NA	NA	No	-
Albersen et al. 2010 [7]	NA	1.3–1.5 times above RDA [1]		NA	NA	-	No	-
Rocha et al. 2012 [34]	Food history from the nutrition appointment	HPA: 1.16 ± 0.53 MPKU: 0.59 ± 0.33 CPKU: 0.59 ± 0.36	HPA: 1.13 ± 0.41 MPKU: 1.38 ± 0.43 CPKU: 1.25 ± 0.53	HPA: 58 ± 5 MPKU: 60 ± 4 CPKU: 58 ± 6	HPA: 30 ± 5 MPKU: 25 ± 4 CPKU: 26 ± 4	HPA: 2260 ± 332 MPKU: 2351 ± 391 CPKU: 2451 ± 316	No	-
Doulgeraki et al. 2014 [35]	NA	NA	NA	NA	NA	No	No	HPA on free diet
Mazzola et al. 2016 [38]	NA	NA	NA	NA	NA	NA	No	-
Evans et al. 2017 [25]	Food diary	0.50 ± 0.18	1.54 ± 0.50	NA	NA	1665 ± 546	Yes (5 patients)	-
Hermida-Ameijeiras et al. 2017 [37]	NA	NA	NA	NA	NA	NA	Yes (7 patients)	-
Couce et al. 2018 [30]	3 day dietary records	1.3–1.5 times above RDA [1]		CPKU: 57.0 ± 8.6 MPKU: 53.5 ± 9.8	NA	NA	Yes (10 patients)	HPA on free diet
Evans et al. 2019 [31]	1 day dietary record	0.43 ± 0.26	2.75 ± 0.39 [2]	60 [2]	25 [2]	1320 [2]	No	-
Azabdaftari et al. 2019 [36]	3 day dietary records	0.19 ± 0.13	0.73 ± 0.21	NA	NA	NA	No	-
Sailer et al. 2020 [18]	24 h dietary recall	0.39 ± 0.31	1.10 ± 0.72	67 ± 9	24 ± 8	2356 ± 620	Yes (4 patients)	-

Abbreviations: BH4: sapropterin; BMR: basal metabolic rate; CPKU: classical PKU; G: grams; HPA: hyperphenylalaninaemia; Kcal: kilocalorie; Kg: kilograms; MPKU: mild–moderate PKU; NA: not available; PE: protein equivalent; PS: protein substitute; RDA: recommended dietary allowances; y: years. [1] Total protein (g/kg/day); [2] at 24 months of age. Examining PKU phenotype, five studies included only patients with classical PKU [7,18,26,32,41], seven mixed phenotypes [30,31,34–38], and three did not specify [25,39,40].

3.3. NutriGrade Assessment

On the basis of the NutriGrade assessment (Supplementary Materials, Section C—Table S5), we found that the quality of the evidence for the meta-analysis using BMI was low, with meta-evidence limited and uncertain. The quality of the evidence for the meta-analysis using body fat percentage was very low, with meta-evidence very limited and uncertain.

3.4. Risk of Bias Assessment

Using the NIH Quality Assessment Tool for Observational Cohort and Cross-Sectional Studies, we found that 4 studies were assessed as fair with moderate risk of bias [26,30,31,34], and 11 as poor with high risk of bias [7,18,25,32,35–41]. Figure 2 presents the percentages of compliance for each tool item across all included studies. The risk of bias summary with review authors' judgments about each item for all included studies can be found in the Supplementary Materials, Section C—Figure S1.

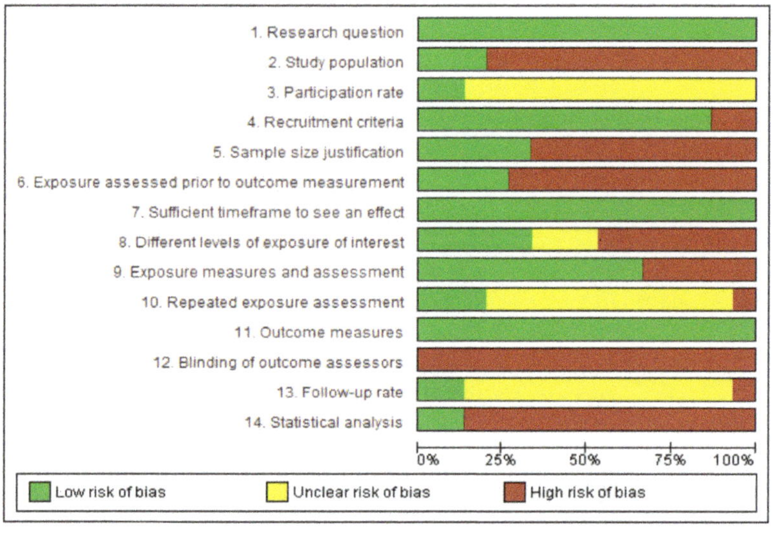

Figure 2. Risk of bias: judgements about each risk of bias item presented as percentages across all included studies.

Visual inspection of the funnel plot did not indicate substantial asymmetry (Supplementary Materials, Section C—Figure S7).

3.5. Synthesis of Results

3.5.1. Patients with PKU vs. Healthy Controls

In the 12 studies included in the meta-analysis, there were no differences for BMI of patients with PKU compared with healthy controls (SMD = 0.12 [−0.04, 0.28], $p = 0.14$; $I^2 = 27\%$, $p = 0.18$; Figure 3).

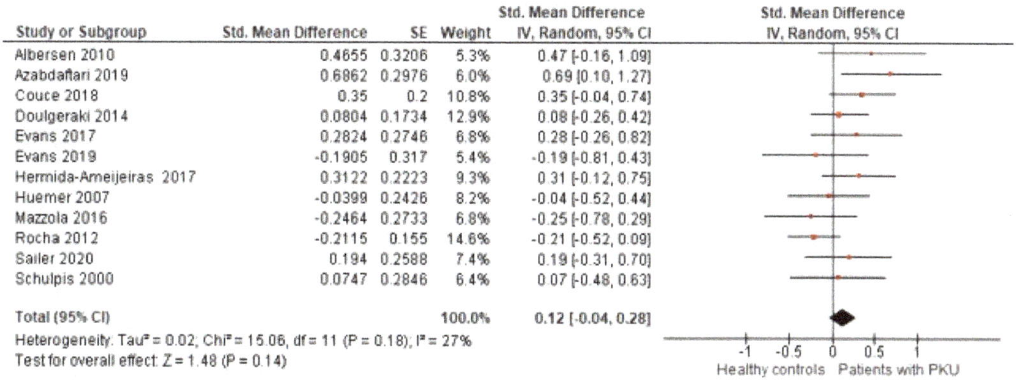

Figure 3. Forest plot comparing the BMI between patients with PKU and healthy controls. Abbreviations: BMI: body mass index; CI: confidence interval; df: degrees of freedom; IV: inverse variance; PKU: phenylketonuria; SE: standard error; Std: standardised. Moderate risk of bias: Couce 2018, Evans 2019, Huemer 2007, and Rocha 2012. High risk of bias: Albersen 2010, Azabdaftari 2019, Doulgeraki 2014, Evans 2017, Hermida-Ameijeiras 2017, Mazzola 2016, Sailer 2020, and Schulpis 2000. Time of diagnosis: Couce 2018 included 70 early and 13 late diagnosed patients, Hermida-Ameijeiras 2017 included both early and late diagnosed patients, Mazzola 2016 included 11 early and 16 late diagnosed patients, and Schulpis 2000 did not provide information on the time of diagnosis. Metabolic control: Azabdaftari 2019 included only one patient with good metabolic control (Phe blood levels < 600 μmol/L). BH4 treatment: Couce 2018 included 10 (12%) patients taking BH4, Evans 2017 included 5 (14%), Hermida-Ameijeiras 2017 included 7 (17%), and Sailer 2020 included 4 (13%).

3.5.2. Moderate vs. Poor Risk of Bias Studies

A subgroup analysis was conducted according to the risk of bias for each study (Supplementary Materials, Section C—Figure S2). Studies assessed as fair with moderate risk of bias [26,30,31,34] found no difference in BMI between patients and healthy controls (SMD = −0.02 [−0.30, 0.27], p = 0.91; I^2 = 43%, p = 0.16). Studies assessed as poor with high risk of bias [7,18,25,32,35–38] found a significantly higher BMI in patients with PKU compared to healthy controls (SMD = 0.20 [0.03, 0.37], p = 0.02; I^2 = 1%, p = 0.42).

3.5.3. Time of Diagnosis

Three studies included late diagnosed patients in their samples [30,37,38], and Schulpis et al. [32] did not provide information on diagnostic age. Thus, a subgroup analysis was conducted according to diagnostic age (Supplementary Materials, Section C—Figure S3). The subgroup of studies including only early diagnosed patients found no differences in BMI between patients and healthy controls (SMD = 0.11 [−0.10, 0.31], p = 0.32; I^2 = 35%, p = 0.15). Moreover, the subgroup of studies including both early and late diagnosed patients found no differences between patients with PKU and healthy controls (SMD = 0.18 [−0.17, 0.52], p = 0.31; I^2 = 43%, p = 0.18). There were no statistical differences between the two subgroups (p = 0.73).

3.5.4. Age

The studies included in the meta-analysis covered a wide patient age. We performed a subgroup analysis (Supplementary Materials, Section C—Figure S4) comparing studies including children and adolescents only [7,18,25,26,31,32,35], adults only [36], and all age groups (children, adolescents, and adults) [30,34,37,38]. We found no differences between the three subgroups (p = 0.15), and a higher heterogeneity in the subgroup of studies that included all age groups (I^2 = 61%). The subgroup that included adults only had one study [36] that identified adult patients with PKU, having a significantly higher BMI when compared to healthy adults.

3.5.5. Sapropterin (BH4) Treatment

Four studies included patients prescribed BH4 in their patient cohort [18,25,30,37]. To understand if there was any difference between studies that included patients taking BH4 (mixed sample) and studies that included only patients on a Phe-restricted diet, we performed a subgroup analysis (Supplementary Materials, Section C—Figure S5).

Studies that included some patients with PKU treated with diet and BH4 [18,25,30,37] found a significantly higher BMI in the overall group than in healthy controls (SMD = 0.30 [0.07, 0.52], $p = 0.01$; $I^2 = 0\%$, $p = 0.97$). Studies that included only patients on a Phe-restricted diet [7,26,31,32,34–36,38] found no differences between the PKU group and healthy controls (SMD = 0.04 [−0.17, 0.24], $p = 0.74$; $I^2 = 35\%$, $p = 0.15$).

3.5.6. Phenotype

Four studies in the meta-analysis included only patients with classical PKU [7,18,26,32]. The remaining studies included patients with different phenotypes and reported their BMI together; therefore, it was not possible to analyse any association between different phenotypes and BMI from these studies [30,31,34–38]. To understand if there were any differences between studies including only patients with classical PKU and studies that included patients with different phenotypes, we performed a subgroup analysis (Supplementary Materials, Section C—Figure S6). In both subgroups, there were no differences between patients with PKU and controls.

3.5.7. Patients with Classical PKU vs. Healthy Controls

Several authors of the included studies provided individual participant data, including disease severity [7,18,31,34–36]. On the basis of this additional data, we conducted a meta-analysis comparing patients with classical PKU only with healthy controls (Figure 4) [7,18,26,32]. In the remaining studies, we calculated the mean BMI of patients with classical PKU [30,31,34–36] and excluded data from patients with other phenotypes. Individual participant data was unavailable from two studies (Hermida-Ameijeiras et al. [37] and Mazzola et al. [38]), and Evans et al. [25] did not include information on the patient phenotype. Therefore, these three studies were excluded from this meta-analysis.

Study or Subgroup	Std. Mean Difference	SE	Weight	Std. Mean Difference IV, Random, 95% CI
Albersen 2010	0.4655	0.3209	8.2%	0.47 [−0.16, 1.09]
Azabdaftari 2019	0.6459	0.3102	8.7%	0.65 [0.04, 1.25]
Couce 2018	0.82	0.24	12.5%	0.82 [0.35, 1.29]
Doulgeraki 2014	0.1338	0.1961	15.9%	0.13 [−0.25, 0.52]
Evans 2019	−0.0648	0.3486	7.2%	−0.06 [−0.75, 0.62]
Huemer 2007	−0.0399	0.2426	12.3%	−0.04 [−0.52, 0.44]
Rocha 2012	0.043	0.2171	14.1%	0.04 [−0.38, 0.47]
Sailer 2020	0.194	0.2588	11.3%	0.19 [−0.31, 0.70]
Schulpis 2000	0.0747	0.2846	9.8%	0.07 [−0.48, 0.63]
Total (95% CI)			**100.0%**	**0.24 [0.04, 0.45]**

Heterogeneity: $Tau^2 = 0.03$; $Chi^2 = 11.63$, $df = 8$ ($P = 0.17$); $I^2 = 31\%$
Test for overall effect: $Z = 2.33$ ($P = 0.02$)

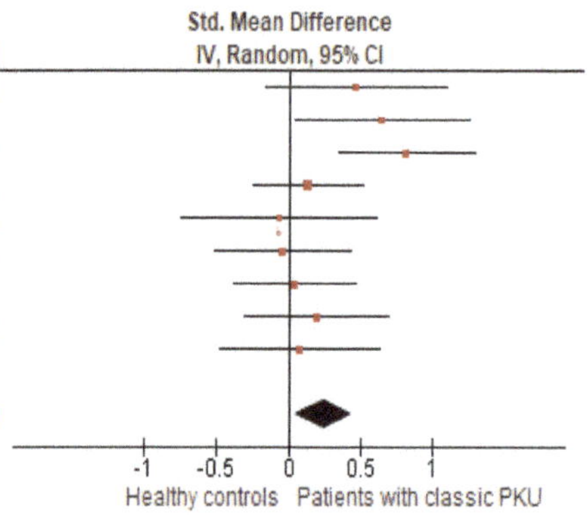

Figure 4. Forest plot comparing the BMI between patients with classical PKU and healthy controls. Abbreviations: BMI: body mass index; CI: confidence interval; df: degrees of freedom; IV: inverse variance; PKU: phenylketonuria; SE: standard error; Std: standardised. Moderate risk of bias: Couce 2018, Evans 2019, Huemer 2007, and Rocha 2012. High risk of bias: Albersen 2010, Azabdaftari 2019, Doulgeraki 2014, Sailer 2020, and Schulpis 2000. Time of diagnosis: Couce 2018 included 70 early- and 13 late-diagnosed patients, and Schulpis 2000 did not provide information on the time of diagnosis. Metabolic control: Azabdaftari 2019 included only one patient with good metabolic control (Phe blood levels < 600 μmol/L). BH4 treatment: Couce 2018 included 1 (3%) patient taking BH4, and Sailer 2020 included 4 (13%) patients.

We found that patients with classical PKU had a significantly higher BMI than healthy controls (SMD = 0.24 [0.04, 0.45], $p = 0.02$; $I^2 = 31\%$, $p = 0.17$).

To reject the hypothesis that this result was due to the removal of the three studies, whose individual participant data is unknown, we performed the first meta-analysis (Figure 3) without them. Removing these three studies did not affect the overall result, compared with the 12 included studies (SMD = 0.12 [−0.07, 0.31], $p = 0.22$; $I^2 = 34\%$, $p = 0.15$).

3.5.8. Sex

Only six studies provided adequate information to establish a comparison on sex, which limits the subsequent interpretation of its effect on overweight. However, when comparing females with PKU and healthy females, all studies found a trend towards a higher BMI in females with PKU (Supplementary Materials, Section C—Table S4).

3.5.9. Metabolic Control

We tried to explore the association between metabolic control and BMI. However, only five studies provided information on metabolic control, and the comparison between patients with poor metabolic control and healthy controls (Supplementary Materials, Section C—Table S4) had substantial heterogeneity ($I^2 = 58\%$, $p = 0.05$); thus, we were unable to present accurate data on metabolic control.

3.5.10. Body Fat Percentage

The methods used to assess body fat percentage across studies were different. This led to a heterogeneous overall result, rendering it unfeasible to present and compare body fat results (Supplementary Materials, Section C—Table S4).

4. Discussion

4.1. Summary of Evidence

To the best of our knowledge, this is the first systematic review with meta-analysis evaluating the association between a Phe-restricted diet and overweight and obesity in patients with PKU. We pooled data from 12 observational studies for the meta-analysis and found no differences between patients with PKU and healthy controls for BMI. The pooled data included diverse patient phenotypes with variable Phe-restriction, with dissimilar contributions from the PS and SLPFs to total protein and energy intake [16,42,43]. Our meta-analysis suggests that dietary Phe-restriction alone is not a risk factor for the development of overweight and obesity.

However, patients with classical PKU had a significantly higher BMI than healthy controls. This observation resulted from nine studies, including only patients with classical PKU and studies whose authors provided additional individual participant data, although these results should be considered with caution. One plausible explanation is that more calories may be given to patients with classical PKU in order to prevent catabolism that causes higher blood Phe levels. This may lead to the development of overweight.

Among the studies included in qualitative synthesis, 4 studies had a moderate risk of bias and 11 had a high risk of bias using the NIH Quality Assessment Tool. The subgroup of studies with moderate risk of bias did not find a higher BMI in patients with PKU. In contrast, studies assessed as poor due to their methodological flaws found a significantly higher BMI in patients with PKU compared to healthy controls. Therefore, this work highlights the fragility of the evidence supporting the idea that a Phe-restricted diet promotes overweight and indicates the need for controlled studies with improved methodology and comprehensive data collection.

Three of the seven most common flaws observed in the studies were limited description of the study population using demographics (who), location (where), and time period (when) (question 2 of the NIH tool) [7,18,25,26,31,32,35,36,38–41]; absence of sample size justification (question 5 of the NIH tool) [18,25,26,30,32,35,37,38,40,41]; and outcome assessors being aware of participants' exposure status (question 12 of the NIH tool) in all included studies. These flaws were not considered fatal, and studies that failed these criteria could still be classified as fair with moderate risk of bias.

Eleven studies were cross-sectional [7,18,30,32,34–40], and the exposure was not assessed prior to outcome measurement (question 6 of the NIH tool). For this reason, it is not possible to establish a relation of causality between the exposure to a Phe-restricted diet and overweight.

For the different levels of exposure assessment (question 8 of the NIH tool), from the 10 studies that included patients with different phenotypes, the use of BH4 with a relaxed Phe-restriction or patients who were late diagnosed with PKU, only five studies considered these factors [25,30,32,34,35]. These different levels of exposure to the Phe-restricted diet renders it difficult to analyse the association between the Phe-restricted diet and overweight. For example, we identified three studies that included patients with HPA [30,34,35] and, in two of three of these studies, patients were on an unrestricted diet [30,35]. The fact that most studies included patients with different phenotypes does not allow for conclusions about the association between phenotype and overweight, as verified in the subgroup analysis by phenotypes (Supplementary Materials, Section C—Figure S6).

In addition, between 20 and 50% of patients with PKU are responsive to the synthetic form of the cofactor (BH4), meaning that a less restricted diet is followed. Evidence suggests that 51% of patients on BH4 therapy completely stop PS intake [44]. In our meta-analysis, the studies that included patients taking both BH4 combined with patients on a traditional Phe-restricted diet only found a significantly higher BMI in the overall group of patients with PKU compared to healthy controls. Although this is an interesting finding, it is unknown as to how many of these patients were overweight before BH4 commencement. A study conducted in Spain, including patients from 13 hospitals, found that patients taking BH4 had significantly higher BMI z-scores than patients on a Phe-restricted diet only, with follow up consistently over 2 years [45]. These results highlight the need for a continuous nutritional monitoring and specialised nutritional care, even in patients under pharmacological treatment. This observation warrants further study.

Of the 12 studies included in the meta-analysis, 4 did not assess patients' dietary intake [7,35,37,38]. In the remaining eight studies, the methods used to assess intake were different, and only four studies [18,31,32,34] provided detailed information on the amount of protein, CHO, fat, and energy patients consumed. This information is central to accurately address our review question and is considered an important omission in studies. Different reimbursement policies in different countries determine access to PS and SLPFs, which ultimately will alter the intake of macronutrients supplied by a Phe-restricted diet [46,47].

We also tried to determine if there was an association between patients' BMI and metabolic control (which may reflect patients' exposure to the Phe-restricted diet). However, most of the studies did not report patients' BMI, nor its comparison with metabolic control. In the literature, some studies have found a positive correlation between mean Phe levels and BMI [3,36,48], and between mean Phe levels and the prevalence of overweight [1,9,34], indicating that good metabolic control is associated with a lower risk of overweight. Conversely, two studies from Spain found a higher prevalence of overweight and BMI in patients with good metabolic control compared to poorly controlled patients [30,49].

Most of the included studies did not adjust for key prognostic variables, such as physical activity, family history, socioeconomic status, parents' weight, and epigenetics, among other determinant factors that may be associated with overweight.

Finally, none of the included studies considered the regular follow-up of patients by a nutritionist. Nutritionists play a crucial role in monitoring the patient's weight while ensuring they meet their complex dietary needs [50]. Consequently, we were not only analysing the influence of the Phe-restricted diet alone on overweight, but also on the quality of the follow-up that the patients receive.

4.2. Strengths and Limitations of This Study

Several limitations in this systematic review should be acknowledged. First, our systematic review included observational studies only. Observational evidence usually provides lower strength evidence than RCTs, due to confounding variables. Nevertheless, RCTs addressing our question have not been conducted, which is unsurprising, given that PKU is a rare disease and the exposure to an unrestricted Phe-diet is clinical and ethically unacceptable. In addition, there was large heterogeneity in the design of observational studies and in the reporting of results.

The diversity of the study populations also contributes to the heterogeneity of the results. For instance, some studies included patients with different disease severities, with variable degrees of Phe-restriction, being diagnosed early and later on, patients on BH4 treatment, and patients with poor metabolic control. Additionally, patients had a wide age range.

The Phe-restricted diet was not always well defined: not all studies reported patients' dietary intake, and some studies did not assess it.

In relation to the comparator, we did not define any inclusion criteria for healthy controls. Most of them were matched for age and gender only, and the number of controls included in our work was less than the number of patients with PKU.

Regarding the outcome, one study [30] only presented the prevalence of overweight, which led us to convert the respective OR to a SMD to include it in the meta-analysis. Although BMI is an important predictor of adiposity and is a tool widely used in clinical practice [23], it may not always identify individuals with increased fat mass percentage [51], which underlines the weakness of the BMI as an indicator of adiposity. Measuring body composition appears to be a better approach to identify individuals with increased fat mass percentage, specifically those at a higher risk of metabolic complications, which is crucial to help prevent the development of comorbidities [51]. Increased abdominal obesity is associated with dyslipidaemia, hypertension, insulin resistance, and inflammation.

Finally, most of the included studies had a high risk of bias according to the NIH tool. On the basis of the NutriGrade assessment, we found that the quality of the meta-analysis comparing all patients with PKU to controls was 'low', and the quality of the meta-analysis comparing patients with classical PKU to controls was 'very low'.

In order to strengthen the conclusions of our systematic review with meta-analysis, we used the best methodology, namely, (1) following the PRISMA guidelines and registering on the PROSPERO database—studies that do appear to be of higher quality [27,52]; (2) clear definition of the aim of our work; (3) clear definition of the inclusion and exclusion criteria, according to the PECO strategy; (4) using several databases for the search and searching reference lists of the retrieved studies; (5) describing the study selection process using a flow diagram; (6) providing the list of the excluded studies and the reasons; (7) providing of the characteristics of individual studies; (8) contacting the correspondence authors to request further information; (9) performing meta-analysis and subgroup analysis; and (10) having two independent authors performing study selection, data extraction, and assessment of the risk of bias and the quality of the evidence.

As the study of risk factors is based on comparisons between exposed and unexposed individuals [53], only studies with a control group were included in our systematic review, which is another strength of this meta-analysis. Indeed, several studies that propose that the Phe-restricted diet promotes overweight did not include a control group.

Finally, our systematic review provides a clear overview of the available evidence on the topic overweight and PKU and will be useful in guideline development. It also identifies the main flaws and pitfalls that should be avoided when designing novel studies to address this question in the future.

5. Conclusions

We found no differences between patients with PKU and healthy controls in BMI. Thus, there is no evidence to support the concept of Phe-restricted diet as a risk factor for the development of overweight. However, a subgroup of patients with classical PKU had a significantly higher BMI than healthy controls. In addition, studies assessed as poor with high risk of bias and studies that included both diet-treated and BH4-treated patients found a significantly higher BMI in patients with PKU compared to healthy controls.

Given the increasing prevalence of overweight in the general population, patients with PKU should remain in long-term follow-up, receiving personalised nutritional advice with systematic nutritional status monitoring by a multidisciplinary team in inherited metabolic disorders. This is essential to prevent overweight, obesity, and its related comorbidities.

Future studies with improved methodology are needed to properly address this question and to help in guiding the clinical practice of health professionals.

Supplementary Materials: The following are available online at https://www.mdpi.com/article/10.3390/nu13103443/s1, Figure S1: Risk of bias summary: Review authors' judgements about each risk of bias item for each included study. Figure S2: Forest plot comparing the BMI between patients with PKU and healthy controls among studies with moderate and high risk of bias. Figure S3: Forest plot comparing the BMI between patients with PKU and healthy controls among studies including only early diagnosed patients and studies including both early and late diagnosed patients. Figure S4: Forest plot comparing the BMI between patients with PKU and healthy controls among studies including only children and adolescents; studies including only adults; and studies including children, adolescents, and adults. Figure S5: Forest plot comparing the BMI between patients with PKU and healthy controls among studies including both patients taking BH4 and patients not taking BH4, as well as studies including only patients not taking BH4. Figure S6: Forest plot comparing the BMI between patients with PKU and healthy controls among studies including patients with mixed phenotypes and studies including only patients with classical PKU. Figure S7: Publication bias plot. The SMD of BMI is plotted on the x-axis and the SE of the SMD is plotted on the y-axis. Table S1: Syntax of Mesh/Emtree terms per database. Table S2: Syntax of title, abstract, and author keyword per database. Table S3: Studies excluded from the systematic review with reasons. Table S4: Summary of between-group meta-analysis results. Table S5: NutriGrade assessment of the quality of the evidence.

Author Contributions: A.M.-R. and J.C.R. conceived and designed the protocol for this systematic review and supervised the study. A.M.-R. designed the methodology and the statistical analysis. A.M.J.v.W. defined the search strategy. A.M. and J.C.R. searched the literature and performed the study selection. C.R. and A.P. extracted the data and applied the risk of bias and NutriGrade assessments tools. C.R. performed the statistical analysis and drafted the manuscript. A.P., A.F., D.T., A.M.J.v.W., K.A., F.F., C.C., A.M., A.M.-R. and J.C.R. reviewed and edited the manuscript. All authors have read and agreed to the published version of the manuscript.

Funding: CINTESIS—UIDB/4255/2020 a program financially supported by Fundação para a Ciência e Tecnologia/Ministério da Educação e Ciência, through national funds is acknowledged. Support from Human Nutrition and Metabolism Master Program from NOVA Medical School, Universidade Nova de Lisboa is ackowledged.

Institutional Review Board Statement: Not applicable.

Informed Consent Statement: Not applicable.

Acknowledgments: We would like to thank the following authors of the studies included in our systematic review for providing their data: Aline Azabdaftari from the Department of Pediatrics, Division of Gastroenterology, Nephrology and Metabolic Diseases, Charité—Universitätsmedizin Berlin, Berlin, Germany; Artemis Doulgeraki from the Department of Bone and Mineral Metabolism, Institute of Child Health, Athens, Greece; Gepke Visser from the Department of Metabolic and Endocrine Diseases, Wilhelmina Children's Hospital, University Medical Centre Utrecht, Utrecht, the Netherlands; Maureen Evans from the Department of Metabolic Medicine, The Royal Children's Hospital, Melbourne, Australia, and the Department of Nutrition and Food Services, Royal Children's Hospital, Melbourne, Australia; Melanie Gillingham from the Departments Molecular and Medical Genetics, Graduate Programs in Human Nutrition at Oregon Health and Science University, Oregon, USA; and Sharon Evans from the Birmingham Women's and Children's Hospital NHS Foundation Trust, Birmingham, United Kingdom.

Conflicts of Interest: A.P. has received an educational grant from Cambrooke Therapeutics and grants from Vitaflo, Nutricia, Merck Serono, Biomarin, and Mevalia to attend scientific meetings. A.M.J.W. received a research grant from Nutricia, honoraria from Biomarin as a speaker, and travel support from Nutricia and Vitaflo. K.A. is a member of the European Nutrition Expert Panel (Biomarin). F.F. has been a board member and received payments for, e.g., lectures/honoraria, and support for travel, accommodations, and/or meeting expenses from BioMarin, Genzyme, Merck-Serono, Nutricia, and Vitaflo. A.M. received research funding and honoraria from Nutricia, Vitaflo International, and Merck Serono. She is a member of the European Nutrition Expert Panel (Biomarin), member of Sapropterin Advisory Board (Biomarin), member of the Advisory Board entitled ELEMENT (Danone-Nutricia), and member of an Advisory Board for Arla and Applied Pharma Research. J.C.R. is a member of the European Nutritionist Expert Panel (Biomarin), the Advisory Board for Applied Pharma Research and Nutricia, and has received honoraria as a speaker from APR, Merck Serono, Biomarin, Nutricia, Vitaflo, Cambrooke, PIAM, and Lifediet.

References

1. Burrage, L.C.; McConnell, J.; Haesler, R.; O'Riordan, M.A.; Sutton, V.R.; Kerr, D.S.; McCandless, S.E. High prevalence of overweight and obesity in females with phenylketonuria. *Mol. Genet. Metab.* **2012**, *107*, 43–48. [CrossRef]
2. Burton, B.K.; Jones, K.B.; Cederbaum, S.; Rohr, F.; Waisbren, S.; Irwin, D.E.; Kim, G.; Lilienstein, J.; Alvarez, I.; Jurecki, E.; et al. Prevalence of comorbid conditions among adult patients diagnosed with phenylketonuria. *Mol. Genet. Metab.* **2018**, *125*, 228–234. [CrossRef] [PubMed]
3. Ozturk, Y.; Gencpinar, P.; Erdur, B.; Tokgöz, Y.; Isik, I.; Akin, S.B. Overweight and obesity in children under phenylalanine restricted diet. *Hong Kong J. Paediatr.* **2018**, *23*, 169–172.
4. Trefz, K.F.; Muntau, A.C.; Kohlscheen, K.M.; Altevers, J.; Jacob, C.; Braun, S.; Greiner, W.; Jha, A.; Jain, M.; Alvarez, I.; et al. Clinical burden of illness in patients with phenylketonuria (PKU) and associated comorbidities—a retrospective study of German health insurance claims data. *Orphanet J. Rare Dis.* **2019**, *14*, 181. [CrossRef] [PubMed]
5. Gokmen Ozel, H.; Ahring, K.; Bélanger-Quintana, A.; Dokoupil, K.; Lammardo, A.M.; Robert, M.; Rocha, J.C.; Almeida, M.F.; van Rijn, M.; MacDonald, A. Overweight and obesity in PKU: The results from 8 centres in Europe and Turkey. *Mol. Genet. Metab. Rep.* **2014**, *1*, 483–486. [CrossRef] [PubMed]
6. McBurnie, M.A.; Kronmal, R.A.; Schuett, V.E.; Koch, R.; Azeng, C.G. Physical growth of children treated for phenylketonuria. *Ann. Hum. Biol.* **1991**, *18*, 357–368. [CrossRef]
7. Albersen, M.; Bonthuis, M.; de Roos, N.M.; van den Hurk, D.A.; Carbasius Weber, E.; Hendriks, M.M.; de Sain-van der Velden, M.G.; de Koning, T.J.; Visser, G. Whole body composition analysis by the BodPod air-displacement plethysmography method in children with phenylketonuria shows a higher body fat percentage. *J. Inherit. Metab. Dis.* **2010**, *33* (Suppl. 3), S283–S288. [CrossRef] [PubMed]
8. Stroup, B.M.; Hansen, K.E.; Krueger, D.; Binkley, N.; Ney, D.M. Sex differences in body composition and bone mineral density in phenylketonuria: A cross-sectional study. *Mol. Genet. Metab. Rep.* **2018**, *15*, 30–35. [CrossRef]
9. Walkowiak, D.; Kaluzny, L.; Bukowska-Posadzy, A.; Oltarzewski, M.; Staszewski, R.; Moczko, J.A.; Musielak, M.; Walkowiak, J. Overweight in classical phenylketonuria children: A retrospective cohort study. *Adv. Med. Sci.* **2019**, *64*, 409–414. [CrossRef]
10. World Health Organization. Obesity and Overweight. Available online: https://www.who.int/news-room/fact-sheets/detail/obesity-and-overweight (accessed on 4 January 2021).
11. Hruby, A.; Hu, F.B. The Epidemiology of Obesity: A Big Picture. *Pharmacoeconomics* **2015**, *33*, 673–689. [CrossRef]
12. WHO Consultation on Obesity & World Health Organization. *Obesity: Preventing and Managing the Global Epidemic: Report of a WHO Consultation*; 0512-3054; World Health Organization: Geneva, Switzerland, 2000; pp. i–xii, 1–253.
13. Blüher, M. Obesity: Global epidemiology and pathogenesis. *Nat. Rev. Endocrinol.* **2019**, *15*, 288–298. [CrossRef]
14. Hruby, A.; Manson, J.E.; Qi, L.; Malik, V.S.; Rimm, E.B.; Sun, Q.; Willett, W.C.; Hu, F.B. Determinants and Consequences of Obesity. *Am. J. Public Health* **2016**, *106*, 1656–1662. [CrossRef]
15. Van Wegberg, A.M.J.; MacDonald, A.; Ahring, K.; Bélanger-Quintana, A.; Blau, N.; Bosch, A.M.; Burlina, A.; Campistol, J.; Feillet, F.; Giżewska, M.; et al. The complete European guidelines on phenylketonuria: Diagnosis and treatment. *Orphanet J. Rare Dis.* **2017**, *12*, 162. [CrossRef]
16. Pena, M.J.; Almeida, M.F.; van Dam, E.; Ahring, K.; Bélanger-Quintana, A.; Dokoupil, K.; Gokmen-Ozel, H.; Lammardo, A.M.; MacDonald, A.; Robert, M.; et al. Special low protein foods for phenylketonuria: Availability in Europe and an examination of their nutritional profile. *Orphanet J. Rare Dis.* **2015**, *10*, 162. [CrossRef] [PubMed]
17. Wood, G.; Evans, S.; Pointon-Bell, K.; Rocha, J.C.; MacDonald, A. Special Low Protein Foods in the UK: An Examination of Their Macronutrient Composition in Comparison to Regular Foods. *Nutrients* **2020**, *12*, 1893. [CrossRef]
18. Sailer, M.; Elizondo, G.; Martin, J.; Harding, C.O.; Gillingham, M.B. Nutrient intake, body composition, and blood phenylalanine control in children with phenylketonuria compared to healthy controls. *Mol. Genet. Metab. Rep.* **2020**, *23*, 100599. [CrossRef] [PubMed]
19. Rocha, J.C.; MacDonald, A. Dietary intervention in the management of phenylketonuria: Current perspectives. *Pediatr. Health Med. Ther.* **2016**, *7*, 155–163. [CrossRef]

20. Moretti, F.; Pellegrini, N.; Salvatici, E.; Rovelli, V.; Banderali, G.; Radaelli, G.; Scazzina, F.; Giovannini, M.; Verduci, E. Dietary glycemic index, glycemic load and metabolic profile in children with phenylketonuria. *Nutr. Metab. Cardiovasc. Dis. NMCD* **2017**, *27*, 176–182. [CrossRef] [PubMed]
21. Daly, A.; Evans, S.; Pinto, A.; Ashmore, C.; MacDonald, A. Protein Substitutes in PKU; Their Historical Evolution. *Nutrients* **2021**, *13*, 484. [CrossRef]
22. Rocha, J.C.; van Spronsen, F.J.; Almeida, M.F.; Ramos, E.; Guimarães, J.T.; Borges, N. Early dietary treated patients with phenylketonuria can achieve normal growth and body composition. *Mol. Genet. Metab.* **2013**, *110*, S40–S43. [CrossRef]
23. Camatta, G.C.; Kanufre, V.C.; Alves, M.R.A.; Soares, R.D.L.; Norton, R.C.; de Aguiar, M.J.B.; Starling, A.L.P. Body fat percentage in adolescents with phenylketonuria and associated factors. *Mol. Genet. Metab. Rep.* **2020**, *23*, 100595. [CrossRef]
24. Dobbelaere, D.; Michaud, L.; Debrabander, A.; Vanderbecken, S.; Gottrand, F.; Turck, D.; Farriaux, J.P. Evaluation of nutritional status and pathophysiology of growth retardation in patients with phenylketonuria. *J. Inherit. Metab. Dis.* **2003**, *26*, 1–11. [CrossRef] [PubMed]
25. Evans, M.; Truby, H.; Boneh, A. The relationship between dietary intake, growth and body composition in Phenylketonuria. *Mol. Genet. Metab.* **2017**, *122*, 36–42. [CrossRef] [PubMed]
26. Huemer, M.; Huemer, C.; Möslinger, D.; Huter, D.; Stöckler-Ipsiroglu, S. Growth and body composition in children with classical phenylketonuria: Results in 34 patients and review of the literature. *J. Inherit. Metab. Dis.* **2007**, *30*, 694–699. [CrossRef] [PubMed]
27. Liberati, A.; Altman, D.G.; Tetzlaff, J.; Mulrow, C.; Gøtzsche, P.C.; Ioannidis, J.P.; Clarke, M.; Devereaux, P.J.; Kleijnen, J.; Moher, D. The PRISMA statement for reporting systematic reviews and meta-analyses of studies that evaluate health care interventions: Explanation and elaboration. *PLoS Med.* **2009**, *6*, e1000100. [CrossRef] [PubMed]
28. Higgins, J.; Thomas, J.; Chandler, J.; Cumpston, M.; Li, T.; Page, M.; Welch, V. *Cochrane Handbook for Systematic Reviews of Interventions*; Version 6.2 (Updated February 2021); Cochrane: London, UK, 2021. Available online: www.training.cochrane.org/handbook (accessed on 1 September 2020).
29. National Institutes of Health. Quality Assessment Tool for Observational Cohort and Cross-Sectional Studies. Available online: https://www.nhlbi.nih.gov/health-topics/study-quality-assessment-tools (accessed on 14 September 2020).
30. Couce, M.L.; Sánchez-Pintos, P.; Vitoria, I.; De Castro, M.J.; Aldámiz-Echevarría, L.; Correcher, P.; Fernández-Marmiesse, A.; Roca, I.; Hermida, A.; Martínez-Olmos, M.; et al. Carbohydrate status in patients with phenylketonuria. *Orphanet J. Rare Dis.* **2018**, *13*, 103. [CrossRef]
31. Evans, S.; Daly, A.; Wildgoose, J.; Cochrane, B.; Chahal, S.; Ashmore, C.; Loveridge, N.; MacDonald, A. Growth, Protein and Energy Intake in Children with PKU Taking a Weaning Protein Substitute in the First Two Years of Life: A Case-Control Study. *Nutrients* **2019**, *11*, 552. [CrossRef]
32. Schulpis, K.H.; Papakonstantinou, E.D.; Tzamouranis, J. Plasma leptin concentrations in phenylketonuric patients. *Horm. Res.* **2000**, *53*, 32–35. [CrossRef] [PubMed]
33. Schwingshackl, L.; Knüppel, S.; Schwedhelm, C.; Hoffmann, G.; Missbach, B.; Stelmach-Mardas, M.; Dietrich, S.; Eichelmann, F.; Kontopantelis, E.; Iqbal, K.; et al. Perspective: NutriGrade: A Scoring System to Assess and Judge the Meta-Evidence of Randomized Controlled Trials and Cohort Studies in Nutrition Research. *Adv. Nutr.* **2016**, *7*, 994–1004. [CrossRef]
34. Rocha, J.C.; van Spronsen, F.J.; Almeida, M.F.; Soares, G.; Quelhas, D.; Ramos, E.; Guimarães, J.T.; Borges, N. Dietary treatment in phenylketonuria does not lead to increased risk of obesity or metabolic syndrome. *Mol. Genet. Metab.* **2012**, *107*, 659–663. [CrossRef]
35. Doulgeraki, A.; Skarpalezou, A.; Theodosiadou, A.; Monopolis, I.; Schulpis, K. Body composition profile of young patients with phenylketonuria and mild hyperphenylalaninemia. *Int. J. Endocrinol. Metab.* **2014**, *12*, e16061. [CrossRef]
36. Azabdaftari, A.; van der Giet, M.; Schuchardt, M.; Hennermann, J.B.; Plöckinger, U.; Querfeld, U. The cardiovascular phenotype of adult patients with phenylketonuria. *Orphanet J. Rare Dis.* **2019**, *14*, 213. [CrossRef]
37. Hermida-Ameijeiras, A.; Crujeiras, V.; Roca, I.; Calvo, C.; Leis, R.; Couce, M.L. Arterial stiffness assessment in patients with phenylketonuria. *Medicine* **2017**, *96*, e9322. [CrossRef]
38. Mazzola, P.N.; Nalin, T.; Castro, K.; van Rijn, M.; Derks, T.G.; Perry, I.D.; Mainieri, A.S.; Schwartz, I.V. Analysis of body composition and nutritional status in Brazilian phenylketonuria patients. *Mol. Genet. Metab. Rep.* **2016**, *6*, 16–20. [CrossRef]
39. Allen, J.R.; McCauley, J.C.; Waters, D.L.; O'Connor, J.; Roberts, D.C.; Gaskin, K.J. Resting energy expenditure in children with phenylketonuria. *Am. J. Clin. Nutr.* **1995**, *62*, 797–801. [CrossRef]
40. Fisberg, R.M.; da Silva-Fernandes, M.E.; Schmidt, B.J.; Fisberg, M. Nutritional evaluation of children with phenylketonuria. *Sao Paulo Med. J.* **1999**, *117*, 185–191. [CrossRef]
41. Allen, J.R.; Baur, L.A.; Waters, D.L.; Humphries, I.R.; Allen, B.J.; Roberts, D.C.; Gaskin, K.J. Body protein in prepubertal children with phenylketonuria. *Eur. J. Clin. Nutr.* **1996**, *50*, 178–186.
42. Pena, M.J.; de Almeida, M.F.; van Dam, E.; Ahring, K.; Bélanger-Quintana, A.; Dokoupil, K.; Gokmen-Ozel, H.; Lammardo, A.M.; MacDonald, A.; Robert, M.; et al. Protein substitutes for phenylketonuria in Europe: Access and nutritional composition. *Eur. J. Clin. Nutr.* **2016**, *70*, 785–789. [CrossRef] [PubMed]
43. Yi, S.H.; Singh, R.H. Protein substitute for children and adults with phenylketonuria. *Cochrane Database Syst. Rev.* **2015**, *2015*, Cd004731. [CrossRef] [PubMed]

44. Ilgaz, F.; Marsaux, C.; Pinto, A.; Singh, R.; Rohde, C.; Karabulut, E.; Gökmen-Özel, H.; Kuhn, M.; MacDonald, A. Protein Substitute Requirements of Patients with Phenylketonuria on BH4 Treatment: A Systematic Review and Meta-Analysis. *Nutrients* **2021**, *13*, 1040. [CrossRef] [PubMed]
45. Aldámiz-Echevarría, L.; Bueno, M.A.; Couce, M.L.; Lage, S.; Dalmau, J.; Vitoria, I.; Andrade, F.; Llarena, M.; Blasco, J.; Alcalde, C.; et al. Tetrahydrobiopterin therapy vs phenylalanine-restricted diet: Impact on growth in PKU. *Mol. Genet. Metab.* **2013**, *109*, 331–338. [CrossRef] [PubMed]
46. Wilson, K.; Charmchi, P.; Dworetzky, B. *State Statutes & Regulations on Dietary Treatment of Disorders Identified Through Newborn Screening*; The Catalyst Center: Denver, CO, USA; Boston University School of Public Health: Boston, MA, USA; Ed. Catalyst Center: New Orleans, LA, USA, 2016. Available online: https://www.ciswh.org/wp-content/uploads/2016/11/State-Statues-and-Leg-re-Dietary-Treatment_Updated-7.24.17.pdf (accessed on 1 February 2021).
47. Belanger-Quintana, A.; Dokoupil, K.; Gokmen-Ozel, H.; Lammardo, A.M.; MacDonald, A.; Motzfeldt, K.; Nowacka, M.; Robert, M.; van Rijn, M.; Ahring, K. Diet in phenylketonuria: A snapshot of special dietary costs and reimbursement systems in 10 international centers. *Mol. Genet. Metab.* **2012**, *105*, 390–394. [CrossRef] [PubMed]
48. Robertson, L.V.; McStravick, N.; Ripley, S.; Weetch, E.; Donald, S.; Adam, S.; Micciche, A.; Boocock, S.; MacDonald, A. Body mass index in adult patients with diet-treated phenylketonuria. *J. Hum. Nutr. Diet.* **2013**, *26* (Suppl. 1), 1–6. [CrossRef]
49. Couce, M.L.; Vitoria, I.; Aldámiz-Echevarría, L.; Fernández-Marmiesse, A.; Roca, I.; Llarena, M.; Sánchez-Pintos, P.; Leis, R.; Hermida, A. Lipid profile status and other related factors in patients with Hyperphenylalaninaemia. *Orphanet J. Rare Dis.* **2016**, *11*, 123. [CrossRef]
50. Giżewska, M.; MacDonald, A.; Bélanger-Quintana, A.; Burlina, A.; Cleary, M.; Coşkun, T.; Feillet, F.; Muntau, A.C.; Trefz, F.K.; van Spronsen, F.J.; et al. Diagnostic and management practices for phenylketonuria in 19 countries of the South and Eastern European Region: Survey results. *Eur. J. Pediatr.* **2016**, *175*, 261–272. [CrossRef]
51. Rocha, J.C.; van Rijn, M.; van Dam, E.; Ahring, K.; Bélanger-Quintana, A.; Dokoupil, K.; Gokmen Ozel, H.; Lammardo, A.M.; Robert, M.; Heidenborg, C.; et al. Weight Management in Phenylketonuria: What Should Be Monitored. *Ann. Nutr. Metab.* **2016**, *68*, 60–65. [CrossRef]
52. Von Elm, E.; Altman, D.G.; Egger, M.; Pocock, S.J.; Gøtzsche, P.C.; Vandenbroucke, J.P. The Strengthening the Reporting of Observational Studies in Epidemiology (STROBE) statement: Guidelines for reporting observational studies. *J. Clin. Epidemiol.* **2008**, *61*, 344–349. [CrossRef]
53. Dekkers, O.M.; Vandenbroucke, J.P.; Cevallos, M.; Renehan, A.G.; Altman, D.G.; Egger, M. COSMOS-E: Guidance on conducting systematic reviews and meta-analyses of observational studies of etiology. *PLoS Med.* **2019**, *16*, e1002742. [CrossRef] [PubMed]

nutrients

MDPI

Article

Provision and Supervision of Food and Protein Substitute in School for Children with PKU: Parent Experiences

Hannah Jones [1], Alex Pinto [2], Sharon Evans [2], Suzanne Ford [3,4], Mike O'Driscoll [5], Sharon Buckley [6], Catherine Ashmore [2], Anne Daly [2] and Anita MacDonald [2,*]

1 Faculty of Health, Education & Life Sciences, Birmingham City University: City South Campus, Westbourne Road, Edgbaston, Birmingham B15 3TN, UK; Hannah.Jones12@mail.bcu.ac.uk
2 Birmingham Women's and Children's NHS Foundation Trust, Steelhouse Lane, Birmingham B4 6NH, UK; alex.pinto@nhs.net (A.P.); sharon.morris6@nhs.net (S.E.); catherine.ashmore@nhs.net (C.A.); a.daly3@nhs.net (A.D.)
3 National Society for Phenylketonuria, Sheffield S12 9ET, UK; suzanne.ford@nspku.org
4 North Bristol NHS Trust, Southmead Road, Bristol BS10 5NB, UK
5 School of Health and Education, Middlesex University, Room WG41A (Williams Building), The Burroughs Hendon, London NW4 4BT, UK; m.odriscoll@mdx.ac.uk
6 Department of Psychology, Faculty of Health, Psychology and Social Care, Manchester Campus, Manchester Metropolitan University, 53 Bonsall Street, Manchester M15 6GX, UK; sharon.j.buckley@stu.mmu.ac.uk
* Correspondence: anita.macdonald@nhs.net

Abstract: Children spend a substantial part of their childhood in school, so provision of dietary care and inclusion of children with phenylketonuria (PKU) in this setting is essential. There are no reports describing the dietary support children with PKU receive whilst at school. The aim of this cross-sectional study was to explore the experiences of the dietary management of children with PKU in schools across the UK. Data was collected using an online survey completed by parents/caregivers of children with PKU. Of 159 questionnaire responses, 92% (n = 146) of children attended state school, 6% (n = 10) private school and 2% (n = 3) other. Fourteen per cent (n = 21/154) were at nursery/preschool, 51% (n = 79/154) primary and 35% (n = 54/154) secondary school. Sixty-one per cent (n = 97/159) said their child did not have school meals, with some catering services refusing to provide suitable food and some parents distrusting the school meals service. Sixty-one per cent of children had an individual health care plan (IHCP) (n = 95/155). Children were commonly unsupervised at lunchtime (40%, n = 63/159), with snacks (46%, n = 71/155) and protein substitute (30%, n = 47/157), with significantly less supervision in secondary than primary school ($p < 0.001$). An IHCP was significantly associated with improved supervision of food and protein substitute administration ($p < 0.01$), and better communication between parents/caregivers and the school team ($p < 0.05$). Children commonly accessed non-permitted foods in school. Therefore, parents/caregivers described important issues concerning the school provision of low phenylalanine food and protein substitute. Every child should have an IHCP which details their dietary needs and how these will be met safely and discreetly. It is imperative that children with PKU are supported in school.

Keywords: PKU; food; protein substitute; school; IHCP; parent/caregiver experiences

Citation: Jones, H.; Pinto, A.; Evans, S.; Ford, S.; O'Driscoll, M.; Buckley, S.; Ashmore, C.; Daly, A.; MacDonald, A. Provision and Supervision of Food and Protein Substitute in School for Children with PKU: Parent Experiences. *Nutrients* **2021**, *13*, 3863. https://doi.org/10.3390/nu13113863

Academic Editor: Shanon L. Casperson

Received: 8 September 2021
Accepted: 28 October 2021
Published: 28 October 2021

Publisher's Note: MDPI stays neutral with regard to jurisdictional claims in published maps and institutional affiliations.

Copyright: © 2021 by the authors. Licensee MDPI, Basel, Switzerland. This article is an open access article distributed under the terms and conditions of the Creative Commons Attribution (CC BY) license (https://creativecommons.org/licenses/by/4.0/).

1. Introduction

In the UK, it is estimated there are approximately 800 children with phenylketonuria (PKU) aged 5 to 16 years [1]; they are expected to attain normal educational achievement and attend mainstream school. Children with classical PKU are treated with a phenylalanine restricted diet only; if they have mild PKU they may be treated with an adjunct therapy, sapropterin. Children with classical PKU usually tolerate < 80% of usual natural protein intake and treatment includes: avoidance of high protein foods, strict measurement and limited intake of moderate protein containing foods, inclusion of special low protein foods (SLPF's) and supplementation with a low phenylalanine protein substitute [2]. Most children will be expected to eat at least one meal and take one dose of protein substitute at school. It is essential that there is safe provision and supervision of dietary treatment with appropriate adjustments that integrates the medical needs of a child with PKU into school life.

Section 100 of the UK Children and Families Act 2014, updated in 2015, states that schools in the UK have a duty to support pupils with medical conditions [3,4]. This act mandates that children with PKU are properly supported, enabling them to have a full and active role in school, remain healthy and achieve their academic potential. It states that school leaders should consult health and social care professionals, pupils, and parents so that the needs of children with medical conditions are accurately understood and effectively met. Schools have a duty to ensure that all relevant staff are trained to provide the support that pupils' need, and that policies, plans, procedures, and systems are implemented. Although not mandatory, each school should have policies to ensure all relevant staff are aware of the child's condition; that there are cover arrangements in case of staff absences or staff turnover, and that risk assessments are conducted for school visits, holidays, and other activities outside the normal timetable. Failure to make reasonable adjustment for a child with a disability is considered discrimination under the UK Equality Act 2010 [5].

Ideally each child with PKU should have an individual health care plan (IHCP) although these are not obligatory by law [4]. These should be developed in partnership between the school, parents, pupils, and relevant healthcare professionals who can advise on individual medical care needs. An IHCP should ensure that schools know how to support children with PKU effectively by providing clarity about what needs to be done, when and by whom. They should be reviewed at least annually or earlier if health care needs change. School governing bodies should ensure that their schools have policies and appoint staff who are responsible for managing IHCP's.

In addition, in UK state-funded schools, every child in reception, year 1 and 2 (children aged 4–7 years) are entitled to a free school lunch [6]. They should have access to a healthy, balanced diet and it is recommended that they have at least one hot meal provided every day. Food and drinks provided by school must comply with certain nutritional standards [7] and reasonable adjustment should be made for children on special diets. The Education Act 1996 requires maintained schools and academies to provide free school meals to disadvantaged pupils aged between 5 to 16 years, with 20.8% of children in England (2020/2021) being entitled to this service [8].

Dietary treatment is expected to have both a physiological and psychological impact on the lives of young people with PKU in school. Whilst consumption of non-permitted foods and poor adherence to protein substitute will lead to elevated blood phenylalanine and neurological dysfunction, teacher/peer insensitivity and exclusion may have an enduring impact on a child's mental health, and attitude and acceptance of PKU. There are no studies examining care provision in school and the opinions and experiences of parents of school children with PKU are unknown. The aim of this study was to explore the views and experiences of parents/caregivers of children with PKU in school and nursery. Additionally, the care of children with and without an IHCP was also studied.

2. Materials and Methods

2.1. Study Design

This was a cross-sectional study using an online survey that collected both qualitative and quantitative data from UK parents of children aged 3 to 16 y with PKU attending school or nursery. Non-UK respondents were excluded.

The questionnaire was built in the Online Surveys platform (https://www.onlinesurveys.ac.uk, accessed on 28 October 2021) to gather quantitative data. This was placed on the UK National Society for Phenylketonuria (NSPKU) website, with additional promotion on the NSPKU Twitter, Instagram and Facebook. The survey was open for five months, from 20 March until 20 August 2020.

2.2. Questionnaire

The non-validated questionnaire contained 22 questions: $n = 17$ multiple choice (with $n = 14$ inviting additional comments), $n = 3$ multiple responses, $n = 1$ Likert scale and $n = 1$ open ended questions (Supplementary Material).

The questionnaire was developed collaboratively by dietitians with expert practical and scientific knowledge of PKU (AP, SE, AM), a colleague from the NSPKU (SF), a researcher (MO) and a student dietitian from Birmingham City University (HJ). It was reviewed amongst colleagues and lay people to ensure its readability and then amended according to feedback.

2.3. Data Collected

The questionnaire was divided into four sections. Information collected included: the age of the child, type of school, school year group, the availability of an IHCP, administration of protein substitute in school, provision and acceptance of lunches provided by school catering services, information about the suitability of school lunches, school staff training and supervision of food and protein substitute. All data that was collected was based on the parents own perception or knowledge about the quality of the care and support provided by the nursery or school.

2.4. Statistics

Quantitative data analysis (inferential and descriptive statistics) was carried out with the Statistical Package for the Social Sciences (SPSS) version 25 (SPSS Inc., Chicago, IL, USA). Multiple response questions were analysed with descriptive statistics only. Statistical significance was set at $p < 0.05$.

Qualitative data analyses of 14 open-ended responses were carried out in NVIVO v 12 PRO. The whole survey dataset was imported into NVIVO, so that coding of open-ended responses could be broken down by attributes of survey questions. All open-ended question responses were analysed thematically.

2.5. Ethics

Ethical approval was obtained from the Birmingham City University ethics committee prior to commencement of the study (Jones/5042/R(A)/2020/Mar/HELS FAEC - Provision of school food for children with PKU: A parent's perspective. Approved 19/3/2020). At the beginning of the online questionnaire, respondents gave consent, and it was emphasized that questionnaire completion was voluntary. Potential respondents were advised that data from the survey may be published in an anonymized form. If names of schools or hospitals were mentioned in verbatim abstracts these were removed from results presented in this manuscript.

3. Results

There were 159 responses. The number of respondents who answered each question was variable (as not all questions were applicable to each respondent). All respondents were parents/caregivers of children with PKU. A description of the school type, school age group and provision of IHCP for children is given in Table 1.

Table 1. School type, age group and provision of IHCP.

School type	%	Number of children/total number of responses
State school	92	146/159

Table 1. *Cont.*

School type	%	Number of children/total number of responses
Private school	6	10/159
Other (e.g., special needs school)	2	3/159
Year group in school	**%**	**Number of children/total number of responses**
Nursery/reception	14	21/154
Primary school Years 1–3 Years 4–6	51 (27) (24)	79/154 (42) (37)
Secondary school Years 7–9 Years 10–11	35 (18) (18)	54/154 (27) (27)
Provision of Individual Health Care Plan	**%**	**Number of children/total number of responses**
Yes	60	95/159
No	33	53/159
Don't know	7	11/159

When considering the provision of written IHCP's, there was no difference between state or private school or between school year groups (Pearson Chi-Square test, $p > 0.5$).

3.1. Uptake of School Meals

Uptake of school lunches and entitlement to free school meals is given in Table 2. Most parents/caregivers (61%, $n = 96/157$) said their children were not eating meals provided by the school catering service.

Table 2. Uptake of school lunches and entitlement to free school lunches.

Numbers of times school lunch is eaten each week prepared by the school	%	Number of children
0	61	96
1	6	10
2–3	7	11
4–5	26	40
Total	100	157
Entitled to free school lunch	**%**	**Number of children**
Yes **Nursery/Reception** **Primary School** **Secondary School**	35 (71) (40) (19)	56 (15) (31) (10)
No **Nursery/Reception** **Primary School** **Secondary School**	64 (29) (60) (81)	96 (6) (47) (43)
Don't know	1	2
Total	100	154

Sixty-two per cent ($n = 73/117$) of parents/caregivers said that they would like their child to have school lunches more often. Only 52% ($n = 29/56$) utilized their free school lunch entitlement. Of those with free school meal entitlement, 41% were eating school lunches 4–5 times a week compared to 16% of those without the entitlement (Pearson Chi-Square test $p = 0.05$). Of the children eating school lunches, 76% ($n = 48/63$) of parents were satisfied with the school lunch service.

Respondents were asked in two open-ended questions, about barriers to accessing school meals more frequently. The main themes which emerged were: school refusing to cater for children with PKU, limited food choice offered by school, child or parent preferring packed lunch, parents did not trust school to prepare appropriate food for their child with PKU, parents were more in control of what their child eats with packed lunches, and children refuse school meals because they openly advertise that they are different. Some parents described how the school or school catering were unwilling or reluctant to cater for children with PKU, particularly in secondary school. They described the inflexibility of catering services, how some parents had to supplement the school lunch with food prepared at home, and exclusion from special occasion meals such as Christmas dinner.

Parents/caregivers verbatim quotes:

- *"The school use an outside catering company who were not prepared to cook any food that was not sourced by them."*
- *"School refused to provide school lunches due to health and safety."*
- *"Not comfortable with someone else having control of portions in case they aren't weighed properly, or wrong foods given by mistake."*
- *"Tried school lunches. Blood phenylalanine levels went too high. Child was not supervised."*

3.2. Food Included in School Lunch Service

The type of school meal plans and variety of low protein foods given are outlined in Table 3.

Table 3. Meal provision within school and type of special low protein foods used.

School Meal Plans	% (Number of patients/total number of responses)
Food chosen from standard school menu	32% ($n = 20/63$) *
Separate low protein meal prepared	51% ($n = 32/63$)
Meals provided by parents/caregivers or standard school menu adapted to make it suitable for children with PKU	17% ($n = 11/63$)
Common low protein foods substituted used when menus were adapted	
Low protein pasta (52%, $n = 33/63$) Low protein pizza (48%, $n = 30/63$) Vegan or 'free from' low protein cheese (46%, $n = 29/63$) Low protein bread (46%, $n = 29/63$) Low protein 'meat'/'fish' substitutes (40%, $n = 25/63$) 'Fishless' fingers (17%, $n = 11/63$)	

'free from': food without one or more specific ingredients, designed for people with food allergies or other intolerances/diseases). * 40% ($n = 8/20$) of children that had food chosen from standard school menu were taking sapropterin and were permitted a higher protein intake.

Parents usually supplied the SLPF's such as pasta and bread which they obtained on prescription; the school usually provided low protein/vegan cheese and 'fishless' fingers purchased from wholesalers. Some parents said the school 'do not provide anything.' Children with an IHCP (68%, $n = 25/37$) were much more likely than those without IHCP (50%, $n = 7/14$) to have alternative meals prepared but the difference was not statistically significant (Pearson Chi Square test $p > 0.05$). Children in private school were more likely to have a separate meal prepared (100%, $n = 5/5$) compared with 58% ($n = 26/45$) of state schools, but the difference did not reach statistical difference due to the small numbers of children in private school. There were no clear differences related to the school year of the child.

Fifty-nine percent (n = 37/63) said catering staff measured or weighed protein exchange foods (e.g., mashed potato or peas) and 2% (n = 1/63) were unaware if foods were measured. Some parents commented that it was unnecessary for the school to weigh protein exchanges because they either provided the food pre-measured, the main meal did not contain protein exchanges, or they did not ask the school catering to weigh exchange foods.

Weighing and measuring of food protein exchanges was most common (80%, n = 12/15) in nursery/reception school compared to other school age groups (57%, n = 24/42) [Pearson Chi-Square test, p = 0.014]. Parents/caregivers were asked to score satisfaction with the school meal service on a scale of 1 (extremely dissatisfied) to 5 (extremely satisfied). They gave a higher satisfaction score (median 5) when the school measured/weighed protein exchanges compared with scoring for schools who did not weigh/measure protein exchanges (median 4) (Mann–Whitney U test, p = 0.003)

There were some parent comments about the quality, variety and presentation of food provided by the school catering service.

Parents/caregivers verbatim quotes:

- *"The dinners came from another school and the presentation when they arrived was not that appetising."*
- *"Would like a wider choice of salads being provided and more attractive fruit at lunches."*
- *"Some of the protein exchanges were noted wrongly and also weighed out incorrectly."*

3.3. *Training and Knowledge about PKU and Diet*

Parents/caregivers said that only 47% (n = 74/159) of their child's class teachers and 54% (n = 33/61) of catering staff (for those receiving school meals) had received PKU training from a health professional. Of the teachers and catering team who had received training, 82% (n = 58/71) of teachers and 85% (n = 35/41) of the catering team received training in the previous 2 years. The training was mainly delivered by the child's dietitian.

3.4. *Supervision of Food in School*

Children were commonly unsupervised at lunchtime (43%, n = 66/154) or snack time (48%, n = 74/155). Lack of meal supervision was significantly more common in secondary schools (61%, n = 33/54) than in primary schools (27%, n = 21/79) (Pearson Chi-Squared test $p < 0.001$).

Those without an IHCP (40%, n = 59/148) were more commonly unsupervised at school at meal and snack time (60%, n = 32/53) compared to those who had a plan (28%, n = 27/95) (Pearson Chi-Square test $p < 0.01$). Of the children supervised at lunchtime, school lunchtime supervisors most commonly did this task (27%, n = 24/88), whereas snacks were mainly supervised by teaching assistants (30%, n = 24/81).

3.5. *Feedback about Food Eaten in School*

Only 36% (n = 57/157) of parents/caregivers said they received feedback about what their children eat in school. Feedback was more common for children with an IHCP in a state school compared with children without one (Pearson Chi-square test $p < 0.05$); and more common for children in nursery/reception and primary school (year 1 to 3) (64%, n = 27/42) than in secondary school (15% n = 8/53) (Pearson Chi-square test $p < 0.001$). It was marginally more common in private school (40%, n = 4/10) compared to state school (35%, 51/144) [Pearson Chi-square test $p > 0.05$].

When feedback was received, 56% (n = 32/57) of parents/caregivers received a written record of food eaten, 25% (n = 14/57) verbal feedback and 11% (n = 6/57) photographs of food eaten via online systems. Nine per cent (n = 5/57) received feedback in 'other' forms such as: lunch wrappers and uneaten food being left in the bag (as evidence of what has been eaten), the online system for monitoring school meal purchases, messages in a schoolbook/homework book, and an email or telephone call from the school.

3.6. Incidents of Eating Foods at School That Were Not Permitted

Parents reported 53 incidents of incorrect foods being given accidentally/purposely to children in school in the previous 6 months. Forty per cent (n = 21/53) of parents/caregivers said that it had happened once; 19% (n = 10/53) said 2 to 3 times, 8% (n = 4/53) said 4 to 5 times and 34% (n = 18/53) said that it had happened more than five times. Respondents were asked to describe incidents of their child eating non permitted food at school, and these responses (n = 39) were thematically analysed. The main themes describing incidents were associated with staff errors (n = 4), other children sharing inappropriate foods (n = 11), child choosing inappropriate foods (n = 5) and trying to fit in with others (n = 4). Two parents mentioned that they felt it was much harder for the school to supervise the child's eating once they were in secondary school.

Parents/caregivers verbatim quotes:

- *"He was given an incorrect lunch when the school cook was on holiday."*
- *"She asked her friend to buy her foods like toast and chocolate from the tuck bar each morning."*
- *"I've saw on 'parent pay' that he purchased baked goods such as flapjacks and cakes."*
- *"Because she felt left out so she was going into the canteen on chip day and buying double her amount."*

Secondary school children were much more likely to have eaten foods which were not permitted as part of a low phenylalanine diet (45% (n = 10/22) of secondary school children (year 10 to 11) compared with 26% (n = 9/35) of primary school children (Year 1 to 3) but the differences were not statistically significant (Pearson Chi-square test $p > 0.05$).

Two-thirds (66%, n = 35/53) of parents/caregivers said that they did not feel adequately informed about food incidents. Parents/caregivers were much more likely to say that they felt adequately informed of the incident if children were in nursery/reception (60%, n = 3/5) and primary school (years 4-6) (58%, n = 7/12) [Pearson Chi-square test $p > 0.05$]. Respondents were asked (open-ended question) to comment about the communication they received from the school staff about food incidents. The main common themes from the 25 responses were: informed by child (n = 7), staff were slow or late in informing us (n = 4), should be greater staff understanding or awareness (n = 4), and staff don't care (n = 3).

Parents/caregivers verbatim quotes:

- *"Well, they were not sure what she really ate. My daughter told me what she ate and at the end they confirmed this."*
- *"I was not informed. Being in a secondary school the PKU diet is hard to monitor for all staff and they are not able to monitor my son's actions."*
- *"The teachers don't understand the condition so she is left to get on with it."*

3.7. School Strategies to Prevent Children Being Given the Incorrect Foods at School

The parents of nursery/reception and primary school (years 1–3) children were much more likely to state that there were strategies in place to prevent incorrect food being eaten at school compared with older children with PKU (Pearson Chi-Square test $p < 0.001$). Thirty-eight (n = 60/158) of respondents said there were no procedures in place to prevent such incidents reoccurring. However, parents gave many examples of strategies used by the school staff to try and ensure children were given the correct food Table 4.

Table 4. All strategies suggested by parents/caregivers to prevent incorrect foods being eaten by children with PKU in school.

Supervision at mealtime
• Wears lanyard at lunch time so he is recognizable. Other children on special diets also do this so he is not the only one. • Poster with his name, picture and instructions on for everyone to see. • Not allowed to self-choose food from canteen. • Teaching assistant watches her, and she is served based on what we put on her lanyard that she can eat each day. • The school have a lunch system where each child's name is typed into a ticket system which then says which lunch they have based on the parents ordering. • He has his own dinner lady on his table that sits with him. • No one is allowed to share their lunch.
Communication/education with school staff
• The teacher talks to me before any occasions or food related activity. • Teachers know to ring parents to organize if they are doing cookery lessons so products can be provided. • They check with me before letting her have anything. • They are all very aware and my child has very good awareness himself. • Talk to school cook every morning. • Have a review meeting every year with the teacher to explain about treatment needs. • Regular staff training. • Regular update of health care plans. • Care plan and pack given by dietitian provide school with information. • My child is not allowed to take money to school so she cannot buy food from the tuck shop. • Child takes packed lunch. Can only eat from lunch box. Teachers sit at his lunch table. • We have a hand over book, if anything off limits was eaten it would be recorded. The teachers and kitchen staff also have the NSPKU booklet, so they know what is allowed and what isn't. I help the chef with the menus and he runs any new ideas by me. • School sends a photo and written comments (and sometimes actual food) to show what has been eaten in a communication book. Breakfast club and after school club use the communication book too.
Communication with previous school/nursery
• School visited the nursery and saw the systems that they had in place there and all the measures that they took which I think helped them visualise them in real terms.

3.8. Exclusion: Feeling and Looking Different in School

Thematic analysis of general comments received about provision of food in school showed that parents/caregivers were concerned that their child was either excluded from activities/school events because of PKU or that they looked different from others in school.

Parents/caregivers verbatim quotes:

- *"My teenage son does not want attention brought to his PKU. Refuses to have special food at school or anyone know about his PKU."*
- *"My child does not want to stay for lunch as she only likes to eat chips and the school would have to measure them out. This would lead to others asking lots of questions which she does not want."*
- *"One day they gave everyone a hot chocolate, but they just gave water to my child."*

3.9. Support with Special Diet by the School

Many parents/caregivers (n = 29) positively described the support they received from the school. However, some outlined the amount of work and liaison they have to do with the school team to receive a better service for their children.

Parents/caregivers verbatim quotes:

- *"I have been extremely lucky with the support we have for my son at school. They will do everything they can to ensure my son is as included as we would like him to be. They have*

gained a lot of knowledge and continue to check in and ask questions or change their 'usual' foods where needed."
- *"When my daughter has been on residential holidays with the school the staff have been excellent arranging catering with staff wherever they have stayed (France and UK)."*

3.10. Negative Comments about School Care for PKU Children

Thematic analysis indicated a further 34 negative experiences with school and management of PKU by respondents.

Parents/caregivers verbatim quotes:

- *"It took a long time to get an initial meeting and then there was a lot of work over a 3 month period to get everything sorted. There was lots of obstacles and a lot of work and organization at school."*
- *"Have had to ask for more appointments to see SENCO teacher to discuss issues. She takes very little action."*

3.11. Secondary School Provision

Parents/caregivers gave 10 comments about the issues for children in secondary school. They described the fear children experience and how they do not want to look different from their peers and the difficulties they experience.

Parents/caregivers verbatim quotes:

- *"In a secondary school it is harder to control your child's diet. You have to try and trust they will do the right thing. You can make a fuss but the children resent you for this."*
- *"In a secondary school there is no supervision."*
- *"Although teacher received training it was one teacher out of many- so really not relevant."*

3.12. Administration of Protein Substitute in School

Protein substitute administration was more commonly unsupervised in children in secondary (77%, $n = 34/44$) than primary school (17%, $n = 11/66$) (Pearson Chi-Square Test $p = 0.001$). Those who did not have an IHCP (57%, $n = 25/44$) were less likely to be supervised compared to those who did have a plan (24%, $n = 18/75$) (Pearson Chi-Square test $p = 0.001$). Any supervision was mostly provided by teaching assistants.

Some parents commented that the school had helped with the transition of protein substitute from a paste to a liquid, others described the measures that the school staff took to ensure that a child took the protein substitute. Some described how they chose not to give protein substitute at school because it was unsupervised and consequently not taken. Others explained there that there was less supervision in secondary school, with one respondent describing a medical room being locked so their child could not gain access to their supply of protein substitute.

Parents/caregivers verbatim quotes:

- *"The school have helped my child with the transition of protein substitute from a paste to a liquid."*
- *"School returns the empty protein substitute pouch each day to evidence that it has all been taken."*
- *"The protein substitute is well supervised by teaching assistants. The dietitian and we as parents have spent a lot of time on this."*
- *"She was telling her teacher that she had drank it when she had not. The teacher just accepted the information from the child. Blood levels went high."*
- *"We decided to not give my son his substitute at school as this was getting missed."*

4. Discussion

This is the first study to explore the views and experiences of parents and caregivers of children with PKU in school and nursery. Additionally, the care of children with and without an IHCP were studied. The responses to this questionnaire represent approximately

20% of school-aged children with PKU in the UK [9]. The experiences of parents/caregivers in relation to schools were highly variable ranging from excellent support, to care that was unsafe, potentially adversely impacting metabolic control of children with PKU. Findings from this questionnaire suggest that pre-admission school planning, health professional training of school team members, and a carefully written IHCP that is reviewed at least annually are all essential components of successful PKU management within schools.

Although every child has the right to a varied and nutritious menu in school, uptake of school meals by parents/caregivers of children with PKU was considerably lower than the general population. Only 39% of children with PKU compared with 58% to 79% of UK school aged children received school meals [10]; and 50% of parents did not utilize their child's entitlement to free school lunches. Some parents/caregivers preferred to give their children packed lunches because of safety concerns, so they could maintain control over their child's food. Others reported that this allowed their child to retain some anonymity about the condition because a low phenylalanine packed lunch looked like a regular packed lunch. Consequently, this situation further penalizes families with PKU by increasing their workload and expenditure on food when they are already managing a stringent and costly dietary treatment.

Parents reported numerous barriers to school meals provided by school catering services including poor food quality, inadequate variety, requirement for extra parental organization and liaison, and operational systems in meal delivery (children having to ask for their special meal, wearing lanyards, child photographs) that brought unwelcome attention to the child. When external catering services provided school lunches, greater difficulty with food provision was reported. They appeared 'rigid' in their approach using allergy concerns with risks of cross-contamination as reasons for not providing school meals, and refusal to use SLPF's supplied via parents for children with PKU, despite being unprepared to purchase SLPF's themselves due to the extra cost and their own operating procedures. This refusal and failure to provide appropriate low phenylalanine school meals is discriminatory [4]. To help children with PKU who are entitled to free school lunches but unable to utilize them, the government should consider issuing money vouchers to assist with extra food costs.

Around 60% of children with PKU had a written IHCP but it is unknown how this compares with use of IHCPs in other chronic health conditions. There is some data that predates the 2014 education act to suggest that only 50% of children with conditions such as diabetes, epilepsy and asthma had an IHCP [11]. Although IHCP's are not mandatory, they helped improve care provision for children with PKU at school. Children with PKU with an IHCP were more likely to have protein substitute administration supervised, have alternative suitable low phenylalanine meals prepared, receive supervision at snack and school lunch time and receive feedback from the school staff. It was also evident that some parents worked very hard with schools, particularly at school entry to establish good care for their children. Some described setbacks, but clear management strategies with regular review of the IHCP plan helped.

IHCP's should include information about PKU and treatment, including protein substitute (dose, time, administration, storage), snack and meal choices, protein exchanges, and the level of support needed (some secondary school children may be able to take responsibility for their own health needs). It is mandatory that schools ensure that written records are kept of all protein substitute that is administered. If a child is self-managing their protein substitute and low phenylalanine diet within secondary school, this should be clearly stated, with appropriate arrangements for monitoring, documenting who will provide any additional support, and their training needs. There should be a clear pathway with named personnel about how and from whom they can obtain help if issues arise at school. All arrangements should generate confidence for parents and pupils. The Department of Health has also produced IHCP templates which healthcare professionals and schools may find useful [3]. PKU specific templates are also available online from the UK National Society of Phenylketonuria [12].

Inadequate staff training and lack of supervision with food was commonly described by parents/caregivers and carried a considerable safety risk for children with PKU. There were several descriptions of children eating or being offered the wrong foods either accidentally or purposely due to inadequate supervision. Better training is needed to enable staff to fully support children at school and this should include all school staff who provide care for children with PKU. Teaching assistants often have an important role in supervising protein substitutes and snacks but are commonly omitted from professional training sessions. Lunch time supervisors are also overlooked for training, but they are central to ensuring that children receive the correct food at mealtimes. Although the parents of a child will often be key in providing relevant information to school staff, training should be provided by a health professional. In addition, availability of online training resources developed by health professionals will help improve the school team's basic knowledge of PKU. In conditions such as diabetes, it is reported that attitudes of teachers and their lack of understanding impact on their ability to manage the condition [13].

Parents/caregivers described some of the school strategies used that led to better management of PKU. Some schools had helped with the transition from a spoonable/paste to a liquid protein substitute. At lunch time, if children were allowed to have a friend queue and visit food counters with them it was considered more discreet and enabled children to feel less special and more supported. Teachers or teaching assistants sitting in the dining room or at the table with the children helped check the correct foods were consumed. Photographing meals pre and post consumption helped parents understand what foods had been offered and eaten by children. Cashless payment systems in secondary schools enabled parents to go online to see what foods their children had purchased. Procedures to cover any transitional arrangements between primary and secondary schools (or nursery and primary school), were also highlighted as important.

Parents/caregivers commonly described their concerns about social exclusion. Children may be unintentionally excluded because of inadequate inclusive opportunities with suitable food provision. Social exclusion frequently causes psychological harm and can have negative outcomes on emotional and mental health, lowering self-esteem, increasing feelings of anxiety, depression and aggression and may even have a detrimental impact on academic performance [14]. Generally, older children with chronic health conditions are almost three times as likely as healthy peers to suffer social exclusion in school [15], as they are seen as different from their peers [14]. This has previously been reported in PKU [16].

The transition into secondary school is naturally associated with greater independence amongst adolescents. Parents reported difficulties with managing a low phenylalanine diet once their child entered secondary school and it was commonly associated with deteriorating blood phenylalanine control [17,18]. Children were self-conscious about their condition and were fearful about mistreatment by peers if their disability became known; dietary management was effectively sacrificed to avoid bullying and harassment by other pupils in school. They commonly avoided any special food that appeared different from regular foods and refused protein substitute administration at school. There was also limited staff training in secondary school, so less teacher empathy and support for the child with PKU. Commonly the position of secondary schools is that children with disability should develop independency with their care needs, but there is a high measure of responsibility on a child as they enter their journey through secondary school. It is important that schools, parents, and school governors work together to help ensure that the secondary school culture is supportive and inclusive and that it encourages acceptance of children with a range of differences. A lack of sensitivity toward people with disabilities is a problem that requires attitude change and training. The impact of children attending secondary school and its association with declining blood phenylalanine control warrants further investigation.

Limitations

There are several limitations to this study. This questionnaire was not validated. Data was not collected about individual protein tolerance or about all food provided by school within the day such as breakfast clubs, after school clubs, tuck shops and celebrations in order to ensure that the questionnaire was not too burdensome to complete. The questionnaires were completed at the start of the Covid 19 pandemic, but respondents were asked to document their usual experience at school. Each questionnaire collected information about one child with PKU in a family; it did not refer/collect information about other children in the family with or without PKU. Data was collected based on parents/caregivers' perception of the service or school incidents, so some answers maybe subjective. The respondents were not randomly selected, and participation was voluntary. Additionally, individuals without internet access may have been unable to participate. The survey was promoted on the NSPKU Twitter and Facebook page, meaning participants were more likely to be NSPKU members who may be more proactive and informed about PKU. Therefore, the survey population may not be representative of the entire PKU population although it is estimated that this questionnaire covers around 20% of the children in school with PKU in the UK.

5. Conclusions

There was disparity in the support given to children with PKU across the UK. They received school meals less commonly than their peers, even when they were entitled to 'free school meals.' Some catering services discriminated against children with PKU by refusing to provide suitable food; some parents distrusted the school meals service. Children were commonly unsupervised with food, leading to the consumption of inappropriate foods. Improved supervision and communication were associated with a written IHCP. We recommend that every child with PKU should have an IHCP, with mandatory training of all staff involved in their care. It is imperative that every child with PKU is supported in school, and their individual dietary and health needs are met safely and discreetly.

Supplementary Materials: The following is available online at https://www.mdpi.com/article/10.3390/nu13113863/s1, Full questionnaire.

Author Contributions: Conceptualization, A.M., A.P. and H.J.; methodology, A.M., A.P., S.E., S.F., M.O. and H.J.; formal analysis, M.O.; writing—original draft preparation, A.M.; writing—review and editing, H.J., A.P., S.E., S.F., M.O., S.B., C.A., A.D. and A.M.; supervision, A.M. and A.P. All authors have read and agreed to the published version of the manuscript.

Funding: This research received no external funding.

Institutional Review Board Statement: The study was conducted according to the guidelines of the Declaration of Helsinki and approved by the Birmingham City University ethics committee prior to commencement of the study (Jones/5042/R(A)/2020/Mar/HELS FAEC—Provision of school food for children with PKU: A parent's perspective. Approved 19/3/2020).

Informed Consent Statement: Informed consent was given by all subjects when filling in the questionnaire.

Data Availability Statement: The data will be made available from the authors upon reasonable request.

Acknowledgments: We would like to acknowledge and thank all the patients and families that have taken their time to fill in this survey.

Conflicts of Interest: The authors declare no conflict of interest.

References

1. Newborn Blood Spot Screening Programme in the UK. Data collection and performance analysis report 2016 to 2017. Public Health England leads the NHS Screening Programmes. Available online: https://assets.publishing.service.gov.uk/government/uploads/system/uploads/attachment_data/file/709367/Newborn_blood_spot_screening_data_collection_and_performance_analysis_report_2016_to_2017.pdf (accessed on 10 August 2021).

2. MacDonald, A.; Van Wegberg, A.M.J.; Ahring, K.; Beblo, S.; Bélanger-Quintana, A.; Burlina, A.; Campistol, J.; Coşkun, T.; Feillet, F.; Giżewska, M.; et al. PKU dietary handbook to accompany PKU guidelines. *Orphanet J. Rare Dis.* **2020**, *15*, 1–21. [CrossRef]
3. Department of Education (2015) Supporting Pupils at School with Medical Conditions: Statutory Guidance for Governing Bod-Ies of Maintained Schools and Proprietors of Academies in England. Available online: https://assets.publishing.service.gov.uk/government/uploads/system/uploads/attachment_data/file/803956/supporting-pupils-at-school-with-medical-conditions.pdf (accessed on 10 August 2021).
4. Children and Families Act 2014 (c 6) Part 5 Pupils with Medical Conditions. Section 100. Available online: http://www.legislation.gov.uk/ukpga/2014/6/section/100/enacted (accessed on 10 August 2021).
5. Equality Act 2010. Available online: https://assets.publishing.service.gov.uk/government/uploads/system/uploads/attachment_data/file/570382/Equality_Act_2010-disability_definition.pdf (accessed on 10 August 2021).
6. Free school Meal Entitlement and Child Poverty in England. Available online: https://assets.publishing.service.gov.uk/government/uploads/system/uploads/attachment_data/file/266587/free-school-meals-and-poverty.pdf (accessed on 10 August 2021).
7. School Standards and Framework Act 1998. Available online: https://www.legislation.gov.uk/ukpga/1998/31/contents (accessed on 10 August 2021).
8. Gov.UK. Academic Year 2020/21. Schools, Pupils and Their Characteristics. Available online: https://explore-education-statistics.service.gov.uk/find-statistics/school-pupils-and-their-characteristics (accessed on 14 August 2021).
9. Georgina Wood, A.P.; Evans, S.; Adams, S.; Costelloe, S.; Eardley, J.; Ellerton, C.; Emm, A.; Firman, S.; Ford, S.; French, M.; et al. Special low protein foods pre-scribed in England for PKU patients: An analysis of prescribing patterns and cost. 2021; Unpublished paper.
10. LACA—The School Food People. School Meal Uptake Research November 2019. Independent Research Conducted by Dewberry Redpoint Ltd June 2019–September 2019. Available online: https://laca.co.uk/sites/default/files/attachment/news/SMU%20Research%20Report%202019.pdf (accessed on 20 August 2021).
11. Wong, I.C.K.; Awolowo, T.; Gordon, K.; Mo, Y.W. Survey of administration of medicines to pupils in primary schools within the London area. *Arch. Dis. Child.* **2004**, *89*, 998–1001. [CrossRef] [PubMed]
12. National Society of Phenylketonuria. Health Care Plan for a Child with PKU at School. Available online: https://www.nspku.org/download/nspku-school-health-care-plan/ (accessed on 24 August 2021).
13. Carroll, A.E.; Marrero, D.G. The Role of Significant Others in Adolescent Diabetes: A Qualitative Study. *Diabetes Educ.* **2006**, *32*, 243–252. [CrossRef] [PubMed]
14. Mulvey, K.L.; Boswell, C.; Zheng, J. Causes and Consequences of Social Exclusion and Peer Rejection Among Children and Adolescents. *Rep. Emot. Behav. Disord. Youth* **2017**, *17*, 71–75.
15. Lindsay, S.; McPherson, A.C. Experiences of social exclusion and bullying at school among children and youth with cerebral palsy. *Disabil. Rehabilitation* **2011**, *34*, 101–109. [CrossRef] [PubMed]
16. Ford, S.; O'Driscoll, M.; MacDonald, A. Living with Phenylketonuria: Lessons from the PKU community. *Mol. Genet. Metab. Rep.* **2018**, *17*, 57–63. [CrossRef] [PubMed]
17. Walter, J.; White, F.J. Blood phenylalanine control in adolescents with phenylketonuria. *Int. J. Adolesc. Med. Heal.* **2004**, *16*, 41–46. [CrossRef] [PubMed]
18. Walter, J.; White, F.; Hall, S.; MacDonald, A.; Rylance, G.; Boneh, A.; Francis, D.; Shortland, G.; Schmidt, M.; Vail, A. How practical are recommendations for dietary control in phenylketonuria? *Lancet* **2002**, *360*, 55–57. [CrossRef]

nutrients

MDPI

Article

Special Low Protein Foods Prescribed in England for PKU Patients: An Analysis of Prescribing Patterns and Cost

Georgina Wood [1,*], Alex Pinto [2], Sharon Evans [2], Anne Daly [2], Sandra Adams [3], Susie Costelloe [4], Joanna Gribben [5], Charlotte Ellerton [6], Anita Emm [7], Sarah Firman [5], Suzanne Ford [8], Moira French [9], Lisa Gaff [10], Emily Giuliano [11], Melanie Hill [12], Inderdip Hunjan [13], Camille Newby [14], Allison Mackenzie [15], Rachel Pereira [16], Celine Prescott [10], Louise Robertson [17], Heidi Seabert [18], Rachel Skeath [19], Simon Tapley [20], Allyson Terry [21], Alison Tooke [22], Karen van Wyk [23], Fiona J. White [23], Lucy White [24], Alison Woodall [25], Júlio César Rocha [26,27,28] and Anita MacDonald [2]

1 Faculty of Health, Education & Life Sciences, Birmingham City University, City South Campus, Westbourne Road, Edgbaston, Birmingham B15 3TN, UK
2 Dietetic Department, Birmingham Women's and Children's Hospital, Steelhouse Lane, Birmingham B4 6NH, UK; alex.pinto@nhs.net (A.P.); evanss21@me.com (S.E.); a.daly3@nhs.net (A.D.); anita.macdonald@nhs.net (A.M.)
3 Royal Victoria Infirmary, Queen Victoria Road, Newcastle upon Tyne NE1 4LP, UK; sandraadams2@nhs.net
4 Royal Devon & Exeter NHS Foundation Trust, Barrack Rd, Exeter EX2 5DW, UK; s.costelloe@nhs.net
5 Guy's and St Thomas' NHS Foundation Trust, London SE1 7EU, UK; Joanna.Eardley@gstt.nhs.uk (J.G.); Sarah.Firman@gstt.nhs.uk (S.F.)
6 University College London Hospitals NHS Foundation Trust, National Hospital for Neurology & Neurosurgery, Queen Square, London WC1N 3BG, UK; c.ellerton@nhs.net
7 University Hospital Southampton NHS Foundation Trust, Tremona Road, Southampton SO16 6YD, UK; anita.emm@uhs.nhs.uk
8 North Bristol NHS Trust, Southmead Road, Bristol BS10 5NB, UK; suzanne.ford@nbt.nhs.uk
9 University Hospitals of Leicester NHS Trust, Infirmary Square, Leicester LE1 5WW, UK; moira.french@uhl-tr.nhs.uk
10 Cambridge University Hospitals NHS Foundation Trust, Hills Road, Cambridge CB2 0QQ, UK; lisa.gaff@addenbrookes.nhs.uk (L.G.); celine.prescott@addenbrookes.nhs.uk (C.P.)
11 Northamptonshire Healthcare NHS Foundation Trust, St Mary's Hospital, London Road, Kettering NN15 7PW, UK; emily.giuliano@nhft.nhs.uk
12 Sheffield Teaching Hospitals NHS Foundation Trust, Herries Road, Sheffield S5 7AU, UK; melanie.hill13@nhs.net
13 Bradford Teaching Hospitals NHS Foundation Trust, Duckworth Lane, Bradford BD9 6RJ, UK; Inderdip.Hunjan@bthft.nhs.uk
14 Bristol Royal Hospital for Children, Bristol BS2 8BJ, UK; Camille.Newby@uhbw.nhs.uk
15 Royal Derby Hospital, Uttoxter Road, Derby DE22 3NE, UK; allison.mackenzie1@nhs.net
16 Norfolk and Norwich University Hospital, Colney Lane, Norwich NR4 7UY, UK; rachel.pereira@nnuh.nhs.uk
17 University Hospitals Birmingham NHS Foundation Trust, Queen Elizabeth Hospital, Birmingham B15 2TH, UK; Louise.robertson@uhb.nhs.uk
18 Somerset NHS Foundation Trust, Parkfield Drive, Taunton TA1 5DA, UK; heidi.seabert@somersetft.nhs.uk
19 Great Ormond Street Hospital for Children NHS Foundation Trust, Great Ormond Street, London WC1N 3JH, UK; rachel.skeath@gosh.nhs.uk
20 University Hospitals Bristol & Weston NHS Foundation Trust, Marlborough St, Bristol BS1 3NU, UK; simon.tapley@uhbw.nhs.uk
21 Alder Hey Children's NHS Foundation Trust, E Prescot Road, Liverpool L12 2AP, UK; allyson.terry@alderhey.nhs.uk
22 Nottingham University Hospitals NHS Trust, Queen's Medical Centre Campus, Derby Road, Nottingham NG7 2UH, UK; alison.tooke@nuh.nhs.uk
23 Royal Manchester Children's Hospital, Oxford Road, Manchester M13 9WL, UK; karen.vanwyk@cmft.nhs.uk (K.v.W.); fiona.white@cmft.nhs.uk (F.J.W.)
24 Sheffield Children's NHS Foundation Trust, Clarkson St, Broomhall, Sheffield S10 2TH, UK; lucy.white19@nhs.net
25 Salford Royal NHS Foundation Trust, Stott Lane, Salford M6 8HD, UK; alison.woodall@srft.nhs.uk
26 Nutrition & Metabolism, NOVA Medical School, Faculdade de Ciências Médicas, Universidade Nova de Lisboa, Campo Mártires da Pátria 130, 1169-056 Lisbon, Portugal; rochajc@nms.unl.pt
27 CINTESIS—Center for Health Technology and Services Research, NOVA Medical School, Campo Mártires da Pátria 130, 1169-056 Lisbon, Portugal

Citation: Wood, G.; Pinto, A.; Evans, S.; Daly, A.; Adams, S.; Costelloe, S.; Gribben, J.; Ellerton, C.; Emm, A.; Firman, S.; et al. Special Low Protein Foods Prescribed in England for PKU Patients: An Analysis of Prescribing Patterns and Cost. *Nutrients* **2021**, *13*, 3977. https://doi.org/10.3390/nu13113977

Academic Editor: Adamasco Cupisti

Received: 11 October 2021
Accepted: 3 November 2021
Published: 8 November 2021

Publisher's Note: MDPI stays neutral with regard to jurisdictional claims in published maps and institutional affiliations.

Copyright: © 2021 by the authors. Licensee MDPI, Basel, Switzerland. This article is an open access article distributed under the terms and conditions of the Creative Commons Attribution (CC BY) license (https://creativecommons.org/licenses/by/4.0/).

[28] Reference Centre of Inherited Metabolic Diseases, Centro Hospitalar Universitário de Lisboa Central, 1169-045 Lisbon, Portugal
* Correspondence: georgiewood93@googlemail.com

Abstract: Patients with phenylketonuria (PKU) are reliant on special low protein foods (SLPFs) as part of their dietary treatment. In England, several issues regarding the accessibility of SLPFs through the national prescribing system have been highlighted. Therefore, prescribing patterns and expenditure on all SLPFs available on prescription in England (n = 142) were examined. Their costs in comparison to regular protein-containing (n = 182) and *'free-from'* products (n = 135) were also analysed. Similar foods were grouped into subgroups (n = 40). The number of units and costs of SLPFs prescribed in total and per subgroup from January to December 2020 were calculated using National Health Service (NHS) Business Service Authority (NHSBSA) ePACT2 (electronic Prescribing Analysis and Cost Tool) for England. Monthly patient SLPF units prescribed were calculated using patient numbers with PKU and non-PKU inherited metabolic disorders (IMD) consuming SLPFs. This was compared to the National Society for PKU (NSPKU) prescribing guidance. Ninety-eight percent of SLPF subgroups (n = 39/40) were more expensive than regular and *'free-from'* food subgroups. However, costs to prescribe SLPFs are significantly less than theoretical calculations. From January to December 2020, 208,932 units of SLPFs were prescribed (excluding milk replacers), costing the NHS £2,151,973 (including milk replacers). This equates to £962 per patient annually, and prescribed amounts are well below the upper limits suggested by the NSPKU, indicating under prescribing of SLPFs. It is recommended that a simpler and improved system should be implemented. Ideally, specialist metabolic dietitians should have responsibility for prescribing SLPFs. This would ensure that patients with PKU have the necessary access to their essential dietary treatment, which, in turn, should help promote dietary adherence and improve metabolic control.

Keywords: special low protein foods; phenylketonuria; England; prescribing patterns; costs

1. Introduction

Phenylketonuria (PKU), an inborn error of amino acid metabolism, is caused by phenylalanine hydroxylase deficiency, an enzyme that converts phenylalanine to tyrosine [1]. This leads to neurotoxicity, causing severe intellectual disability if untreated [2]. It is managed by a life-long phenylalanine-restricted diet supplemented with a phenylalanine free/low phenylalanine protein substitute, although adjunct pharmacological therapies may also be prescribed to some patients [2,3]. In particular, patients with classical PKU require severe restrictions of natural protein, commonly tolerating ≤25% of a normal protein intake [1,2]. Regular protein containing foods e.g., bread, flour and pasta, are replaced with special low protein foods (SLPFs) that contain minimal protein [2,3]. These deliver a substantial source of energy, providing up to 50% of daily energy intake [4–6], fibre [7], they offer essential bulk, add variety and so help to sustain dietary adherence and ultimately aid metabolic control [8–10].

The cost of SLPFs to patients in England is reimbursed by the National Health Service (NHS), as these foods are considered borderline substances and are available on NHS prescription [11–13]. Borderline substances are nutritional or dermatological products specifically formulated to manage a medical condition [12]. There are around 150 SLPFs available on borderline substance prescription in England [13]. Each SLPF is approved by the United Kingdom (UK) Advisory Committee on Borderline Substances (ACBS) [12–14], which considers the clinical need of a product, its efficacy and the total price to the NHS [15]. Manufacturers/suppliers of SLPFs provide the ACBS with a statement outlining the proposed NHS list price and any distribution costs charged to dispensers [15]. For SLPFs that are broadly similar to existing products, the ACBS recommends a maximum benchmark cost to the NHS for that category [15]. When a company chooses to increase their NHS list price and maintain 'ACBS status', price increases are benchmarked against a standard inflation comparator [15].

General Practitioners (GPs) issue prescriptions for SLPFs monthly on request, which are then dispensed through local pharmacists or specialist home delivery companies linked to the suppliers of SLPFs [16]. The NHS then pays pharmacists or dispensing doctors a fee for each item they dispense [17,18]. The National Society for PKU (NSPKU) has produced a guide outlining the maximum monthly number of units of SLPFs (e.g., 1 unit = 1 pack of pasta up to 500 g—see Appendix A for full list of definitions for each product) which can be prescribed [19,20]. This guide considers patient age and circumstances to support GPs in prescribing these products and to ensure that expenditure on SLPFs is controlled. This guide has been widely adopted by GPs. In England, NHS prescriptions are free of charge for patients in the following categories: under 16 years of age; aged 16–18 years if in full time education; over 60 years of age; pregnant; receive income support or in other specific circumstances [21]. All other patients must pay a set fee per item, or they can purchase a three-monthly or annual prescription prepayment certificate which covers all of their NHS prescriptions [21].

However, there are many challenges in accessing SLPFs with the current prescribing system [16,22]. Some patients with PKU report that they have had their prescription requests refused; some describe how their GPs advise that they should purchase these foods rather than obtain them on prescription [16]. Others report that their GPs refuse to prescribe the appropriate range of products, as they consider some foods luxury items (e.g., cake mix or cereal bars) or the quantity of SLPFs is reduced due to their costs [16]. In a study by MacDonald et al., 2019, 43% (n = 25/58) of caregivers and parents said they needed more SLPFs for their children than they had been prescribed [22]. These challenges will impact on nutritional intake, directly affecting nutritional status and ultimately metabolic control.

Although studies have considered the cost of SLPFs, the majority were conducted outside the UK, where different reimbursement systems exist [23–26]. One study compared the theoretical costs in 10 international centres, where costs of SLPFs in the UK appeared to be higher than in many other countries [11]. Two nonpeer reviewed articles also discussed the theoretical cost of SLPFs in the UK and suggested that some SLPFs are expensive, but emphasised they are essential in the management of PKU [27,28]. Several papers have discussed costs when looking at the challenges of living with PKU in the UK, but this has not been the single focus of their work [3,16,22,29,30]. No study has compared the costs of SLPFs with regular foods or foods used in other therapeutic diets. Furthermore, no study has considered the prescribing pattern of SLPFs for low protein diets in England, or the UK as a whole.

This study therefore aimed to:

(1) examine the cost of all SLPFs on NHS prescription in England and compare these with similar regular equivalent protein containing and *'free-from'* dietary foods available in the supermarkets; and
(2) determine NHS expenditure on SLPFs and examine the number of SLPF units prescribed annually in England

2. Materials and Methods

2.1. Cost of SLPFs in England in Comparison to Regular Foods and 'Free-From' Foods

Data was collected from August to October 2020 on the price of all individual SLPFs available on ACBS prescription in England using British National Formulary (BNF) resources (Website, mobile phone app and book) and from the following suppliers or manufacturers websites if prices were stated:

1. Promin—https://prominpku.com/shop (accessed on 3 October 2020) [31]
2. Taranis—https://prominpku.com/shop (accessed on 3 October 2020) [31]
3. Metax—https://prominpku.com/shop (accessed on 3 October 2020) [31]

When individual prices of items were unavailable or unclear, companies were contacted directly via email. The cost per kg of each SLPF was calculated. SLPFs were divided into 40 subgroups of equivalent food product types, e.g., low protein burgers, sausages, cookies/biscuits, cake mixes. The mean and range costs across subgroups of similar products were calculated.

The mean and range cost per kg were collected and calculated for at least two regular protein-containing comparable foods and at least two *'free-from'* comparable foods, from major supermarkets in England with data available online (ASDA, Morrisons, Sainsburys, Tesco, Waitrose, Ocado and Marks & Spencer). A *'free-from'* food was defined as a food made without one or more specific ingredients, designed for people with food allergies or other intolerances/diseases e.g., coeliac disease. If data was unavailable from a supermarket's website, it was obtained from alternative online shops or directly from the manufacturer. Where prices differed between supermarkets for the same regular protein-containing food or *'free-from'* food, the mean value was recorded. Percentage differences between SLPFs and regular/*'free-from'* food subgroups for all mean costs were determined. Variations within ± 10% were considered comparable.

2.2. NHS Prescribing Patterns for SLPFs and Expenditure in England

One of the authors (A.P.) was given approval to access and extract prescribing data about SLPFs from the NHS Business Service Authority (NHSBSA) ePACT2 (electronic Prescribing Analysis and Cost Tool 2) for the costs and quantity of SLPFs prescribed in total and for each subgroup in England. This tool provided access to prescription data from the NHSBSA from January to December 2020. An ePACT2 bespoke training session was arranged with NHSBSA to ensure that all data was obtained and interpreted correctly. NSPKU prescribing guidance describing the definition of one unit for each SLPF was used to calculate the number of units of SLPFs prescribed in total and for each subgroup (Appendix A) [19,20].

In order to estimate the number of patients with PKU cared for by NHS centres in England, all NHS centres known to treat and monitor PKU patients were contacted in order to determine the number of patients with PKU (paediatric and adult), the number on dietary treatment (defined as those receiving prescribed protein substitutes and therefore potentially SLPFs), the number of shared care patients and the number of non-PKU inherited metabolic disorders (IMD) patients accessing SLPFs. Information was supplied by dietitians working in n = 26 NHS England hospitals/centres who care for patients with PKU. These data were used to calculate how many units of SLPFs were being prescribed per patient per month and the cost to the NHS per patient per month in England. This was then compared to NSPKU prescribing guidance.

3. Results

3.1. SLPFs, Regular Foods and Free-From Foods Costing Comparison

One hundred and forty-six SLPFs were identified as being available on ACBS prescription in England, with these products grouped and further subcategorised for comparison with at least two regular food products per subgroup. Regular and *'free-from'* comparators for four SLPFs (Calogen neutral, Calogen banana, Calogen strawberry and Duocal—Nutricia) were unavailable. Thus, 142 SLPFs were available for comparison with 182 regular products and 135 *'free-from'* products. Table 1 displays all SLPF, regular product and *'free-from'* food subgroups (n = 40), the mean cost per kg of products within each subgroup and % differences between costs.

Sixty-eight of 142 SLPFs (48%) were unavailable on BNF resources at the time of data collection (August to October 2020), and therefore, their costs had to be obtained directly from the manufacturer or supplier's website or through email contact with the manufacturer/supplier.

When analysed by subgroup, all SLPFs were more expensive than regular foods and *'free-from'* foods, except for regular eggs and *'free-from'* flavour puddings, where their cost per kg was comparable to low protein equivalents.

Table 1. Cost of low protein, regular and *'free-from'* food products for each subgroup and the % differences between costs.

Subgroup	SLPFs *n*	SLPFs Cost (£/kg)	Regular Protein-Containing Foods *n*	Regular Protein-Containing Foods Cost (£/kg)	*'Free-From'* Foods *n*	*'Free-From'* Foods Cost (£/kg)	% Difference between SLPFs and Regular Foods	% Difference between SLPFs and *'Free-From'* Foods
				Bread/pizza bases				
Bread	12	11.11 (8.23–16.13)	12	2.67 (1.31–5.00)	11	6.30 (3.27–11.40)	316%	76%
Pizza base	1	19.80	2	5.17 (4.00–6.33)	2	9.93 (9.86–10.00)	283%	99%
				Pasta/rice/noodles				
Pasta/rice/ noodles	33	15.28 (8.80–19.10)	23	2.60 (1.20–5.04)	16	3.65 (1.20–7.50)	488%	319%
Pasta and sauces (prepared)	5	16.16 (8.82–26.25)	10	2.61 (1.11–4.98)	6	9.36 (7.50–13.32)	519%	73%
Risotto	1	22.00	2	6.82 (6.49–7.14)	2	7.50 (7.50–7.50)	223%	193%
Xpots/pot noodles	4	92.50 (92.50–92.50)	8	7.44 (4.00–9.09)	4	24.32 (16.67–40.32)	1143%	280%
				Flour/mixes				
Bread mix	1	11.96	2	1.64 (1.28–2.00)	2	1.72 (1.69–1.75)	629%	595%
Cake mix	4	15.64 (13.94–19.36)	4	4.27 (1.20–5.29)	4	6.95 (4.57–9.97)	266%	125%
Flour/All Purpose Mix	5	14.80 (11.90–18.02)	2	1.37 (1.21–1.54)	2	1.60 (1.50–1.70)	980%	825%
Pancake/ waffle mix	1	15.33	2	5.14 (5.00–5.28)	2	8.34 (7.00–9.68)	198%	84%
				Egg/replacers				
Egg (prepared)	3	3.01 (1.89–4.08)	2	3.24 (2.46–4.02)	2	1.46 (1.36–1.55)	-7%	106%
Egg whites (powder)	1	108.10	2	49.92 (40.00–59.83)	2	16.02 (15.00–17.04)	117%	575%
				Milk/replacers				
Milk (liquid)	5	5.84 (4.05–6.75)	2	0.48 (0.48–0.48)	2	1.06 (0.59–1.53)	1117%	451%
Milk (powder)	1	22.38	2	7.64 (5.89–9.39)	2	17.56 (15.16–19.96)	193%	27%
				Meat/replacers				
Burgers (prepared)	3	16.88 (8.82–20.91)	4	6.04 (5.02–7.35)	4	7.44 (4.02–10.00)	179%	127%
Fish (prepared)	1	18.07	2	10.03 (8.25–11.81)	2	11.78 (11.67–11.88)	80%	53%
Sausages (prepared)	3	23.72 (23.72–23.72)	6	5.10 (3.06–6.88)	4	8.47 (6.67–9.26)	365%	180%
				Breakfast and cereal bars				
Breakfast bar	4	42.08 (42.08–42.08)	8	11.30 (6.42–15.28)	4	13.83 (8.57–18.18)	272%	204%
Breakfast cereal (dry)	3	23.35 (23.07–23.92)	6	4.84 (2.37–6.17)	6	6.18 (4.50–10.56)	382%	278%
Fruit bar	1	37.60	2	6.63 (4.28–8.99)	2	13.23 (11.25–15.20)	467%	184%
Hot breakfast cereal (dry)	4	25.11 (25.00–25.45)	4	10.61 (6.00–20.52)	4	11.45 (8.33–14.55)	137%	119%

Table 1. *Cont.*

Subgroup	SLPFs *n*	SLPFs Cost (£/kg)	Regular Protein-Containing Foods *n*	Regular Protein-Containing Foods Cost (£/kg)	*'Free-From'* Foods *n*	*'Free-From'* Foods Cost (£/kg)	% Difference between SLPFs and Regular Foods	% Difference between SLPFs and *'Free-From'* Foods
Snacks								
Biscuits/cookies	7	43.37 (33.60–68.52)	10	8.03 (1.05–25.00)	8	10.39 (6.50–17.86)	440%	317%
Breadsticks	1	41.87	2	8.20 (5.60–10.79)	2	14.69 (12.76–16.62)	411%	185%
Cake	3	26.00 (26.00–26.00)	2	5.58 (5.41–5.75)	2	13.04 (11.58–14.49)	366%	99%
Chocolate	2	52.32 (49.10–55.54)	2	7.62 (7.44–7.81)	2	12.08 (11.30–12.86)	587%	332%
Crackers	3	25.38 (24.00–26.07)	6	7.07 (3.25–9.56)	4	12.58 (12.00–13.81)	259%	102%
Crisps	4	37.50 (37.50–37.50)	8	8.46 (6.67–10.33)	4	16.05 (14.71–17.39)	343%	134%
Crispbread crackers	1	32.80	2	2.66 (1.33–3.98)	2	8.93 (8.89–8.98)	1133%	267%
French toast crackers	1	20.00	2	6.35 (6.25–6.45)	2	11.24 (10.80–11.67)	215%	78%
Hazelnut spread	1	35.43	2	5.16 (2.88–7.43)	2	10.55 (9.30–11.80)	587%	236%
Desserts								
Dessert pot	2	20.30 (20.30–20.30)	4	5.99 (4.69–7.14)	2	7.71 (2.93–12.50)	239%	163%
Flavoured pudding (powder)	4	30.68 (30.68–30.68)	7	9.79 (6.65–11.43)	2	29.00 (13.00–45.00)	213%	6%
Jelly (unprepared)	2	25.59 (25.59–25.59)	2	4.18 (4.16–4.19)	2	15.88 (15.88–15.88)	512%	61%
Rice pudding	4	24.35 (24.35–24.35)	6	3.26 (2.17–3.86)	2	8.25 (8.00–8.50)	647%	195%
Yogurt	1	7.19	2	2.68 (2.30–3.05)	2	3.13 (2.50–3.75)	168%	130%
Other snacks/meals								
Cheese sauce	1	24.18	2	7.47 (6.58–8.36)	2	13.02 (10.77–15.27)	224%	86%
Croutons	1	42.94	2	10.26 (10.00–10.52)	2	25.84 (18.51–33.17)	319%	66%
Potato cakes	1	8.68	2	3.07 (2.68–3.45)	2	4.37 (1.33–7.41)	183%	99%
Potato pots/dehydrated potato	3	87.25 (87.25–87.25)	4	9.21 (6.25–12.62)	2	23.46 (20.00–26.93)	847%	272%
Soup	4	53.85 (48.57–59.18)	8	13.03 (9.26–16.29)	4	26.67 (15.88–34.10)	313%	102%

Abbreviations: *n* = number of products; SLPFs = special low protein foods. Values displayed as mean (range).

Low protein crispbread crackers, Xpots (low protein equivalent of a pot noodle) and milk replacements (liquid) had the highest percentage cost difference, being 1117% to 1143% more expensive than the regular food comparator. When compared to *'free-from'* foods, low protein flour, bread mix and egg whites had the highest percentage differences (575% to 825%) in costs. In contrast, low protein milk powder, fish substitute and jelly were only 27% to 61% more expensive than their *'free-from'* food comparators. Basic SLPFs, including bread, pasta, rice, noodles and milk replacers (liquid), were 76% to 451% more expensive than *'free-from'* equivalent foods.

3.2. NHS Prescribing and Costing Data in England for SLPFs

Table 2 displays the prescribing and costing data for SLPFs from January–December 2020.

Table 2. Number of units, actual cost of prescribing SLPFs, and percentage of total units and total actual costs of all SLPFs by subgroup from January to December 2020 by the NHS for England.

Subgroup	Number of Units Prescribed from January to December 2020		Actual Costs * from January to December 2020 (£)		For the Year of January to December 2020	
	Total	Monthly Average	Total	Monthly Average	% of Total Units of SLPFs Prescribed	% of Total Actual Cost of SLPFs Prescribed
Bread/pizza bases						
Bread (n = 12)	42,171	3514	232,873	19,406	20.2%	10.8%
Pizza base (n = 1)	3382	282	38,566	3214	1.6%	1.8%
Pasta/rice/noodles						
Pasta/rice/noodles (n = 33)	39,043	3254	295,619	24,635	18.7%	13.7%
Pasta and sauces (prepared) (n = 5)	3574	298	37,592	3133	1.7%	1.7%
Risotto (n = 1)	258	22	2758	230	0.1%	0.1%
Xpots (n = 4)	1682	140	36,023	3002	0.8%	1.7%
Flour/mixes						
Bread mix (n = 1)	2111	176	11,780	982	1.0%	0.5%
Cake mix (n = 4)	6790	566	53,697	4475	3.2%	2.5%
Flour/All Purpose Mix (n = 5)	32,720	2727	239,559	19,963	15.7%	11.1%
Pancake/waffle mix (n = 1)	700	58	3565	297	0.3%	0.2%
Egg replacers						
Egg replacer (n = 3)	1312	109	16,412	1368	0.6%	0.8%
Egg white replacer (n = 1)	334	28	3398	283	0.2%	0.2%
Milk replacers						
Milk replacer (liquid) (n = 5)	n/a	n/a	655,437	54,620	n/a	30.5%
Milk replacer (powder) (n = 1)	n/a	n/a	1623	135	n/a	0.1%
Meat/fish replacers						
Burger replacements (n = 3)	4601	383	53,038	4420	2.2%	2.5%
Fish replacement (n = 1)	358	30	4069	339	0.2%	0.2%
Sausage replacements (n = 3)	7591	633	59,545	4962	3.6%	2.8%
Breakfast and cereal bars						
Breakfast bar (n = 4)	1595	133	16,876	1406	0.8%	0.8%
Breakfast cereal (dried) (n = 3)	6073	506	50,533	4211	2.9%	2.3%
Fruit bar (n = 1)	6424	535	28,863	2405	3.1%	1.3%
Hot breakfast cereal (n = 4)	3264	272	27,511	2293	1.6%	1.3%
Snacks						
Biscuits/cookies (n = 7)	9841	820	65,126	5427	4.7%	3.0%
Breadsticks (n = 1)	653	93 **	3928	561 **	0.3%	0.2%
Cake (n = 3)	3827	319	26,619	2218	1.8%	1.2%
Chocolate (n = 2)	7299	608	46,714	3893	3.5%	2.2%
Crackers (n = 3)	12,331	1028	50,952	4246	5.9%	2.4%
Crisps (n = 4)	1015	85	7528	627	0.5%	0.3%
Crispbread crackers (n = 1)	180	15	920	77	0.1%	0.0%
French toast crackers (n = 1)	270	23	1402	117	0.1%	0.1%
Hazelnut spread (n = 1)	812	68	7219	602	0.4%	0.3%
Desserts						
Dessert pot (n = 2)	1548	129	14,782	1232	0.7%	0.7%
Flavoured pudding (dried) (n = 4)	3188	266	21,439	1787	1.5%	1.0%
Jelly (dried) (n = 2)	196	16	1728	144	0.1%	0.1%
Rice pudding (n = 4)	1156	96	7961	663	0.6%	0.4%
Yogurt substitute (n = 1)	203	17	3855	321	0.1%	0.2%
Other snacks/meals						
Cheese sauce (n = 1)	288	24	1716	143	0.1%	0.1%
Croutons (n = 1)	328	27	2292	191	0.2%	0.1%
Potato cakes (n = 1)	311	26	2002	167	0.1%	0.1%
Potato pots (n = 3)	676	56	11,677	973	0.3%	0.5%
Soup (n = 4)	827	69	4776	398	0.4%	0.2%
TOTAL	**208,932**	**17,451**	**2151,973**	**179,566**	**100%**	**100%**

Abbreviations: n = number of products; SLPFs = special low protein foods * Actual Costs on ePACT2 is calculated as the Net Ingredient Cost of the item(s) supplied, less the National Average Discount Percentage (NADP) plus Payment for Consumables, Out of Pocket Expenses and Payment for Containers. ** Data from June 2020–December 2020 only.

In total, 208,932 units of SLPFs (monthly mean of 17,451 units) were prescribed from January to December 2020. This equated to a total actual cost of £2,151,973 (monthly mean cost of £179,566). The most frequently prescribed subgroups were bread, pasta/rice and flour, in total equating to 54.6% of all SLPFs prescribed. Milk replacers accounted for the highest percentage (30.5%) of the total actual cost of these products. There is not a definition for a unit of milk replacer, as the amount prescribed should be determined on an individual patient basis (Appendix A) [19,20]. Flour, pasta/rice and bread each accounted for just over 10% of total actual cost of SLPFs from January to December 2020 (11.1%, 13.7% and 10.8%, respectively).

Other expenses included payment for containers, consumables and out of pocket expenses, contributing 4.4% (£94,669) of the annual SLPFs costs to the NHS in England. Out of pocket expenses reimbursed to the pharmacy may include: postage and packaging costs; handling costs; and the cost of phone calls to manufacturers or suppliers to order products [32]. Payment at a rate of 10p for every prescription item is paid for containers where the quantity of a prescription item is ordered outside of the pack size or a multiple of the pack size (except for those granted 'special container status' where it is not practical to split a pack) [33]. An additional payment of 1.24p is made for all prescriptions including SLPFs in case additional consumables may need to be dispensed by the pharmacist (e.g., oral syringes, measuring spoons), although SLPFs usually do not need additional consumables. [33]. Also, a dispensing fee of £1.29 is allocated for each item prescribed [18].

3.3. NHS Patient Prescribing and Costing Data for SLPFs in England Compared to NSPKU Guidelines

Patients with PKU are the major consumers of SLPFs. It is estimated that there were 2359 patients with PKU in hospital follow-up in England (1436 adult patients, 923 paediatric patients), with n = 1814 (77%) on dietary treatment (Table 3). There were a further 422 patients using SLPFs with other inherited metabolic disorders of protein metabolism in England, suggesting that approximately 2236 patients in total were accessing SLPFs. On average, 93 units were prescribed per patient per year, which equates to approximately 8 units per month per patient. This is significantly less than the recommended maximum number of units per patient that could be prescribed each month as outlined by the NSPKU (Table 4). Actual cost data suggest that it costs a monthly mean of £80 per patient.

For the 877 paediatric patients with PKU on full or partial diet, it was estimated that 20% were aged 4 months–3 years (n = 175), 20% 4–6 years (n = 175), 20% 7–10 years (n = 175) and 40% 11–18 years (n = 352). Therefore, if all of these children, combined with adults with PKU on a full or partial diet (n = 937) were receiving the maximum number of low protein items on prescription each month, as per NSPKU guidance (Table 4), this would equate to 77,575 units each month. This is much higher than the average monthly prescribed units of 17,451 (excluding milk replacers) for the calendar year of 2020.

Table 3. Number of patients in England with PKU and/or using SLPFs under the care of an NHS hospital/centre.

Centre	Number of PKU Paediatric Patients ***	Number of PKU Adult Patients ***	Number of Patients on Full/Partial Phe-Restricted Diet	Number of Non-PKU Inherited Metabolic Disorder Patients Using SLPFs
Birmingham Women's and Children's Hospital	110	0	110	15
Evelina London Children's Healthcare—part of Guy's and St Thomas' NHS Foundation Trust	168	0	144	55
Guy's and St Thomas' NHS Foundation Trust—Adult IMD service	0	195	145	10
Great Ormond Street Hospital	163	0	159	53
University Hospitals Birmingham NHS Foundation Trust—Queen Elizabeth Hospital	0	153	134	30
University College London Hospitals NHS Foundation Trust	0	378	235	30
Bradford Teaching Hospitals NHS Foundation Trust	58	0	58	21

Table 3. *Cont.*

Centre	Number of PKU Paediatric Patients ***	Number of PKU Adult Patients ***	Number of Patients on Full/Partial Phe-Restricted Diet	Number of Non-PKU Inherited Metabolic Disorder Patients Using SLPFs
Royal Manchester Children's Hospital	96	0	96	27
Bristol Royal Hospital for Children	71	0	67	18
North Bristol NHS Trust	0	58	41	1
Alder Hey Children's NHS Foundation Trust	54	0	54	17
Salford Royal NHS Foundation Trust	0	334	186	58
Cambridge University Hospitals NHS Foundation Trust	14	36	47—of which 14 are paediatric patients	3
Sheffield Children's NHS Foundation Trust	52	0	42	21
Sheffield Teaching Hospitals NHS Foundation Trust	0	160	90	20
University Hospitals of Leicester NHS Trust	13	0	10	12
Nottingham University Hospitals NHS Trust	24	0	24	9
Great North Children's Hospital—within the Royal Victoria Infirmary	64	0	63	9
Royal Victoria Infirmary—Adult IMD services	0	74	43	5
Norfolk and Norwich University Hospital	15	0	15	-
Royal Derby Hospital	6	6	6—all of which are paediatric patients	-
Somerset NHS Foundation Trust	(1)	8 (+1)	5 (+1)	3
Royal Devon & Exeter NHS Foundation Trust	1 (+1)	9	5 (+1)—1 of which is a paediatric patient and not shared care	2
University Hospital Southampton NHS Foundation Trust	4 (+7)	23 (+1)	23 (+8)—4 of which are paediatric patients and not shared care	3
Northamptonshire Healthcare NHS Foundation Trust	10	2	12—10 of which are paediatric patients	0
University Hospitals Bristol & Weston NHS Foundation Trust	0	(21)	(20)	0
TOTAL	**923**	**1436**	**1814—877 of which are paediatric patients**	**422**

Abbreviations: SLPFs = special low protein foods; PKU = phenylketonuria; Phe = phenylalanine. ( ) shared care with another unit so numbers not included in totals. *** This includes patients with mild PKU/hyperphenylalaninaemia who maintain phenylalanine levels within target therapeutic range without dietary treatment.

Table 4. NSPKU guideline for recommended amounts of special low protein products per month [19] compared with monthly average per patient estimated in the current study which does not include milk replacers.

Age of Patient with PKU	Recommended Maximum Number of SLPFs to Prescribe Each Month (Not Including Milk Replacers)	Estimated Number of SLPFs Prescribed Per Person Each month (Not Including Milk Replacers)
4 months–3 years	20 units	8 units
4–6 years	25 units	
7–10 years	30 units	
11–18 years	50 units	
Adults	50 units	
Pre-conception/Pregnancy	50 units	

Abbreviations: SLPFs = special low protein foods; PKU = phenylketonuria; NSPKU = The National Society for Phenylketonuria.

4. Discussion

This is the first study to examine the cost of all SLPFs available on prescription in England compared to regular and *'free-from'* foods available in supermarkets. It is also the first study to examine the number and type of low protein items prescribed and expenditure on individual SLPFs and total SLPFs prescribed by the NHS in England over 1 year. There is a lower than expected volume of SLPFs prescribed in England, meaning that the costs to prescribe these products are significantly less than theoretically calculated [11,28], with a total of 17,451 units per month, costing £179,566. This equates to an estimated annual cost to the NHS per person with PKU in England of £962 with just 8 units (excluding low protein milk) prescribed per person per month, indicating that patients are receiving significantly less than the upper NSPKU prescribing guidance [16,19,20].

Over half (54.6%) of the units of SLPFs prescribed from January to December 2020 were basic foods such as bread, flour/mixes and pasta/rice. This accounted for just over one-third (35.6%) of the total annual costs. Just under a third (30.5%) of the costs were attributed to prescribing special low protein milks (liquid). It is likely that it is primarily children accessing SLPFs, as recent research suggested that it is mainly children aged <10 years with PKU who use prescribed special low protein milks [6]. There was previous concern that there may be over prescription of sweet SLPFs [8]. In Scotland, a 2014 survey found that special low protein pasta/rice/couscous, biscuits and flour were most commonly ordered by children, whereas adults with PKU mainly ordered pasta/rice/couscous, flour and bread [8]. In contrast, the amount of special low protein snacks and desserts (n = 14/40 subgroups including low protein chocolate, cookies, biscuits, cakes, and crisps) prescribed in England was minimal, with each subgroup only accounting for 0.1–5.9% of all SLPFs prescribed and contributing just 0.1–3.0% of the total NHS expenditure on SLPFs from January to December 2020. This is consistent with research reporting that special low protein cakes, biscuits and chocolate provide minimal contributions to daily energy intake in children with PKU [6]. It is clear that the expenditure on prescribing SLPFs is limited, particularly for sweet foods.

Overall, very little is known about SLPFs usage by adults with PKU in England. Our study suggests that 35% of adults with PKU were not following a phenylalanine restricted diet (Table 3). Although some adult patients may use SLPFs, others may not attempt to access them due to the complexity of the access system or the costs of the prescription fee for every food item ordered, unless the individual is entitled to free prescriptions. In one UK survey, 15% of patients with PKU stated that recurrent access problems with SLPFs was frustrating, and even led them to abandon their dietary treatment [16]. GP administration staff have been described as unhelpful, judgemental or obstructive when ordering SLPFs [8,16]; home delivery services are complex and sometimes unreliable, and SLPFs may arrive out of date or damaged, or of poor quality [16]. Some children with PKU were not on dietary treatment or not accessing SLPFs; this was associated with mild PKU, a higher natural protein tolerance, using sapropterin as an adjunct therapy, young infants not yet on solids or a dislike of SLPFs.

It is understandable that SLPFs cost more than regular and *'free-from'* foods. The demand for SLPFs is small in a limited global market. Few companies manufacture or distribute SLPFs in the UK [13]. Production runs are small scale with high staffing ratios, leading to increased costs. Some of the raw ingredients and packaging materials are purchased in low volumes, increasing productions costs. Packaging may be subject to frequent label changes due to alterations in legislation. Manufacturing wastage may be high if final products do not meet the necessary standards. Manufacturers also need to make some profit to allow them to invest in research and development to improve and expand their SLPF range.

The availability, accessibility and cost of SLPFs vary between countries [5,7,8,11,13,23–25,34]. Comparisons are challenging due to differences in currency, age of patients, degree of dietary adherence and study methodology. China reported a mean cost of $573 (American dollars or approximately £415) a year per patient for SLPFs [25], whereas the United States of America found a mean cost of $1615 (approximately £1171) for children aged 0–17 years for SLPFs and just $967 (approximately £701) for adults [23]. The Netherlands reported a mean annual cost of €680 (approximately £576) on SLPFs, whereas the Czech Republic found this value to be significantly higher at €1560 (approximately £1321) [24,26].

The overall use of SLPFs is affected by the national access system and any consequential economic burden [11,23–26]. Some countries do not reimburse SLPFs costs; but may be funded by insurance coverage [11,24]. When national reimbursement schemes do not exist, families have to self-finance the purchase of SLPFs [11,23,25,26]. This is a huge financial burden for patients, which influences their ability to adhere to dietary treatment [11,23,25,26].

For patients with PKU to have better access to SLPFs through the NHS, several recommendations should be implemented. Consistent with previous suggestions by MacDonald et al. and Ford et al. [16,22], specialist metabolic dietitians should play a key role in prescribing SLPFs, as they control dietary management and oversee any dietary changes according to the individual patient's metabolic control, nutritional needs, growth and overall nutritional status. This would be more efficient, minimise administration time and professional and patient confusion and enable patients with PKU to have minimal contact with healthcare professionals/prescribers who know very little about their condition and how it is managed. Instead, their SLPF prescriptions would be managed by those who are most equipped to support them in meeting their dietary needs and maintaining good metabolic control.

This study has some limitations. When obtaining the cost of each SLPF in August–October 2020, 68 products were not visible on any BNF resource, and therefore, prices were obtained directly from the manufacturer or supplier of SLPFs. The selection of protein-containing foods and *'free-from'* foods as comparators, and how the products were grouped, was subjective. Certain powdered/dried SLPF products e.g., burger mix, had to be compared to a prepared regular protein-containing or *'free-from'* product e.g., cooked burger; therefore, the cost of the SLPF in its prepared form per kg was estimated. This study only examined products accessible on prescription in England compared with protein-containing products and *'free-from'* foods available from supermarket websites in England. Also, NHS prescribing and costing data were only available for England and not the whole of the UK, and were only collected from January to December 2020. From March 2020 onwards, England experienced multiple 'lockdowns' due to the coronavirus pandemic, and it is possible that this may have affected food behaviours and, consequently, the number and/or types of SLPFs that patients were requesting on prescription. However, there was no evidence from clinical practice that use or supplies of SLPFs were affected in England.

When calculating the number of units of SLPF and the costs per person with PKU in England, the numbers of patients on dietary treatment were estimated. However, dietetic colleagues throughout England provided representative and recent data from their clinics. It is difficult to state exactly how many patients were requesting SLPFs, as we did not examine individual prescribing data for each patient. On ePACT2, there were nine occasions in 2020 where a SLPF appeared on a prescription, but the quantity prescribed was unclear. Consequently, these data were removed from our spreadsheet. It is possible that there may be under-reporting of SLPFs by the NHSBSA ePACT2. The NHSBSA ePACT2 trainers/help team stated that there was a small possibility that data can be incorrectly processed, but that data is scanned from each prescription form directly, so the NHSBSA ePACT2 should accurately reflect all the prescriptions issued in England.

5. Conclusions

The annual cost to the NHS in England to prescribe SLPFs is £962 per patient with PKU and non-PKU IMD conditions. Surveys have repeatedly shown that patients or caregivers have access difficulties with current systems. If patients with PKU are expected to adhere to their dietary treatment for life, they must be able to easily access all SLPFs on prescription in a timely manner via the NHS. Given how little is currently being spent on prescribing

SLPFs in England in comparison to the upper NSPKU guidance, cost should not be given as a reason to restrict a patient's access to their essential dietary treatment. A review of how SLPFs are prescribed, supplied and controlled is warranted to improve the system, which, in turn, could lead to increased dietary adherence and improved patient outcomes.

Author Contributions: Conceptualization, A.M. (Anita MacDonald) and A.P.; methodology, G.W., A.M. (Anita MacDonald) and A.P.; formal analysis, G.W. and S.E.; investigation, all authors except J.C.R. and A.D.; data curation, G.W., A.P. and S.E.; writing—original draft preparation, G.W.; writing—review and editing, all authors; visualization, G.W., A.P., S.E. and A.M. (Anita MacDonald); supervision, A.M (Anita MacDonald). All authors have read and agreed to the published version of the manuscript.

Funding: This research received no external funding.

Institutional Review Board Statement: Not applicable.

Informed Consent Statement: Not applicable.

Data Availability Statement: NHSBSA prescribing data on special low protein foods in England was obtained from ePACT2.

Acknowledgments: Thank you to the manufacturers and suppliers of special low protein foods for providing data on their products. Thank you to the NHSBSA for giving A.P. permission to access prescribing data on special low protein foods and providing support and guidance with running our reports on ePACT2.

Conflicts of Interest: A.M. (Anita MacDonald) is a member of, the advisory board ELEMENT Danone-Nutricia, the advisory board for Arla and Applied Pharma Research, and received research funding and honoraria from Nutricia, Vitaflo International, Biomarin, MetaHealth, Metax and Merck Serono. S.E. receives research funding from Nutricia, and has received financial support and honoraria from Nutricia and Vitaflo to attend/speak at study days and conferences. A.P. received an educational grant from Cambrooke Therapeutics and grants from Vitaflo International, Nutricia, Merck Serono, Biomarin and Mevalia to attend scientific meetings. A.D. received research funding from Vitaflo International, financial support from Nutricia, Mevalia and Vitaflo International to attend study days and conferences. J.C.R. is a member of the European Nutritionist Expert Panel (Biomarin), the Advisory Board for Applied Pharma Research and Nutricia, and has received honoraria as a speaker from APR, Merck Serono, Biomarin, Nutricia, Vitaflo, Cambrooke, PIAM, and Lifediet. S.F. (Suzanne Ford) is a member of the advisory board for Nutricia, and MetaHealth and has received financial support and honoraria from Cambrooke and Vitaflo. R.S. has received sponsorship to attend conferences and study days, payment to present at conferences from Nutricia Metabolics, Vitaflo Internation and Mevalia. C.E. has received honoraria and educational grants to attend events from Vitaflo, Nutricia, Meta Healthcare and SOBI and is a member of the Advisory Commitee on Borderline Substances. S.F. (Sarah Firman) has received funding to attend conferences and study days from Nutricia, Vitaflo International and Dr. Schär UK Ltd., and consulting fees from Vitaflo International and Meta Healthcare Ltd. C.N. has received financial support and honoraria from Nutricia and Vitaflo to attend/speak at study days and conferences. L.R. is a member of the Nutricia adult advisory board and received honorarium from Nutricia and Vitaflo in the past. F.J.W. receives honoraria, educational and travel grants from Nutricia and Vitaflo. M.H. is a member of advisory board Nutricia Danone and Applied Pharma and has received financial support and honoraria from Nutricia, Vitaflo, Cambrooke, Mevalia, Promin for attendance at conferences/meetings and speakers fee. A.T. (Allyson Terry) has received payment from Vitaflo International for speaking at a patient event. K.vW. receives honoraria and educational grants from Nutricia, Vitaflo and Meta Healthcare. S.T. is expected to receive honoraria from Vitaflo International for providing feedback and reviewing guidelines for PKU Sphere. G.W., M.F., L.G., C.P., S.A., A.M. (Allison Mackenzie), A.T. (Alison Tooke), A.W., E.G., L.W., I.H., H.S., J.G., A.E., S.C., R.P. have no conflicts of interest to declare.

Appendix A

Table A1. Definition of 1 unit for each SLPF (table adapted slightly from NSPKU special low protein foods on prescription document) [20].

Burger Mixes, Sausage Mixes	Pack Size	Number of Units
Firstplay Dietary Foods:		
Promin Low Protein Burger Mix: -		
Lamb & Mint Flavour	4 × 62 g	1
Original Flavour	4 × 62 g	1
Promin Low Protein Sausage Mix: -		
Apple and Sage	4 × 30 g	1
Original	4 × 30 g	1
Tomato and Basil	4 × 30 g	1
Mevalia		
Low Protein Burger Mix	350 g	1
Taranis		
Low Protein Fish Substitute	4 × 62 g sachets	1
Bread, Flour and Mixes	**Pack Size**	**Number of Units**
Fate Special Foods		
Fate Low Protein All-Purpose Mix	500 g	1
Fate Low Protein Cake Mix	2 × 250 g	1
Fate Low Protein Chocolate Flavour Cake Mix	2 × 250 g	1
Firstplay Dietary Foods		
Promin Low Protein All Purpose Baking Mix	1000 g	2
Promin Low Protein Fresh Baked Bread Buns	6 × 75 g	1
Promin Low Protein Fresh Baked Sliced Brown Bread	4 × 400 g loaves	4
Promin Low Protein Fresh Baked Sliced White Bread	4 × 800 g loaves	4
Promin Low Protein Potato Cake Mix	300 g	1
Taranis:		
Taranis Low Protein Natural Cake Mix	300 g	1
Taranis Low Protein Pancakes & Waffles Mix	300 g	1
Gluten Free Foods (PK Foods):		
PK Foods All Purpose Low Protein Flour Mix	750 g	1.5
PK Foods Low Protein White Sliced Bread	300 g	1
Juvela		
Juvela Low protein Bread Rolls	5 rolls	1
Juvela Low Protein Loaf—sliced	400 g	1
Juvela Low Protein Mix	500 g	1
Mevalia		
Mevalia Low Protein Bread Mix	500 g	1
Mevalia Low Protein Ciabattine	4 × 65 g	0.5
Mevalia Low Protein Grissini (Breadsticks)	150 g (3 × 50 g)	1
Mevalia Low Protein Mini Baguette	2 × 100 g	0.5
Mevalia Low Protein Pan Carre	2 × 200 g	0.5
Mevalia Low Protein Pan Rustico	2 × 200 g	0.5
Mevalia Low Protein Pane Casereccio	220 g	0.5
Mevalia Low Protein Pizza Base	2 × 150 g	0.5
Nutricia:		
Loprofin Low Protein Part-Baked Sliced Bread	400 g	1
Loprofin Low Protein Mix	500 g	1
Loprofin Low Protein Chocolate Cake Mix	500 g	1

Table A1. *Cont.*

Pasta and Rice	Pack Size	Number of Units
Firstplay Dietary Foods		
Promin Low Protein Pasta:		
Low Protein Alphabets	500 g	1
Low Protein Elbows	500 g	1
Low Protein Flat Noodles	500 g	1
Low Protein Macaroni	500 g	1
Low Protein Shells	500 g	1
Low Protein Short Cut Spaghetti	500 g	1
Low Protein Spirals	500 g	1
Promin Low Protein Tricolour:		
Low Protein Alphabets	500 g	1
Low Protein Elbows	500 g	1
Low Protein Shells	500 g	1
Low Protein Spirals	500 g	1
Promin Low Protein Specialty Pasta:		
Low Protein Couscous	500 g	1
Low Protein Lasagne Sheets	200 g	0.5
Low Protein Pastameal	500 g	1
Low Protein Rice	500 g	1
Promin Pasta in Sauce:		
Low Protein Cheese and Broccoli	4 × 66 g	1
Low Protein Moroccan Flavour Tomato,	4 × 72 g	1
Low Protein Tomato Pepper & Herb	4 × 72 g	1
Promin Mac Pots:		
Low Protein Macaroni Cheese	4 × 61 g	1
Low Protein Tomato Macaroni	4 × 61 g	1
Promin Low Protein Pasta Plus (with fibre):		1
Promin Plus Low Protein Flat Noodles	500 g	1
Promin Plus Low Protein Macaroni	500 g	1
Promin Plus Low Protein Spaghetti	500 g	1
Promin Plus Low Protein Spirals	500 g	
Promin Potato Pots:		
Low Protein Cabbage & Bacon Flavour and croutons	4 × 50 g	1
Low Protein Onion Flavour and croutons	4 × 50 g	1
Low Protein Sausage Flavour and croutons	4 × 50 g	1
Promin X-Pots:		
Low Protein All Day Scramble	4 × 60 g	1
Low Protein Beef & Tomato	4 × 60 g	1
Low Protein Chip Shop Curry	4 × 60 g	1
Low Protein Rogan Style Curry	4 × 60 g	1
Taranis		
Low Protein Risotto Substitute	4 × 300 g	2.5
Mevalia		
Mevalia Low Protein Ditali	500 g	1
Mevalia Low Protein Fusilli	500 g	1
Mevalia Low Protein Penne	500 g	1
Mevalia Low Protein Rice	400 g	1
Mevalia Low Protein Spaghetti	500 g	1
Nutricia		
Loprofin Low Protein Animal pasta	500 g	1
Loprofin Low Protein Fusilli	500 g	1
Loprofin Low Protein Lasagne	250 g	0.5
Loprofin Low Protein Long Spaghetti	500 g	1
Loprofin Low Protein Macaroni Elbows	250 g	0.5
Loprofin Low Protein Penne	500 g	1
Loprofin Low Protein Rice	500 g	1
Loprofin Low Protein Tagliatelle	250 g	0.5
Gluten Free Foods (PK Foods):		
PK Foods Pasta spirals	250 g	0.5

Table A1. *Cont.*

Breakfast Cereals	Pack Size	Number of Units
Firstplay Dietary Foods		
Low Protein Hot Breakfast:		
Low Protein Apple and Cinnamon Flavour	6 × 57 g	1
Low Protein Banana Flavour	6 × 57 g	1
Low Protein Chocolate Flavour	6 × 57 g	1
Low Protein Original Flavour	6 × 56 g	1
Low Protein Breakfast Bars:		
Low Protein Apple & Cinnamon	6 × 40 g	1
Low Protein Banana	6 × 40 g	1
Low Protein Chocolate & Cranberry	6 × 40 g	1
Low Protein Cranberry	6 × 40 g	1
Nutricia		
Low Protein Loprofin Cereal Loops	375 g	1
Low Protein Loprofin Flakes—Chocolate	375 g	1
Low Protein Loprofin Flakes—Strawberry	375 g	1
Biscuits/Crackers	**Pack Size**	**Number of Units**
Gluten Free Foods (PK Foods):		
PK Foods Low Protein Crispbread	75 g	0.5
Mevalia:		
Low Protein Cookies	200 g	1
Low Protein Frollini	200 g	1
Low Protein Fruit Bar	5 × 25 g	1
Nutricia:		
Loprofin Low Protein Crackers (Savoury)	150 g	1
Loprofin Low Protein Herb Crackers	150 g	1
Taranis:		
Taranis Chocolate Chip Cookies	135 g	1
Taranis Shortbread Biscuits	120 g	1
Taranis Raspberry Shortbread Biscuits	120 g	1
Taranis Chocolate Chip Biscuits	120 g	1
Biscuits with caramel shards	130 g	1
Taranis French Toasts	250 g	1
Vitaflo:		
Vitaflo Choices Low Protein Mini Crackers	40 g (15 × 40 g)	3
Puddings, Desserts & Cakes	**Pack Size**	**Number of Units**
Firstplay Dietary Foods:		
Metax Low Protein YoguMaxx (yoghurt substitute)	400 g (23 servings)	1
Promin Low Protein Desserts:		
Caramel Dessert	6 × 36.5 g	1
Chocolate and Banana Dessert	6 × 36.5 g	1
Custard Dessert	6 × 36.5 g	1
Strawberry and Vanilla Dessert	6 × 36.5 g	1
Promin Low Protein Rice Pudding Mix:		
Low Protein Apple	4 × 69 g	1
Low Protein Banana	4 × 69 g	1
Low Protein Original	4 × 69 g	1
Low Protein Strawberry	4 × 69 g	1
Taranis Low Protein Cakes:		
Taranis Low Protein Apricot Cake	6 × 40 g	1
Taranis Low Protein Lemon Cake	6 × 40 g	1
Taranis Low Protein Pear Cake	6 × 40 g	1
Taranis Low Protein Desserts:		
Taranis Low Protein Pause Caramel Dessert	pack of four pots (×125 g)	1
Taranis Low Protein Pause Strawberry Dessert	pack of four pots (× 125 g)	1
Gluten Free Foods:		
PK Foods Low Protein Cherry Jelly Mix	4 × 80 g	1
PK Foods Low Protein Orange Jelly Mix	4 × 80 g	1

Table A1. *Cont.*

Low Protein Energy Bars	Pack Size	Number of Units
Mevalia:		
Low Protein Chocotino	100 g	1
Vitaflo:		
Low Protein Vitabite	7 × 25 g	1
Miscellaneous Foods	**Pack Size**	**Number of Units**
Promin Low Protein Salted Croutons	4 × 40 g	1
Promin Low Protein Cheese Sauce Mix	225 g	1
Promin Low Protein Snax:		
4 flavours: Ready Salted, Jalapeno, Cheese & Onion and Salt & Vinegar in a mixed box	12 × 25 g	1.5
Promin Low Protein Soups:		
Low Protein Chicken Flavour with Croutons	4 × 28 g	1
Low Protein Creamy Tomato with Croutons	4 × 23 g	1
Low Protein Minestrone with Croutons	4 × 28 g	1
Low Protein Pea & Mint with Croutons	4 × 23 g	1
Taranis:		
Taranis Low Protein Hazelnut Spread	230 g tub	1
Low Protein Drinks	**Pack Size**	**Number of Units**
Taranis:		
Taranis Dalia Liquid milk	24 × 200 mL	n/a
Taranis Dalia powder milk	400 g	n/a
Mevalia:		
Low Protein Lattis	500 mL	n/a
Nutricia:		
Loprofin PKU Milk	27 × 200 ml	n/a
Sno-Pro	27 × 200 ml	n/a
Vitaflo:		
ProZero Protein Free Drink	18 × 250 mL or 6 × 1 L	n/a

Abbreviations: SLPFs = special low protein foods; NSPKU = The National Society for Phenylketonuria.

References

1. Van Wegberg, A.M.J.; MacDonald, A.; Ahring, K.; Bélanger-Quintana, A.; Blau, N.; Bosch, A.M.; Burlina, A.; Campistol, J.; Feillet, F.; Giżewska, M.; et al. The complete European guidelines on phenylketonuria: Diagnosis and treatment. *Orphanet J. Rare Dis.* **2017**, *12*, 162. [CrossRef]
2. MacDonald, A.; Van Wegberg, A.M.J.; Ahring, K.; Beblo, S.; Bélanger-Quintana, A.; Burlina, A.; Campistol, J.; Coşkun, T.; Feillet, F.; Giżewska, M.; et al. PKU dietary handbook to accompany PKU guidelines. *Orphanet J. Rare Dis.* **2020**, *15*, 171. [CrossRef]
3. MacDonald, A.; van Rijn, M.; Feillet, F.; Lund, A.M.; Bernstein, L.; Bosch, A.M.; Gizewska, M.; van Spronsen, F.J. Adherence issues in inherited metabolic disorders treated by low natural protein diets. *Ann. Nutr. Metab.* **2012**, *61*, 289–295. [CrossRef]
4. Moretti, F.; Pellegrini, N.; Salvatici, E.; Rovelli, V.; Banderali, G.; Radaelli, G.; Scazzina, F.; Giovannini, M.; Verduci, E. Dietary glycemic index, glycemic load and metabolic profile in children with phenylketonuria. *Nutr. Metab. Cardiovasc. Dis.* **2017**, *27*, 176–182. [CrossRef]
5. Pena, M.J.; Almeida, M.F.; van Dam, E.; Ahring, K.; Bélanger-Quintana, A.; Dokoupil, K.; Gokmen-Ozel, H.; Lammardo, A.M.; MacDonald, A.; Robert, M.; et al. Special low protein foods for phenylketonuria: Availability in Europe and an examination of their nutritional profile. *Orphanet J. Rare Dis.* **2015**, *10*, 162. [CrossRef] [PubMed]
6. Daly, A.; Evans, S.; Pinto, A.; Ashmore, C.; Rocha, J.C.; Macdonald, A. A 3 year longitudinal prospective review examining the dietary profile and contribution made by special low protein foods to energy and macronutrient intake in children with phenylketonuria. *Nutrients* **2020**, *12*, 3153. [CrossRef]
7. Wood, G.; Evans, S.; Pointon-Bell, K.; Rocha, J.C.; MacDonald, A. Special Low Protein Foods in the UK: An Examination of Their Macronutrient Composition in Comparison to Regular Foods. *Nutrients* **2020**, *12*, 1893. [CrossRef]
8. Cochrane, B.; Schwahn, B.; Galloway, P.; Robinson, P.; Gerasimidis, K. A questionnaire survey on the usage of low protein staple foods by people with phenylketonuria in Scotland. *J. Hum. Nutr. Diet.* **2014**, *27*, 533–541. [CrossRef] [PubMed]
9. Rocha, J.C.; MacDonald, A. Dietary intervention in the management of phenylketonuria: Current perspectives. *Pediatr. Health Med. Ther.* **2016**, *7*, 155–163. [CrossRef]
10. Handoom, B.; Megdad, E.; Al-Qasabi, D.; Al Mesned, M.; Hawary, R.; Al-Nufiee, S.; Al-Hassnan, Z.; Alsayed, M.D.; Eldali, A. The effects of low protein products availability on growth parameters and metabolic control in selected amino acid metabolism disorders patients. *Int. J. Pediatr. Adolesc. Med.* **2018**, *5*, 60–68. [CrossRef] [PubMed]
11. Belanger-Quintana, A.; Dokoupil, K.; Gokmen-Ozel, H.; Lammardo, A.M.; MacDonald, A.; Motzfeldt, K.; Nowacka, M.; Robert, M.; van Rijn, M.; Ahring, K. Diet in phenylketonuria: A snapshot of special dietary costs and reimbursement systems in 10 international centers. *Mol. Genet. Metab.* **2012**, *105*, 390–394. [CrossRef] [PubMed]

12. Department of Health and Social Care. Advisory Committee on Borderline Substances. Available online: https://www.gov.uk/government/groups/advisory-committee-on-borderline-substances (accessed on 26 August 2021).
13. Wood, G.; Evans, S.; Pointon-Bell, K.; MacDonald, A. Special low protein foods: The challenges in obtaining accurate information about the availability of special low protein foods (SLPFs) and their nutritional composition. *Netw. Health Dig.* **2020**, *153*, 36–38.
14. Department of Health and Social Care. Appendix 1: Categorisation of Nutritional Products Submitted for Approval by the Advisory Committee on Borderline Substances. Available online: https://assets.publishing.service.gov.uk/government/uploads/system/uploads/attachment_data/file/792993/Appendix1.pdf (accessed on 18 August 2020).
15. Department of Health and Social Care. Information on the Pricing of ACBS Products. Available online: https://www.gov.uk/government/publications/guidelines-on-the-pricing-of-acbs-products/information-on-the-pricing-of-acbs-products (accessed on 18 August 2020).
16. Ford, S.; O'Driscoll, M.; MacDonald, A. Prescribing issues experienced by people living with phenylketonuria in the UK. *Mol. Genet. Metab. Rep.* **2019**, *21*, 100527. [CrossRef]
17. NHS Prescription Services. NHS and LA Reforms Factsheet 4: Payment, Charging and Apportioning Arrangements of Costs Associated with NHS Pharmaceutical Service Provision from 1 April 2013. Available online: https://www.nhsbsa.nhs.uk/sites/default/files/2017-07/NHS_Reforms_factsheet_4_v2.0.pdf (accessed on 25 October 2020).
18. Pharmaceutical Services Negotiating Committee. Fees and Allowances. Available online: https://psnc.org.uk/dispensing-supply/endorsement/fees-allowances (accessed on 25 October 2020).
19. The National Society for Phenylketonuria (NSPKU). The Prescription of Low Protein Foods for Patients with PKU. Available online: https://www.nspku.org/download/unit-guidelines-for-the-prescription-of-low-protein-foods (accessed on 19 September 2021).
20. The National Society for Phenylketonuria (NSPKU). Special Low Protein/Low Phenylalanine Foods on Prescription. Available online: https://www.nspku.org/download/low-protein-foods-available-on-prescription-for-patients-with-pku/#.pdf (accessed on 19 September 2021).
21. NHS Business Services Authority. Are you Claiming Free Prescriptions? Available online: https://www.nhsbsa.nhs.uk/sites/default/files/2018-09/Freeprescriptionsbooklet%28V3%2909.2018.pdf (accessed on 19 September 2021).
22. MacDonald, A.; Pinto, A.; Evans, S.; Ashmore, C.; MacDonald, J.; Daly, A. Home delivery service of low protein foods in inherited metabolic disorders: Does it help? *Mol. Genet. Metab. Rep.* **2019**, *19*, 100466. [CrossRef]
23. Rose, A.M.; Grosse, S.D.; Garcia, S.P.; Bach, J.; Kleyn, M.; Simon, N.J.E.; Prosser, L.A. The financial and time burden associated with phenylketonuria treatment in the United States. *Mol. Genet. Metab. Rep.* **2019**, *21*, 100523. [CrossRef]
24. Eijgelshoven, I.; Demirdas, S.; Smith, T.A.; van Loon, J.M.T.; Latour, S.; Bosch, A.M. The time consuming nature of phenylketonuria: A cross-sectional study investigating time burden and costs of phenylketonuria in the Netherlands. *Mol. Genet. Metab.* **2013**, *109*, 237–242. [CrossRef]
25. Wang, L.; Zou, H.; Ye, F.; Wang, K.; Li, X.; Chen, Z.; Chen, J.; Han, B.; Yu, W.; He, C.; et al. Household financial burden of phenylketonuria and its impact on treatment in China: A cross-sectional study. *J. Inherit. Metab. Dis.* **2017**, *40*, 369–376. [CrossRef]
26. Mlčoch, T.; Puda, R.; Ješina, P.; Lhotáková, M.; Šterbová; Doležal, T. Dietary patterns, cost and compliance with low-protein diet of phenylketonuria and other inherited metabolic diseases. *Eur. J. Clin. Nutr.* **2018**, *72*, 87–92. [CrossRef] [PubMed]
27. Kidd, C. Low-protein foods: Cost-effective prescribing. *Netw. Health Dig.* **2019**, *145*, 40–43.
28. Ford, S. The costs of a special diet. *Netw. Health Dig.* **2019**, *141*, 35–39.
29. MacDonald, A.; Smith, T.A.; de Silva, S.; Alam, V.; van Loon, J.M.T. The personal burden for caregivers of children with phenylketonuria: A cross-sectional study investigating time burden and costs in the UK. *Mol. Genet. Metab. Rep.* **2016**, *9*, 1–5. [CrossRef] [PubMed]
30. Ford, S.; O'Driscoll, M.; MacDonald, A. Living with Phenylketonuria: Lessons from the PKU community. *Mol. Genet. Metab. Rep.* **2018**, *17*, 57–63. [CrossRef] [PubMed]
31. Promin Web Shop. Available online: https://prominpku.com/shop (accessed on 3 October 2020).
32. Pharmaceutical Services Negotiating Committee. Out of Pocket Expenses. Available online: https://psnc.org.uk/dispensing-supply/endorsement/fees-allowances/oop (accessed on 27 October 2021).
33. Pharmaceutical Services Negotiating Committee. Consumables and Container Allowance. Available online: https://psnc.org.uk/dispensing-supply/endorsement/fees-allowances/consumables-containers (accessed on 24 September 2021).
34. Camp, K.M.; Lloyd-Puryear, M.A.; Huntington, K.L. Nutritional treatment for inborn errors of metabolism: Indications, regulations, and availability of medical foods and dietary supplements using phenylketonuria as an example. *Mol. Genet. Metab.* **2012**, *107*, 3–9. [CrossRef] [PubMed]

nutrients MDPI

Article

Dietetic Management of Adults with Phenylketonuria (PKU) in the UK: A Care Consensus Document

Louise Robertson [1,*], Sarah Adam [2], Charlotte Ellerton [3], Suzanne Ford [4,5], Melanie Hill [6], Gemma Randles [7], Alison Woodall [8], Carla Young [2] and Anita MacDonald [9]

1 Department of Inherited Metabolic Disorders, University Hospitals Birmingham NHS Foundation Trust, Queen Elizabeth Hospital, Birmingham B15 2TH, UK
2 NHS Greater Glasgow and Clyde, Royal Hospital for Children, Glasgow G51 4TF, UK; Sarah.Adam@ggc.scot.nhs.uk (S.A.); Carla.Young@ggc.scot.nhs.uk (C.Y.)
3 Charles Dent Metabolic Unit, University College London Hospitals NHS Foundation Trust, National Hospital for Neurology & Neurosurgery, Queen Square, London WC1N 3BG, UK; c.ellerton@nhs.net
4 North Bristol NHS Trust, Southmead Road, Bristol BS10 5NB, UK; suzanne.ford@nbt.nhs.uk
5 National Society for Phenylketonuria, Richard House, Winckley Square, Preston PR1 3HP, UK
6 Adult Inherited Metabolic Disorders Service at Sheffield Teaching Hospitals NHS Foundation Trust, Herries Road, Sheffield S5 7AU, UK; melanie.hill13@nhs.net
7 Guys and St Thomas NHS Foundation Trust, London SE1 7EU, UK; gemmarandles@gmail.com
8 Salford Care Organisation, Northern Care Alliance NHS Foundation Trust, Stott Lane, Salford M6 8HD, UK; alison.woodall@nca.nhs.uk
9 Birmingham Women's and Children's Hospital, Steelhouse Lane, Birmingham B4 6NH, UK; Anita.MacDonald@nhs.net
* Correspondence: louise.robertson@uhb.nhs.uk; Tel.: +44-121-3716989

Citation: Robertson, L.; Adam, S.; Ellerton, C.; Ford, S.; Hill, M.; Randles, G.; Woodall, A.; Young, C.; MacDonald, A. Dietetic Management of Adults with Phenylketonuria (PKU) in the UK: A Care Consensus Document. *Nutrients* **2022**, *14*, 576. https://doi.org/10.3390/nu14030576

Academic Editor: J. Mark Brown

Received: 22 December 2021
Accepted: 21 January 2022
Published: 28 January 2022

Publisher's Note: MDPI stays neutral with regard to jurisdictional claims in published maps and institutional affiliations.

Copyright: © 2022 by the authors. Licensee MDPI, Basel, Switzerland. This article is an open access article distributed under the terms and conditions of the Creative Commons Attribution (CC BY) license (https://creativecommons.org/licenses/by/4.0/).

Abstract: There is an increasing number of adults and elderly patients with phenylketonuria (PKU) who are either early, late treated, or untreated. The principal treatment is a phenylalanine-restricted diet. There is no established UK training for dietitians who work with adults within the specialty of Inherited Metabolic Disorders (IMDs), including PKU. To address this, a group of experienced dietitians specializing in IMDs created a standard operating procedure (SOP) on the dietetic management of adults with PKU to promote equity of care in IMD dietetic services and to support service provision across the UK. The group met virtually over a period of 12 months until they reached 100% consensus on the SOP content. Areas of limited evidence included optimal blood phenylalanine reporting times to patients, protein requirements in older adults, management of weight and obesity, and management of disordered eating and eating disorders. The SOP does not include guidance on maternal PKU management. The SOP can be used as a tool for training dietitians new to the specialty and to raise the standard of education and care for patients with PKU in the UK.

Keywords: phenylketonuria; adult phenylketonuria; standard operating procedure; inherited metabolic disorders; dietary management; phenylalanine; protein substitute

1. Introduction

Phenylketonuria (PKU) is an autosomal recessive disorder of protein metabolism that is caused by a deficiency of phenylalanine hydroxylase, the enzyme which metabolizes the amino acid phenylalanine to tyrosine. The incidence in the UK is 1 in 10,000 [1], with regional variations. Individuals are recommended to follow a lifelong phenylalanine-restricted diet, supplemented with a low-phenylalanine protein substitute [2,3] to protect the brain from the toxic effect of elevated phenylalanine. In the UK, PKU is detected through neonatal screening, which began in 1969.

Neonatal screening and subsequent early diagnosis and initiation of treatment have changed the outcome of PKU [4], enabling the affected individuals to reach their full cognitive and intellectual potential. The healthcare and social-care savings are highly significant, as individuals do not need institutional care. Those with late-treated PKU are more likely to require special community care packages [5]. The burden of dietary treatment to individuals and carers cannot be underestimated [6–9].

A range of cognitive sequelae are seen in some patients with PKU [10–12]; however, the impact of current phenylalanine levels compared to historical childhood control is still uncertain [3]. There are variations in reported psychosocial outcomes for adults with PKU and indications that partial adherence to treatment negatively impacts on quality of life [8,10].

Dietitians play an important role in helping patients access and achieve effective treatment for PKU. There are several established metabolic centers across the UK that are dedicated to supporting adults living with an inherited metabolic disorder (IMD), including PKU. The needs of adults living with PKU are considerably different from those of pediatric patients, and these change over time as individuals become older. Research has indicated that transition of patients with PKU to adult services is successful with maintenance of metabolic control and high levels of patient engagement [13,14]. Adult clinics also support up to 23% of patients who are not following dietary treatment [15], usually because they maintain phenylalanine levels within target range without treatment (hyperphenylalaninemia) or the dietary treatment was discontinued in childhood by medical teams prior to life-long treatment recommendations. There are adult patients who recognize the benefits of maintaining lower phenylalanine levels but find it too challenging and impractical to sustain dietary treatment. Maintaining contact with this group of patients is important to monitor clinical outcome; to ensure good overall nutritional status; and to keep them informed of any treatment recommendation changes, new research, and developments. A number of adults with PKU choose (and are supported) to restart dietary treatment after a period of discontinuation in adolescence and/or adulthood [16]. Adults with PKU are a highly heterogeneous patient group in terms of treatment history, which includes late diagnosed and late treated, untreated, early treated who have stopped treatment at different stages in childhood, and early and continuously treated patients. This variability in treatment exposure may be reflected in a spectrum of different cognitive, co-morbidities, and life outcomes in adults with PKU attending metabolic clinics.

Dietitians working in the field of adult IMD have scarce access to formal specialty training. Few rotational or dietetic training posts exist within the UK, and therefore identifying the need for and creating a Standard Operating Procedure (SOP) forms part of the standardization of training and dietetic care for adults with PKU. Within the British Inherited Metabolic Disease Group-dietitians' group, there is a subgroup for adult dietitians. The adult dietitians group meets to specifically discuss dietetic management, develop resources, and arrange adult-focused education and training events to support learning and development within the specialty.

The publication of the first European PKU guidelines in 2017 set out clear standards for care, including for adults with PKU [2,3]. The guidelines explicitly state the need for adult metabolic services that are staffed by healthcare professionals with training in this specialty.

Standard Operating Procedures set out clear guidance about what needs to be achieved to support best practice, ensure transparency, and reduce ambiguity [17]. The aim of this dietetic SOP is to outline the role of the dietetic team in treating adults with PKU. The dietitian is an autonomous practitioner, and this SOP does not replace the dietitian's decision-making about the care of each individual patient, using evidence and his or her clinical judgment [18]. Dietitians have unique skills to counsel regarding dietary care. This document defines the standards of care that should be offered to all adults with PKU attending specialist care in the UK to ensure equity. This was guided by the first publication of the European PKU guidelines [2,3].

2. Materials and Methods

Eight experienced Dietitians specializing in the care of adults with IMDs in the UK met regularly over 12 months (September 2020–September 2021) to discuss the best practice in PKU care in the UK and to create the SOP. The SOP was based on the European PKU guidelines [3] and clinical expertise. Meetings were held virtually for one hour every 1–2 months, with a total of nine over one year. After each meeting, the draft SOP was emailed to all group members who reviewed and commented on this before the next meeting. All ideas and opinions were discussed at the following meting and adjustments made to SOP after 100% verbal consensus at each stage.

This SOP was based on existing SOPs at individual centers which were reviewed and further developed, and then a 100% consensus gained within the group in the meetings. The core group consisted of experienced IMD dietitians working in England and Scotland, and comments were sought from dietitians working in Wales and Northern Ireland to ensure that the whole of the UK was represented.

Once written, the SOP was reviewed by seven adults with PKU via an anonymous online survey, the British Inherited Metabolic Diseases (BIMDG) dietitians' group, the BIMDG committee, and the National Society for PKU (NSPKU). Feedback was provided and the SOP adapted as required.

The following areas were discussed: (1) glossary, (2) scope, (3) clinical SOP introduction, (4) aims and objectives of the SOP, (5) duties of the adult IMD dietitian, (6) SOP delivery and implementation, and (7) monitoring and assurance. In Appendix A, the section on SOP delivery and implementation examines dietetic assessment and interventions for adults with PKU. These sections include additional adult-specific areas, such as weight management and obesity, eating disorders or disordered eating, and patients who have discontinued dietary treatment.

A separate SOP for maternal PKU will be developed in the future.

3. Results

The full SOP is given in Appendix A.

This SOP addresses the standards of dietetic care and intervention for adults with PKU. The aspects of care described in the SOP include the following:

- Aims of dietetic care.
- Dietetic assessment for patient on and off treatment.
- Interventions, including the following:
    - Protein substitutes.
    - Avoidance of foods high in phenylalanine.
    - Prescribed special low-protein foods, e.g., low-protein bread or pasta.
    - Importance of including naturally low-protein foods, such as fruits and vegetables.
    - Specific considerations for females.
    - Adults not on treatment.
    - Those returning to diet.
    - Late treated PKU starting back on diet.
    - Weight management/obesity.
    - Eating disorders.
    - Blood phenylalanine monitoring.
    - Nutritional blood biochemistry/nutritional status.
- Patient follow-up.
- Dietetic contact with patients between outpatient appointments.
- Signposting to other services.
- Discharge or transfer from service.
- Outcome measures.
- Resources.

Variance from the European guidelines [3] occurred where differences in practice across the centers was evident or barriers existed to implementation of the guidelines. Areas requiring further consideration and research included the timescale of informing patients of their phenylalanine blood results, protein requirements, and the inclusion of the assessment and management of disordered eating and eating disorders.

It was agreed that dietitians should report blood phenylalanine results within three days of receipt from the hospital laboratory. All members of the group shared their experience of managing patients with PKU who described disordered eating behaviors. The SOP therefore includes guidance on the identification of disordered eating and eating disorders, provision of support, and signposting to other services if an overt eating disorder was suspected.

It is recommended that the SOP is reviewed every 3 years or is updated within 6 months if any new evidence or guidance is published that necessitates a change in practice. The authors also recommend that all services should perform an annual audit by using a representative sample of patients, using this SOP as a benchmark.

4. Discussion

This PKU Adult dietetic SOP is a practical interpretation of the European PKU guidelines [3]. It helps the adult IMD dietitian to translate and further develop the guidance into care in the UK. This document is the first consensus SOP for the dietetic management of an IMD in adults in the UK. Its purpose is to promote care equity for patients with PKU, followed up in IMD dietetic services across the UK and to support service provision. It can be used as a tool for training dietitians new to the specialty.

Patient-centered care is important to build positive dietitian–patient relationships. These relationships enable problem-solving, engagement in care, and earning of patient trust [19]. Working in collaboration with patients and carefully considering their beliefs and values will help guide shared decision-making between the dietitian and the patient [18]. The World Health Organization defines patient-centered care as care that which "meets people's expectations and respects their wishes" [20]. The dietitian can use the SOP as a treatment guide whilst maintaining patient-centered care at the forefront of management.

To provide holistic nutritional care, the SOP examines aspects of care specific to adults with PKU, including protein intake, weight management and obesity, eating disorders or disordered eating, non-dietary treatment, and patients lost to the service and co-morbidities.

Calculation of protein requirements

The calculation of protein requirements for adults with PKU was considered (Table 1). There are two components: (1) calculation of total protein requirements and (2) calculation of the dose of protein substitute required (which usually provides 52–80% of the total protein intake for a person with PKU treated with a phenylalanine restriction only [21]). The level and type of physical activity undertaken by individuals when calculating their protein requirements should also be considered.

The European PKU guidelines propose "*providing an additional 20% of L-amino acids to compensate for the 'digestible indispensable amino acid score' and also a further 20% of L-amino acids to optimize their impact on blood Phenylalanine control*" [3]. The incremental factors serve to compensate for the reduced uptake and utilization of amino acids from protein substitutes and offer metabolic benefits from the large neutral amino acid (LNAA) content. The above refers to protein substitutes derived from L-amino acids, and there may be differences in protein utilization with casein-glycomacropeptide (C-GMP) protein substitutes [22].

Minimum protein requirements are commonly derived from "safe levels" of protein intake [23] that are age-specific until the age of 19 years and then remain constant over the adult lifespan. In a recent review paper, Firman et al. [24] suggests that this may not be suitable for older adults with PKU with higher demands for protein associated with ageing. More research is needed to understand optimal protein needs for adults at different life stages and to investigate the body composition of older adults with PKU.

Given the awareness of overweight and obesity amongst adults with PKU [25], it is recommended that protein requirements be based on ideal body weight [3,26]. It is also important to consider patient tolerance of higher doses of protein substitute and the energy balance implications of additional calories supplied at higher prescribed doses of protein substitute.

Table 1. Outlining different ways of calculating protein requirements in adults.

	Parameter	Evidence Supporting	Protein Requirement Recommendations
1	Safe protein intake per kilogram of body weight per day	[23]	0.83 g/kg/day
2	Reference nutrient intake for protein	[26]	0.75 g/kg/day
3	Use of Indicator Amino Acid Oxidation Method for protein requirement calculation	[27]	0.93–1.2 g/kg/day
4	Appropriate protein requirements for older adults (> 65 years)	[26,28]	1.2–1.5 g/kg/day
5	Protein requirements for injury and disease (adults)	[28]	1–1.5 g/kg/day
6	Appropriate protein requirement adjustments for obesity	[3,28]	BMI > 30–75% of calculated requirements for actual body weight BMI > 50–65% of calculated requirements for actual body weight

Weight management and Obesity

In 1982, White et al. [29] observed an increased likelihood of an increased body mass index (BMI) in children with PKU. Since then, several studies have found the female PKU population (both adults and children) to have increased levels of overweight and obesity in comparison to the general population [25,30,31]. In a recent systematic review, Rodrigues et al. [32] conducted a meta-analysis and found that the BMI of patients with PKU was similar to their healthy controls; however, a subgroup of patients with classical PKU had a significantly higher BMI. The meta-analysis dataset included both adults and children; the age range was between 0.2 and 52 years. The authors also noted a trend towards a higher BMI in females with PKU in all studies with male and female datasets.

Interestingly, it has been noted that LDL cholesterol and other biomarkers of increased cardiovascular risk that may be increased in obesity are not elevated in patients with PKU. In fact, studies have shown biomarkers of cardiovascular risk, including LDL cholesterol, were reduced in healthy participants with PKU [33,34]. It is not currently known if the decreased levels of cardiovascular biomarkers in PKU confers a protective effect against cardiovascular events in the PKU population.

The likelihood of a patient with PKU being overweight or obese does not correlate with choice of protein substitute [35] and may be associated with treatment adherence. Cammatta et al. [31] observed no correlation between treatment adherence and prevalence of obesity in Brazilian patients with PKU. However, in UK patients over 16 years old, high phenylalanine levels were found to correlate with obesity [36]. Cammatta et al. [31] also observed that 94% of patients with PKU were sedentary.

It is important that the need for weight-management advice, including advice around exercise and activity, is considered within the dietetic-assessment process for all patients with PKU. Further work is needed to monitor the incidence of overweight and obesity and identify the underlying causes in all patients with PKU. Referral to specialist weight-management services (with appropriate support from the IMD dietitian) may be indicated.

Bariatric surgery is also possible for adults with PKU who meet the referral criteria; however, careful consideration is needed for both pre- and post-operative management to ensure that a phenylalanine-restricted diet can be maintained.

Disordered eating and eating disorders

Disordered eating and eating disorders occur in adults with PKU. Disordered eating is described as eating behaviors that are lower in severity and intensity than that of an eating disorder. However, both can have an impact of everyday life of the adult with PKU.

The occurrence of eating disorders is recognized in the European Guidelines for PKU [3], but due to the paucity of the literature, they could only recommend that this area required further study. The prevalence of eating disorders self-reported in the PKU patient population is significantly higher than in the general population [37]. Patients with disordered eating are also at a greater risk of developing eating disorders and should have early referral to specialists in psychology and dietetics [20].

Studies also suggest that patients with poor metabolic control are more likely to exhibit symptoms of disordered eating and may be more at risk of developing eating disorders [21,38]. In adolescents and adults with PKU, the occurrence of eating disorders has not been systematically reviewed and is under-reported, so it may not be detected and treated [3].

Disordered eating patterns may be common in patients with PKU without their having an overt eating disorder; regular health-professional support, especially from a psychologist, may provide some measure of protection [3]. Contact with the patient's general physician and signposting to local support agencies may be warranted as appropriate.

Diagnosing an eating disorder in a patient with PKU is challenging. Existing validated tools for eating disorders may not be appropriate for in individuals with PKU, as they often answer questions differently, due to their prescribed dietary treatment. This can produce false positive or low sensitivity at identifying an eating disorder [38]. Another challenge is the treatment of PKU versus the treatment of the eating disorder. The treatment of PKU involves a low-protein diet which restricts foods high in protein. This is at odds with the treatment of eating disorders such as anorexia nervosa, where the aim of treatment is to remove the self-imposed restriction of food. Regarding referral and treatment of an overt eating disorder, appropriate national guidelines [39,40] and/or local policies should be followed.

It is important that IMD dietitians support individuals with PKU diagnosed with an eating disorder and work in close liaison with dietitians specializing in eating disorders and the wider MDT in a shared care approach. The eating-disorders team is unlikely to have any experience in managing PKU.

Reporting Blood Phenylalanine Concentrations

The NHS England Specialist Services Quality Dashboard for IMD Services [41] directs laboratories to report results within three days of receipt. The European guidelines [3] advise that the ideal standard for time between blood sampling and receiving results should be no more than five days. Barriers to reporting results within five days of the sample being taken include delays in postal service and samples not being posted/given to the laboratory immediately after the procedure is completed. The Australasian PKU Guidelines do not suggest any specific timeframe but advise that dietitians should report results to patients as soon as possible once received from the laboratory [42]. It is important that blood phenylalanine results are reported promptly so that patients can recall how they managed their PKU in the immediate period prior to the blood test, and timely changes can be advised to maintain metabolic control. The European PKU guidelines also recommend that adults should have their phenylalanine concentrations measured monthly [3]. The current group acknowledged that dietitians can only be responsible for the time between results being reported by laboratories to the patient receiving their results. Therefore, for the purposes of this SOP, a realistic standard for patients receiving their results from the dietitian was agreed at three days from receipt of blood results from the laboratory.

The best practice is to report the phenylalanine result as soon as possible, but the group acknowledges that this is not always practical, due to inadequate staffing levels. Three days was agreed on an arbitrary basis and is a pragmatic goal for the timeframe of blood phenylalanine reporting.

Non-Dietary Treatments for PKU

Currently there is only one non-dietary adjunct treatment, sapropterin, that has recently been funded by NHS England only for treating adults with PKU. Dietitians will adjust natural protein and protein substitute intake, as well as (potentially) sapropterin dose, for patients who are responsive to this therapy. In Northern Ireland and Wales, sapropterin is routinely available for people with PKU up to the age of 22, and it is hoped that access will be extended to adults. Scottish healthcare has not yet commissioned sapropterin for routine use as a treatment for PKU. Sapropterin management protocols are currently being agreed.

Maintaining Patient Engagement and Avoidance of Patients Being "Lost to Follow Up"

Adult patients vary greatly in their neurocognitive abilities, from having profound learning disabilities and high levels of dependence on nursing care for engagement with treatment (associated with late treated PKU) to complete independence with the dietary regimen. Adults with PKU can present with levels of functioning in between these points, with subtler executive function deficits.

Patients' variable neurocognitive abilities or executive function deficits that are associated with their heterogeneous treatment experiences and disorder severity need consideration when organizing adult clinics. Impairment of working memory, planning, cognitive flexibility, and sustained attention [8,10] is likely to impact on consistent clinic attendance.

The European Guidelines recommendation is that all adults with PKU should be under systematic follow-up at specialist metabolic clinics and organization of clinics should support adults' continued engagement [3]. Mechanisms such as reminders to attend just prior to appointments, additional telephone or text messages prompting attendance, and removal of barriers to re-access clinics after missing appointments support better outcomes than systems which discharge patients after a one- or two-time non-attendance. Transition of patients from pediatric to adult clinics is a point in care when patients might be "lost to follow up" for a variety of reasons. Robust transition arrangements will reduce this [21]. Finally, remote clinic appointments using video or telephone calls may support patient attendance if (independent) travel is a barrier to attending adult clinics.

Shared Care

Services caring for people with long-term conditions need to consider the holistic needs of patients with co-morbidities, particularly if this affects dietary management. Co-morbidities may include diabetes mellitus, cancer, inflammatory bowel disorders, irritable bowel syndrome, and dysphagia (late treated) [43]. Collaboration with other medical teams is necessary to advocate for and support PKU treatment alongside concurrent treatments and management of co-morbidities. Additionally, awareness of the impact of PKU management on concurrent conditions or illnesses is essential to adequately support adults with PKU. Although PKU is not a decompensating metabolic disorder, during any hospital admission, provision of a phenylalanine restricted diet supplemented with a low-phenylalanine protein substitute should be organized and supplied. If there is a requirement for enteral feeding, a modular feed using the protein substitute, a natural protein source, fat and carbohydrate modules, and electrolytes can be designed. IMD dietitians should work collaboratively with services supporting hospital admissions and consider any comorbidities to ensure that the requirements of PKU are considered alongside their treatment.

Monitoring and assurance of the SOP

It is important that the SOP is reviewed regularly (every 3 years) to ensure that it remains up to date and informed by clinical practice. If any new evidence or guidance is published which necessitates a change in practice, the SOP will be revised within 6 months of publication. Adult IMD dietitians can use this SOP as a benchmark to audit their service. By providing agreed and defined national guidance for dietetic treatment of PKU in the UK, this SOP will allow all Adult Inherited Metabolic Disorders (AIMD) services to audit provision of care against an agreed national standard. This will also promote consistency of care between services. The SOP will be disseminated via the BIMDG dietitians' group. This provides an exciting opportunity for services to collaborate on a national audit or future research, with the SOP defining agreed outcomes of dietetic care.

Limitations

The SOP is based on the consensus opinion drawn from the experience of the authors and their interpretation of a scarce evidence base. As the authors are UK-based dietitians working within the UK National Health Services, there is a primary focus on UK services. Official methodology was not used to reach consensus, but 100% consensus was reached in all aspects of the SOP.

There are minimal outcome data on early and continuous treated adults and late treated adults with PKU. The provision of dietary care within adult IMD services (in the UK) is variable due to lack of funding and limited dietetic staffing, which may prevent the recommendations in the SOP from being incorporated into practice.

This SOP document is for the dietetic care only; it does not include the role of the rest of the IMD team in management of the adult with PKU. As more non-dietary treatments become available and adults are at increased risk of other co-morbidities, e.g., diabetes and metabolic syndrome, then future work on the SOP should include the role of the whole team.

5. Conclusions

This is the first dietetic SOP for adults with PKU in the UK. The SOP outlines the role of the dietetic team in treating adults with PKU. The SOP and this supporting publication aim to strengthen service provision and achieve equity in the dietetic management of patients in the UK with PKU. The SOP is a consensus based on experience in an area where there is a limited or minimal evidence base to support dietetic management at the present time.

As further non-dietary treatments are expected to become available in the UK, the SOP will be updated to reflect this. Future work is needed, especially in key areas where current evidence is scarce. These include determining protein requirements across the adult lifespan, developing strategies to effectively prevent and manage obesity, and improving the understanding of etiology and optimal treatment approaches with regard to eating disorders. Research focused on adults with PKU remains a high priority to ensure optimal care throughout the lifespan.

Author Contributions: Conceptualization, L.R., S.A., C.E., S.F., M.H., G.R., A.W. and C.Y.; funding acquisition, L.R. and S.F.; project administration, L.R.; supervision, A.M.; writing—original draft, L.R., S.A., C.E., S.F., M.H., G.R., A.W. and C.Y.; writing—review and editing, L.R., S.A., C.E., S.F., M.H., G.R., A.W., C.Y. and A.M. All authors have read and agreed to the published version of the manuscript.

Funding: This research received no external funding. The publication cost was funded by the National Society for Phenylketonuria (NSPKU) and University Hospitals Birmingham Charity.

Institutional Review Board Statement: Not applicable.

Informed Consent Statement: Not applicable.

Data Availability Statement: Not applicable.

Acknowledgments: The development of the PKU SOP was guided by the SOP standards from the Nutrition and Dietetic Department at Guy's and St Thomas' Hospital. We thank all who reviewed the SOP, including Sarah Bailey, Clinical Lead Dietitian, The All Wales Inherited Metabolic Disease Service; Nicola McStravick, Advanced Practitioner Dietitian in IMD, The Royal Hospitals Belfast,

Belfast Health and Social Care Trust; Sarah Firman, Specialist Dietitian Guy's & St Thomas' NHS Foundation Trust; British Inherited Metabolic Disorders (BIMDG) dietitians' group, BIMDG committee, National Society for Phenylketonuria (NSPKU) committee; Adults with PKU. The National Society for Phenylketonuria (NSPKU) and University Hospitals Birmingham Charity for funding publication costs.

Conflicts of Interest: L.R. is a member of the Nutricia Danone advisory board and received honorarium from Nutricia and Vitaflo. S.A. is a member of the Nutricia Danone Advisory Board and for Meta Healthcare, has received honoraria/speaker's fees from Nutricia and Vitaflo and financial support for attendance at meetings from Nutricia, Vitaflo and Cambrooke. C.E. has received honoraria and educational grants to attend events from Vitaflo, Nutricia, Meta Healthcare and SOBI and is a member of the Advisory Committee on Borderline Substances. S.F. is a member of the advisory board for Nutricia Danone and Meta Healthcare and has received financial support and honoraria from Cambrooke and Vitaflo. M.H. is a member of advisory board Nutricia Danone and Applied Pharma and has received financial support and honoraria from Nutricia, Vitaflo, Cambrooke, Mevalia, and Promin for attendance at conferences/meetings and speaker's fees. A.W. has received speakers fee from Nutricia and honoraria from Travere. C.Y. has received educational grants and financial support for attendance at conferences/meetings from Nutricia.

Appendix A

SOP for the Dietetic Management of Adults with Phenylketonuria (PKU) in the UK

Written by Louise Robertson, Sarah Adam, Charlotte Ellerton, Suzanne Ford, Melanie Hill, Gemma Randles, Alison Woodall, and Carla Young on December 2021.

Contents

1. Glossary

- Patient: patient or patient advocate.
- Adult Inherited Metabolic Disorders Dietitian (AIMD dietitian): a dietitian who works with adults who have an inherited metabolic disorder.
- Patient-centered approach: treating the patient as an individual and an equal partner in the healthcare management.
- Psychosocial: how social conditions affect mental health or how someone copes with PKU.
- Neurocognitive: the ability to think and reason. This includes the ability to concentrate, remember things, process information, and understand.
- Capacity: to use and understand information to make and communicate a decision.
- Protein substitutes: a medical food containing all amino acids, except/very small amount of phenylalanine.
- Prescribed low-protein foods: foods manufactured to be very low in protein only found on prescription in the UK.

- Phenylalanine exchange: one exchange is the amount of food that contains 1 g of protein or 50 mg phenylalanine.
- NSPKU: The National Society for Phenylketonuria—patient society in the UK.
- BIMDG: The British Inherited Metabolic Diseases Group—health-professionals interest group in the UK.
- ACBS: Advisory Committee on Borderline Substances—The ACBS is responsible for advising on the prescribing of borderline substances for use in the NHS primary care. Borderline substances are nutritional or dermatological products that have been specially formulated to manage medical conditions.

2. Scope

- This SOP outlines the dietitian care pathway for adult patients (16+ years) with Phenylketonuria (PKU) under the care of Adult Inherited Metabolic Disease (AIMD) teams in the UK.
- This SOP does not cover management of maternal and preconception patients with PKU.
- This SOP does not cover the management of acute inpatient admissions
- The role of the AIMD dietitian in PKU care is highlighted.
- This document is to be used with the clinical judgment of the dietitian to tailor it to the adult with PKU.
- The roles of other healthcare professionals are noted, although the entirety of their role in the pathway has not been included.

3. Introduction to the clinical SOP

- PKU is the most common inborn error of protein metabolism with an incidence of approximately 1 in 10,000 births, with varying incidence across the UK. Management by restricted dietary intake of phenylalanine (natural protein), along with supplemented phenylalanine-free amino acids (L-AA) or glycomacropeptide (GMP) [3,21], remains the mainstay of management in the UK.
- The current European Phenylketonuria Guidelines [3] recommend that all patients with PKU remain on treatment/restricted diet for life if phenylalanine is >600 µmol/L without treatment.
- Across the UK, there are recognized adult metabolic centers.
- This procedure is necessary to achieve the following:
    - Standardize the dietetic care of all adult patients with PKU across the UK.
    - Provide a framework to support the dietitian's decision-making around treatment of patients with PKU.
    - Assist development and supervision of AIMD dietitians in the UK and ensure that all AIMD dietitians are providing equal standards of care to patients with PKU.

4. Aim and Objectives of this SOP

- To outline the role of the AIMD dietetic team in provision of care to adults with PKU.
- To ensure equity of patient care throughout the UK.
- To agree on standards for patient care to ensure patient safety and optimal care provision by referring to European PKU guidelines 2017 [3].
- To outline the service provision required to provide optimal care.

5. Duties of the AIMD dietitian

- It is the responsibility of the AIMD dietetic service to implement the procedures and provide best practice care, as outlined in this document.
- All members of the AIMD dietetic team have a role in advocating for adults with PKU to receive the care that is aligned with this document, unless this is otherwise indicated in the course of the review.
- All members of the AIMD dietetic team are required to escalate any patient management not within the dietetic scope of practice to an appropriate IMD team member.

6. SOP Delivery and Implementation

6.1 Key stakeholders in the SOP

- The AIMD Dietitians in AIMD centers across the UK.
- The metabolic team, including physician, clinical nurse specialist, and dietetic assistant to support implementation of the SOP.
- Patients, their families, carers and advocates.
- The British Inherited Metabolic Diseases Group (BIMDG).
- National Society for Phenylketonuria (NSPKU).

6.2 Dietetic Assessment and Interventions for Adults with PKU

Please also refer to Appendix B Pathway for Dietetic Management of Adults with Phenylketonuria (PKU) in the UK.

Aims of dietetic care

- To optimize normal neurocognitive and psychosocial functioning for the patient.
- To support adults with PKU identify their personal aims and goals through the lifecycle.
- To ensure the patient is fully informed on best practice management of PKU in accordance with European PKU Guidelines 2017 [3] and to support the patient to make informed treatment decisions.
- To educate on how to maintain phenylalanine levels between 120 and 600 μmol/L [1].
- To encourage lifelong PKU management.
- To ensure the diet is nutritionally adequate.
- To help promote a healthy weight.

6.2.1 Dietetic Assessment

For patient on a phenylalanine restricted diet

- Check identification of the patient and seek consent for assessment.
- Medical/surgical history.
- Psychosocial considerations, e.g., change of living circumstances.
- Anthropometry: weight, height, and body mass index.
- Current clinical issues.
- Relevant medications, including protein substitutes and prescribed low-protein foods.
- Biochemistry: nutritional status bloods (refer to European PKU guidelines [3]) and history of blood phenylalanine monitoring.
- Diet history, including the following:
    - Total protein intake (including food sources) and distribution over the day; prescribed and actual intake.
    - Quantity and timing of protein substitute, prescribed and actual intake.
    - How much low-protein food is being used and confidence with incorporating low-protein foods in the diet.
    - Menu planning and cooking skills.
    - Home delivery/local dispensing of protein substitutes and low-protein foods.
    - Discussion regarding patient's regulation of protein intake, e.g., if he or she is using phenylalanine exchange system/counting grams of protein.
    - Meal timings.
    - Additional vitamin and mineral, omega 3 supplementation, and history of nutritional deficiencies.
    - Overall dietary adequacy, including assessment of total energy intake.
- Patient- and non-patient-related factors affecting treatment management and any specific concerns the patient has relating to his or her PKU.
- Discussion regarding prescription charges (if appropriate).
- To explore relationships with food if concerns are raised.
- Calculated protein requirements; refer to the European PKU guidelines [3].

For patients not on treatment

- Check identification of the patient and seek consent for assessment.

- Medical/surgical history.
- Psychosocial issues.
- Anthropometry: weight, height, and body mass index.
- Current clinical issues.
- Medication (including non-prescribed medications, e.g., herbal remedies and probiotics).
- Biochemistry: nutritional bloods (and history of blood phenylalanine monitoring).
- Diet history, including the following:
    - Total protein intake and distribution.
    - Meal timings
    - Any extra nutritional supplementation of vitamins and minerals, trace elements, and omega 3.
    - Overall dietary adequacy.
    - Protein aversion.
- Patient- and non-patient-related factors affecting treatment management and any specific concerns the patient has relating to his or her PKU.
- Exploring barriers to being on treatment.
- Patient education/update on the management of PKU.

6.2.2 Interventions

Protein substitutes

- Advise on adequate dose.
- Consider nutritional composition of prescribed protein substitute intake; is it nutritionally complete or is additional micronutrient supplementation required?
- Advise on when it should be taken.
- Advise on any new alternative protein substitutes—amino acid/GMP substitutes.
- Offer to arrange patient samples.
- Discuss tolerability and/or barriers to management adherence.

Patient switching protein substitute or starting new protein substitute

- Discuss new substitute regimen with patient (if required).
- Send prescription request letter to GP.
- Discuss collection options with patient, e.g., pharmacy or home delivery.
- Seek verbal or written permission to contact home-delivery company to register patient and update the company on the patient's current prescription if appropriate.
- Advise GP that a home-delivery company will manage prescription requests on behalf of the patient.

Avoidance of food high in phenylalanine

- Educate patients about the practicalities of a low-phenylalanine diet, considering individual phenylalanine tolerance and patient preferences.
- Education should include the following:
    - Avoiding high-phenylalanine foods.
    - Suitable natural low-phenylalanine foods.
    - Measuring and counting phenylalanine exchanges.
    - Avoidance of aspartame and discuss suitable phenylalanine-free sweeteners.
    - Appropriate alcohol consumption.
    - Provide sufficient resources to prepare low-phenylalanine meals.
    - Ensure patient understands how to read food labels.

Prescribed low-protein foods

- Ensure patients receive enough supplies via ACBS prescription to meet calorie requirements and to allow variety in the diet.
- Advise patients on the availability of new special low-protein foods.
- Provide a list of special low-protein foods available on ACBS prescription.

- Advise on the NSPKU guidance—up to 50 units per month (excluding low-protein milk alternatives and protein substitutes).
- Arrange special low-protein food samples if requested.
- Outline the system on how to obtain regular supply of special low-protein foods on prescription/home delivery, as above.

Females

- If appropriate, discuss the importance of the strict low-phenylalanine diet for pregnancy and planning a pregnancy.
- Signpost to obtaining contraception if appropriate.
- Ensure patient knows what to do if she finds out that she is pregnant.

Adults not on treatment

- Ensure adequate intakes of macro- and micronutrients.
- Discuss benefits of lifelong PKU management.
- Advise on support available.
- Discuss importance of attending annual appointments and keeping in touch.

Patients returning to a low-phenylalanine diet

- An appropriate step-by-step patient-centered approach should be used if a patient wishes to return to dietary treatment.
- Discuss with the patient the responsibilities of the dietitian and the patient.

Late Treated PKU patients starting back on diet
Determine:

- If the patient has capacity.
- Baseline behaviors and functions, communication limitations, support needs, and support/care package.
- Possible previous experience of the diet, number of phenylalanine exchanges, and protein substitute used.
- Tolerance of any monitoring, i.e., finger-prick blood taking (including blood spot and capillary).

Identify:

- Number of phenylalanine exchanges, items needed on prescription.
- Key personnel/carers/cooks to teach concepts of low-phenylalanine diet.
- Devise practical menu plans for care homes or equivalent.
- Anthropometric monitoring—weekly weight charts by carers.
- Review dates for carers' feedback on any behavior changes (improvements).
- 3-month trial to identify if the diet is helping or not.

Weight Management/Obesity

- Identify any history of previous strategies used to manage weight or restrict diet. Discuss with the patient the efficacy of these previous strategies from the perspective of weight loss. Explore impact on mental and physical health, and quality of life.
- Assess information on previous attempts at weight loss which were unsuccessful or not sustainable. Use this information to inform the current weight-management strategy.
- Identify health issues or current medical treatment which may impact on the weight-management strategy, e.g., mental-health issues, medical conditions and treatments, socioeconomic factors, age, gender, culture, ethnicity, and personal support mechanisms.
- Assess risk of comorbidities by monitoring lipid profile, blood pressure, and HbA1c (44).
- Assess understanding of the wide range of dietary and nutritional information available. This can be overwhelming and may be a barrier to a weight-management plan.
- Tailor the education and supporting information provided to aid understanding and reduce any barriers which may have formed.
- Individually tailor patient education to help the patient identify realistic goals. This should include no more than two or three diet and lifestyle changes at a time.

- Focus education to promote understanding on food choices to support a low-protein-food diet. For example, at meals, fill half the plate with vegetables, which are naturally low in protein and calories.
- Advise on the lower calorie protein substitute options and also ensure an adequate intake of the protein substitute and micronutrients. This will help ensure nutritional balance which is essential to a healthy weight loss to ensure nutritional to and supporting weight loss.
- Regular physical activity of a moderate intensity is recommended to help support and maintain health and weight loss. NICE (2014) [44] recommends 45 to 60 min exercise, for example brisk walking of cycling, per day as part of a weight loss program.
- If there are significant barriers in place towards weight loss then discuss the possibility of delaying the weight management until a more appropriate time.
- Consider referral or signposting to other services or organizations for support if this is indicated. This could include referring to a weight-management service or program that the AIMD dietitian can then adapt to suit a low-protein diet for patients with PKU.

Eating Disorders

- Identify any history of eating disorder from referral into adult service/liaise with pediatric service for more information and if any treatment received or ongoing.
- Identify common behaviors of eating disorders, i.e., missing meals; food avoidance; bingeing behaviors; and compensatory behaviors, including laxative or diet pill misuse, vomiting, or excessive exercise. Identify any issues with body image, irregular meal pattern.
- Identify any physical signs of eating disorders, e.g., excessive tiredness, feeling cold, dizzy, digestive problems, or dental problems unrelated to PKU.
- Identify if not having periods (females) unless due to contraception method or other medical conditions.
- Identity an unusually low or high body mass index (BMI).
- Any rapid weight loss.
- Whether they take part in activities associated with a high risk of eating disorders (for example, professional sport, fashion, dance, or modeling).
- Other mental-health problems.
- If appropriate, educate on effects of low-calorie intake on Phe control/adverse effect on phenylalanine levels.
- Signpost to eating disorders other local resources and charities, e.g., https://www.beateatingdisorders.org.uk/ (accessed on 24 August 2021).
- Refer to appropriate local services, e.g., GP, community mental-health team. Support patient with PKU diet if going through treatment for eating disorder alongside a specialist eating disorders Dietitian.

Phenylalanine monitoring

- Review blood phenylalanine control with patient.
- Monthly blood phenylalanine sampling is recommended to support dietetic management and understanding of blood phenylalanine control [1]. Tailored plans can be discussed with patient, e.g., increased frequency, whilst dietary changes are being made.
- Ensure patient has sufficient blood-sampling equipment.
- Support independence with taking blood samples or ensuring an appropriate plan for those unable to take their own blood samples.
- Encourage the patient to take blood samples at the same time of day so that results are comparable, i.e., fasting.
- Clinical nurse specialist can help trouble shoot issues with taking blood samples.
- Agree patient preference to receiving blood phenylalanine results as per local data governance, e.g., telephone call, text, and email.
- Aim to report the blood phenylalanine result back within 3 working days of receipt of the blood result from the lab.

Nutritional biochemical blood tests

- Discuss with medical consultant if these are required and refer to the European PKU guidelines [3].

Advances in research/developments

- To inform and discuss any new research, treatments, or guidelines as appropriate.

6.2.3 Follow-up

- Agree to follow up with patient.
- 6–12 month appointment if on treatment.
- 12 month appointment if not on treatment.
- Less frequent follow-up arrangements may be agreed upon if appropriate (e.g., male patients with hyperphenylalaninaemia).
- The option of video or telephone appointment to be considered if appropriate.

Follow-up contact in between clinic appointments

- Contact details provided for queries or help between appointments.
- Encourage patient-led approach to seek support and information as required.

Guiding patient to access support from other agencies or other Healthcare Professionals (HCP)

- Referral to other HCPs might be needed, e.g., psychologist.
- Signposting to services outside of the NHS, e.g., IAPT (Improving Access to Psychological Therapies (England)/mental health/GP (Northern Ireland).
- Supporting letters may be needed, e.g., for travel, applying for benefits (for example PIP), employers, etc.

Discharge/transfer arrangements if appropriate

- Discharge arrangements will vary between services.
- PKU is a long-term condition which should have lifelong treatment and metabolic-specialist follow-up.
- Some patients may have neurological and cognitive impairments, e.g., poor working memory, meaning that they may need extra support/reminders to attend appointments.
- Dietitians will facilitate patients transfer to another center if they relocate, e.g., university students.

6.2.4 Potential outcome measures

- Patient experience feedback.
- Knowledge and skills of managing diet.
- Attending appointments.
- Frequency of blood phenylalanine monitoring.
- Blood phenylalanine concentrations.
- Adherence to protein substitute.
- Variety in diet.
- Healthy body weight.
- No nutritional deficiencies.
- Good quality of life, e.g., PKU Quality of Life Survey.

6.2.5 Resources

- NSPKU diet booklet.
- Relevant center resources.
- Picture booklets on the NSPKU website.
- Company literature/websites and recipe books.
- Apps for smart phones or tablets.

7. Monitoring and Assurance

SOP group: The SOP working group will review this document every three years to ensure it remains up to date and informed by clinical guidance and evidence. The SOP was written in December 2021, and the review date will be in December 2024. If any new evidence or guidance is published which requires a change in practice, it will be updated within 6 months of publication. The working group will meet to update this.

Service level: AIMD dietetic services should use this SOP as a benchmark to audit provision of services to patients, to highlight gaps in services, and to identify changes in service provision required to conform to the latest guidelines and requirements and the in development of business cases.

It is recommended that each AIMD dietetic service complete an annual audit on a representative sample of patients on key outcomes outlined in this document (such as frequency of consultations and time take to report blood phenylalanine results) and act on the findings of the audit appropriately.

A suggestion for an audit tool which could be used on a representative sample of the patient group is outlined in Appendix C.

Appendix B

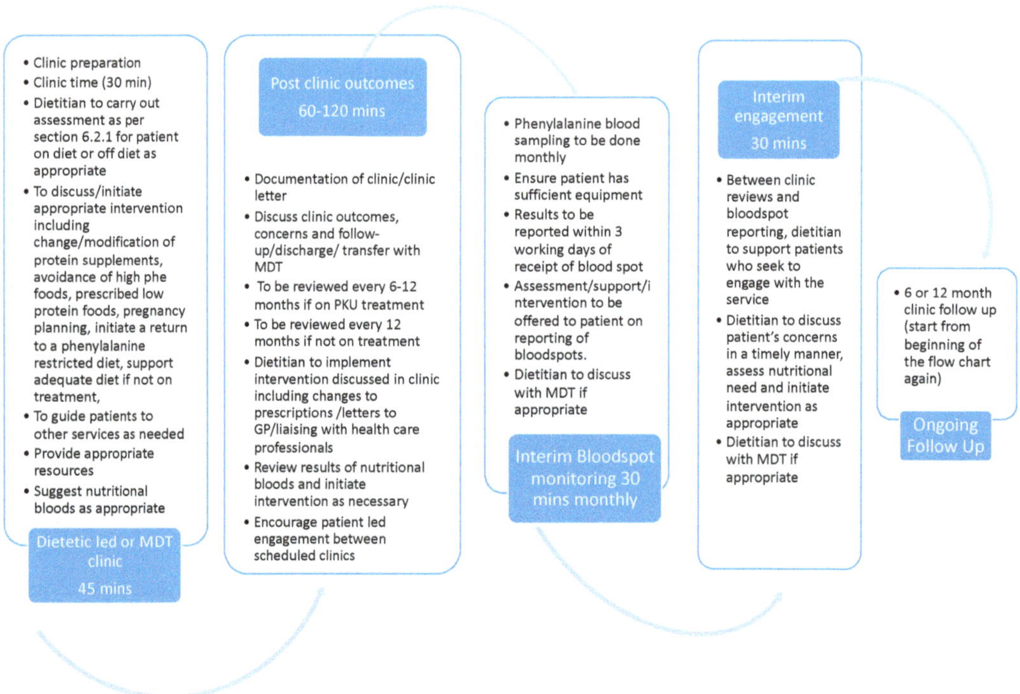

Figure A1. Pathway for Dietetic Management of Adults with Phenylketonuria (PKU) in the UK.

Appendix C

Table A1. Suggested Audit tool for the SOP.

Auditing adherence to the SOP for the Dietetic Management of Adults with Phenylketonuria (PKU) in the UK						
Date audit completed:						
Time period covered from:		To:				
Frequency of dietetic clinic consultation (MDT or Dietetic led)						
Patient identifier	Date of most recent clinic appointment offered (D1)	Date of previous clini-cappointment offered (D2)	Is the most recent appointment (D1) a rescheduled appointment due to the patient requesting a change of appointment? Y/N	If yes, what was the date of the appointment offered prior to D2 (D3)	Time (weeks) betweenap-pointments offered to patient (D2–D1 or D3–D1)	Recommended timeframe achieved? Y/N
Frequency of dietetic clinic consultation (MDT- or Dietetic-led)—simplified table						
Patient identifier	D1	D2	Is D1 pt requested reschedule? Y/N	If Y, D3	Time b/w D2–D1 or D3–D1	Recommendation achieved? Y/N?
Time taken to report phenylalanine blood sampling						
Patient identifier	Date result reported by lab (DA)	Date result reported to patient (DB)	Time (days) between date result reported by lab (DA) and date result reported to patient (DB)	Recommendation achieved? Y/N?		

References

1. Pandor, A.; Eastham, J.; Beverley, C.; Chilcott, J.; Paisley, S. Clinical Effectiveness and Cost-Effectiveness of Neonatal Screening for Inborn Errors of Metabolism Using Tandem Mass Spectrometry: A Systematic Review. In *NIHR Health Technology Assessment Programme: Executive Summaries*; NIHR Journals Library: Southampton, UK, 2004; Volume 8.
2. van Spronsen, F.J.; van Wegberg, A.M.J.; Ahring, K.; Bélanger-Quintana, A.; Blau, N.; Bosch, A.M.; Burlina, A.; Campistol, J.; Feillet, F.; Giżewska, M.; et al. Key European Guidelines for the Diagnosis and Management of Patients with Phenylketonuria. *Lancet Diabetes Endocrinol.* **2017**, *5*, 743–756. [CrossRef]
3. van Wegberg, A.M.J.; MacDonald, A.; Ahring, K.; Bélanger-Quintana, A.; Blau, N.; Bosch, A.M.; Burlina, A.; Campistol, J.; Feillet, F.; Giżewska, M.; et al. The Complete European Guidelines on Phenylketonuria: Diagnosis and Treatment. *Orphanet J. Rare Dis.* **2017**, *12*, 162. [CrossRef] [PubMed]
4. Enns, G.M.; Koch, R.; Brumm, V.; Blakely, E.; Suter, R.; Jurecki, E. Suboptimal Outcomes in Patients with PKU Treated Early with Diet Alone: Revisiting the Evidence. *Mol. Genet. Metab.* **2010**, *101*, 99–109. [CrossRef] [PubMed]
5. Brown, M.C.J.; Guest, J.F. Economic Impact of Feeding a Phenylalanine-Restricted Diet to Adults with Previously Untreated Phenylketonuria. *J. Intellect. Disabil. Res.* **1999**, *43*, 30–37. [CrossRef] [PubMed]
6. Eijgelshoven, I.; Demirdas, S.; Smith, T.A.; van Loon, J.M.T.; Latour, S.; Bosch, A.M. The Time Consuming Nature of Phenylketonuria: A Cross-Sectional Study Investigating Time Burden and Costs of Phenylketonuria in the Netherlands. *Mol. Genet. Metab.* **2013**, *109*, 237–242. [CrossRef] [PubMed]
7. MacDonald, A.; Smith, T.A.; de Silva, S.; Alam, V.; van Loon, J.M.T. The Personal Burden for Caregivers of Children with Phenylketonuria: A Cross-Sectional Study Investigating Time Burden and Costs in the UK. *Mol. Genet. Metab. Rep.* **2016**, *9*, 1–5. [CrossRef]
8. Palermo, L.; MacDonald, A.; Limback, E.; Robertson, L.; Howe, S.; Geberhiwot, T.; Romani, C. Emotional Health in Early-Treated Adults with Phenylketonuria (PKU): Relationship with Cognitive Abilities and Blood Phenylalanine. *J. Clin. Exp. Neuropsychol.* **2020**, *42*, 142–159. [CrossRef]
9. Rose, A.M.; Grosse, S.D.; Garcia, S.P.; Bach, J.; Kleyn, M.; Simon, N.-J.E.; Prosser, L.A. The Financial and Time Burden Associated with Phenylketonuria Treatment in the United States. *Mol. Genet. Metab. Rep.* **2019**, *21*, 100523. [CrossRef]

10. Aitkenhead, L.; Krishna, G.; Ellerton, C.; Moinuddin, M.; Matcham, J.; Shiel, L.; Hossain, S.; Kiffin, M.; Foley, J.; Skeath, R.; et al. Long-Term Cognitive and Psychosocial Outcomes in Adults with Phenylketonuria. *J. Inherit. Metab. Dis.* **2021**, *44*, 1353–1368. [CrossRef]
11. Burlina, A.P.; Lachmann, R.H.; Manara, R.; Cazzorla, C.; Celato, A.; van Spronsen, F.J.; Burlina, A. The Neurological and Psychological Phenotype of Adult Patients with Early-Treated Phenylketonuria: A Systematic Review. *J. Inherit. Metab. Dis.* **2019**, *42*, 209–219. [CrossRef]
12. Hofman, D.L.; Champ, C.L.; Lawton, C.L.; Henderson, M.; Dye, L. A Systematic Review of Cognitive Functioning in Early Treated Adults with Phenylketonuria. *Orphanet J. Rare Dis.* **2018**, *13*, 150. [CrossRef] [PubMed]
13. Mütze, U.; Thiele, A.G.; Baerwald, C.; Ceglarek, U.; Kiess, W.; Beblo, S. Ten Years of Specialized Adult Care for Phenylketonuria—A Single-Centre Experience. *Orphanet J. Rare Dis.* **2016**, *11*, 27. [CrossRef] [PubMed]
14. Peres, M.; Almeida, M.F.; Pinto, É.J.; Carmona, C.; Rocha, S.; Guimas, A.; Ribeiro, R.; Martins, E.; Bandeira, A.; MacDonald, A.; et al. Implementing a Transition Program from Paediatric to Adult Services in Phenylketonuria: Results after Two Years of Follow-up with an Adult Team. *Nutrients* **2021**, *13*, 799. [CrossRef] [PubMed]
15. Wood, G.; Pinto, A.; Evans, S.; Daly, A.; Adams, S.; Costelloe, S.; Gribben, J.; Ellerton, C.; Emm, A.; Firman, S.; et al. Special Low Protein Foods Prescribed in England for PKU Patients: An Analysis of Prescribing Patterns and Cost. *Nutrients* **2021**, *13*, 3977. [CrossRef]
16. Ford, S.; O'Driscoll, M.; MacDonald, A. Living with Phenylketonuria: Lessons from the PKU Community. *Mol. Genet. Metab. Rep.* **2018**, *17*, 57–63. [CrossRef] [PubMed]
17. NHS Creative. Guidance for Standard Operating Procedure (SOP) for Management of COVID-19 Risk Assessments. Available online: https://www.england.nhs.uk/south-east/wp-content/uploads/sites/45/2021/05/Standard-Operating-Process-for-Management-and-COVID.pdf (accessed on 4 December 2021).
18. British Dietetic Association. Model and Process for Nutrition and Dietetic Practice. The Model and Process. Available online: https://www.bda.uk.com/practice-and-education/nutrition-and-dietetic-practice/professional-guidance/model-and-process-for-dietetic-practice.html (accessed on 4 December 2021).
19. Sladdin, I.; Ball, L.; Bull, C.; Chaboyer, W. Patient-Centred Care to Improve Dietetic Practice: An Integrative Review. *J. Hum. Nutr. Diet.* **2017**, *30*, 453–470. [CrossRef]
20. World Health Organization. *People-Centred and Integrated Health Services: An Overview of the Evidence. Services Deliver and Safety*; World Health Organization: Geneva, Switzerland, 2015.
21. MacDonald, A.; van Wegberg, A.M.J.; Ahring, K.; Beblo, S.; Bélanger-Quintana, A.; Burlina, A.; Campistol, J.; Coşkun, T.; Feillet, F.; Giżewska, M.; et al. PKU Dietary Handbook to Accompany PKU Guidelines. *Orphanet J. Rare Dis.* **2020**, *15*, 171. [CrossRef]
22. Daly, A.; Evans, S.; Chahal, S.; Santra, S.; Pinto, A.; Gingell, C.; Rocha, J.C.; van Spronsen, F.; Jackson, R.; MacDonald, A. The Effect of Glycomacropeptide versus Amino Acids on Phenylalanine and Tyrosine Variability over 24 Hours in Children with PKU: A Randomized Controlled Trial. *Nutrients* **2019**, *11*, 520. [CrossRef]
23. Joint WHO/FAO/UNU Expert Consultation. *Protein and Amino Acid Requirements in Human Nutrition*; WHO Technical Report Series, No. 935; World Health Organization: Geneva, Switzerland, 2007; pp. 1–265.
24. Firman, S.; Witard, O.C.; O'Keeffe, M.; Ramachandran, R. Dietary Protein and Protein Substitute Requirements in Adults with Phenylketonuria: A Review of the Clinical Guidelines. *Clin. Nutr.* **2021**, *40*, 702–709. [CrossRef]
25. Gokmen Ozel, H.; Ahring, K.; Bélanger-Quintana, A.; Dokoupil, K.; Lammardo, A.M.; Robert, M.; Rocha, J.C.; Almeida, M.F.; van Rijn, M.; MacDonald, A. Overweight and Obesity in PKU: The Results from 8 Centres in Europe and Turkey. *Mol. Genet. Metab. Rep.* **2014**, *1*, 483–486. [CrossRef]
26. Dietary Reference Values for Food Energy and Nutrients for the United Kingdom. Report of the Panel on Dietary Reference Values of the Committee on Medical Aspects of Food Policy. *Rep. Health Soc. Subj.* **1991**, *41*, 1–210.
27. Elango, R.; Humayun, M.A.; Ball, R.O.; Pencharz, P.B. Evidence That Protein Requirements Have Been Significantly Underestimated. *Curr. Opin. Clin. Nutr. Metab. Care* **2010**, *13*, 52–57. [CrossRef] [PubMed]
28. Parenteral and Enteral Nutrition Group. *A Specialist Group of the British Dietetic Association. A Pocket Guide to Clinical Nutrition*, 5th ed.; Parenteral and Enteral Nutrition Group: South Yorkshire, UK, 2018.
29. White, J.E.; Kronmal, R.A.; Acosta, P.B. Excess Weight among Children with Phenylketonuria. *J. Am. Coll. Nutr.* **1982**, *1*, 293–303. [CrossRef] [PubMed]
30. Burrage, L.C.; McConnell, J.; Haesler, R.; O'Riordan, M.A.; Sutton, V.R.; Kerr, D.S.; McCandless, S.E. High Prevalence of Overweight and Obesity in Females with Phenylketonuria. *Mol. Genet. Metab.* **2012**, *107*, 43–48. [CrossRef] [PubMed]
31. Camatta, G.C.; de Cássia Kanufre, V.; Alves, M.R.A.; Soares, R.D.L.; de Carvalho Norton, R.; de Aguiar, M.J.B.; Starling, A.L.P. Body Fat Percentage in Adolescents with Phenylketonuria and Associated Factors. *Mol. Genet. Metab. Rep.* **2020**, *23*, 100595. [CrossRef] [PubMed]
32. Rodrigues, C.; Pinto, A.; Faria, A.; Teixeira, D.; van Wegberg, A.M.J.; Ahring, K.; Feillet, F.; Calhau, C.; MacDonald, A.; Moreira-Rosário, A.; et al. Is the Phenylalanine-Restricted Diet a Risk Factor for Overweight or Obesity in Patients with Phenylketonuria (PKU)? A Systematic Review and Meta-Analysis. *Nutrients* **2021**, *13*, 3443. [CrossRef]
33. Williams, R.A.; Hooper, A.J.; Bell, D.A.; Mamotte, C.D.S.; Burnett, J.R. Plasma Cholesterol in Adults with Phenylketonuria. *Pathology* **2015**, *47*, 134–137. [CrossRef]

34. Mazzola, P.N.; Teixeira, B.C.; Schirmbeck, G.H.; Reischak-Oliveira, A.; Derks, T.G.J.; van Spronsen, F.J.; Dutra-Filho, C.S.; Schwartz, I.V.D. Acute Exercise in Treated Phenylketonuria Patients: Physical Activity and Biochemical Response. *Mol. Genet. Metab. Rep.* **2015**, *5*, 55–59. [CrossRef]
35. Daly, A.; Evans, S.; Pinto, A.; Jackson, R.; Ashmore, C.; Rocha, J.C.; MacDonald, A. The Impact of the Use of Glycomacropeptide on Satiety and Dietary Intake in Phenylketonuria. *Nutrients* **2020**, *12*, 2704. [CrossRef]
36. Robertson, L.V.; McStravick, N.; Ripley, S.; Weetch, E.; Donald, S.; Adam, S.; Micciche, A.; Boocock, S.; MacDonald, A. Body Mass Index in Adult Patients with Diet-Treated Phenylketonuria. *J. Hum. Nutr. Diet.* **2013**, *26* (Suppl. S1), 1–6. [CrossRef]
37. Cazzorla, C.; Bensi, G.; Biasucci, G.; Leuzzi, V.; Manti, F.; Musumeci, A.; Papadia, F.; Stoppioni, V.; Tummolo, A.; Vendemiale, M.; et al. Living with Phenylketonuria in Adulthood: The PKU ATTITUDE Study. *Mol. Genet. Metab. Rep.* **2018**, *16*, 39–45. [CrossRef] [PubMed]
38. Luu, S.; Breunig, T.; Drilias, N.; Kuhl, A.; Scott Schwoerer, J.; Cody, P. A Survey of Eating Attitudes and Behaviors in Adolescents and Adults with Phenylalanine Hydroxylase Deficiency. *WMJ* **2020**, *119*, 37–43. [PubMed]
39. NICE 2020. Eating Disorders: Recognition and Treatment NICE Guideline NG69. 2020. Available online: https://www.nice.org.uk/guidance/ng69 (accessed on 4 November 2021).
40. SIGN Evidence-Based Guidelines. Eating Disorders, A National Clinical Guideline, Consultation Draft May 2021. 2021. Available online: https://www.sign.ac.uk/media/1849/20210511-ed-draft-v115-peer-review.pdf (accessed on 10 December 2021).
41. NHS England. Highly Specialised Services Quality Dashboards—Women and Children Metric Definitions for 2020/21. 2020. Available online: https://www.england.nhs.uk/publication/highly-specialised-services-quality-dashboards-women-and-children-metric-definitions-for-2020-21 (accessed on 30 November 2021).
42. Inwood, A.C.; Balasubramaniam, S.; Wiley, V.; Kreis, C.; Harrigan, K. Australasian Consensus Guidelines for the Management of Phenylketonuria (PKU) throughout the Lifespan. Available online: https://www.mdda.org.au/wp-content/uploads/2017/10/Australasian-PKU-Lifespan-Guidelines-FINAL-Endorsed-5.10.2017.pdf (accessed on 4 November 2021).
43. MacDonald, A.; Ahring, K.; Almeida, M.F.; Belanger-Quintana, A.; Blau, N.; Burlina, A.; Cleary, M.; Coskum, T.; Dokoupil, K.; Evans, S.; et al. The Challenges of Managing Coexistent Disorders with Phenylketonuria: 30 Cases. *Mol. Genet. Metab.* **2015**, *116*, 242–251. [CrossRef] [PubMed]
44. Obesity: Identification, Assessment and Management Clinical Guideline. Available online: https://www.nice.org.uk/guidance/cg189/resources/obesity-identification-assessment-and-management-pdf-35109821097925 (accessed on 4 November 2021).

nutrients

MDPI

Article

Hungry for Change: The Experiences of People with PKU, and Their Caregivers, When Eating Out

Grace Poole [1], Alex Pinto [2], Sharon Evans [2], Suzanne Ford [3,4], Mike O'Driscoll [5], Sharon Buckley [6], Catherine Ashmore [2], Anne Daly [2] and Anita MacDonald [2,*]

1 Faculty of Health, Education & Life Sciences, Birmingham City University City South Campus, Westbourne Road, Edgbaston, Birmingham B15 3TN, UK; grace.poole@mail.bcu.ac.uk
2 Birmingham Women's and Children's NHS Foundation Trust, Steelhouse Lane, Birmingham B4 6NH, UK; alex.pinto@nhs.net (A.P.); sharon.morris6@nhs.net (S.E.); catherine.ashmore@nhs.net (C.A.); a.daly3@nhs.net (A.D.)
3 National Society for Phenylketonuria, P.O. Box 6046, Sheffield S12 9ET, UK; suzanne.ford@nspku.org
4 North Bristol NHS Trust, Southmead Road, Bristol BS10 5NB, UK
5 School of Health and Education, Middlesex University, Room WG41A (Williams Building), The Burroughs Hendon, London NW4 4BT, UK; m.odriscoll@mdx.ac.uk
6 Department of Psychology, Faculty of Health, Psychology and Social Care, Manchester Campus, Manchester Metropolitan University, 53 Bonsall Street, Manchester M15 6GX, UK; sharon.j.bucley@stu.mmu.ac.uk
* Correspondence: anita.macdonald@nhs.net

Citation: Poole, G.; Pinto, A.; Evans, S.; Ford, S.; O'Driscoll, M.; Buckley, S.; Ashmore, C.; Daly, A.; MacDonald, A. Hungry for Change: The Experiences of People with PKU, and Their Caregivers, When Eating Out. *Nutrients* **2022**, *14*, 626. https://doi.org/10.3390/nu14030626

Academic Editor: Jose Lara

Received: 18 December 2021
Accepted: 27 January 2022
Published: 31 January 2022

Publisher's Note: MDPI stays neutral with regard to jurisdictional claims in published maps and institutional affiliations.

Copyright: © 2022 by the authors. Licensee MDPI, Basel, Switzerland. This article is an open access article distributed under the terms and conditions of the Creative Commons Attribution (CC BY) license (https://creativecommons.org/licenses/by/4.0/).

Abstract: For patients with phenylketonuria (PKU), stringent dietary management is demanding and eating out may pose many challenges. Often, there is little awareness about special dietary requirements within the hospitality sector. This study's aim was to investigate the experiences and behaviours of people with PKU and their caregivers when dining out. We also sought to identify common problems in order to improve their experiences when eating outside the home. Individuals with PKU or their caregivers residing in the UK were invited to complete a cross-sectional online survey that collected both qualitative and quantitative data about their experiences when eating out. Data were available from 254 questionnaire respondents (136 caregivers or patients with PKU < 18 years and 118 patients with PKU ≥ 18 years (n = 100) or their caregivers (n = 18)). Fifty-eight per cent dined out once per month or less (n = 147/254) and the biggest barrier to more frequent dining was 'limited choice of suitable low-protein foods' (90%, n = 184/204), followed by 'no information about the protein content of foods' (67%, n = 137/204). Sixty-nine per cent (n = 176/254) rated their dining experience as less than satisfactory. Respondents ranked restaurant employees' knowledge of the PKU diet as very poor with an overall median rating of 1.6 (on a scale of 1 for extremely poor to 10 for extremely good). Forty-four per cent (n = 110/252) of respondents said that restaurants had refused to prepare alternative suitable foods; 44% (n = 110/252) were not allowed to eat their own prepared food in a restaurant, and 46% (n = 115/252) reported that restaurants had refused to cook special low-protein foods. Forty per cent (n = 101/254) of respondents felt anxious before entering restaurants. People with PKU commonly experienced discrimination in restaurants, with hospitality staff failing to support their dietary needs, frequently using allergy laws and concerns about cross-contamination as a reason not to provide suitable food options. It is important that restaurant staff receive training regarding low-protein diets, offer more low-protein options, provide protein analysis information on all menu items, and be more flexible in their approach to cooking low-protein foods supplied by the person with PKU. This may help people with PKU enjoy safe meals when dining out and socialising with others.

Keywords: phenylketonuria; eating out; low protein food; restaurants

1. Introduction

Eating out, defined as eating foods that are prepared by others and consumed out of the home in food establishments such as restaurants, cafes, canteens, and fast-food outlets, is a growing trend. It is a well-established core social activity among people in the UK [1,2]. Eating similar foods is a cue for social connection, providing an avenue for people to communicate and relate to each other and many people prefer to gather to share a meal rather than eat alone [3,4]. People with phenylketonuria (PKU), an inherited metabolic disorder, characterised by the inability to hydrolyse the amino acid phenylalanine, are treated with a low-phenylalanine and aspartame-free diet. Whilst this dietary treatment is critical to avoid neurological damage, it is complex, with the natural protein intake of patients with classical PKU being decreased to as low as 20% of regular intake when prescribed dietary treatment only. Eating outside the home may be uncomfortable for people with PKU as they must constantly navigate social situations in which they are unable to eat what others eat, with most of the regular meal items being excluded.

There is an expectation in society that people can eat out at any time, any place, anywhere. Food and drinks are at the heart of consumer culture, increasing the pressure and desire on people with PKU to eat outside the home. According to the Kantar Worldpanel survey, in 2018, 98% of people in the UK reported eating or drinking 'out', with overall UK expenditure on food and drink reaching £49 billion a year [5]. Also in 2018, in an English survey of 2241 people aged 16 years and over, 68% had eaten in a restaurant in the last month, while 41% had eaten in a pub, bar or nightclub. Restaurants, takeaway food and cafes or coffee shops were the most popular options for eating out in the UK [6]. The Office for National Statistics (ONS) (2019) estimated that a UK household spent on average £38.80/week on food prepared out of the home, including £18.60 on restaurants and cafés. In a Food Standards Survey (2018), 85% of respondents ate out for dinner, 70% for lunch and 38% for breakfast; this was more common among young people (aged 16–34 years) and men tended to eat out more than women for breakfast, lunch and dinner.

Eating out in restaurants presents many challenges for individuals with PKU. Menu choices in restaurants usually do not state what ingredients are added to dishes or give their protein content, leaving a person with PKU the difficult choice of non-participation or choosing inappropriate foods, intensifying dietary adherence issues that may lead to poor metabolic control. They may lack self-confidence skills to seek the necessary help to secure appropriate food choices. Although there is legislation (The Food Information Regulations 2014 ("FIR") [7] and The Food Information (Amendment) (England) Regulations 2019) [8] requiring all operators to disclose food allergens, there is no mandatory catering training for special dietary provision. Evidence suggests that there are significant knowledge gaps regarding special diets among the employees of the UK hospitality industry [9–11]. The workforce in restaurants often consists of young employees, some of whom are undertaking their first job, and there may be high employee turnover with low engagement. When training is initiated, it is usually for new employees and there may be infrequent training updates [10].

Therefore, it is important to explore factors that contribute towards experiences of people with PKU when eating out. This will help to characterise the main issues encountered, any social impacts and the effect on their ability to follow their dietary treatment. Thus, this study was designed to investigate the experiences of patients with PKU, and their caregivers, in eating establishments. The aim was to identify common problems of eating out in order to improve their dining experiences in the future.

2. Materials and Methods

2.1. Methods

This was a cross-sectional study using an online survey that collected both qualitative and quantitative data from adults with PKU and caregivers of children and adults. Respondents were excluded if they did not reside in the UK.

The questionnaire was built in the Online Surveys platform (https://www.onlinesurveys.ac.uk, accessed on the 2 November 2020) to gather quantitative data. This was placed on the UK National Society for Phenylketonuria (NSPKU) website, with additional promotion on the NSPKU Twitter, Instagram and Facebook pages. The questionnaire was open for 7 months, from April until October 2020.

2.2. Questionnaire

The non-validated questionnaire contained 20 questions (Table S1). Eight questions were multiple choice, $n = 8$ multiple responses, $n = 2$ Likert scale and $n = 2$ open ended questions. Thirteen questions invited additional comments.

The questionnaire was developed by dietitians with expert practical and scientific knowledge of PKU (AP, SE, CA, AD, AM), a colleague from the NSPKU (SF), a researcher (MO) and a student dietitian from Birmingham City University (GP). It was reviewed by colleagues and lay people to ensure its readability and then amended according to feedback.

2.3. Data Collected

The questionnaire was divided into three sections, collecting information on patient age, frequency of eating out, factors that prevented the individual from eating out, impact of low protein diet, factors that affected the choice of restaurant, and influences that affected meal choice in restaurants. Information on the perception of knowledge about a low-protein diet by restaurant staff, descriptions, and characteristics of good restaurants for patients with PKU, and opinion of restaurant chains was also requested. All data collected were based on the patients/caregiver's experiences when eating out.

2.4. Statistics

Quantitative data analysis (inferential and descriptive statistics) was carried out with the Statistical Package for the Social Sciences (SPSS) version 25 (SPSS Inc., Chicago, IL, USA). For multiple response questions, only descriptive statistics were used (inferential statistics are not normally used with such questions). For testing differences between two categorical variables, chi square was used. Statistical significance was set at $p < 0.05$.

Qualitative data analyses of open-ended responses were carried out in NVIVO version 12 PRO (QSR International Pty Ltd.). The whole survey dataset was imported into NVIVO so that the coding of open-ended responses could be broken down by survey questions. All open-ended questions responses were analysed thematically.

2.5. Ethics

Ethical approval was obtained from the Birmingham City University ethics committee prior to commencement of the study (Poole/6128/R(A)/2020/Mar/HELS FAEC: What knowledge and attitudes do restaurateurs have about provision of the phenylketonuria (PKU) diet?/What are the experiences of people with PKU, and their caregivers, when eating out in restaurants or cafes?). At the beginning of the online questionnaire, respondents gave consent, and it was emphasised that the questionnaire completion was voluntary. Potential respondents were advised that data from the survey may be published in an anonymized form. If names or hospitals were mentioned in verbatim abstracts these were removed from results presented in this manuscript.

3. Results

Data were available from 254 participants (whole or partial completions of the questionnaire). The number of respondents for each question varied, as not all respondents answered all questions. Fifty-four per cent ($n = 136/254$) of responses were related to people with PKU under 18 years of age. Forty-six per cent ($n = 118/254$) of responses were from people aged $\geq$18 years of age 100 adults with PKU and 18 caregivers of adults with PKU aged $\geq$18 years.

3.1. Frequency of Dining Out

Most respondents of the questionnaire dined out only once per month or less (n = 147/254; 58%). Eighteen per cent (n = 46/254) reported doing so 'once per week', 18% (n = 45/254) said they did so 'once per fortnight', and 6% (n = 15/254) did so '2–3 times per week.' Furthermore, most participants (n = 204; 80%) expressed the desire to dine out more often; and reported factors which prevent this (Table 1). The biggest barrier overall was 'limited choice of suitable low protein foods' (90%, n = 184/204) followed by 'no information about the protein content of foods' (67%, n = 137/204). More adults with PKU (n = 27, 30%) said they 'Have no choice but to eat foods that are not permitted in the PKU diet' compared to the responses of children's caregivers (n = 12, 10%). More caregivers of children compared with adults with PKU described issues such as 'restaurants refusing to prepare low protein foods they provided' e.g., pasta (41%, n = 47 children vs. 33%, n = 29 adults); 'feeling hungry after eating out due to limited food choice' (34%, n = 39 children vs. 24%, n = 21 adults); and 'no information about the protein content of foods (72%, n = 83 children vs. 60%, n = 53).

Table 1. Factors that prevent people with phenylketonuria (PKU) from eating out (n = 204) *.

Factors That Prevent People with PKU from Eating Out	Number of Responses n = 204	% Responses
Limited choice of suitable low protein foods	183	90
No information about the protein content of foods	136	67
Restaurant have limited knowledge about PKU	124	61
Feels like too much effort	108	53
Restaurants refuse to use low protein foods e.g., pasta	76	37
Embarrassed when explaining about PKU diet	69	34
The restaurant does not offer aspartame free drinks	60	29
Still feel hungry after eating out due to limited choice	60	29
Do not want to look different	56	28
Unhelpful restaurant staff	46	23
Have no choice but to eat foods that are not permitted	39	19
Restaurant staff often get my food order wrong	31	15
Other	22	11

* Multiple response question.

Twenty-two responses answered "other". Several responses indicated that the cost of dining out was higher or of poor value for people with PKU e.g., 'often it ends up costing quite a lot of money for what is actually eaten'. They said there was more wasted food, or they provided low-protein ingredients for the restaurant to cook without a price reduction or they had to pay more than they received if sharing the bill with people who do not have PKU. Other issues identified by respondents included: 'if no information is provided about the food's protein content, I tend to go over my daily allowance and suffer migraines and I do not feel 100% the next day;' and 'I will not ask for low-protein food to be cooked, as too many people are within earshot. Usually, staff taking orders are very young'.

3.2. *Choice of Restaurant*

Eighty-nine per cent (n = 227/254) said the choice of restaurant was influenced by the need to follow a low-protein diet for the person with PKU. Factors that influenced the choice of restaurant are given in Table 2. Parents of children < 18 years of age were more likely to choose a restaurant if 'catering staff were happy to cook with low-protein foods', (46%, n = 63 vs. 34%, n = 40 of those aged ≥18 years). Parents of children < 18 years of age were less likely than adults with PKU to say 'Like to socialise with friends/family regardless of food choice' (n = 30, 22% vs. adults n = 50, 42%), and 'good choice of low protein foods on the menu' (parents of children aged <18 years: n = 93, 68% vs. adults: n = 93, 79%).

Table 2. Factors that influence the choice of restaurant/café when the person with PKU is eating out (n = 254) *.

Factors That Influence the Choice of Restaurant/Café When the Person with PKU is Eating Out	Number of Responses n = 254	% Responses
Good choice of low protein foods on the menu	186	73
Restaurant staff are happy to help	163	64
Catering staff will prepare a suitable meal independent of menu choice	121	48
Unlimited access to vegetables	120	47
Catering staff are happy to cook with low protein foods	103	41
Information about protein content of foods provided	102	40
Good choice of aspartame-free drinks	101	40
Like to socialize with family/friends regardless of food choice	80	32
Restaurant staff are discreet about the dietary needs for PKU	51	20
Restaurant staff have good knowledge about of the PKU diet	33	13
Other	17	7

* Multiple response question.

Respondents added 17 verbatim comments describing factors that influenced their restaurant choice. These included: 'My daughter goes to places she's tried before just so she has the information she needs about protein content in food'; 'she will always Google the menu to see if there is anything on the menu, if nothing available she will make an excuse to her friends to decline going'. Other comments included: 'there are limited places to go and even then, the same food is eaten every time'; and 'the majority of restaurants will not cook food I supply for my 5-year-old daughter so we can't go very far'.

3.3. *Practices When Eating Out*

Seventy-four per cent (n = 188/254) of respondents said that they ordered from the menu and chose something that may be suitable for PKU. Respondents for children under 18 years of age were more likely than adults with PKU to bring in some low protein food from home and ask the restaurant/cafe to cook it or to prepare an alternative meal (Table 3). Differences by age were statistically significant ($p < 0.001$). There were 20 other comments about food choices when eating out which included: 'we usually feed our child with PKU before going out and then choose either chips or olives in the restaurant'; 'I call ahead to discuss suitable food choices'; and 'I do a combination of ordering low-protein options, taking low-protein bread with me, sometimes pasta too'.

Table 3. What people with PKU normally do when eating out divided by age of respondents (n = 254).

Practices by Respondents	Respondents Aged < 18 Years	Respondents Aged ≥ 18 Years	Total
Just order from the menu and choose some-thing that may be suitable for PKU	64.7%	84.7%	74%
Ask the restaurant/cafe to prepare something different	8.8%	2.5%	5.9%
Bring in some pre-prepared low protein food from home	14.0%	1.7%	8.3%
Other	7.4%	8.5%	7.9%
Total responses	136	118	254

3.4. *Views on Restaurant Brands*

Respondents rated a series of popular chain restaurants regarding the suitability of meal choices and the customer services they received to help them with their dietary needs. The scale ran from 'very poor' to 'very good.' The results are summarised in Table 4. Only one restaurant scored more than 50% of ratings as good or very good (Hungry Horse, 53%, n = 82/154). Many high street chain restaurants had less than 25% of users saying they were good or very good at helping provide suitable food or supporting patients with PKU.

Table 4. Percentage of UK restaurant chains scored by adult patients or parents/caregivers of children with PKU scoring "good or very good" for their provision of low protein foods.

Restaurant	Number and % Who Scored Good or Very Good		Total Nunber of Answers for Each Restaurant
	n	%	Count
Hungry Horse	82	53%	154
Pizza Express	84	46%	181
McDonalds	84	46%	184
Wetherspoons	66	44%	149
Toby Carvery	31	39%	79
Las Iguanas	36	38%	94
Ask Italian	57	33%	173
Pizza Hut	48	29%	163
Wagamama	32	29%	112
Stonehouse Carvery	31	27%	114
Nandos	40	27%	149
Beefeater	25	26%	95
Chiquito	23	25%	93
Prezzo	19	24%	78
Frankie and Bennys	29	20%	144
Zizzi	19	20%	97
Bella Italia	20	18%	109
Harvester	20	18%	109
Greggs	13	18%	74
Brewers Fayre	13	15%	86
KFC	10	9%	107
Five Guys	14	8%	171
Café Rouge	13	7%	179
Giraffe	5	7%	72
Burger King	6	4%	155

3.5. Overall Satisfaction When Eating Out

The overall dining experience was unsatisfactory for most respondents. The median overall satisfaction rating was 4 (n = 254) (on a scale of 1 for extremely poor to 10 for extremely good). Sixty-nine per cent (n = 176/254) of respondents rated overall satisfaction as 5 or less.

3.6. Rating of Restaurant/Café Employee Staff Knowledge about Phenylketonuria (PKU)

Knowledge of PKU and dietary management was rated as very poor by respondents with an overall median rating of 1.6 from 254 responses (on a scale of 1 for extremely poor to 10 for extremely good). There were 100 free text comments to this question from which the themes given in Table 5 were derived.

Table 5. Open-ended responses to the questionnaire rating restaurant/café employee staff knowledge about PKU.

Theme	Examples of Verbatim Comments by Questionnaire Respondents
Low staff awareness of PKU (n = 68)	• *'most staff don't even know what PKU is! When we explain it, many people seem to think we're just being awkward for the sake of it.'* • *'the few times I tried to explain it, the waiter made fun of me and said I was 'being picky.'*
PKU gets confused/conflated with food allergies or vegetarianism (n = 14)	• *'they get it confused with food allergies and some don't even try to understand when we explain.'* • *'they just think you are a picky veggie/vegan.'*
Did not expect staff to be aware of PKU (n = 8)	• *'the employee cannot be expected to know about every condition.'* • *'I think it's poor but the waitress should not need a medical exam to earn a minimum wage.'*
Staff rudeness/unhelpfulness (n = 8):	• *'nobody ever knows anything about PKU and people are sometimes very rude.'* • *'most of the time they think it's made up and I'm being awkward.'*
Staff are sometimes helpful (n = 5):	• *'no one has ever heard of it, but some places are willing to try and make something work.'* • *'one time the restaurant did cook our own pizza base.'*
We do not discuss PKU in restaurants (n = 2)	• *'I would just find the experience not enjoyable if I had to keep asking questions about the menu.'*

Verbatim comments are presented in italic.

3.7. Helpfulness of Restaurants/Cafes in Finding a Solution to Cater for PKU

Sixty-three per cent (n = 159/254) of respondents said that they had at least one positive experience when dining out, particularly at local/ independent restaurants and non-chain restaurants ('after repeated visits, they went out of their way to cater for PKU') and it was considered particularly helpful when restaurants provided a full list of ingredients with their protein content. However, only one third of respondents (33%, n = 83/254) considered that restaurants/cafes were always or often helpful, 39% (n = 100/254) felt that they were 'sometimes' helpful, and 21% (n = 54/254) thought that they were rarely or never helpful.

Forty-four per cent (n = 110/252) of respondents said that they had experienced restaurants refusing to prepare alternative foods; 44% (n = 110/252) said that they had not been allowed to eat their own prepared food in a restaurant; and 46% (n = 115/252) said that a restaurant had refused to cook low-protein pasta, burger mix or pizzas. The lack of low-protein food choices and inflexibility was considered unhelpful.

3.8. Changes That Would Encourage People with PKU to Dine Out

Seventy-nine per cent (n = 200/254) of respondents said changes would help improve their experience dining outside the home but 21% (n = 54/254) said changes would not help. There were 200 free text responses. The main themes are shown below and illustrated through a selection of verbatim quotes in Table 6.

Table 6. Open ended responses to the questionnaire describing the changes that would help people with PKU dine out.

Theme	Examples of Verbatim Comments by Questionnaire Respondents
More low protein choices on the menu (n = 69)	• *'There should be at least one low protein menu choice that isn't just vegetables and potato, with one or two flavour options (e.g., spices or sauce).'* • *'Cafes could offer soups, jacket potatoes or salads that don't have added protein ingredients.'* • *'There should be more vegetarian options on the menu with the choice of exchanging ingredients such as low protein cheese and cream. Restaurants should also be able to cook e.g., low protein rice or pasta for people on a different diet.'* • *'Allow different ingredients on the menu to be mixed. For example, if mushrooms and grilled tomatoes are served on a steak—can they be bought as a portion on their own and served with a salad.'*
Educating and raising awareness amongst staff (catering/food retail) n = 64	• *'Staff more should be educated about PKU and adapt the restaurant menus.'* • *'Ensure the employees of restaurants know which foods contain protein and how important it is to have the nutritional information of their food readily available to customers.'* • *'I think it would help for people to know why phenylalanine/aspartame is often highlighted on drinks and why there is a need to stock aspartame free drinks—more so for smaller places like cafes.'* • *'Generally, more awareness is needed of PKU and the diet in the catering/dining sector as I think most staff now are prepared to accommodate dietary needs however simply don't have the knowledge about it.'* • *'Training on how to specifically not make the customer feel like a nuisance via hospitality training.'* • *'An explanation that PKU is not an allergy.'* • *'Willingness to listen and not just say different foods cannot be used.'* • *'More understanding of very rare conditions—would they stop a blind person eating.'* • *'Have a card briefly explaining the details of the PKU diet for catering staff. I often find that they assume it's like a peanut allergy'.* • *'All servers should ask all customers if anyone has any special dietary needs.'* • *'Restaurants should provide nutritional info about the food so customers can make informed decisions.'* • *'More friendly staff to make you feel confident and helpful.'* • *'Provide all staff with a fact sheet or information pack on what PKU is and what we can eat'*

Table 6. *Cont.*

Theme	Examples of Verbatim Comments by Questionnaire Respondents
Publishing protein content of menu items (n = 36)	• *'Having nutritional information in a booklet so anyone can read the protein level in different foods.'* • *'Cafes and restaurants should list the amount of protein in their foods and drinks.'* • *'More nutritional info for sauces and vegan cheeses. Ability to get the chef to weigh foods too.'* • *'Nutritional information to be available for every dish offered on the menu so that people with PKU can make an informed choice about what they eat.'*
Staff should be able to adapt/tailor recipes (n = 15)	• *'Making a main meal up out of side dishes where choices are limited.'* • *'Restaurants more flexible in making meals with replacement ingredients to suit low protein diets.'* • *'Bring out the dish at the same time as other meals are being served so the person with PKU doesn't feel singled out or different in anyway.'* • *'Being flexible with the menu'.* • *'Cooking something from scratch with suitable ingredients. Being happy to use prescription foods in their kitchen.'* • *'Talk about requirements away from table so child with PKU does not have to sit and listen to all the negotiations that have to go on before they can eat something.'*
Make it more normal/acceptable for people to bring their own food to be cooked (n = 11)	• *'They should cook something using your own pasta and rice.'* • *'Microwaves in food courts so can heat up our own food.'*
Publicise restaurants that are PKU friendly (n = 4)	• *'Give praise and positive reviews for the good restaurants, and shame those that have offered bad experiences.'* • *'More opportunities and advertising that they are happy to cater for special diets.'* • *'A sign on the restaurant to say we cater for all diets or be willing to help would be a start.'*

Verbatim comments are presented in italic.

3.9. Emotions around Dining Out

Respondents' feelings and emotions before dining out are presented in Table 7.

Table 7. Emotions of adults with PKU/caregivers of children before dining out (n = 254) from multiple response question.

	Under 18 Years of Age		18 Years of Age or Over		Total Number of Respondents	
	n	%	n	%	n	%
Anxious	42	31%	59	50%	101	40%
Excited	54	40%	37	31%	91	36%
Hungry	37	27%	42	36%	79	31%
Happy	47	35%	29	25%	76	30%
Uneasy	32	24%	42	36%	74	29%
Concerned	22	16%	47	40%	69	27%
Pleasure	11	8%	12	10%	23	9%
Other	9	7%	11	9%	20	8%
Not applicable	12	9%	5	4%	17	7%
Total	136		118		254	

When leaving a restaurant/café, only 35% (n = 88/254) of respondents said they were satisfied, with only 31% (n = 79/252) saying they were happy. Twenty-eight per cent (n = 71/254) left disappointed, 26% (n = 66/254) frustrated and 22% (n = 57/254) were still hungry. Adults with PKU (n = 43/118, 36%) were more than twice as likely to feel frustrated post-meal than caregivers of children under the age of 18 years (n = 23/136, 17%).

4. Discussion

This research is the first to purposefully investigate the eating out experiences, behaviours and concerns of people with PKU or their caregivers. Although eating out is a routine activity enjoyed by the general population, people with PKU chose not to do this regularly. While it is expected that people dining outside the home should derive social and psychological enjoyment [12], with satisfaction of appetite, and respite from low-protein meal preparation, our results suggest that people with PKU or their caregivers were unable to enjoy stress-free and spontaneous meals. In fact, 40% said eating out was associated with anxiety, only 9% derived any pleasure from it, with over one quarter of survey participants leaving restaurants feeling frustrated, disappointed, and still hungry.

Individuals with PKU or their caregivers were eager to find restaurants that were willing to accommodate their dietary needs. Personalisation of menu choices with unlimited access to vegetables was considered almost mandatory for people with PKU. They commonly favoured familiar, non-chain/independent eating out venues that they had visited previously, with a proven track-record of preparing appropriate low-protein foods. Most preferred restaurants who cooked with fresh ingredients onsite rather than those who used pre-assembled meals that could not be modified. Some used eating establishments that had 'build-your-own options' (e.g., brands such as Subway or salad bars) allowing for more customization. Many found food-chain restaurants inflexible scoring disappointingly when rated by people with PKU or their caregivers. Restaurants often used pre-prepared foods, with some vegetable options being coated in wheat flour. Although vegan meal choices are now common in restaurants, they are usually high in protein.

Overall, incompatibility of menu choice with low-protein diets, inadequate food choice, uncertainty about the protein content of meals, and limited suitable drink options were all concerns of people with PKU or their caregivers. Consumers with PKU need transparency around meal ingredients, protein content and food portion size. Some restaurants only sell aspartame-containing soft drinks to avoid extra costs associated with sugar taxes. There was frustration that some restaurants would not agree to cook or even allow people with PKU to eat their own special low-protein foods e.g., low-protein bread, pasta and pizza bases prescribed by their general practitioner on their premises, even though the restaurant staff were unable to supply these foods themselves. Although some restaurants could offer gluten-free equivalents, these foods were often too high in protein for most people with PKU.

Written information about the protein content of food provided on a website that could be studied in advance of a restaurant booking was considered helpful as it enabled the person with PKU or parents/caregivers to assess the suitability of food choices without the need for conversations with restaurant staff. Although most restaurants post their menus online, not all give their nutritional content and food portion sizes may differ if unweighted. Some fast-food chains post online the protein content of meals, but this information may be difficult to locate and given in small print tables. It was requested that restaurant food nutritional analysis and portion sizes should also be available by mobile app, with written reviews about special diet provision. There are currently no mandatory labelling requirements for any unpackaged products sold by catering businesses to state the protein content or list all the ingredients (except allergens, some additives and aspartame) [13]. The UK Government plans to introduce a new menu-labelling requirements law, which will enforce major foodservice operators to include a calorie count on the food items of both their digital and physical menus by April 2022, but it does not specify other nutrients or require provision of a full list of ingredients [14].

The results of this survey indicated that some people with PKU were reluctant to eat outside the home and experienced a spike in anxiety when visiting a restaurant because they anticipate it will not be a pleasurable experience. In another study on PKU, families reported avoiding eating out in restaurants, to prevent children from feeling excluded [15]. In our study, there was commonly social embarrassment, discomfort, and much sensitivity in the behaviours associated with social eating. The respondents experienced food worries about how others perceive them based on what they eat. To avoid causing others (e.g., staff or social companions) inconvenience, some respondents deliberately downplayed or did not mention their low-protein dietary requirements in conversations and opted for food options that were lower in protein and safe such as a baked potato, potato chips or a side salad. If they asked for alternative food choices, they felt that they were making unreasonable and excessive demands on staff. Some even felt they were being difficult when asking restaurant staff about the ingredients added to foods and the protein content of dishes. Others feared that the food venue would refuse to serve them after they had explained their dietary needs. Generally, people with PKU did not like drawing extra attention to their dietary needs within restaurants and any public discussions about their condition were commonly unwelcome.

The quality of the relationship or interaction that people with PKU or their parents/caregivers experience with food venues is important. They should be able to comfortably communicate with restaurant staff regarding their dietary needs. However, many perceive themselves as being made to feel as though they were a 'fussy customer' or a 'nuisance' so it constrained any conversation about food risks associated with incorrect food choices being served. Restaurant staff rarely proactively ask customers about special dietary needs, therefore leaving consumers to initiate any communication with staff regarding their requirements [16,17]. If the restaurant team genuinely listened to the dietary issues through taking the time to speak to the person and paying attention to what they said, the customer would be more forthcoming to discuss their dietary needs. This could lead to a willingness to modify food choices on a 'plate' in order to accommodate consumers' needs and discretion whilst still holding conversations regarding dietary requirements. These actions are signs of extra care and respect. Commonly the waiter/waitress fail to understand the requests for low-protein food as there is no/low awareness of PKU, and people with PKU say 'it is sometimes like talking to a brick wall'. The lack of knowledge leads to a customer perception of poor-quality provision. People with PKU might be more candid with staff whom they consider caring and trustworthy. The readiness of food establishments to adapt the dishes whilst respecting consumers' food preferences and desire to try out different foods was also highly valued by patients with foods allergies [18,19].

A large proportion of the hospitality industry possess no or a very limited knowledge of special diets and may be unable to respond adequately to low-protein requests and this was clear from the results of the survey. However, ignorance of special diets by those people involved in delivering special dietary menus is not a defense for failing to meet the customer's needs and expectations. Any current mandatory training predominantly focuses on food safety and technical preparation skills only, with an absence of education on special dietary requirements [20]. There should be mandatory special diet training for all employees who work in catering establishments. Special diet training has been shown to be effective. A short training programme on allergies was found to increase the knowledge and awareness of employees from all restaurants in one UK town as well as encouraging more information to be available for customers [21]. Furthermore, a survey that included 861 restaurant staff and members of the general public, found high levels of awareness of allergies and coeliac disease among trained chefs, in comparison to the general public and untrained staff, demonstrating the effectiveness of training [22].

Limitations

Recruitment of participants for this online survey was via the NSPKU website and promoted on PKU social media sites, so respondents were limited to any individuals who had access to the internet using the appropriate technology. It is likely that respondents were people who accessed social media sites frequently, were not randomly selected, and the extent to which the sample matched the demographic characteristics of the general PKU population is unknown. However, the sample size was large, so this factor is likely to have had minimal impact on the overall results. In addition, caregivers acted as proxy respondents on behalf of children and described what they perceived to be their child's feelings when eating out, so their answers may have been inaccurate. We did not distinguish between male and female respondents. Also, the number of respondents to scaled questions varied which may added errors to the results. Additionally, the questionnaire was non-validated, and the respondent's level of understanding was unknown. Protein tolerance was not reported, and this may have influenced the respondents dining experiences.

Furthermore, research to compare dining out experiences of patients with PKU and those with other conditions requiring dietary management may be useful to give additional insight into this practical issue.

5. Conclusions

In summary, there is a considerable lack of awareness and inability to successfully meet the needs of people with PKU on low-protein diets in restaurants and catering establishments in the UK. Reputation, revenue and customer relationships may be jeopardized if hospitality businesses do not meet the dietary needs of their customers. There is a need to better understand the knowledge and practices of restaurant and food-service establishment personnel toward the management of special diets in order to improve consumer experiences when eating out. Changes to staff training, flexibility to adapt menus, provision of more low-protein options, and a change in the law to enforce better availability of nutritional information in restaurants should be implemented. It is necessary to improve the experience of people with PKU and end the barriers they continually face in trying to enjoy a basic human social activity (dining out together) that most people can take for granted.

Supplementary Materials: The following supporting information can be downloaded at: https://www.mdpi.com/article/10.3390/nu14030626/s1, Table S1: Full Questionnaire.

Author Contributions: Conceptualization, A.M., A.P. and G.P.; methodology, A.M., A.P., S.E., S.F., M.O., S.B. and G.P.; formal analysis, M.O.; writing original draft preparation, A.M.; writing, review and editing, G.P., A.P., S.E., S.F., M.O., S.B., C.A., A.D., S.B. and A.M.; supervision, A.M. and A.P. All authors have read and agreed to the published version of the manuscript.

Funding: This research received no external funding.

Institutional Review Board Statement: The study was conducted according to the guidelines of the Declaration of Helsinki and approved by the Birmingham City University ethics committee prior to commencement of the study (Poole/6128/R(A)/2020/Mar/HELS FAEC: What knowledge and attitudes do restaurateurs have about provision of the phenylketonuria (PKU) diet?; and What are the experiences of people with PKU, and their caregivers, when eating out in restaurants or cafes?).

Informed Consent Statement: Informed consent was given by all subjects when filling inthe questionnaire.

Data Availability Statement: The data will be made available from the authors upon reasonable request.

Acknowledgments: We would like to acknowledge and thank all the patients and families that have taken their time to fill in this survey.

Conflicts of Interest: The authors declare no conflict of interest.

References

1. Ratner, R.K.; Hamilton, R.W. Inhibited from bowling alone. *J. Consum. Res.* **2015**, *42*, 266–283. [CrossRef]
2. Rozin, P. The Meaning of Food in Our Lives: A Cross-Cultural Perspective on Eating and Well-Being. *J. Nutr. Educ. Behav.* **2005**, *37*, S107–S112. [CrossRef]

3. Douglas, M. *Meal Formats, Meal Cycles and Menu Negotiation in the Maintenance of an Italian-American Community. Food in the Social Order: Studies of Food and Festivities in Three American Communities*, 1st ed.; Douglas, M., Ed.; Taylor & Francis Group: London, UK, 1984; pp. 143–218.
4. Mintz, S.W.; Du Bois, C.M. The Anthropology of Food and Eating. *Annu. Rev. Anthr.* **2002**, *31*, 99–119. [CrossRef]
5. Kantar Worldpanel. Out of Home, Out of Mind? Understanding Food and Drink Habits. Available online: https://www.kantarworldpanel.com/global/News/New-report-out-Out-of-home-out-of-mind (accessed on 13 December 2021).
6. Reported Eating Out Behavior in the Last Month in the United Kingdom (UK) in 2018, by Venue. Available online: https://www.statista.com/statistics/328443/eating-out-in-the-united-kingdom-uk-by-venue (accessed on 11 December 2021).
7. UK Government. *The Food Information Regulations*; 2014. Available online: https://www.legislation.gov.uk/uksi/2014/1855/contents (accessed on 10 December 2021).
8. UK Government. *The Food Information (Amendment) (England) Regulations*; 2019. Available online: https://www.legislation.gov.uk/uksi/2019/1218/made (accessed on 10 December 2021).
9. Common, L.A.R.; Corrigan, C.J.; Smith, H.; Bailey, S.; Harris, S.; Holloway, J.A. How safe is your curry? Food allergy awareness of restaurant staff. *J. Allergy Ther.* **2013**, *4*, 4. [CrossRef]
10. Lee, Y.M.; Sozen, E. Who knows more about food allergies—Restaurant managerial staff or employees? *Br. Food J.* **2018**, *120*, 876–890. [CrossRef]
11. Soon, J. 'Food allergy? Ask before you eat': Current food allergy training and future training needs in food services. *Food Control* **2020**, *112*, 107129. [CrossRef]
12. Dunbar, R.I.M. Breaking Bread: The Functions of Social Eating. *Adapt. Hum. Behav. Physiol.* **2017**, *3*, 198–211. [CrossRef] [PubMed]
13. UK Government. Food Labelling and Packaging. Available online: https://www.gov.uk/food-labelling-and-packaging (accessed on 13 December 2021).
14. UK Government. *The Calorie Labelling (Out of Home Sector) (England) Regulations*; 2021. Available online: https://www.legislation.gov.uk/ukdsi/2021/9780348223538 (accessed on 10 December 2021).
15. Roberts, R.M.; Muller, T.; Sweeney, A.; Bratkovic, D.; Gannoni, A.; Morante, B. Relationships Between Childhood Experiences and Adulthood Outcomes in Women with PKU: A Qualitative Analysis. *JIMD Rep.* **2016**, *32*, 41–49. [CrossRef] [PubMed]
16. Wen, H.; Kwon, J. Food allergy risk communication in restaurants. *Food Prot. Trends* **2016**, *36*, 372–383.
17. Wen, H.; Kwon, J. Food allergy information sharing and communication strategies in full-service restaurants in the U.S. *J. Foodserv. Bus. Res.* **2018**, *22*, 50–65. [CrossRef]
18. Barnett, J.; Vasileiou, K.; Lucas, J.S. Conversations about food allergy risk with restaurant staff when eating out: A customer perspective. *Food Control* **2019**, *108*, 106858. [CrossRef]
19. Begen, F.M.; Barnett, J.; Payne, R.; Roy, D.; Gowland, M.H.; Lucas, J. Consumer Preferences for Written and Oral Information about Allergens When Eating Out. *PLoS ONE* **2016**, *11*, e0156073. [CrossRef] [PubMed]
20. Towers, N.; Pratten, J. Managing special dietary needs for food allergies in the hospitality industry; A survey based on the Coeliac disorder. *Int. J. Hosp. Manag.* **2003**, *22*, 341–351. [CrossRef]
21. Young, I.; Thaivalappil, A. A systematic review and meta-regression of the knowledge, practices, and training of restaurant and food service personnel toward food allergies and Celiac disease. *PLoS ONE* **2018**, *13*, e0203496. [CrossRef] [PubMed]
22. Simpson, S.; Lebwohl, B.; Lewis, S.; Tennyson, C.; Sanders, D.; Green, P. Awareness of gluten-related disorders: A survey of the general public, chefs and patients. *e-SPEN Eur. e-J. Clin. Nutr. Metab.* **2011**, *6*, e227–e231. [CrossRef]

nutrients

MDPI

Review

Glycomacropeptide in PKU—Does It Live Up to Its Potential?

Anne Daly *, Alex Pinto, Sharon Evans and Anita MacDonald

Birmingham Women's and Children's Hospital NHS Foundation Trust, Steelhouse Lane, Birmingham B4 6NH, UK; alex.pinto@nhs.net (A.P.); evanss21@me.com (S.E.); anita.macdonald@nhs.net (A.M.)
* Correspondence: a.daly3@nhs.net

Abstract: The use of casein glycomacropeptide (CGMP) as a protein substitute in phenylketonuria (PKU) has grown in popularity. CGMP is derived from κ casein and is a sialic-rich glycophosphopeptide, formed by the action of chymosin during the production of cheese. It comprises 20–25% of total protein in whey products and has key biomodulatory properties. In PKU, the amino acid sequence of CGMP has been adapted by adding the amino acids histidine, leucine, methionine, tyrosine and tryptophan naturally low in CGMP. The use of CGMP compared to mono amino acids (L-AAs) as a protein substitute in the treatment of PKU promises several potential clinical benefits, although any advantage is supported only by evidence from non-PKU conditions or PKU animal models. This review examines if there is sufficient evidence to support the bioactive properties of CGMP leading to physiological benefits when compared to L-AAs in PKU, with a focus on blood phenylalanine control and stability, body composition, growth, bone density, breath odour and palatability.

Keywords: glycomacropeptide; PKU; protein substitute; amino acids

Citation: Daly, A.; Pinto, A.; Evans, S.; MacDonald, A. Glycomacropeptide in PKU—Does It Live Up to Its Potential?. *Nutrients* **2022**, *14*, 807. https://doi.org/10.3390/nu14040807

Academic Editors: Shanon L. Casperson and Adamasco Cupisti

Received: 31 December 2021
Accepted: 7 February 2022
Published: 14 February 2022

Publisher's Note: MDPI stays neutral with regard to jurisdictional claims in published maps and institutional affiliations.

Copyright: © 2022 by the authors. Licensee MDPI, Basel, Switzerland. This article is an open access article distributed under the terms and conditions of the Creative Commons Attribution (CC BY) license (https://creativecommons.org/licenses/by/4.0/).

1. Introduction

It is estimated there are 0.45 million people worldwide with the inherited metabolic disorder phenylketonuria (PKU) [1], which causes irreversible neurological damage if untreated. Although pharmaceutical therapies are being actively developed, a phenylalanine restricted diet remains the only effective treatment. In classical PKU, protein substitutes (low phenylalanine protein replacements) provide up to 80% of dietary protein requirements and are essential to ensure metabolic stability and growth. Protein substitutes are derived from either phenylalanine free amino acids (L-AAs) or a combination of low phenylalanine peptides with added amino acids (casein glycomacropeptide: CGMP). They are usually supplemented with vitamins, minerals and trace elements, and may contain essential and/or long chain fatty acids and prebiotics. In 1953, the first protein substitute was made using a low phenylalanine hydrolysed casein [2,3]; subsequently, the number and type of manufactured preparations have exponentially increased [4]. In 2008, CGMP, a by-product of whey from the manufacture of cheese, was introduced as an alternative protein substitute to L-AAs, but it is still unclear if this protein source has any advantage over conventional L-AAs in the dietary management of PKU. Overall, their composition, bioavailability and long term impact on metabolic efficacy has received limited systematic investigation in PKU.

This review examines the evidence of using the bioactive protein substitute CGMP compared to L-AAs in the treatment of PKU, focusing on benefits to blood phenylalanine stability, body composition, bone mass, density and geometry and the influence of protein substitutes on breath malodour and palatability.

2. Protein Substitutes Pharmacological Benefits

Protein substitutes meet the protein requirement for cellular function and growth and have several pharmacological and physiological functions (Table 1). They improve phenylalanine tolerance and optimise metabolic control by suppressing blood phenylalanine concentrations. This is particularly important during illness and trauma, where protein substitutes have a protective role by counteracting protein catabolism [5–8]. Irrespective of their nitrogen source, each protein substitute has a different amino acid profile consisting of essential and non-essential amino acids, and around 40% large neutral amino acids (LNAAs). They provide the principal source of tyrosine, although there is no consensus on the optimal amount required [9]. Similarly, there is no agreement on the quantity and ratio of branched chain amino acids, and there is also limited data about the absorption and retention of amino acids [10–13].

Table 1. Functional properties of protein substitutes in PKU.

Functional Properties	Action	References
Large neutral amino acids (LNAAs)	Phenylalanine transport from the plasma into the brain is via the LNAA transporter (LAT1). Competition at the blood brain barrier using LNAAs for LAT1 prevents excess phenylalanine from entering the brain, preventing neurocognitive damage	[14–16]
	LNAAs and cationic amino acids cross the intestinal mucosa via a carrier protein system. The affinity of the amino acids for the intestinal carrier is higher than at the blood brain barrier. By providing LNAAs, there is a decreased entry of phenylalanine across the intestinal mucosa	[17–19]
Normal growth and cellular function	Provide nitrogen to maintain and improve muscle mass and promote growth	[20,21]
Provide a source of nitrogen for the synthesis of nitrogen containing compounds	Nitrogen is necessary for the manufacture of small molecular substances, e.g., nitric oxide	[22]
Provide tyrosine	Phenylalanine to tyrosine conversion is severely limited or absent in classical PKU. Tyrosine becomes a surrogate essential amino acid, and adequate amounts must be provided by protein substitutes to prevent deficiency. Tyrosine is important for the biosynthesis of neurotransmitters, thyroxine and melanin	[8,9]
Optimise blood phenylalanine control	Protein substitutes support stabilisation of blood phenylalanine concentrations by providing a complement of amino acids (except phenylalanine) allowing protein anabolism and nitrogen retention. For maximum effectiveness, they must be given frequently throughout the day	[7,23]
Prevent nutritional deficiencies	Most protein substitutes are supplemented with vitamins, minerals and trace elements. Adherence with separate vitamin and mineral supplements is poor in patients with PKU	[24]

3. The Role of Functional Amino Acids in Protein Substitutes

Amino acids in protein substitutes have several nutritional, biochemical and physiological roles linked to growth, health and disease prevention [22,25]. Functional amino acids (essential or non-essential) regulate key metabolic pathways. They provide nitrogen, hydrocarbon skeletons and sulphur [26]; both nitrogen and sulphur are unable to be synthesised de novo. Some roles of functional amino acids include regulation of body composition and bone health, others include modulating bacterial flora, glucose homeostasis and inflammatory responses. Amino acids are also involved in cell signalling (including mammalian target of rapamycin complex 1 (mTORC1), and the interaction and generation of small peptides, glucagon-like peptide 1 (GLP-1), peptide-YY (PYY), serotonin and insulin. Insulin plays a key regulatory role in amino acid metabolism, and amino acids alter insulin action by regulating glucose and protein metabolism [27,28]. The composition of a protein substitute affects the rate at which amino acids are delivered into the systemic system, changing their cellular uptake and biological utilisation. Different rates of absorption have been reported when amino acids are ingested as free amino acids, peptides or bound to proteins [13,26,29]. Free amino acids appear in the peripheral plasma more quickly than those from an intact protein source [10]. Any protein substitute that can maximise amino acid absorption will increase anabolism and subsequently alter phenylalanine metabolism.

4. What Is a Casein Glycomacropeptide (CGMP)?

In 1954 while working on a variant of *lactobacillus bifidus*, György et al. [30] found evidence of protein bound sialic acid (N acetylneuraminic acid) in cow's milk. In 1965, Delfour et al. [31] established that this milk bound sialic acid protein was called κ casein and reported that CGMP was formed by separation of κ casein by the action of chymosin during cheese production. CGMP is found in the soluble whey elute [32] and constitutes 20–25% of total proteins in whey products manufactured from cheese whey. It is a 64 amino acid phosphoglycoprotein [33]. Five oligosaccharides (glycans) have been identified as part of the glycomacropeptide structure [32].

In its pure form, the glycophosphopeptide has an unusual amino acid sequence containing no aromatic amino acids (tryptophan, tyrosine, phenylalanine) or the sulphur amino acid cysteine [34]. Of the five glycan structures common to bovine CGMP, the one of most interest is the nine-carbon sugar molecule, sialic acid, which forms 7–9% of CGMP. This is a component of human milk oligosaccharides and neural tissues and is an integral part of brain gangliosides and glycoproteins. The glycan chains are attached via two types of glycosylation: *N*-linked when the glycan chain is attached to the amide side chain of the asparagine residue, and *O*-linked when the glycan is attached to the oxygen of a serine or threonine residue [35,36]. Around 60% of CGMP is glycosylated [37] with exclusively *O*-linking glycans. There is evidence to suggest glycosylation is a controlled hierarchical process that influences the associated biological activities of CGMP [38,39]. These bioactive properties provide a functional ingredient for the food and pharmaceutical industry.

5. Potential Clinical Properties of CGMP

Carbohydrates, whether free or bound to proteins or lipids, are essential communication molecules in inter and intracellular processes. The biological properties associated with CGMP include immunomodulatory, antimicrobial and prebiotic [32,35,40]. CGMP interacts with cholera toxins through the glycan chains [41,42], and bind to *E. coli* and *Salmonella enteritidis* [43]. It also has an important role in anticariogenesis; CGMP inhibits adherence of oral bacteria, preventing tooth decay [44,45]. In animal experiments, a CGMP enriched infant formula increased learning ability, which was linked to an increase in sialioprotein in the frontal brain cortex [46]. These findings need further investigation.

6. Potential Commercial Use of CGMP

CGMP is an acidic peptide, highly soluble and heat stable [35]. It also has a wide pH range and solubility, and has emulsifying, gel and foaming properties, making it desirable in the food and nutritional products industry as it alters the structural matrix of foods and improves the texture and mouth feel.

7. Adaptation of CGMP for Use as a Low Phenylalanine Protein Substitute in PKU

Isolating CGMP from cheese whey is difficult and expensive, with residual phenylalanine remaining in the final product [32]. CGMP has inadequate amounts of five indispensable amino acids: histidine, leucine, methionine, tryptophan, and tyrosine, but supplementation with these amino acids enables it to be used as an alternative to L-AAs [47].

The first case study using CGMP [47] was reported in a 29-year-old male with PKU. Over 15 weeks, CGMP and L-AA protein substitutes were compared. CGMP was supplemented with histidine, leucine and tryptophan providing 130% and tyrosine at 150% of the USA 2002 recommendation [48]. Added vitamins, minerals and trace elements were supplemented when taking CGMP. An additional 500 mg of tyrosine was taken orally twice daily, providing the same tyrosine intake as that from L-AAs. Significant increases in plasma glutamine, isoleucine, proline and threonine, with an overall increase in the LNAAs and a 16% increase in the BCAAs were noted. CGMP is naturally higher in threonine and isoleucine, explaining the observed increases. In a subsequent study in 2009, van Calcar et al. [49] compared the effects of L-AAs and CGMP in 11 subjects with PKU over 8 days. The CGMP product was supplemented with histidine, leucine, methionine and tryptophan, but the additional supplement of 1000 mg/day of tyrosine was omitted. This led to a mean fasting tyrosine concentration below the normal reference range in the CGMP group, with an expected increase in isoleucine and threonine consistent with the higher concentration in CGMP. After an overnight fast, plasma blood concentration of arginine, a conditionally essential amino acid, was significantly lower. The limiting amino acids added to the CGMP, histidine, leucine methionine and tryptophan, remained within the normal biochemical reference ranges, but tyrosine and arginine concentrations required further supplementation. Methionine supplementation was stopped as there was an adequate amount in the CGMP to meet the new lower requirements as suggested by Humayun et al. [50].

8. The Impact of CGMP on Blood Phenylalanine Control in PKU

Ten published studies have investigated the effect of CGMP compared to L-AAs on blood phenylalanine control. The majority ($n = 7/10$, 70%) have suggested no significant alteration in blood phenylalanine concentrations despite residual phenylalanine being present in CGMP [49,51–54]. Nine of ten studies reported higher blood phenylalanine concentrations when using CGMP, but only three studies demonstrated a statistically significant increase. All three studies were in children from one centre, but this included two long term longitudinal studies over 6 and 12 months [55,56], and one randomised controlled study over 6 weeks [57]. Four other studies collected data mainly in adults for a minimal period of 8 to 21 days, with suboptimal blood phenylalanine concentration at study baseline; some subjects were taking adjunctive sapropterin treatment that improved phenylalanine tolerance. Two studies were retrospective reviews in 11 teenagers and adults, with follow up at 20 and 29 months [58,59]. One study [54] examined CGMP as a food (GMP soft cheese) supplement in children; it was consumed 3 times daily over 9 weeks. No information was provided on its residual phenylalanine content or amino acid profile. The supplement was provided in combination with L-AAs and provided 50% of the total protein substitute intake.

It is difficult to interpret the effectiveness of results from short-term studies. One of the earliest studies [49] suggested that the residual phenylalanine in the CGMP was too high at 0.4 g/100 g of product. This was only given to three subjects, all with high phenylalanine tolerance. In the remaining nine subjects, the CGMP composition was refined, with a phenylalanine content of 0.2 g/100 g of product. A statistically significant increase in blood phenylalanine was only evident in the longitudinal studies in children, with blood phenylalanine being maintained within a narrow therapeutic target range of 120 to 360 μmol/L. This suggests caution is necessary when using CGMP that contains residual phenylalanine, particularly in children with classical PKU. Table 2 lists the PKU studies using CGMP and their outcomes. The impact of residual phenylalanine may be less important in patients using adjunct drug management that improves phenylalanine tolerance or in teenagers and adults who maintain blood phenylalanine levels under a higher upper therapeutic target. Further studies are needed in adults and in pregnancy when CGMP is the only protein substitute source.

Table 2. Studies using CGMP compared to L-amino acid protein substitutes in PKU.

Author/ Year	Country	Study Design Age (Range)	Nos of Subjects/ Gender	PKU Phenotype	Study Intervention	Mean/Median Phenylalanine Concentrations in L-AAs Compared to CGMP (μmol/L)
Van Calcar [49] 2009	United States	Cross-sectiona 123 y ± 7 (11–31)	11 4 F, 7 M	10 Classical 1 Variant	100% L-AAs vs. 100% CGMP 4 days on each product	L-AAs = 619 CGMP = 676, p = ns
MacLeod [60] 2010	United States	Cross-sectional 23 y ± 7 (11–31)	11 4 F, 7 M	11 Classical	100% L-AAs vs. 100% CGMP 4 days on each product	L-AAs = 619 CGMP = 676, p = ns
Ney [52] 2016	United States	Randomised crossover clinical study (15–49)	301 8 F,12 M	20 Classical 10 Variant	21 days: 100% CGMP or 100% L-AAs	L-AAs = 655 CGMP = 777, p = ns
Zaki [54] 2016	Egypt	Clinical study 6.7 y (5.0–11.8)	10 4 F, 6 M	10 Classical	9 weeks: 50% CGMP + 50% L-AA 9 weeks: 100% L-AA	100% L-AA s = 490 CGMP 50% + 50% L-AAs = 376, p = ns
Pinto [59] 2017	Portugal	Retrospective longitudinal study 27 y ± 10 (13–42)	11 8 F, 3 M	6 Classical 4 Mild 1 HPA	Median 20 months: n = 11 CGMP, n = 11 L-AAs	L-AAs = 516 CGMP = 540, p = ns
Daly [55] 2017	UK	Prospective clinical study 11 y (6–16)	21 9 F, 12 M	20 Classical 1 Mild	6 months n = 12 CGMP n = 9 L-AAs	L-AAs: pre study 325, end of study 280, p = ns CGMP: pre study 275, end of study 317, $p < 0.02$
Ahring [51] 2018	Denmark	Randomised crossover clinical study. 4 PS given over 4 visits 33.3 y ± 11.2 (15–48)	8 7 F, 1 M	8 Classical	PS1 = CGMP, PS2 = L-AAs PS1 and PS2 same AA profilePS3 = CGMP + L-AAs, PS4 = L-AAs PS3 and PS4 same L-AA profile but no Phe	L-AAs = 688 CGMP = 819, p = ns
Daly [56] 2019	UK	Prospective clinical study over 12 months 9.2 y (5–16)	48 21 F, 27 M	46 Classical 2 Mild	12 months n = 29 CGMP n = 19 L-AAs	L-AAs pre study 315, 52 weeks 340, p = 0.236 CGMP pre study 270, 52 weeks 300, p = 0.001
Daly [57] 2019	UK	Randomised control study (RCT) 10 y (6–16)	18 11 F, 7 M	17 Classical 2 Mild	6-week RCT • 2 weeks CGMP 100% no dietary changes (R1) • 2 weeks CGMP 100% minus dietary • phenylalanine contributed from CGMP (R2) • 2 weeks L-AAs nodietary changes (R3)	Median phenylalanine R1: 290 (30–580) R2: 220 (10–670) R3: 165 (10–640) R1 vs. R2, R1 vs. R3 $p < 0.0001$ R2 vs. R3, $p = 0.0009$
Pena [58] 2021	Portugal	Retrospective longitudinal study 28 y (15–43)	11 8 F, 3 M	3 Classical 3 Late diagnosed 3 Mild 2 HPA	29 months CGMP 66%, L-AAs 34% n = 4 CGMP 100% n = 4 CGMP 50 < 100% n = 2 CGMP < 50%	Pre study on L-AAs: 562 ± 289 Post study L-AAs and CGMP 628 ± 317, p = ns

Legend: PKU phenylketonuria; L-AA, amino acid protein substitute; CGMP, caseinglycomacropeptide; PS, protein substitute; ns, not significant; HPA, hyperphenylalaninemia; F, female; M, male; y, years; m, months; vs, versus.

9. Kinetic Properties of Protein Substitutes

There is evidence from animal studies that protein substitutes engineered to slowly release amino acids have improved physiological functions, but proving this remains a challenge in PKU [61]. The speed of absorption of dietary amino acids by the gut varies according to the type of ingested dietary protein. Whey protein is established as a 'fast' protein and casein as a 'slow' protein, the latter provides greater nitrogen retention and whole-body protein anabolism [62,63]. L-AAs are incapable of replicating the physiological actions of whole protein being directly absorbed from the small intestine [22]. Amino acids from L-AAs are rapidly absorbed, peak but then fall rapidly compared to amino acids slowly released from whole protein, and this influences their utilization [12,64,65]. Herrmann et al. [66] demonstrated that ingestion of large doses of L-AAs increased amino acid oxidation and nitrogen excretion, decreasing their availability for cellular functioning. For effective protein synthesis, all essential amino acids must be available to the tissues in appropriate amounts simultaneously [29]. There is circumstantial evidence to suggest that CGMP lowers the rate of amino acid absorption and improves nitrogen retention. Van Calcar et al. studied 11 subjects with PKU over 4 days and reported lower blood phenylalanine after an overnight fast using CGMP compared to L-AAs, implying a slower release of amino acids in CGMP. Two-hour post prandial blood urea nitrogen concentrations were lower, and insulin concentrations were marginally but significantly higher in the CGMP group, suggesting lower nitrogen excretion and improved amino acid utilisation. Any protein substitute that will imitate the physiological absorption of whole protein will theoretically improve growth, body composition and bone density, and may possibly influence inflammatory responses and appetite.

There are no kinetic studies reviewing the action of L-AAs versus CGMP on blood urea nitrogen, insulin or amino acid absorption. Until studies are reported, it cannot be concluded that CGMP improves amino acid utilisation. However, CGMP does influence phenylalanine and tyrosine variability over a 24-h period. In a randomised controlled crossover study [57], children with PKU were randomised to three groups taking CGMP or L-AAs as a protein substitute: group R1 (no dietary adjustment with CGMP), group R2 (dietary adjustment with phenylalanine from CGMP deducted from the dietary phenylalanine allowance) and group R3 (no dietary adjustment with L-AAs). Each arm of the study was for 14 days, and on the last 2 days, subjects had 4-hourly day and night blood spots measuring blood phenylalanine and tyrosine. All median phenylalanine concentrations were within recommended target ranges, there was a significant difference in median phenylalanine at each time point between R1 and R2 ($p = 0.0027$) and R1 and R3 ($p < 0.0001$), but no differences between R2 and R3. Tyrosine was significantly higher in the CGMP groups. This work shows two main findings: the residual phenylalanine given in R1 increased blood phenylalanine concentrations (in this group, 18% had phenylalanine concentrations greater than the target reference range compared to none in the R3 group), and secondly, CGMP appears to give less blood phenylalanine variability when compared to L-AAs. Any mechanism that permits a constant delivery of amino acids would allow a steady state of protein synthesis, improving body protein balance and skeletal muscle protein synthesis.

In a preliminary investigation [67] to review if CGMP compared to L-AAs altered pre and post prandial amino acid profiles in children with PKU, quantitative amino acids were measured after an overnight fast and 2 h post prandially after consuming breakfast and 20 g protein equivalent from the allocated protein substitute. CGMP was provided as CGMP1, in which the amino acid profile met WHO recommendations, or CGMP2, which had higher concentrations of histidine, tyrosine, tryptophan and valine. Forty-three children, median age 9 years (range 5–16 years) were studied; 11 took CGMP1, 18 CGMP2 and 14 L-AAs. The results showed, regardless of the protein substitute source, there was a significant increase in post prandial amino acids. In CGMP2, post prandial histidine ($p < 0.001$), leucine ($p < 0.001$) and tyrosine ($p < 0.001$) were higher than in CGMP1 (reflecting the additional amounts in this formulation), and leucine ($p < 0.001$), threonine ($p < 0.001$) and tyrosine ($p = 0.003$) were higher in CGMP2 than in L-AAs, reflecting the amino acid composition of the three different protein substitute formulations. There is a suggestion that CGMP does alter amino acid absorption, leading to a greater stability of phenylalanine over 24 h, but controlled kinetic studies are necessary.

10. The Impact of CGMP on Growth and Body Composition in Children with PKU

In PKU, the impact of using a phenylalanine-restricted diet on physical growth was first reported in the late 1970s, and despite improvements in dietary treatment, contradictory findings on growth outcome are reported [68–71]. Early studies [72] demonstrated that children had improved growth if they were prescribed a protein equivalent from protein substitute that exceeded the WHO/FAO/UNU 1973 [73] safe levels of protein intake. Smith et al. [74] showed that even if amino acids are efficiently absorbed from the intestinal tract, there is a higher loss of nitrogen as urea when compared to natural protein. McBurnie et al. and Holm et al. [75,76] assessed height, weight and head circumference in two prospective collaborative studies, evaluating 133 and 124 children with PKU over 8 and 4 years, respectively. In both studies, weight and height increased similarly to that of control groups.

In contrast, three European studies [77–79] found children with PKU had reduced height growth when compared to control subjects. Protein substitute intake was not always reported, but typical total protein intake only provided safe recommended intakes [73]. It is possible that phenylalanine deficiency may have occurred but was not described. Dhondt et al. [77] reported normal height and weight were achieved after dietary relaxation at 8 years of age. Schaefer et al. [78] reported negative weight and height in the first 2 years with catch up by 3 years of age. A recent systematic and meta-analysis examining growth in subjects with PKU [70] reported normal growth at birth and during infancy, but children were significantly shorter and had lower weight for age compared with reference populations during the first four years of life. Linear growth was reduced until the end of adolescence. These findings were not identified in patients with mild hyperphenylalaninemia on no dietary restrictions.

Overall, optimal growth was noted in studies where total protein intake (a combined protein intake from natural protein and protein substitute) was higher [80–83]. Nitrogen balance is regulated by urea production [63,84], which is produced linearly in response to plasma amino acid concentrations. Ney et al. and Calcar et al. [49,85] suggested that CGMP may induce a slower and more sustained release of amino acids, leading to decreased urea and greater availability of amino acids for protein synthesis, possibly leading to improved growth.

In PKU, it is important to monitor lean and fat mass, but there are no long-term prospective studies or systematic/meta-analyses describing body composition in PKU. Of eleven studies reported in children (Table 3), any comparison is challenging due to an absence of national reference standards, different body composition techniques, variable pubertal status and different PKU phenotypes. Of six controlled studies, compared with healthy controls, four showed no statistically significant differences in body composition. One study demonstrated a correlation with increased blood phenylalanine concentrations and higher fat mass in male subjects with PKU only [86]. Albersen et al. [87] showed body fat was significantly higher in subjects with PKU, and higher in females >11 years. Long-term associated comorbidities such as type II diabetes and cardiometabolic diseases may be linked to altered body composition, with evidence suggesting an association between abdominal obesity, increased insulin resistance and cardiovascular disease. Therefore, the composition of a protein substitute needs careful formulation as this may alter body composition and possibly long-term health outcomes [88–90].

Table 3. Studies measuring body composition in children with PKU.

Author/Year	Number/Age of Subjects Body Composition Measurement Technique	Parameters Measured	Main Outcome	Limitations
Allen 1996 [91] Australia	n = 30 PKU (classical) Mean age: 9.6 y n = 65 control Mean age: 11.2 y Skinfold thickness	Body fat Resting energy expenditure	No differences in body fat compared to controls No difference in resting energy expenditure	Skinfold measurements provide no information on lean mass.

Table 3. *Cont.*

Author/Year	Number/Age of Subjects Body Composition Measurement Technique	Parameters Measured	Main Outcome	Limitations
Dobbelaere 2003 [68] France	n = 20 PKU (classical) n = 20 control Mean age: 4.5 y Age- and gender-matched Skinfold thickness Bioelectrical impedance	Weight, height, body mass index (BMI) head circumference Skin folds triceps, biceps, subscapular and suprailiac measurement Body density, body fat, lean mass Blood tyrosine and phenylalanine concentrations Zinc, selenium, thyroid, insulin, growth factor Weighed 4-day dietary intake	No differences in body composition compared with controls Growth was significantly different from that of reference population $p < 0.05$ No correlation with phenylalanine biochemical bloods or calorie intake	Body mass index measures nutritional status, not body composition Impedance associated with poor accuracy for individuals and groups
Huemer 2007 [92] Study over 12 months Austria	n = 34 PKU (classical) n = 34 control Mean age: 8.7 y Age/gender-matched Total body electrical conductivity (TOBEC)	Weight, height, BMI % fat, fat-free mass Blood phenylalanine concentrations	No differences between groups for the measured parameters Significant correlation between natural protein g/kg/d and fat-free mass	TOBEC rarely used and unknown accuracy compared to other body composition measurements
Albersen 2010 [87] The Netherlands	n = 20 PKU (classical) n = 20 control Mean age: 10 y Age/gender-matched BodPod/whole-body air displacement plethysmograph	Weight, height, BMI % body fat Blood phenylalanine concentrations	No difference for weight, height, BMI Body fat significantly higher in PKU despite similar BMI to that of controls $p = 0.002$ Body fat higher in girls >11 y, $p = 0.027$ Body fat increased with weight only in PKU No correlation with blood phenylalanine	4/20 PKU children were from different ethnic background
Adamczyk 2011 [93] Poland	n = 45 PKU (classical) Mean age: 13.8 y Group 1 = 15 prepubertal Group 2 = 18 pubertal good control Group 3 = 12 pubertal poor control Dual X-ray absorptiometry (DXA)	Weight, height, BMI Lean body mass Fat mass Total bone density Bone mineral content Ratio of bone mineral content/lean body mass Bone markers Data compared with Polish DXA reference values	Normal body fat and lean body mass Statistically significant differences for ratio of bone mineral content/lean body mass between groups Blood phenylalanine negatively affected bone status	No control group DXA radiation exposure, whole-body bias dependent on size, gender and amount of fat
Douglas 2013 [94] USA	n = 59 PKU (classical and mild) Mean age: 14.4 y BodPod/whole-body air displacement plethysmograph Tricep, subscapular, suprailiac, thigh skinfold	Weight, height, BMI Body fat	Normal body fat Lean mass not evaluated Inverse relationship between age and body fat $p = 0.016$	Mixed PKU phenotype No control group Agreement between skinfold depends on equations used to convert measurement to body fat

Table 3. *Cont.*

Author/Year	Number/Age of Subjects Body Composition Measurement Technique	Parameters Measured	Main Outcome	Limitations
Rocha 2012 Rocha 2013 [95,96] Portugal	*n* = 89 PKU (classical, mild, hyperphenylalaninemia) Mean age: 14.4 y *n* = 78 controls Mean age: 15.9 y Bioelectrical impedance analysis (BIA)	Weight, height, BMI Fat mass Lean body mass Body cell mass Muscular mass Phase angle	No differences in fat mass No differences in lean body mass No differences in body cell, muscular mass or phase angle All classical PKU negative height z-score No differences in height compared to controls in children aged <19 y In PKU group, aged >19 y, height statistically significantly worse than that of controls $p = 0.017$	Impedance is associated with poor accuracy for individuals and groups Mixed PKU phenotype
		Blood pressure, amino acids Glucose, insulin Total cholesterol, high-density cholesterol Triglycerides, C- reactive protein, uric acid Assessment of protein substitute and natural protein intake	Anthropometric parameters no differences to controls Higher triglycerides/high density cholesterol in PKU group Metabolic syndrome no difference compared with controls In PKU subjects, those with central obesity had significantly higher triglycerides/high-density cholesterol compared to those without central obesity	
Doulgeraki 2014 [97] Greece	*n* = 48 PKU (classical) Mean age: 10.9 y 32 HPA (mild hyperphenylalaninemia) Mean age: 10.9 y *n* = 57 control Age/gender-matched Dual X-ray absorptiometry (DXA)	Lean body mass Fat mass Bone mineral density	No differences in body composition Weight and BMI significantly different between mild PKU and classical PKU Bone mineral density lower in classical PKU compared to mild and controls Fat mass significantly higher in PKU teenagers with poor phenylalanine control Positive correlation between bone, muscle and fat mass in both groups and fat mass and phenylalanine concentrations	Mixed PKU phenotype Control group not reported in study DXA radiation exposure, whole-body bias dependent on size, gender and amount of fat
Mazzola 2016 [98] Brazil	*n* = 27 PKU *n* = 11 early diagnosed *n* = 16 late-diagnosed (classical and mild) *n* = 27 control Mean age: 12 y Age/gender-matched Bioelectrical impedance analysis (BIA)	Weight, height, BMI Fat mass Lean body mass Extracellular mass/body cell mass ratio Phase angle (PA)	No differences in body fat No differences in lean body mass No effect on time of diagnosis or PKU phenotype	Age at diagnosis variable, some early and late-treated PKU Mixed PKU phenotype

Table 3. *Cont.*

Author/Year	Number/Age of Subjects Body Composition Measurement Technique	Parameters Measured	Main Outcome	Limitations
Sailer 2020 [86] USA	n = 30 PKU n = 30 control Mean age: 11.6 y Age/gender-matched 4 subjects on Kuvan Dual X-ray absorptiometry (DXA)	Weight, height, BMI Fat mass Lean body mass 24 h dietary recall	Male subjects with PKU had significantly lower lean body mass and more fat mass compared to controls $p = 0.024$ No differences for females and controls when measuring same parameters Age/fat mass positively correlated with blood phenylalanine, $p = 0.02$ Protein substitute negatively correlated with blood phenylalanine $p = 0.04$ Males with PKU had significantly lower height compared with controls $p < 0.05$ No difference in energy intake between the groups	Mixed PKU phenotype 13% on sapropterin DXA radiation exposure, whole-body bias dependent on size, gender and amount of fat
Daly 2021 [99] UK	n = 48 PKU Mean age: 9.2 y (5–16) 3 groups taking different protein substitutes n = 19 L-AAs only n = 16 CGMP and L-AAs (CGMP50) n = 13 CGMP only (CGMP100) Dual X-ray absorptiometry (DXA)	Weight, height, BMI Fat mass Lean body mass % body fat	No correlation or statistically significant differences (after adjusting for age, gender, puberty and blood phenylalanine concentrations) were found between the groups for fat mass, % body fat or lean body mass The change in height z-scores: L-AAs 0, CGMP50 +0.4, CGMP100 +0.7 showed a trend that children in the CGMP100 group were taller, had improved lean body mass with decreased fat mass and % body fat	DXA radiation exposure, whole body bias dependent on size, gender and amount of fat No control non PKU group Different intake of CGMP protein substitute

Legend: Sapropterin, drug treatment for PKU; BMI, body mass index; PKU, Phenylketonuria; L-AA, amino acid protein substitute; CGMP, caseinglycomacropeptide protein substitute; PS, protein substitute; ns, not significant; F, female; M, male; HPA, hyperphenylalaninemia; y, years

To date, only two studies have examined the role of CGMP compared to L-AAs on body composition and growth in PKU: one three-year prospective study [99] in children, and a retrospective review in adults by Pena et al. in 2021. In the three-year study, n = 19 children (median age 11 years; range 5–15 years) took L-AAs only, n = 16 (median age 7.3 years; range 5–15 years) took a combination of CGMP and L-AAs (CGMP50), and n = 13 (median age 9.2 years; range 5–16 years) took CGMP only (CGMP100). A dual-energy X-ray absorptiometry (DXA) scan at enrolment and 36 months measured lean body mass (LBM), % body fat (%BF) and fat mass (FM). Height was measured at enrolment, 12, 24 and 36 months. No correlation or statistically significant differences (after adjusting for age, gender, puberty and phenylalanine blood concentrations) were found between the three groups. The change in height z-scores (L-AAs 0, CGMP50 +0.4 and CGMP100 +0.7) showed a trend that children in the CGMP100 group were taller, had improved LBM with decreased FM and %BF, although this did not reach statistical significance. We can only speculate about this suggested trend shown in the CGMP100 group. One possibility is that the branched-chain amino acids leucine and isoleucine (the latter is naturally higher in CGMP) modulate protein turnover, as both are potent modulators of insulin and glucose metabolism [100]. If insulin sensitivity is enhanced, it is possible that growth could be improved. Further long term studies are needed to confirm these findings.

11. Impact of CGMP Compared to L-AAs on Bone Mass, Density and Geometry in Children with PKU

Bone mass is maintained by a complex and dynamic process involving resorption of bone by the osteoclast and formation of bone by the osteoblast. In children, this is a dynamic continuous process of modelling and remodelling [101]. Peak bone mass, which programmes the future risk of osteoporosis, is established in childhood and adolescence [102,103]. Factors that influence bone mass include genetics, lean mass, adiposity, adipocytokines, physical activity and nutrition. The relationship between fat and bone is contentious. Evidence [103] suggests that in early childhood, obesity confers a structural advantage, but with age this relationship is reversed, and excessive fat is detrimental. Clark et al. [104] in 3082 healthy children, reported a positive relationship between adiposity and bone mass accrual. Others have reported conflicting findings [105,106]. Lean body mass is the strongest significant predictor of bone mineral content [107,108] and relates to bone mass and skeletal development in children.

Dietary protein promotes peripubertal bone growth and slows bone loss [109]. Protein is necessary for optimal bone metabolism during growth, positively influencing bone mass, density and strength [109–111]. In children and adults with PKU, bone density is inconsistently reported [112–118]. Four systematic and three meta-analysis studies report mixed results. Enns et al. reported nine suboptimal bone health outcomes. The scope of this review was on general health problems in PKU, and therefore it failed to interpret the results on bone health in depth. Hansen et al. described a lower spine bone mineral density, but this review had methodological errors and assessment bias. Demirdas et al. [119] reported bone mineral density (BMD) was within the normal range; although it was lower than normal, it was not clinically significant. There was no correlation with phenylalanine concentrations, vitamin D, parathyroid hormone and individual nutrients. De Castro et al. [120] supported the findings from Demirdas et al., showing BMD was lower than that of the reference groups but within the normal range. They also demonstrated an imbalance between bone formation and resorption, favouring bone removal.

Solverson et al. [121] studied the effect of three different diets on bone strength in mice with or without PKU. They were given a low-protein diet with (a) CGMP, (b) L-AAs or (c) a normal (casein) diet. The PKU mice fed either CGMP or L-AAs had a lower BMD compared with non-PKU mice. In PKU mice fed the L-AAs, the femur length independent of gender was significantly shorter compared to that of the PKU mice given CGMP or a normal diet. Skeletal fragility (brittle and weak femora) was a consistent finding in the PKU mice regardless of gender or diet. The reduction of BMD and bone mineral content (BMC) of the femora measured by DXA was more pronounced in the mice receiving L-AAs compared to those receiving CGMP. This group concluded that the type of protein influenced bone outcome in mice, with CGMP giving better results compared to L-AAs. However, careful consideration is needed to determine the impact of CGMP or L-AAs on bone growth. In humans, bone growth is a slow, multifaceted process affected by hormonal patterns, gender, obesity, dietary intake and physical activity.

Only one three-year longitudinal study [122] in children with PKU has compared the impact of CGMP and L-AAs on bone mass, density and geometry (comparing the same group of children who participated in the body composition study previously described). Measurements were taken by DXA and peripheral quantitative computer tomography (pQCT), in addition to blood biochemistry and bone turnover markers. No statistical significance was evident between the three study groups (L-AAs, CGMP50 or CGMP100). In all three groups, there was a strong positive correlation between bone resorption and formation markers: type 1 collagen cross-linked C telopeptide (β CTX) and procollagen type 1 terminal propeptide (P1NP), and there was evidence of an increased PINP in the CGMP100 group independent of age compared to the L-AA group ($p = 0.04$). The synergy between bone formation and resorption shows active bone turnover and reflects appropriate bone growth since these markers are derived from physiological processes. Bone density was clinically normal, although the median z-scores were below the population mean and agreed with the findings of systematic reviews by Demirdis et al. and de Castro et al. Bone remodelling processes appeared active in children with PKU taking either L-AAs or CGMP, but it was unknown why the median z-scores were below the population norm.

12. Does Glycomacropeptide Improve Palatability of Protein Substitutes?

A potential advantage of using a peptide-based protein substitute is the altered taste profile. L-AAs are generally bitter tasting, and both children and adults dislike the aftertaste they leave post consumption [123]. In a blind sensory study, Lim et al. 2007 evaluated the acceptability of CGMP compared to L-AAs and found CGMP was rated favourable for odour and taste. This improved taste profile has been observed by other researchers [49,51,52,54,55,59,124]. Pena et al. [53] highlighted the lack of uniformity in the methods used to evaluate palatability, with some studies evaluating food and others liquid based CGMP protein substitutes. The improved taste profile may improve concordance with a lifelong rigorous diet.

13. Impact of CGMP on Breath Malodour in Children with PKU

In clinical practice, caregivers of children with PKU report their children have breath malodour, particularly after protein substitute consumption. This may increase non adherence by lowering self-esteem and affect interpersonal communication, leading to social isolation. No study has quantitatively measured breath odour in children with PKU. In a randomised, crossover study using gas chromatography ion mobility spectrometry (GS-IMS), exhaled volatile organic compounds were measured in children taking CGMP or L-AAs over the course of 10 h [123]

Forty children (20 PKU; 20 healthy non-PKU controls) were recruited; the children with PKU took either L-AAs or CGMP exclusively for one week in a randomised order. On the seventh day, seven exhaled breath samples were collected over a 10-h period. Subjects than transferred to the alternative protein substitute for a week, and the breath sampling process was repeated. In the PKU group, the aim was to collect breath samples 30 min after consuming their protein substitute; this happened in all but three cases, when breath samples were collected 5 min after protein substitute consumption. In all three groups (L-AAs, CGMP and controls), fasting breath samples contained similar numbers of volatile organic compounds (VOCs) (10–12). Similarly, post prandial samples showed no significant differences in the number of exhaled VOCs (12–18) between L-AAs/CGMP and controls, or between L-AAs and CGMP. A different breath signature occurred in the three subjects who had breath measurements 5 min post completing their protein substitute. In this subset, a higher number of VOCs (25–30) were detected; however, these were no longer detectable at 30 min post consumption. This study demonstrated that protein substitutes have a transient effect on exhaled breath, and after 30 min post consumption, VOCs in children with PKU were no different to those of controls. Timing food and drink with protein substitute consumption may potentially reduce or eliminate the immediate unpleasant protein substitute breath odour.

14. Summary

In PKU, evidence suggests that the use of a bioactive CGMP protein substitute does not show any overwhelming benefit compared to L-AAs on post prandial amino acid absorption, body composition, bone mineral density or breath odour. It is clear that CGMP increases blood phenylalanine concentrations, particularly in children with a low phenylalanine tolerance. However, there is a trend that children taking CGMP as their sole source of protein substitute are taller, with improved lean body mass and decreased fat mass. Overall, the residual phenylalanine content in CGMP appears to be a limitation, particularly for those with minimal or no phenylalanine hydroxylase activity. The full clinical potential of CGMP in PKU has not yet been determined, and its role in gut microbiota and potential brain development awaits further investigation.

Author Contributions: Conceptualization, A.D. and A.M.; investigation, A.D.; supervision, A.M. writing—original draft, A.D.; writing—review and editing, A.D., A.P., S.E. and A.M. All authors have read and agreed to the published version of the manuscript.

Funding: No funding has been provided.

Data Availability Statement: Data available at the request of the corresponding author.

Conflicts of Interest: A.D., research funding from Vitaflo International, financial support from Nutricia, Mevalia and Vitaflo International to attend study days and conferences. S.E., research funding from Nutricia, financial support from Nutricia and Vitaflo International to attend study days and conferences. A.P. has received an educational grant from Cambrooke Therapeutics and grants from Vitaflo International, Nutricia, Merck Serono, Biomarin and Mevalia to attend scientific meetings. A.M. received research funding and honoraria from Nutricia, Vitaflo International and Merck Serono, and is a member of the European Nutrition Expert Panel (Merck Serono International), the Sapropterin Advisory Board (Merck Seronointernational) and the Advisory Board Element (Danone-Nutricia).

References

1. Hillert, A.; Anikster, Y.; Belanger-Quintana, A.; Burlina, A.; Burton, B.K.; Carducci, C.; Chiesa, A.E.; Christodoulou, J.; Đorđević, M.; Desviat, L.R.; et al. The Genetic Landscape and Epidemiology of Phenylketonuria. *Am. J. Hum. Genet.* **2020**, *107*, 234–250. [CrossRef] [PubMed]
2. Bickel, H.; Gerrard, J.; Hickmans, E.M. Influence of phenylalanine intake on phenylketonuria. *Lancet* **1953**, *265*, 812–813. [CrossRef]
3. Woolf, L.I.; Griffiths, R.; Moncrieff, A. Treatment of Phenylketonuria with a Diet Low in Phenylalanine. *Br. Med. J.* **1955**, *1*, 57–64. [CrossRef] [PubMed]
4. Daly, A.; Evans, S.; Pinto, A.; Ashmore, C.; MacDonald, A. Protein Substitutes in PKU: Their Historical Evolution. *Nutrients* **2021**, *13*, 484. [CrossRef]
5. Gardiner, K.; Barbul, A. Review: Intestinal Amino Acid Absorption during Sepsis. *J. Parenter. Enter. Nutr.* **1993**, *17*, 277–283. [CrossRef]
6. MacDonald, A.; Chakrapani, A.; Hendriksz, C.; Daly, A.; Davies, P.; Asplin, D.; Hall, K.; Booth, I.W. Protein substitute dosage in PKU: How much do young patients need? *Arch. Dis. Child.* **2006**, *91*, 588–593. [CrossRef]
7. Macdonald, A.; Rylance, G.; Davies, P.; Asplin, D.; Hall, S.K.; Booth, I.W. Administration of protein substitute and quality of control in phenylketonuria: A randomized study. *J. Inherit. Metab. Dis.* **2003**, *26*, 319–326. [CrossRef]
8. Van Wegberg, A.M.J.; Macdonald, A.; Ahring, K.; BéLanger-Quintana, A.; Blau, N.; Bosch, A.M.; Burlina, A.; Campistol, J.; Feillet, F.; Giżewska, M.; et al. The complete European guidelines on phenylketonuria: Diagnosis and treatment. *Orphanet J. Rare Dis.* **2017**, *12*, 162. [CrossRef]
9. Van Spronsen, F.J.; Smit, P.G.A.; Koch, R. Phenylketonuria: Tyrosine beyond the phenylalanine-restricted diet. *J. Inherit. Metab. Dis.* **2001**, *24*, 1–4. [CrossRef]
10. Elango, R.; Ball, R.O.; Pencharz, P.B. Amino acid requirements in humans: With a special emphasis on the metabolic availability of amino acids. *Amino Acids* **2009**, *37*, 19–27. [CrossRef]
11. Gropper, S.S.; Acosta, P.B. Effect of Simultaneous Ingestion of L-Amino Acids and Whole Protein on Plasma Amino Acid and Urea Nitrogen Concentrations in Humans. *J. Parenter. Enter. Nutr.* **1991**, *15*, 48–53. [CrossRef]
12. Gropper, S.S.; Gropper, D.M.; Acosta, P.B. Plasma Amino Acid Response to Ingestion of L-Amino Acids and Whole Protein. *J. Pediatr. Gastroenterol. Nutr.* **1993**, *16*, 143–150. [CrossRef]
13. Metges, C.C.; El-Khoury, A.E.; Selvaraj, A.B.; Tsay, R.H.; Atkinson, A.; Regan, M.M.; Bequette, B.J.; Young, V.R. Kinetics of L-[1-(13)C]leucine when ingested with free amino acids, unlabeled or intrinsically labeled casein. *Am. J. Physiol. Endocrinol. Metab.* **2000**, *278*, E1000–E1009. [CrossRef]
14. Jahja, R.; Huijbregts, S.; de Sonneville, L.; van der Meere, J.J.; van Spronsen, F.J. Neurocognitive Evidence for Revision of Treatment Targets and Guidelines for Phenylketonuria. *J. Pediatr.* **2014**, *164*, 895–899.e2. [CrossRef]
15. Scala, I.; Riccio, M.P.; Marino, M.; Bravaccio, C.; Parenti, G.; Strisciuglio, P. Large Neutral Amino Acids (LNAAs) Supplementation Improves Neuropsychological Performances in Adult Patients with Phenylketonuria. *Nutrients* **2020**, *12*, 1092. [CrossRef]
16. Van Vliet, D.; Bruinenberg, V.M.; Mazzola, P.N.; van Faassen, M.H.; de Blaauw, P.; Kema, I.P.; Heiner-Fokkema, M.R.; van Anholt, R.D.; van der Zee, E.A.; van Spronsen, F.J. Large Neutral Amino Acid Supplementation Exerts Its Effect through Three Synergistic Mechanisms: Proof of Principle in Phenylketonuria Mice. *PLoS ONE* **2015**, *10*, e0143833.
17. Hidalgo, I.J.; Borchardt, R.T. Transport of a large neutral amino acid (phenylalanine) in a human intestinal epithelial cell line: Caco-2. *Biochim. Biophys. Acta* **1990**, *1028*, 25–30. [CrossRef]
18. Matalon, R.; Michals-Matalon, K.; Bhatia, G.; Grechanina, E.; Novikov, P.; McDonald, J.D.; Grady, J.; Tyring, S.K.; Güttler, F. Large neutral amino acids in the treatment of phenylketonuria (PKU). *J. Inherit. Metab. Dis.* **2006**, *29*, 732–738. [CrossRef]
19. Wapnir, R.A.; Hawkins, R.L.; Lifshitz, F. Local and Systemic Effects of Phenylalanine on Intestinal Transport of Tyrosine and Tryptophan. *Proc. Soc. Exp. Biol. Med.* **1972**, *140*, 661–666. [CrossRef]

20. Schuck, P.F.; Malgarin, F.; Cararo, J.H.; Cardoso, F.; Streck, E.L.; Ferreira, G.C. Phenylketonuria Pathophysiology: On the Role of Metabolic Alterations. *Aging Dis.* **2015**, *6*, 390–399.
21. Thiele, A.G.; Gausche, R.; Lindenberg, C.; Beger, C.; Arelin, M.; Rohde, C.; Mütze, U.; Weigel, J.F.; Mohnike, K.; Baerwald, C.; et al. Growth and Final Height among Children with Phenylketonuria. *Pediatrics* **2017**, *140*, e20170015. [CrossRef]
22. Wu, G. Amino acids: Metabolism, functions, and nutrition. *Amino Acids* **2009**, *37*, 1–17. [CrossRef]
23. MacDonald, A.; Rylance, G.; Hall, S.K.; Asplin, D.; Booth, I.W. Factors affecting the variation in plasma phenylalanine in patients with phenylketonuria on diet. *Arch. Dis. Child.* **1996**, *74*, 412–417. [CrossRef]
24. Macdonald, A.; Lee, P.; Davies, P.; Daly, A.; Lilburn, M.; Ozel, H.G.; Preece, M.A.; Hendriksz, C.; Chakrapani, A. Long-term compliance with a novel vitamin and mineral supplement in older people with PKU. *J. Inherit. Metab. Dis.* **2008**, *31*, 718–723. [CrossRef]
25. Li, P.; Yin, Y.-L.; Li, D.; Kim, S.W.; Wu, G. Amino acids and immune function. *Br. J. Nutr.* **2007**, *98*, 237–252. [CrossRef]
26. Wu, G. Dietary protein intake and human health. *Food Funct.* **2016**, *7*, 1251–1265. [CrossRef]
27. Patti, M.E.; Brambilla, E.; Luzi, L.; Landaker, E.J.; Kahn, C.R. Bidirectional modulation of insulin action by amino acids. *J. Clin. Investig.* **1998**, *101*, 1519–1529. [CrossRef]
28. Patti, M.-E.; Kahn, C.R. The Insulin Receptor—A Critical Link in Glucose Homeostasis and Insulin Action. *J. Basic Clin. Physiol. Pharmacol.* **1998**, *9*, 89–110. [CrossRef]
29. Millward, D.J.; Layman, D.K.; Tomé, D.; Schaafsma, G. Protein quality assessment: Impact of expanding understanding of protein and amino acid needs for optimal health. *Am. J. Clin. Nutr.* **2008**, *87*, 1576S–1581S. [CrossRef]
30. György, P.; Kuhn, R.; Rose, C.S.; Zilliken, F. Bifidus factor. II. Its occurrence in milk from different species and in other natural products. *Arch. Biochem. Biophys.* **1954**, *48*, 202–208. [CrossRef]
31. Delfour, A.; Jollès, J.; Alais, C.; Jolles, P. Caseino-glycopeptides: Characterization of a methionine residue and of the N-terminal sequence. *Biochem. Biophys. Res. Commun.* **1965**, *19*, 452–455. [CrossRef]
32. Brody, E.P. Biological activities of bovine glycomacropeptide. *Br. J. Nutr.* **2000**, *84* (Suppl. S1), 39–46. [CrossRef] [PubMed]
33. Idota, T.; Kawakami, H.; Nakajima, I. Growth promoting effects of N-actylneuraminic acid containing substances on Bifidobacteria. *Biosci. Biotechnol. Biochem.* **1994**, *58*, 413–416. [CrossRef]
34. Sharma, R.; Rajput, Y.S.; Mann, B. Chemical and functional properties of glycomacropeptide (GMP) and its role in the detection of cheese whey adulteration in milk: A review. *Dairy Sci. Technol.* **2013**, *93*, 21–43.
35. Thomä-Worringer, C.; Søensen, J.; López-Fandiño, R. Health effecs and technological features of caseingmacropeptide. *Int. Dairy J.* **2006**, *16*, 1324–1333. [CrossRef]
36. Vance, B.A.; Wu, W.; Ribaudo, R.K.; Segal, D.M.; Kearse, K.P. Multiple Dimeric Forms of Human CD69 Result from Differential Addition of N-Glycans to Typical (Asn-X-Ser/Thr) and Atypical (Asn-X-Cys) Glycosylation Motifs. *J. Biol. Chem.* **1997**, *272*, 23117–23122. [CrossRef]
37. Vreeman, H.J.; Visser, S.; Slangen, C.J.; Van Riel, J.A. Characterization of bovine kappa-casein fractions and the kinetics of chymosin-induced macropeptide release from carbohydrate-free and carbohydrate-containing fractions determined by high-performance gel-permeation chromatography. *Biochem. J.* **1986**, *240*, 87–97. [CrossRef]
38. Holland, J.W.; Deeth, H.C.; Alewood, P.F. Analysis of O-glycosylation site occupancy in bovine kappa-casein glycoforms separated by two-dimensional gel electrophoresis. *Proteomics* **2005**, *5*, 990–1002. [CrossRef]
39. O'Riordan, N.; Kane, M.; Joshi, L.; Hickey, R.M. Structural and functional characteristics of bovine milk protein glycosylation. *Glycobiology* **2014**, *24*, 220–236. [CrossRef]
40. Ntemiri, A.; Chonchúir, F.N.; O'Callaghan, T.; Stanton, C.; Ross, R.; O'Toole, P.W. Glycomacropeptide Sustains Microbiota Diversity and Promotes Specific Taxa in an Artificial Colon Model of Elderly Gut Microbiota. *J. Agric. Food Chem.* **2017**, *65*, 1836–1846. [CrossRef]
41. Kawasaki, Y.; Isoda, H.; Tanimoto, M.; Dosako, S.I.; Idota, T.; Ahiko, K. Inhibition by lactoferrin and kappa-casein glycomacropeptide of binding of Cholera toxin to its receptor. *Biosci. Biotechnol. Biochem.* **1992**, *56*, 195–198. [CrossRef]
42. Oh, S.; Worobo, R.W.; Kim, B.; Rheem, S.; Kim, S. Detection of the cholera toxin-binding activity of kappa-casein macropeptide and optimization of its production by the response surface methodology. *Biosci. Biotechnol. Biochem.* **2000**, *64*, 516–522. [CrossRef]
43. Nakajima, H.; Yamazaki, M.; Kariya, H.; Ohata, R. Isolation of enteroaggregative escherichia coli (EAggEC) from patients with diarrheal disease in an outbreak case of overseas tour. *Kansenshogaku Zasshi* **2005**, *79*, 314–321. [CrossRef]
44. Aimutis, W.R. Bioactive Properties of Milk Proteins with Particular Focus on Anticariogenesis. *J. Nutr.* **2004**, *134*, 989S–995S. [CrossRef]
45. Rose, R. Binding Characteristics of Streptococcus mutans for Calcium and Casein Phosphopeptide. *Caries Res.* **2000**, *34*, 427–431. [CrossRef]
46. Wang, H.X.; Chen, Y.; Haque, Z.; de Veer, M.; Egan, G.; Wang, B. Sialylated milk oligosaccharides alter neurotransmitters and brain metabolites in piglets: An In vivo magnetic resonance spectroscopic (MRS) study. *Nutr. Neurosci.* **2019**, *24*, 885–895. [CrossRef]
47. Ney, D.M.; Hull, A.K.; Van Calcar, S.C.; Liu, X.; Etzel, M.R. Dietary Glycomacropeptide Supports Growth and Reduces the Concentrations of Phenylalanine in Plasma and Brain in a Murine Model of Phenylketonuria. *J. Nutr.* **2008**, *138*, 316–322. [CrossRef]
48. Institute of Medicine, F.N.B. *Dietary Reference Intakes for Energy, Carbohydrates, Fibre, Fat, Protein and Amino Acids*; National Academy Press: Washington, DC, USA, 2002.

49. Van Calcar, S.C.; MacLeod, E.L.; Gleason, S.T.; Etzel, M.R.; Clayton, M.K.; Wolff, J.A.; Ney, D.M. Improved nutritional management of phenylketonuria by using a diet containing glycomacropeptide compared with amino acids. *Am. J. Clin. Nutr.* **2009**, *89*, 1068–1077. [CrossRef]
50. Humayun, M.A.; Turner, J.M.; Elango, R.; Rafii, M.; Langos, V.; Ball, R.O.; Pencharz, P.B. Minimum methionine requirement and cysteine sparing of methionine in healthy school-age children. *Am. J. Clin. Nutr.* **2006**, *84*, 1080–1085. [CrossRef]
51. Ahring, K.K.; Lund, A.; Jensen, E.; Jensen, T.G.; Brøndum-Nielsen, K.; Pedersen, M.; Bardow, A.; Holst, J.J.; Rehfeld, J.F.; Møller, L. Comparison of Glycomacropeptide with Phenylalanine Free-Synthetic Amino Acids in Test Meals to PKU Patients: No Significant Differences in Biomarkers, Including Plasma Phe Levels. *J. Nutr. Metab.* **2018**, *2018*, 1–11. [CrossRef]
52. Ney, D.M.; Stroup, B.M.; Clayton, M.K.; Murali, S.G.; Rice, G.M.; Rohr, F.; Levy, H.L. Glycomacropeptide for nutritional management of phenylketonuria: A randomized, controlled, crossover trial. *Am. J. Clin. Nutr.* **2016**, *104*, 334–345. [CrossRef]
53. Pena, M.J.; Pinto, A.; Daly, A.; Macdonald, A.; Azevedo, L.; Rocha, J.C.; Borges, N. The Use of Glycomacropeptide in Patients with Phenylketonuria: A Systematic Review and Meta-Analysis. *Nutrients* **2018**, *10*, 1794. [CrossRef]
54. Zaki, O.K.; El-Wakeel, L.; Ebeid, Y.; Elarab, H.E.; Moustafa, A.; Abdulazim, N.; Karara, H.; Elghawaby, A. The Use of Glycomacropeptide in Dietary Management of Phenylketonuria. *J. Nutr. Metab.* **2016**, *2016*, 2453027. [CrossRef]
55. Daly, A.; Evans, S.; Chahal, S.; Santra, S.; MacDonald, A. Glycomacropeptide in children with phenylketonuria: Does its phenylalanine content affect blood phenylalanine control? *J. Hum. Nutr. Diet* **2017**, *30*, 515–523. [CrossRef]
56. Daly, A.; Evans, S.; Chahal, S.; Santra, S.; Pinto, A.; Jackson, R.; Gingell, C.; Rocha, J.; Van Spronsen, F.J.; Macdonald, A. Glycomacropeptide: Long-term use and impact on blood phenylalanine, growth and nutritional status in children with PKU. *Orphanet J. Rare Dis.* **2019**, *14*, 44. [CrossRef]
57. Daly, A.; Evans, S.; Chahal, S.; Santra, S.; Pinto, A.; Gingell, C.; Rocha, J.C.; Van Spronsen, F.; Jackson, R.; Macdonald, A. The Effect of Glycomacropeptide versus Amino Acids on Phenylalanine and Tyrosine Variability over 24 Hours in Children with PKU: A Randomized Controlled Trial. *Nutrients* **2019**, *11*, 520. [CrossRef]
58. Pena, M.J.; Pinto, A.; de Almeida, M.F.; Barbosa, C.D.S.; Ramos, P.C.; Rocha, S.; Guimas, A.; Ribeiro, R.; Martins, E.; Bandeira, A.; et al. Continuous use of glycomacropeptide in the nutritional management of patients with phenylketonuria: A clinical perspective. *Orphanet J. Rare Dis.* **2021**, *16*, 1–10. [CrossRef]
59. Pinto, A.; Almeida, M.; Ramos, P.C.; Rocha, S.; Guimas, A.; Ribeiro, R.; Martins, E.; Bandeira, A.; MacDonald, A.; Rocha, J.C. Nutritional status in patients with phenylketonuria using glycomacropeptide as their major protein source. *Eur. J. Clin. Nutr.* **2017**, *71*, 1230–1234. [CrossRef] [PubMed]
60. Macfarlane, G.T.; Macfarlane, S. Bacteria, colonic fermentation, and gastrointestinal health. *J. AOAC Int.* **2012**, *95*, 50–60. [CrossRef] [PubMed]
61. Giarratana, N.; Gallina, G.; Panzeri, V.; Frangi, A.; Canobbio, A.; Reiner, G. A New Phe-Free Protein Substitute Engineered to Allow a Physiological Absorption of Free Amino Acids for Phenylketonuria. *J. Inborn Errors Metab. Screen.* **2018**, *6*. [CrossRef]
62. Boirie, Y.; Dangin, M.; Gachon, P.; Vasson, M.-P.; Maubois, J.-L.; Beaufrère, B. Slow and fast dietary proteins differently modulate postprandial protein accretion. *Proc. Natl. Acad. Sci. USA* **1997**, *94*, 14930–14935. [CrossRef]
63. Dangin, M.; Boirie, Y.; Garcia-Rodenas, C.; Gachon, P.; Fauquant, J.; Callier, P.; Ballèvre, O.; Beaufrère, B. The digestion rate of protein is an independent regulating factor of postprandial protein retention. *Am. J. Physiol. Endocrinol. Metab.* **2001**, *280*, E340–E348. [CrossRef]
64. Mönch, E.; Herrmann, M.E.; Brösicke, H.; Schöffer, A.; Keller, M. Utilisation of amino acid mixtures in adolescents with phenylketonuria. *Eur. J. Pediatr.* **1996**, *155* (Suppl. S1), S115–S120. [CrossRef]
65. Silk, D.B.A.; Fairclough, P.D.; Clark, M.L.; Hegarty, J.E.; Addison, J.M.; Burston, D.; Clegg, K.M.; Matthews, D.M. Use of a Peptide Rather Than Free Amino Acid Nitrogen Source in Chemically Defined "Elemental" Diets. *J. Parenter. Enter. Nutr.* **1980**, *4*, 548–553. [CrossRef]
66. Herrmann, M.E.; Brösicke, H.G.; Keller, M.; Monch, E.; Helge, H. Dependence of the utilization of a phenylalanine-free amino acid mixture on different amounts of single dose ingested. A case report. *Eur. J. Pediatr.* **1994**, *153*, 501–503. [CrossRef]
67. Daly, A.; Evans, S.; Pinto, A.; Jackson, R.; Ashmore, C.; Rocha, J.C.; Macdonald, A. Preliminary Investigation to Review If a Glycomacropeptide Compared to L-Amino Acid Protein Substitute Alters the Pre- and Postprandial Amino Acid Profile in Children with Phenylketonuria. *Nutrients* **2020**, *12*, 2443. [CrossRef]
68. Dobbelaere, D.; Michaud, L.; DeBrabander, A.; Vanderbecken, S.; Gottrand, F.; Turck, D.; Farriaux, J.P. Evaluation of nutritional status and pathophysiology of growth retardation in patients with phenylketonuria. *J. Inherit. Metab. Dis.* **2003**, *26*, 1–11. [CrossRef]
69. Dokoupil, K.; Gokmen-Ozel, H.; Lammardo, A.M.; Motzfeldt, K.; Robert, M.; Rocha, J.C.; van Rijn, M.; Ahring, K.; Bélanger-Quintana, A.; MacDonald, A. Optimising growth in phenylketonuria: Current state of the clinical evidence base. *Clin. Nutr.* **2011**, *31*, 16–21. [CrossRef]
70. Ilgaz, F.; Pinto, A.; Gökmen-Özel, H.; Rocha, J.C.; Van Dam, E.; Ahring, K.; Bélanger-Quintana, A.; Dokoupil, K.; Karabulut, E.; Macdonald, A. Long-Term Growth in Phenylketonuria: A Systematic Review and Meta-Analysis. *Nutrients* **2019**, *11*, 2070. [CrossRef]
71. Zaffanello, M.; Zamboni, G.; Tatò, L. Growth parameters in newborns with hyperphenylalaninaemia. *Paediatr. Périnat. Epidemiol.* **2002**, *16*, 274–277. [CrossRef]

72. Kindt, E.; Motzfeldt, K.; Halvorsen, S.; Lie, S.O. Protein requirements in infants and children: A longitudinal study of children treated for phenylketonuria. *Am. J. Clin. Nutr.* **1983**, *37*, 778–785. [CrossRef]
73. FAO; WHO. *Energy and Protein Requirements*; Report of a Joint FAO/WHO Ad Hoc Expert Committee, in Technical Report Series 1973; FAO: Quebec City, QC, Canada; WHO: Geneva, Switzerland, 1973; p. 118.
74. Smith, J.L.; Arteaga, C.; Heymsfield, S.B. Increased Ureagenesis and Impaired Nitrogen Use during Infusion of a Synthetic Amino Acid Formula. *N. Engl. J. Med.* **1982**, *306*, 1013–1018. [CrossRef]
75. Holm, V.A.; Kronmal, R.A.; Williamson, M.; Roche, A.F. Physical growth in phenylketonuria: II. Growth of treated children in the PKU collaborative study from birth to 4 years of age. *Pediatrics* **1979**, *63*, 700–707. [CrossRef]
76. McBurnie, M.; Kronmal, R.; Schuett, V.; Koch, R.; Azeng, C. Physical growth of children treated for phenylketonuria. *Ann. Hum. Biol.* **1991**, *18*, 357–368. [CrossRef]
77. Dhondt, J.L.; Moreno, L.; Farriaux, J.P. Physical growth in patients with phenylketonuria. *J. Inherit. Metab. Dis.* **1995**, *18*, 135–137. [CrossRef]
78. Schaefer, F.; Burgard, P.; Batzler, U.; Rupp, A.; Schmidt, H.; Gilli, G.; Bickel, H.; Bremer, H.J. Growth and skeletal maturation in children with phenylketonuria. *Acta Paediatr.* **1994**, *83*, 534–541. [CrossRef]
79. Verkerk, P.H.; van Spronsen, F.J.; Smit, G.P.; Sengers, R.C. Impaired prenatal and postnatal growth in Dutch patients with phenylketonuria. The National PKU Steering Committee. *Arch. Dis. Child.* **1994**, *71*, 114–118. [CrossRef]
80. Belanger-Quintana, A.; Martínez-Pardo, M. Physical development in patients with phenylketonuria on dietary treatment: A retrospective study. *Mol. Genet. Metab.* **2011**, *104*, 480–484. [CrossRef]
81. Couce, M.L.; Guler, I.; Anca-Couce, A.; Lojo, M.; Mirás, A.; Leis, R.; Pérez-Muñuzuri, A.; Fraga, J.M.; Gude, F. New insights in growth of phenylketonuric patients. *Eur. J. Pediatr.* **2015**, *174*, 651–659. [CrossRef]
82. Evans, M.; Truby, H.; Boneh, A. The relationship between dietary intake, growth and body composition in Phenylketonuria. *Mol. Genet. Metab.* **2017**, *122*, 36–42. [CrossRef]
83. Hoeksma, M.; Van Rijn, M.; Verkerk, P.H.; Bosch, A.M.; Mulder, M.F.; De Klerk, J.B.C.; De Koning, T.J.; Rubio-Gozalbo, E.; De Vries, M.; Sauer, P.J.J.; et al. The intake of total protein, natural protein and protein substitute and growth of height and head circumference in Dutch infants with phenylketonuria. *J. Inherit. Metab. Dis.* **2005**, *28*, 845–854. [CrossRef] [PubMed]
84. Young, V.R.; El-Khoury, A.E.; Raguso, C.A.; Forslund, A.; Hambraeus, L. Rates of Urea Production and Hydrolysis and Leucine Oxidation Change Linearly over Widely Varying Protein Intakes in Healthy Adults. *J. Nutr.* **2000**, *130*, 761–766. [CrossRef] [PubMed]
85. Ney, D.M.; Etzel, M.R. Designing medical foods for inherited metabolic disorders: Why intact protein is superior to amino acids. *Curr. Opin. Biotechnol.* **2017**, *44*, 39–45. [CrossRef] [PubMed]
86. Sailer, M.; Elizondo, G.; Martin, J.; Harding, C.O.; Gillingham, M.B. Nutrient intake, body composition, and blood phenylalanine control in children with phenylketonuria compared to healthy controls. *Mol. Genet. Metab. Rep.* **2020**, *23*, 100599. [CrossRef] [PubMed]
87. Albersen, M.; Bonthuis, M.; De Roos, N.M.; Hurk, D.A.M.V.D.; Weber, E.C.; Hendriks, M.M.W.B.; Velden, M.G.M.D.S.-V.D.; De Koning, T.J.; Visser, G. Whole body composition analysis by the BodPod air-displacement plethysmography method in children with phenylketonuria shows a higher body fat percentage. *J. Inherit. Metab. Dis.* **2010**, *33* (Suppl. S3), 283–288. [CrossRef] [PubMed]
88. Despres, J.P. Is visceral obesity the cause of the metabolic syndrome? *Ann. Med.* **2006**, *38*, 52–63. [CrossRef]
89. Després, J.-P.; Lemieux, I. Abdominal obesity and metabolic syndrome. *Nature* **2006**, *444*, 881–887. [CrossRef]
90. Forte, N.; Fernández-Rilo, A.C.; Palomba, L.; Di Marzo, V.; Cristino, L. Obesity Affects the Microbiota-Gut-Brain Axis and the Regulation Thereof by Endocannabinoids and Related Mediators. *Int. J. Mol. Sci.* **2020**, *21*, 1554. [CrossRef]
91. Allen, J.R.; Baur, L.; Waters, D.L.; Humphries, I.R.; Allen, B.J.; Roberts, D.C.; Gaskin, K.J. Body protein in prepubertal children with phenylketonuria. *Eur. J. Clin. Nutr.* **1996**, *50*, 178–186.
92. Huemer, M.; Huemer, C.; Möslinger, D.; Huter, D.; Stöckler-Ipsiroglu, S. Growth and body composition in children with classical phenylketonuria: Results in 34 patients and review of the literature. *J. Inherit. Metab. Dis.* **2007**, *30*, 694–699. [CrossRef]
93. Adamczyk, P.; Morawiec-Knysak, A.; Płudowski, P.; Banaszak, B.; Karpe, J.; Pluskiewicz, W. Bone metabolism and the muscle-bone relationship in children, adolescents and young adults with phenylketonuria. *J. Bone Miner. Metab.* **2011**, *29*, 236–244. [CrossRef]
94. Douglas, T.D.; Kennedy, M.J.; Quirk, M.E.; Yi, S.H.; Singh, R.H. Accuracy of Six Anthropometric Skinfold Formulas Versus Air Displacement Plethysmography for Estimating Percent Body Fat in Female Adolescents with Phenylketonuria. *JIMD Rep.* **2013**, *10*, 23–31. [CrossRef]
95. Rocha, J.C.; MacDonald, A.; Trefz, F. Is overweight an issue in phenylketonuria? *Mol. Genet. Metab.* **2013**, *110*, S18–S24. [CrossRef]
96. Rocha, J.C.; van Spronsen, F.J.; Almeida, M.; Soares, G.; Quelhas, D.; Ramos, E.; Guimarães, J.T.; Borges, N. Dietary treatment in phenylketonuria does not lead to increased risk of obesity or metabolic syndrome. *Mol. Genet. Metab.* **2012**, *107*, 659–663. [CrossRef]
97. Doulgeraki, A.; Skarpalezou, A.; Theodosiadou, A.; Monopolis, I.; Schulpis, K. Body Composition Profile of Young Patients With Phenylketonuria and Mild Hyperphenylalaninemia. *Int. J. Endocrinol. Metab.* **2014**, *12*, e16061. [CrossRef]
98. Mazzola, P.N.; Nalin, T.; Castro, K.; van Rijn, M.; Derks, T.G.; Perry, I.D.; Mainieri, A.S.; Schwartz, I.V.D. Analysis of body composition and nutritional status in Brazilian phenylketonuria patients. *Mol. Genet. Metab. Rep.* **2016**, *6*, 16–20. [CrossRef]

99. Daly, A.; Högler, W.; Crabtree, N.; Shaw, N.; Evans, S.; Pinto, A.; Jackson, R.; Strauss, B.; Wilcox, G.; Rocha, J.; et al. Growth and Body Composition in PKU Children—A Three-Year Prospective Study Comparing the Effects of L-Amino Acid to Glycomacropeptide Protein Substitutes. *Nutrients* **2021**, *13*, 1323. [CrossRef]
100. Duan, Y.; Li, F.; Tan, K.; Liu, H.; Li, Y.; Liu, Y.; Kong, X.; Tang, Y.; Wu, G.; Yin, Y. Key mediators of intracellular amino acids signaling to mTORC1 activation. *Amino Acids* **2015**, *47*, 857–867. [CrossRef]
101. Frost, H.; Schönau, E. The "Muscle-Bone Unit" in Children and Adolescents: A 2000 Overview. *J. Pediatr. Endocrinol. Metab.* **2000**, *13*, 571–590. [CrossRef]
102. Bachrach, L.K. Acquisition of optimal bone mass in childhood and adolescence. *Trends Endocrinol. Metab.* **2001**, *12*, 22–28. [CrossRef]
103. Dimitri, P. Fat and bone in children—Where are we now? *Ann. Pediatr. Endocrinol. Metab.* **2018**, *23*, 62–69. [CrossRef]
104. Clark, E.M.; Ness, A.; Bishop, N.J.; Tobias, J. Association between Bone Mass and Fractures in Children: A Prospective Cohort Study. *J. Bone Miner. Res.* **2006**, *21*, 1489–1495. [CrossRef]
105. Rauch, F.; Bailey, D.A.; Baxter-Jones, A.; Mirwald, R.; Faulkner, R. The 'muscle-bone unit' during the pubertal growth spurt. *Bone* **2004**, *34*, 771–775. [CrossRef]
106. Russell, M.; Mendes, N.; Miller, K.K.; Rosen, C.J.; Lee, H.; Klibanski, A.; Misra, M. Visceral Fat Is a Negative Predictor of Bone Density Measures in Obese Adolescent Girls. *J. Clin. Endocrinol. Metab.* **2010**, *95*, 1247–1255. [CrossRef]
107. Carbone, J.W.; Pasiakos, S.M. Dietary Protein and Muscle Mass: Translating Science to Application and Health Benefit. *Nutrients* **2019**, *11*, 1136. [CrossRef]
108. Crabtree, N.; Kibirige, M.; Fordham, J.; Banks, L.; Muntoni, F.; Chinn, D.; Boivin, C.; Shaw, N. The relationship between lean body mass and bone mineral content in paediatric health and disease. *Bone* **2004**, *35*, 965–972. [CrossRef] [PubMed]
109. Rizzoli, R.; Bianchi, M.L.; Garabédian, M.; McKay, H.A.; Moreno, L.A. Maximizing bone mineral mass gain during growth for the prevention of fractures in the adolescents and the elderly. *Bone* **2010**, *46*, 294–305. [CrossRef] [PubMed]
110. Bonjour, J.-P. Calcium and Phosphate: A Duet of Ions Playing for Bone Health. *J. Am. Coll. Nutr.* **2011**, *30* (Suppl. S1), 438S–448S. [CrossRef] [PubMed]
111. Bonjour, J.-P.; Ammann, P.; Chevalley, T.; Rizzoli, R. Protein Intake and Bone Growth. *Can. J. Appl. Physiol.* **2001**, *26*, S153–S166. [CrossRef]
112. Al-Qadreh, A.; Schulpis, K.; Athanasopoulou, H.; Mengreli, C.; Skarpalezou, A.; Voskaki, I. Bone mineral status in children with phenylketonuria under treatment. *Acta Paediatr.* **2007**, *87*, 1162–1166. [CrossRef]
113. Ambroszkiewicz, J.; Gajewska, J.; Chełchowska, M.; Oltarzewski, M.; Laskowska-Klita, T.; Nowacka, M.; Milanowski, A. Concentration of osteoprotegerin, bone formation and resorption markers in patients with phenylketonuria. *Pol. Merkur. Lekarski.* **2008**, *25*, 57–60.
114. Barat, P.; Barthe, N.; Redonnet-Vernhet, I.; Parrot, F. The impact of the control of serum phenylalanine levels on osteopenia in patients with phenylketonuria. *Eur. J. Pediatr.* **2002**, *161*, 687–688. [CrossRef]
115. De Groot, M.J.; Hoeksma, M.; van Rijn, M.; Slart, R.H.; van Spronsen, F.J. Relationships between lumbar bone mineral density and biochemical parameters in phenylketonuria patients. *Mol. Genet. Metab.* **2012**, *105*, 566–570. [CrossRef]
116. Greeves, L.; Magee, A. Fractures and phenylketonuria. *Acta Paediatr.* **1997**, *86*, 242–244. [CrossRef]
117. McMurry, M.P.; Chan, G.M.; Leonard, C.O.; Ernst, S.L. Bone mineral status in children with phenylketonuria—Relationship to nutritional intake and phenylalanine control. *Am. J. Clin. Nutr.* **1992**, *55*, 997–1004. [CrossRef]
118. Zeman, J.; Bayer, M.; Stepan, J. Bone mineral density in patients with phenylketonuria. *Acta Paediatr.* **1999**, *88*, 1348–1351. [CrossRef]
119. Demirdas, S.; Coakley, K.E.; Bisschop, P.H.; Hollak, C.E.M.; Bosch, A.M.; Singh, R.H. Bone health in phenylketonuria: A systematic review and meta-analysis. *Orphanet J. Rare Dis.* **2015**, *10*, 17. [CrossRef]
120. De Castro, M.J.; de Lamas, C.; Sánchez-Pintos, P.; González-Lamuño, D.; Couce, M.L. Bone Status in Patients with Phenylketonuria: A Systematic Review. *Nutrients* **2020**, *12*, 2154. [CrossRef]
121. Solverson, P.; Murali, S.G.; Litscher, S.J.; Blank, R.D.; Ney, D.M. Low Bone Strength Is a Manifestation of Phenylketonuria in Mice and Is Attenuated by a Glycomacropeptide Diet. *PLoS ONE* **2012**, *7*, e45165. [CrossRef]
122. Daly, A.; Högler, W.; Crabtree, N.; Shaw, N.; Evans, S.; Pinto, A.; Jackson, R.; Ashmore, C.; Rocha, J.; Strauss, B.; et al. A Three-Year Longitudinal Study Comparing Bone Mass, Density, and Geometry Measured by DXA, pQCT, and Bone Turnover Markers in Children with PKU Taking L-Amino Acid or Glycomacropeptide Protein Substitutes. *Nutrients* **2021**, *13*, 2075. [CrossRef]
123. Tiele, A.; Daly, A.; Hattersley, J.; Pinto, A.; Evans, S.; Ashmore, C.; Macdonald, A.; Covington, J.A. Investigation of paediatric PKU breath malodour, comparing glycomacropeptide with phenylalanine free L-amino acid supplements. *J. Breath Res.* **2019**, *14*, 016001. [CrossRef]
124. MacLeod, E.L.; Clayton, M.K.; van Calcar, S.C.; Ney, D.M. Breakfast with glycomacropeptide compared with amino acids suppresses plasma ghrelin levels in individuals with phenylketonuria. *Mol. Genet. Metab.* **2010**, *100*, 303–308. [CrossRef]

nutrients

MDPI

Review

The Impact of the Quality of Nutrition and Lifestyle in the Reproductive Years of Women with PKU on the Long-Term Health of Their Children

Maria Inês Gama [1], Alex Pinto [2], Anne Daly [2], Júlio César Rocha [1,3,4] and Anita MacDonald [2,*]

1 Nutrition & Metabolism, NOVA Medical School, Faculdade de Ciências Médicas, Universidade Nova de Lisboa, Campo dos Mártires da Pátria 130, 1169-056 Lisboa, Portugal; inesrgama@gmail.com (M.I.G.); rochajc@nms.unl.pt (J.C.R.)
2 Birmingham Women's and Children's Hospital, Birmingham B4 6NH, UK; alex.pinto@nhs.net (A.P.); a.daly3@nhs.net (A.D.)
3 Reference Centre of Inherited Metabolic Diseases, Centro Hospitalar Universitário de Lisboa Central, 1169-045 Lisboa, Portugal
4 CINTESIS—Center for Health Technology and Services Research, NOVA Medical School, Campo dos Mártires da Pátria 130, 1169-056 Lisboa, Portugal
* Correspondence: anita.macdonald@nhs.net

Abstract: A woman's nutritional status before and during pregnancy can affect the health of her progeny. Phenylketonuria (PKU), a rare disorder causing high blood and brain phenylalanine (Phe) concentrations, is associated with neurocognitive disability. Lifelong treatment is mainly dietetic with a Phe-restricted diet, supplemented with a low-Phe protein substitute. Treatment adherence commonly decreases in adolescence, with some adults ceasing dietary treatment. In maternal PKU, elevated blood Phe is harmful to the fetus so a strict Phe-restricted diet must be re-established preconception, and this is particularly difficult to achieve. A woman's reproductive years introduces an opportunity to adopt healthier behaviours to prepare for successful pregnancies and positive health outcomes for both themselves and their children. Several factors can influence the health status of women with PKU. Political, socioeconomic, and individual food and lifestyle choices affect diet quality, metabolic control, and epigenetics, which then pre-condition the overall maternal health and long-term health of the child. Here, we reflect on a comprehensive approach to treatment and introduce practical recommendations to optimize the wellbeing of women with PKU and the resultant health of their children.

Keywords: adherence; epigenetics; health; phenylketonuria; preconception; women

Citation: Gama, M.I.; Pinto, A.; Daly, A.; Rocha, J.C.; MacDonald, A. The Impact of the Quality of Nutrition and Lifestyle in the Reproductive Years of Women with PKU on the Long-Term Health of Their Children. *Nutrients* **2022**, *14*, 1021. https://doi.org/10.3390/nu14051021

Academic Editor: Yugo Shibagaki

Received: 3 February 2022
Accepted: 24 February 2022
Published: 28 February 2022

Publisher's Note: MDPI stays neutral with regard to jurisdictional claims in published maps and institutional affiliations.

Copyright: © 2022 by the authors. Licensee MDPI, Basel, Switzerland. This article is an open access article distributed under the terms and conditions of the Creative Commons Attribution (CC BY) license (https://creativecommons.org/licenses/by/4.0/).

1. Introduction

Phenylketonuria (PKU, OMIM 261600) is an inherited metabolic disorder caused by mutations in the phenylalanine hydroxylase (PAH) enzyme that impairs phenylalanine (Phe) metabolism, leading to high blood and brain Phe concentrations. It is managed with a lifelong Phe-restricted diet and an adjunct pharmacological treatment, such as sapropterin or pegvaliase [1]. In maternal phenylketonuria (MPKU), it is established that Phe crosses the placenta's blood membrane through a concentration gradient [2,3] and elevated blood Phe levels have a well-recognised teratogenic effect on the developing fetus, particularly in the early stages of pregnancy [4]. MPKU syndrome is characterized by foetal intrauterine growth retardation, facial dysmorphism, microcephaly, congenital heart disease, infant low birth weight, developmental delay, and intellectual disabilities [4]. There is also an increased risk of miscarriage, usually associated with poor maternal metabolic control [5]. Although there are several reports of pregnancy in women with PKU, little is known about the conception rates compared with the general population, though one recent UK/PKU study reported that 37% of 300 women aged ≥18 years had one or more children [6]. MPKU syndrome is preventable if women achieve rigorous blood Phe control by adhering to a Phe-restricted diet that is commenced preconception and continued throughout pregnancy. A considerable amount of professional health time and support is given to women during this challenging time.

In women with PKU, less consideration is given to the overall quality of nutritional care in the reproductive years (spanning from mid-adolescence until mid-adulthood) and interpregnancy. There is mounting evidence in all women of reproductive age that poor maternal and pregnancy health leads to a higher risk of disease in their children as they age [7]. The nutritional health of many women with PKU at the time of conception is likely to be sub-optimal, particularly if a strict dietary treatment has not been maintained through adult life. Some may have adopted an unhealthy eating pattern even if they are able to maintain optimal metabolic control. Furthermore, unplanned pregnancies at any point in time may increase the risk of nutrient imbalances. In England, 45% of all pregnancies are unplanned [7], and similar figures are observed in women with PKU, despite active health professional education to avoid unplanned pregnancy [5].

Therefore, the lifestyle choices of all women in reproductive years can have an enduring influence on the lifetime health of their children, and a clear focus on interventions before conception is necessary. Cohort studies have shown that improving dietary patterns for up to three years prior to conception can influence pregnancy outcomes, including lowering the risk of preterm birth [8]. Preconception environmental and nutritional factors that may affect the foetal outcome in women with PKU are presented in Figure 1. This review aims to highlight the importance of optimal nutrition, lifestyle, and environment in women with PKU in their reproductive years and offers proposals for pragmatic interventions that may improve the outcome of their children.

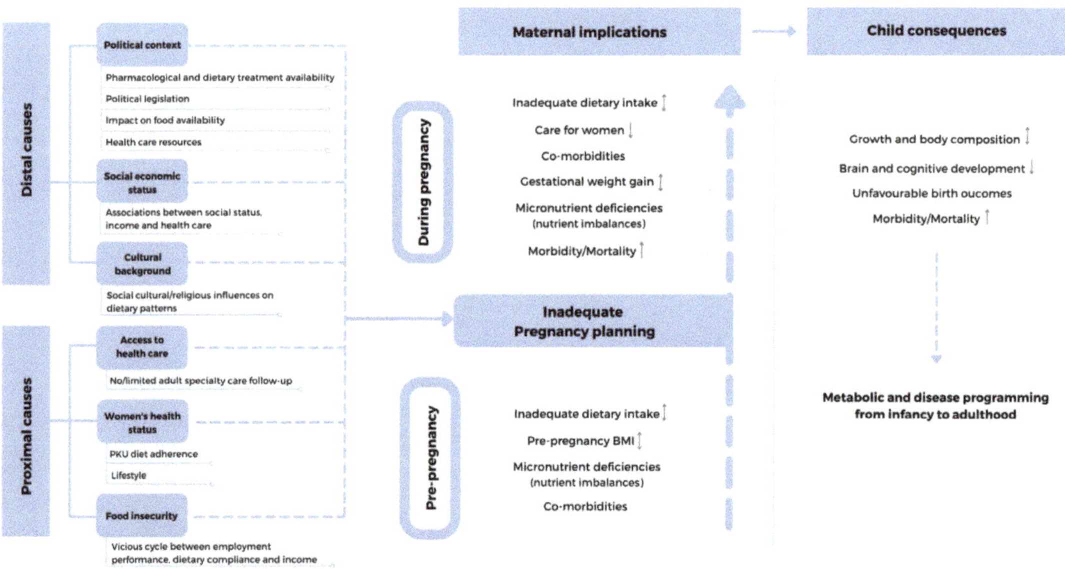

Figure 1. Preconception environmental and nutritional factors that may affect the foetal outcome in women with PKU. ↓-lower; ↓-higher; ↕-lower/higher; ↓↑ broken line-arrows-potentially lower/higher.

2. Nutritional Vulnerability of Women with PKU in Their Reproductive Years

2.1. Distal, Social, and Economic Causes of Nutritional Vulnerability in Adult Women with PKU

There are many economic and political factors that may lead to suboptimal nutritional outcomes associated with the availability of treatment for women with PKU. Health provision varies around the world, and some women with PKU have limited access to 'free' health care from public funding, while few hospitals provide PKU health care teams that provide expertise in the management of adult patients. Low Phe protein substitutes and special low protein foods (SLPFs) are an essential part of treatment but are expensive and may be unaffordable unless provided by insurance or state health care systems. Pharmaceutical treatments may be unavailable or even ineffectual (e.g., sapropterin) for adult patients with classical PKU without residual enzyme activity [4]. Many adult women may be unemployed, receive low earnings due to part-time work, or have minimal earning capacity due to impaired cognitive functioning, affecting their economic security, life quality, and ability to afford their dietary treatment. Political legislation that aims to improve the health of the entire population, e.g., food labelling laws and sugar taxes, may indirectly create additional treatment challenges because of further unintentional dietary restrictions for people with PKU.

Women who do adhere to dietary treatment are dependent on a Phe restricted diet and, if they have classical PKU, usually tolerate <500 mg/day Phe (equivalent to 10 g/natural protein) supplemented with protein substitutes. The protein substitutes are mainly comprised of Phe-free L-amino acids (AA) or low-Phe glycomacropeptide (GMP) and may potentially supply up to 80% of protein intake. Although they usually contain added tyrosine, micronutrients including vitamins, minerals, and long-chain fatty acids, such as docosahexaenoic acid (DHA), the lifetime outcome of habitually taking an artificial protein source is unknown. Amino acid supplements, compared with natural protein, are associated with less efficient utilization and early oxidation, and they may alter insulin release, glycaemic control, and endocrine regulation [9]. The impact on gut microbiota and long-term renal health is undetermined. SPLFs are high in carbohydrates [1,10,11] and contain isolated starches that are more refined or have a higher glycaemic index than equivalent foods made from wheat flour [12,13].

2.2. Proximal Causes Directly Related to Nutritional Vulnerability in Adult Women with PKU

Dietary adherence becomes increasingly challenging with age and metabolic control commonly deteriorates from adolescence [14–18]; it is estimated that 25% to 40% of adults who remain in clinical follow up discontinue treatment [19]. Most adults have difficultly re-establishing dietary control after a period 'off diet' or dietary relaxation [20]. Although more natural protein is consumed than prescribed, clinical practice suggests that the quality of foods eaten is poor, potentially leading to nutritional inadequacy [21,22]. Women may have a low IQ (associated with poor blood Phe control during childhood) and poor executive functioning and possibly have left home and lost the practical support of their parents. This affects their ability to self-manage a Phe restricted diet owing to the daily organisation and planning required [18,23]. Low mood or denial of the condition may also obstruct the ability of people to comply by reducing self-control or motivation. Poor knowledge of diet and food suitability, limited cooking skills and meal choices, the inability to read and interpret protein amounts on food labels, being unable to estimate protein exchanges, and difficulty accessing supplies of protein substitutes/SPLFs also influence the ability to adhere to the diet [24].

2.3. Health of Women with PKU

Obesity: The prevalence of overweight and obesity in all women of childbearing age is high, and approximately 39% of the world's adult population is overweight, with 13% being obese [25]. Although a recent systematic review and meta-analysis of women with PKU [26] found that the body mass index (BMI) of patients with PKU was similar to their healthy controls, a subgroup of patients with classical PKU had a significantly higher BMI. The authors also noted a trend towards a higher BMI in females with PKU in all studies with male and female datasets. The BMI was also higher in an uncontrolled study in women with PKU, particularly if they had poor blood Phe control [27]. Adolescence is a critical period for the development of overweight and obesity [28], with a recent study illustrating that 28% (n = 101) of adolescents with PKU were overweight or obese [29].

Eating disorders: There is increasing evidence of eating disorders, food neophobia, and adverse attitudes towards food in adults with PKU [24,30–32]. Disordered eating refers to abnormal behaviours focused on eating or feeding, but it does not fit the pattern of a specific eating disorder [33]. It can manifest in restrictive, emotional, or uncontrolled eating. It is lower in severity and intensity than that of an eating disorder but impacts everyday life.

Fourteen percent of adults (n= 40/286) self-reported disordered eating in a survey reported by the UK National Society for PKU, with 4% receiving therapy for eating disorders. Individual patient stories described how they had an unpleasant relationship with food; others described how they used food as a reward [24]. Bilder et al. reported that 3.4% of patients (n = 128/3714) with PKU had an eating disorder compared with 0.9% in the general population [31]. Viau et al. discussed that 53% of adults (n= 9/18) on pegvaliase therapy had food neophobia with low enjoyment of food which did not appear to improve with a relaxed protein intake [32]. Luu et al. [33] found that in a group of adults with PKU (n = 15) aged 12–35 y, patients with poor metabolic control had symptoms of disordered eating at a higher frequency than those with good metabolic control. They were more likely to have been overweight, and there was an association between dieting and dissatisfaction with body image.

Food neophobia in adults with PKU may have its origins in childhood [34–38] and is likely to impede long-term dietary patterns, alter food selection, and lower nutritional quality later in life. Intransient feeding problems are very challenging to change, and diagnosing an eating disorder in a patient with PKU is difficult. Existing validated tools for the assessment of eating disorders may not be appropriate for individuals with PKU on a prescribed dietary treatment [33,39].

Dietary pattern quality: There are many concerns about the quality of diets consumed by women who have stopped dietary treatment, potentially causing nutritional fragility in reproductive years. Some patients remain on a self-imposed low-protein diet, avoiding protein-rich foods such as meat, fish, and milk for many years. If they eat higher protein foods, it is commonly only intermittently as many report guilt and having less food enjoyment if they eat foods contraindicated in their dietary treatment [24]. The discontinuation of a protein substitute, supplemented with vitamins and minerals, intensifies the risk of micronutrient deficiencies [18,22]. Women may have unpleasant memories of the taste, smell, and texture of protein substitute from childhood, or they may associate it with causing gastrointestinal symptoms such as reflux and constipation [24]. The absence of protein substitute intake may lead to the thinning of hair and poor skin condition associated with inadequate nutritional status [32]. There are reports of reduced or low normal serum urea levels [40]. In patients on a partial or minimal dietary treatment, a protein [41] and amino acid deficiency, particularly tyrosine [42] with low normal free carnitine values [43], are described.

Overall, there is little qualitative data discussing the dietary patterns of adults with PKU, and it is undetermined if they consume an adequate intake of fruit and vegetables. The habitual intake of meat, fish, dairy products, wholegrain cereals, and nuts and seeds is unknown but thought to be minimal. It is established that teenagers commonly eat high amounts of carbohydrates with a limited intake of fruit and vegetables [44], despite extensive dietary education.

Nutrient deficiency: Women may be at particular risk of iron deficiency due to menstruation and the low intake of Phe-free/low-Phe protein substitutes. In a group of non-adherent UK adult patients with PKU (n= 14) who did not take protein substitute as prescribed, dietary intakes of iron, zinc, vitamin D3, magnesium, calcium, selenium, iodine, vitamin C, vitamin A, and copper were significantly lower than adherent patients (n = 16) and were below the UK Reference Nutrient Intakes [21]. Rohde et al. demonstrated that in 67 patients with PKU who consumed a $\leq$0.5 g/kg protein equivalent from a protein substitute that calcium and vitamin D intake was low, and the majority had low plasma 25-OH- vitamin D levels [22]. Vitamin B12 [41,45], zinc [21,46], and selenium [21,41,47] inadequacies are also reported in adult patients. Lower dietary adherence is associated with mild iodine deficiency and lower urinary selenium levels [48]. Pregnancy also increases the requirements for several macro- and micro-nutrients, compounding the risk of nutritional imbalance in women. The influence on maternal and foetal outcome of genetics, foetal programing, dietary management, and lifestyle of women with PKU are presented in Figure 2.

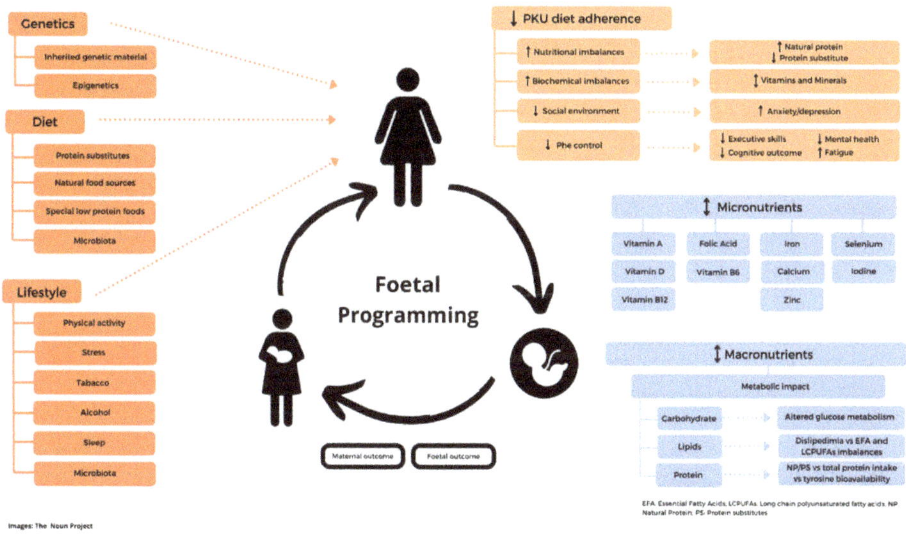

Figure 2. Foetal metabolic programming in women with PKU: influence of genetics, dietary management, and lifestyle on maternal and foetal outcome. ↓-lower; ↓-higher; ↕-lower/higher; ⇣⇡ broken line-arrows-potentially lower/higher.

2.4. Nutrition, Foetal Metabolic Programming, and Epigenetics

The foetal programming concept suggests that maternal nutritional imbalance may have a persistent effect on the health of their children. It may pre-condition for metabolic syndrome and lead to long-term, irreversible changes in the organs and metabolism [49]. Poor maternal nutrition has been linked with early embryogenesis and foetal growth abnormalities, cardiovascular disease risk, and metabolic and renal dysfunction [50,51]. The Dutch famine studies clearly demonstrated how poor nutritional intake affects foetal outcomes. Children from pregnancies influenced by famine in early gestation had increased disease and metabolic risk in adulthood [52]. Even second-generation children of women who experienced famine in pregnancy were at increased metabolic risk, creating a transgenerational effect. Foetal epigenetic programming could play a key role in foetal metabolic programming [53,54].

Epigenetics is defined as changes that modify gene expression and cellular function; they do not change the DNA nucleotide sequence. Unlike genetic changes, these are reversible [53,55]. Epigenetic changes occur when environmental conditions, such as malnutrition or stress during critical periods in early life, modify metabolic and developmental pathways, in turn leading to alterations in their function [55–57] and the predisposition of individuals to disease in later adulthood [58]. Barker [59] first suggested that environmental events occurring during pregnancy could have consequences in adult life, leading to cardiometabolic disease. Thus, the quality of nutrition and nutritional imbalances, dietary restriction, eating behaviors, lifestyle, and nutritional supplementation may affect nutritional programming before, during, and between maternal PKU pregnancies [49,57,60,61].

Micronutrients, including iron, zinc, folic acid, and other vitamins, contribute to epigenetic modifications during organogenesis in early pregnancy [58,62]. Methyl-donor groups, such as folate and vitamin B12, are vital for embryo and early foetal development [62]. Preconception zinc deficiency compromises foetal and placental growth and neural tube closure [63]. Folate, vitamin B12, methionine, choline, and betaine can affect DNA methylation and histone methylation. Folic acid, vitamin B12, and zinc participate in brain DNA and RNA synthesis, which begins early in gestation. Decreased vitamin B12 in the first trimester, associated with raised levels of folate, predicts increased central obesity and insulin resistance in the offspring [62]. Vitamin B12 has also been shown to affect myelination, which begins during gestation, and may affect cognitive functioning.

Folic acid and vitamin B12 participate in the folate–methionine cycle [64]. They are essential in the remethylation of homocysteine into methionine, which, consequently, generates S-adenosylmethionine, a methyl-donor molecule and folic acid essential in the prevention of neural tube defects (NTDs) [65]. There is evidence of inadequate intakes of folate and vitamin B12 in adult patients with PKU [41,66–68]. Many countries have a folic acid food fortification policy to decrease the incidence of NTDs or recommend folic acid supplementation during preconception and early pregnancy. However, regular foods fortified with folic acid (e.g., bread, pasta, and flour) are unsuitable for people with PKU. Protein substitutes are supplemented with folic acid, but reports of inadequate folic acid intake are described in non-adherent adults. In women with PKU, 400 μg/day of folic acid supplementation is recommended during preconception and the first 12 weeks of gestation [4]. Vitamin B12 is obtained from animal foods, which are excluded in a Phe-restricted diet, and acceptable intake is usually only associated with adherence to a nutritionally fortified protein substitute.

There is also evidence from animal and clinical studies that maternal overnutrition can lead to epigenetically mediated alterations in different physiological homeostatic regulatory systems and is associated with increases in the cardiometabolic risk in infants [56]. Observational evidence suggests that metabolic changes due to parental overweight/obesity affect epigenetic markers in oocytes and sperm alike and may influence epigenetic programming and reprogramming processes during embryogenesis [69]. However, mechanisms underlying overweight development and foetal adipogenic programming through influences of early-life stages are still poorly understood.

2.5. Role of Key Micronutrients in Reproductive Nutrition

Iron: A major public health problem that affects all women of reproductive age is anaemia, and in 2019 the global prevalence of anaemia in women of reproductive age (15–49 years) was 29.9% [70]. Anaemia has been associated with an increased risk of poor birth outcomes (low birth weight, preterm births, being small for gestational age, stillbirth, and perinatal and neonatal mortality) and adverse maternal outcomes (maternal mortality, postpartum haemorrhaging, and preeclampsia [71,72]. Perinatal iron deficiency is associated with long-term cognitive abnormalities as iron plays an important role in normal neurodevelopment through enzymes controlling neurotransmitter synthesis, cell division, neuronal energy metabolism, and myelination [73].

Preconception iron status is critical [65], and in women with PKU, the main sources are protein substitutes; women are particularly at risk of deficiency if adherence to this nutrition source is low. Several studies have reported an inadequate micronutrient status, including iron, particularly in non-adherent patients [21,22,74]. Green et al. identified that off-diet individuals with PKU with a blood Phe $\geq$600 μmol/L had iron intakes below the country-specific recommendations [74]. In a further two studies, patients with PKU who had stopped dietary treatment had significantly lower iron intake compared to adherent patients [21,22].

Iodine: Iodine is important in early foetal development and is associated with its involvement in thyroid function and foetal brain development [65]. Due to an increase in the iodine requirement for brain development in early pregnancy, iodine deficiency in the preconception period increases the risk of developmental delay in a child [65]. A meta-analysis by Levie et al. showed that a lower urinary iodine-to-creatine ratio during pregnancy was associated with a lower verbal IQ [75]. In women with PKU, iodine status is strongly influenced by a dietary adherence to protein substitutes supplemented with micronutrients, the main dietary source of iodine [21,22,48,74,76].

Zinc: In an in vivo model, acute dietary zinc deficiency before conception compromised oocyte epigenetic programming and disrupted embryonic development [77]. It is also important for immune function, foetal growth and neurological development, and potentially lowers the risk of preterm birth [65]. Low zinc intakes are commonly observed in women with PKU [21,74].

Long-chain polyunsaturated fatty acids (LC-PUFAs): These play an important role in the inflammatory response as eicosanoid precursors, as well as an important role in foetal–infant brain development in the later stage of pregnancy and early infancy. It is crucial that adequate maternal LC-PUFAs reserves are maintained early in pregnancy and for foetal use in later stages of development [78]. The placenta relies on fatty acids as a major energy source and disturbances in nutritional status could cause placental dysfunction, such as angiogenesis occurring in the first trimester and, consequently, compromise of foetal development [78].

The placental transport of LC-PUFAs is altered in maternal obesity and diabetes, which consequently has implications for foetal metabolic status [78]. Low DHA concentrations are reported in patients with PKU and during pregnancy [79–82] if women do not receive a supply from a protein substitute supplemented with DHA. Pregnant women should be supplemented with an additional supply of ≥200 mg DHA/day, over and above the intake recommended for an adult's general health, and usually achieves a total intake of ≥300 mg DHA/day [83]. This should be given to all women with PKU considering pregnancy and throughout pregnancy [4,83].

Over-nutrition: Obesity is associated with an increased risk of most major adverse maternal and perinatal outcomes, including infertility, miscarriages, complications during pregnancy (pre-eclampsia and gestational diabetes) and delivery (macrosomia), congenital anomalies, stillbirth, unsuccessful breastfeeding, and even maternal death [65,84–88]. A higher BMI before pregnancy is associated with a more significant fat mass gain during pregnancy and is correlated with fat retention postpartum. It is also a strong predictor for increased birth weight, as well as for childhood overweight and obesity [69].

Obesity in pregnancy has been shown to significantly alter glucose metabolism leading to impaired fasting glucose reduction in early pregnancy and a considerable increase of peripheral and hepatic insulin resistance [56]. Any obesity-related, pre-pregnancy insulin resistance is associated with an increase of gestational diabetes and, consequently, a higher risk of foetal glucose metabolism impairment, hyperinsulinemia, and type 2 diabetes.

Maternal gut microbiome: Maternal health and diet play a critical role in the foundation of a child's gut microbiome with long-lasting health implications. The rise in oestrogen and progesterone during pregnancy alters the gut function and microbiome composition, increasing vulnerability to pathogens. Throughout pregnancy, the gut microbiota progressively changes, with the greatest change occurring in the ratio of specific key bacteria (e.g., Firmicutes/Bacteroidetes ratio) mimicking the higher levels of Firmicutes seen in obesity [89]. Gut microbiota [90] can interact and be modulated by dietary factors. Prebiotics, such as fructooligosaccharides and galactooligosaccharides, have a positive influence on the gut microbiota composition. Little is known about the carbohydrate intake of adults with PKU. In a Phe-restricted diet, many of the carbohydrate sources allowed are based on simple sugars, e.g., sucrose and fructose, and this may cause rapid deregulation in the composition of the gut microbiota and, hence, metabolic dysfunction in the host [91]. Although some SPLFS contain added fibre, it is usually in the form of hydrocolloids to help their structure rather than provide nutritional benefits [11,44]. There is evidence that patients with PKU may have dysbiosis with less variety of bacteria, which may interfere with an optimal metabolism [92]. As well as the quality of carbohydrate intake, the high consumption of snacks, late-night eating, and skipping breakfast can also affect the gut microbiota composition [91].

Sleep hygiene: Sleep patterns may be disturbed in adult patients with PKU [93]. Quantity and quality of sleep play important roles in metabolic regulation and homeostasis [57,94]. A good night's sleep is associated with improved glucose, lipid, and energy metabolism, cardiovascular risk, inflammatory response, neurocognitive function, and mental health status [94,95].

2.6. Interventions to Improve Nutritional Health in the Reproductive Years of Women with PKU

Preconception care has been defined as "any intervention provided to women of childbearing age, regardless of pregnancy status or desire, before pregnancy, to improve health outcomes for women, newborns and children" [96]. In MPKU, it is important to identify any opportunities for improving nutrition prior to pregnancy using evidence informed interventions. It should be accepted that improving women's nutritional status may take several years and may be particularly challenging to maintain due to the high levels of food neophobia, maladaptive feeding behaviours, and limited food choices. In addition, individual motivations to engage with improving preconception nutrition will differ according to age, mental health, cognitive ability, and executive function. Understanding and harnessing these motivations will be key to successful intervention. Interventions to improve the nutritional status of PKU patients during their reproductive years are presented in Table 1.

Table 1. Interventions to improve nutritional health in women with PKU in their reproductive years.

Intervention	Recommendation/Action by Individual Women with PKU or Health Care Teams
Prevention of overweightand obesity	Substantial weight loss is particularly difficult in women with PKU due to the catabolic effect of lowering energy intake on PKU and impact on metabolic control and may take months and even years to achieve. Ideally, healthy weight should be established before or during adolescence and pre-pregnancy. Undertake regular preconception assessments of weight, BMI, nutritional monitoring, dietary patterns/intake, and lifestyle. Women with PKU should try and maintain an adequate balance between energy intake and expenditure. Increase amount and range of fruits, vegetables, and plant foods whilst limiting the intake of total fats, free sugars, and sodium. Decrease snacks and late-night eating. Encourage breakfast. Reduce sedentary activity such as television or computer viewing.

Table 1. *Cont.*

Intervention	Recommendation/Action by Individual Women with PKU or Health Care Teams
Regular exercise	Higher levels of preconception physical activity are associated with a lower risk of gestational diabetes and pre-eclampsia [97]. Address sedentary lifestyles early in life by promoting physical activity. Encourage 10,000 steps daily of unstructured activity in the light-to-moderate intensity range that are usually part of daily living (e.g., cycling, climbing stairs, and walking). Sports and structured activities: encourage 150 min per week of structured activities (that range from a moderate to vigorous intensity) [97]. Pedometers or similar apps can be used as forms of motivational support.
Improve quality of Phe-restricted diet	Promote adherence to dietary treatment and explore individual resistance to maintaining a Phe-restricted diet. Promote dietary diversification within the limits of dietary restriction. Encourage at least 400 g/day of fruit and vegetables, equivalent to 5/daily portions [98]. A range of different fruits and vegetables will provide different nutrients, phytochemicals, and fibre [90]. The EFSA recommends 25 g/day of fibre [99]. To help achieve this, a high intake of fruit and vegetables and wholegrain cereals (within natural protein allowance) is necessary. Focus on fat quality rather than quantity; monounsaturated and polyunsaturated fat sources provide health benefits associated with triglyceride and cholesterol metabolism. Avoid trans fats and lower saturated fat intake [90]. Give careful guidance on the choice of SLPF's as some may contain increased amounts of saturated and trans fats when compared to regular foods [11]. Ensure an adequate intake of essential fatty acids such as omega-3 and omega-6, with an emphasis on the optimal ratio of omega-3/omega-6. Encourage less added salt at the meal table and in cooking. Replace salt with herbs and spices. Provide social support to adults with PKU to help attain financial assistance to help purchase basic foods.
Encourage a healthy gut/gut microbiota	Assess gut health (particularly check for presence of gastro-intestinal reflux and constipation) at least annually. Fibre sources, including fruit and vegetables, augment microbiota diversity and are beneficial for gut health [90]. Fibre fermentation end-products and short-chain fatty acids (SCFA) have a role in preventing gut dysbiosis associated with metabolic dysfunction and immune response. SCFA, acetate, propionate, and butyrate are important modulators of gut microbiota [100]. Probiotic foods or supplements may offer additional protection. No controlled supplement trials of probiotics have been conducted in women with PKU.

Table 1. *Cont.*

Intervention	Recommendation/Action by Individual Women with PKU or Health Care Teams
Ensure a vitamin/mineral enriched protein substitute is taken in prescribed amounts	Explore any patient barriers to taking a protein substitute as prescribed. Give protein substitute in at least 3/daily doses and spread evenly throughout the day to minimise blood Phe fluctuations and to aid bioavailability of nutrient absorption. Protein substitutes help ensure that many macro- and micro-nutrient requirements are met. Meta-analyses confirm that supplementation or fortification with the 'big four' micronutrients (vitamin A, iron, zinc, and iodine) is efficacious to reduce the risk of infectious disease and improves growth and cognitive outcome in infants.
Give nutrition supplementsin the peri-conceptual period	Give 400 mg/day of folic acid in the periconceptual period to reduce the risk of neural tube defects by up to 72% [4,101]. Folic acid supplementation will also decrease the risk of pre-eclampsia, miscarriage, low-birth weight, being small for gestational age, a stillbirth, neonatal death, and autism in children [61,102]. A minimum of 4–6 weeks of folic acid supplementation is required to reach adequate levels before neurulation begins three weeks after conception. There is no information about adherence with folic acid supplementation in women with MPKU.
General lifestyle factors	Discourage smoking. While there are no published trials showing that reducing smoking before conception improves outcomes, indirect evidence suggests that smoke-free legislation in different countries has been associated with substantial reductions in preterm births [8]. Encourage moderate alcohol consumption in case of unplanned pregnancy. Maternal alcohol consumption can result in a range of foetal alcohol spectrum disorders [8]. Evaluate sources and perceived levels of stress, mood, and support systems. Offer psychological support and counselling. Encourage attendance of 'online' group mindfulness/support sessions.
Use of sapropterin	Sapropterin can liberate a woman's diet and increase natural food sources and nutrient intake in sub-groups of responsive women, but education and careful monitoring is needed, as changes in food patterns may have a negative impact on nutrient adequacy [32,103].
Maintain regular nutritional monitoring	Monitor nutritional intake at each dietetic review. Assess food patterns and check for any disordered eating or maladaptive eating practices. Monitor weight, BMI, and abdominal circumference at each face-to-face review. Review the condition of the hair, skin, and nails. Assess patients' biochemical nutritional status at least once a year. Assess for risk of comorbidities by monitoring lipid profile, blood pressure, and HbA1c [104]. Monitor blood Phe levels (according to European PKU guidelines) [4].

Table 1. *Cont.*

Intervention	Recommendation/Action by Individual Women with PKU or Health Care Teams
Encourage good sleep hygiene	Evaluate sleep patterns. Adults should aim to sleep at least 7 h per night to maintain optimal health [105]. Eating at late hours in the day has a negative effect on glucose, lipid, and energy metabolism [94], although a late-night dose of protein substitute may help decrease Phe fluctuations.

Abbreviations: PKU, Phenylketonuria; MPKU, Maternal Phenylketonuria; BMI, Body mass index; HbA1c, hemoglobin A1c; EFSA, European Food Safety Authority; Phe, Phenylalanine.

3. Conclusions

The health of a mother and her children cannot be completely separated, and a heightened awareness of the importance of preconception health, particularly diet and nutrition, is essential in women with PKU. Birth outcomes are influenced by the long-term interaction of a woman's biology, behaviour, social and environmental factors, and quality of diet. Therefore, the optimal health status of women with PKU before and inter-conception is essential. It is important that there is attention to dietary adequacy, healthy weight, and lifestyle. Women should be encouraged to maintain dietary and pharmaceutical treatments for PKU for optimal neuropsychological functioning and the provision of self-care during their reproductive years. In addition, the attainment of optimal nutrition should be the goal of health professionals. Any approach that improves the long-term nutritional health of women with PKU will help enhance the well-being of their future children.

Author Contributions: Conceptualization, A.M., M.I.G., A.P. and J.C.R.; methodology, M.I.G., A.P., J.C.R. and A.M.; writing—original draft preparation, M.I.G. and A.M.; writing, review, and editing—M.I.G., A.M., A.P., A.D. and J.C.R.; supervision, A.M. and J.C.R. All authors have read and agreed to the published version of the manuscript.

Funding: This research received no external funding.

Institutional Review Board Statement: Not applicable.

Informed Consent Statement: Not applicable.

Data Availability Statement: Not applicable.

Conflicts of Interest: A.P. has received an educational grant from Cambrooke Therapeutics and grants from Vitaflo, Nutricia, MerckSerono, Biomarin, and Mevalia to attend scientific meetings. A.D. has received research funding and financial support from Nutricia and Vitaflo to attend study days and conferences; Vitaflo has funded a PhD advisory member for APR., and J.C.R. is a member of the European Nutritionist Expert Panel (Biomarin), the Advisory Board for Applied Pharma Research and Nutricia, and received honoraria as a speaker from APR, MerckSerono, Biomarin, Nutricia, Vitaflo, Cambrooke, PIAM, and Lifediet. A.M. has received research funding and honoraria from Nutricia, Vitaflo International, Metax, Applied Pharma Research, and Biomarin. She is a member of the advisory board entitled ELEMENT (Danone-Nutricia) and Applied Pharma Research.

References

1. MacDonald, A.; Van Wegberg, A.M.J.; Ahring, K.; Beblo, S.; Bélanger-Quintana, A.; Burlina, A.; Campistol, J.; Coşkun, T.; Feillet, F.; Giżewska, M.; et al. PKU Dietary Handbook to Accompany PKU Guidelines. *Orphanet J. Rare Dis.* **2020**, *15*, 171. [CrossRef]
2. Levy, H.L.; Lenke, R.R.; Koch, R. Lack of Fetal Effect on Blood Phenylalanine Concentration in Maternal Phenylketonuria. *J. Pediatr.* **1984**, *104*, S0022–S3476. [CrossRef]
3. Davidson, D.C. Maternal Phenylketonuria. *Postgrad. Med. J.* **1989**, *65* (Suppl. 2), S10–S20. [PubMed]
4. Van Wegberg, A.M.J.; MacDonald, A.; Ahring, K.; Bélanger-Quintana, A.; Blau, N.; Bosch, A.M.; Burlina, A.; Campistol, J.; Feillet, F.; Giżewska, M.; et al. The Complete European Guidelines on Phenylketonuria: Diagnosis and Treatment. *Orphanet J. Rare Dis.* **2017**, *12*, 162. [CrossRef]
5. Murphy, E. Pregnancy in Women with Inherited Metabolic Disease. *Obstet. Med.* **2015**, *8*, 61–67. [CrossRef]

6. Ford, S.; O'Driscoll, M.; MacDonald, A. Reproductive Experience of Women Living with Phenylketonuria. *Mol. Genet. Metab. Rep.* **2018**, *17*, 64–68. [CrossRef]
7. Public Health England. Health Matters: Reproductive Health and Pregnancy Planning. Available online: https://www.gov.uk/government/publications/health-matters-reproductive-health-and-pregnancy-planning/health-matters-reproductive-health-and-pregnancy-planning (accessed on 1 February 2022).
8. Stephenson, J.; Heslehurst, N.; Hall, J.; Schoenaker, D.A.J.M.; Hutchinson, J.; Cade, J.E.; Poston, L.; Barrett, G.; Crozier, S.R.; Barker, M.; et al. Before the Beginning: Nutrition and Lifestyle in the Preconception Period and Its Importance for Future Health. *Lancet* **2018**, *391*, 1830–1841. [CrossRef]
9. MacDonald, A.; Singh, R.H.; Rocha, J.C.; van Spronsen, F.J. Optimising Amino Acid Absorption: Essential to Improve Nitrogen Balance and Metabolic Control in Phenylketonuria. *Nutr. Res. Rev.* **2019**, *32*, 70–78. [CrossRef]
10. Pena, M.J.; Almeida, M.F.; van Dam, E.; Ahring, K.; Bélanger-Quintana, A.; Dokoupil, K.; Gokmen-Ozel, H.; Lammardo, A.M.; MacDonald, A.; Robert, M.; et al. Special Low Protein Foods for Phenylketonuria: Availability in Europe and an Examination of Their Nutritional Profile. *Orphanet J. Rare Dis.* **2015**, *10*, 162. [CrossRef]
11. Wood, G.; Evans, S.; Pointon-Bell, K.; Rocha, J.C.; MacDonald, A. Special Low Protein Foods in the UK: An Examination of Their Macronutrient Composition in Comparison to Regular Foods. *Nutrients* **2020**, *12*, 1893. [CrossRef]
12. Couce, M.L.; Sánchez-Pintos, P.; Vitoria, I.; De Castro, M.-J.; Aldámiz-Echevarría, L.; Correcher, P.; Fernández-Marmiesse, A.; Roca, I.; Hermida, A.; Martínez-Olmos, M.; et al. Carbohydrate Status in Patients with Phenylketonuria. *Orphanet J. Rare Dis.* **2018**, *13*, 103. [CrossRef] [PubMed]
13. Moretti, F.; Pellegrini, N.; Salvatici, E.; Rovelli, V.; Banderali, G.; Radaelli, G.; Scazzina, F.; Giovannini, M.; Verduci, E. Dietary Glycemic Index, Glycemic Load and Metabolic Profile in Children with Phenylketonuria. *Nutr. Metab. Cardiovasc. Dis.* **2017**, *27*, 176–182. [CrossRef] [PubMed]
14. Levy, H.; Lamppu, D.; Anastosoaie, V.; Baker, J.L.; DiBona, K.; Hawthorne, S.; Lindenberger, J.; Kinch, D.; Seymour, A.; McIlduff, M.; et al. 5-Year Retrospective Analysis of Patients with Phenylketonuria (PKU) and Hyperphenylalaninemia Treated at Two Specialized Clinics. *Mol. Genet. Metab.* **2020**, *129*, 177–185. [CrossRef]
15. Kanufre, V.; Almeida, M.F.; Barbosa, C.S.; Carmona, C.; Bandeira, A.; Martins, E.; Rocha, S.; Guimas, A.; Ribeiro, R.; MacDonald, A.; et al. Metabolic Control of Patients with Phenylketonuria in a Portuguese Metabolic Centre Comparing Three Different Recommendations. *Nutrients* **2021**, *13*, 3118. [CrossRef] [PubMed]
16. Brown, C.S.; Lichter-Konecki, U. Phenylketonuria (PKU): A Problem Solved? *Mol. Genet. Metab. Rep.* **2016**, *6*, 8–12. [CrossRef]
17. Weglage, J.; Fromm, J.; van Teeffelen-Heithoff, A.; Möller, H.E.; Koletzko, B.; Marquardt, T.; Rutsch, F.; Feldmann, R. Neurocognitive Functioning in Adults with Phenylketonuria: Results of a Long Term Study. *Mol. Genet. Metab.* **2013**, *110*, S44–S48. [CrossRef]
18. Cazzorla, C.; Bensi, G.; Biasucci, G.; Leuzzi, V.; Manti, F.; Musumeci, A.; Papadia, F.; Stoppioni, V.; Tummolo, A.; Vendemiale, M.; et al. Living with Phenylketonuria in Adulthood: The PKU ATTITUDE Study. *Mol. Genet. Metab. Rep.* **2018**, *16*, 39–45. [CrossRef]
19. Wood, G.; Pinto, A.; Evans, S.; Daly, A.; Adams, S.; Costelloe, S.; Gribben, J.; Ellerton, C.; Emm, A.; Firman, S.; et al. Special Low Protein Foods Prescribed in England for PKU Patients: An Analysis of Prescribing Patterns and Cost. *Nutrients* **2021**, *13*, 3977. [CrossRef]
20. Bik-Multanowski, M.; Didycz, B.; Mozrzymas, R.; Nowacka, M.; Kaluzny, L.; Cichy, W.; Schneiberg, B.; Amilkiewicz, J.; Bilar, A.; Gizewska, M.; et al. Quality of Life in Noncompliant Adults with Phenylketonuria after Resumption of the Diet. *J. Inherit. Metab. Dis.* **2008**, *31* (Suppl. 2), S415–S418. [CrossRef]
21. Green, B.; Browne, R.; Firman, S.; Hill, M.; Rahman, Y.; Kaalund Hansen, K.; Adam, S.; Skeath, R.; Hallam, P.; Herlihy, I.; et al. Nutritional and Metabolic Characteristics of UK Adult Phenylketonuria Patients with Varying Dietary Adherence. *Nutrients* **2019**, *11*, 2459. [CrossRef]
22. Rohde, C.; von Teeffelen-Heithoff, A.; Thiele, A.G.; Arelin, M.; Mütze, U.; Kiener, C.; Gerloff, J.; Baerwald, C.; Schultz, S.; Heller, C.; et al. PKU Patients on a Relaxed Diet May Be at Risk for Micronutrient Deficiencies. *Eur. J. Clin. Nutr.* **2014**, *68*, 119–124. [CrossRef] [PubMed]
23. MacDonald, A.; Gokmen-Ozel, H.; van Rijn, M.; Burgard, P. The Reality of Dietary Compliance in the Management of Phenylketonuria. *J. Inherit. Metab. Dis.* **2010**, *33*, 665–670. [CrossRef] [PubMed]
24. Ford, S.; O'Driscoll, M.; MacDonald, A. Living with Phenylketonuria: Lessons from the PKU Community. *Mol. Genet. Metab. Rep.* **2018**, *17*, 57–63. [CrossRef]
25. Corrales, P.; Vidal-Puig, A.; Medina-Gómez, G. Obesity and Pregnancy, the Perfect Metabolic Storm. *Eur. J. Clin. Nutr.* **2021**, *75*, 1723–1734. [CrossRef]
26. Rodrigues, C.; Pinto, A.; Faria, A.; Teixeira, D.; van Wegberg, A.M.J.; Ahring, K.; Feillet, F.; Calhau, C.; MacDonald, A.; Moreira-Rosário, A.; et al. Is the Phenylalanine-Restricted Diet a Risk Factor for Overweight or Obesity in Patients with Phenylketonuria (PKU)? A Systematic Review and Meta-Analysis. *Nutrients* **2021**, *13*, 3443. [CrossRef] [PubMed]
27. Robertson, L.V.; McStravick, N.; Ripley, S.; Weetch, E.; Donald, S.; Adam, S.; Micciche, A.; Boocock, S.; MacDonald, A. Body Mass Index in Adult Patients with Diet-Treated Phenylketonuria. *J. Hum. Nutr. Diet. Off. J. Br. Diet. Assoc.* **2013**, *26* (Suppl. 1), 1–6. [CrossRef] [PubMed]

28. Rocha, J.C.; van Rijn, M.; van Dam, E.; Ahring, K.; Bélanger-Quintana, A.; Dokoupil, K.; Gokmen Ozel, H.; Lammardo, A.M.; Robert, M.; Heidenborg, C.; et al. Weight Management in Phenylketonuria: What Should Be Monitored. *Ann. Nutr. Metab.* **2016**, *68*, 60–65. [CrossRef]
29. Silveira, A.M.; Lima, P.L.; Alves, M.R.A.; Soares, R.D.L.; de Kanufre, V.C.; de Rodrigues, V.M.; Starling, A.L.P.; de Norton, R.C.; de Aguiar, M.J.B. Overweight/Obesity in Adolescents with Phenylketonuria: Protective and Predisposing Factors. *J. Pediatr.* **2021**, *98*, 104–110. [CrossRef]
30. Tonon, T.; Martinez, C.; Poloni, S.; Nalin, T.; Macdonald, A.; Schwartz, I.V.D. Food Neophobia in Patients with Phenylketonuria. *J. Endocrinol. Metab.* **2019**, *9*, 108–112. [CrossRef]
31. Bilder, D.A.; Kobori, J.A.; Cohen-Pfeffer, J.L.; Johnson, E.M.; Jurecki, E.R.; Grant, M.L. Neuropsychiatric Comorbidities in Adults with Phenylketonuria: A Retrospective Cohort Study. *Mol. Genet. Metab.* **2017**, *121*, 1–8. [CrossRef]
32. Viau, K.; Wessel, A.; Martell, L.; Sacharow, S.; Rohr, F. Nutrition Status of Adults with Phenylketonuria Treated with Pegvaliase. *Mol. Genet. Metab.* **2021**, *133*, 345–351. [CrossRef] [PubMed]
33. Luu, S.; Breunig, T.; Drilias, N.; Kuhl, A.; Scott Schwoerer, J.; Cody, P. A Survey of Eating Attitudes and Behaviors in Adolescents and Adults with Phenylalanine Hydroxylase Deficiency. *WMJ* **2020**, *119*, 37–43. [PubMed]
34. Evans, S.; Daly, A.; Chahal, S.; Ashmore, C.; MacDonald, J.; MacDonald, A. The Influence of Parental Food Preference and Neophobia on Children with Phenylketonuria (PKU). *Mol. Genet. Metab. Rep.* **2018**, *14*, 10–14. [CrossRef]
35. Evans, S.; Daly, A.; Chahal, S.; MacDonald, J.; MacDonald, A. Food Acceptance and Neophobia in Children with Phenylketonuria: A Prospective Controlled Study. *J. Hum. Nutr. Diet. Off. J. Br. Diet. Assoc.* **2016**, *29*, 427–433. [CrossRef] [PubMed]
36. MacDonald, A.; Harris, G.; Rylance, G.; Asplin, D.; Booth, I.W. Abnormal Feeding Behaviours in Phenylketonuria. *J. Hum. Nutr. Diet.* **1997**, *10*, 163–170. [CrossRef]
37. MacDonald, A.; Rylance, G.W.; Asplin, D.A.; Hall, K.; Harris, G.; Booth, I.W. Feeding Problems in Young PKU Children. *Acta Paediatr. Suppl.* **1994**, *407*, 73–74. [CrossRef]
38. De Cosmi, V.; Scaglioni, S.; Agostoni, C. Early Taste Experiences and Later Food Choices. *Nutrients* **2017**, *9*, 107. [CrossRef]
39. Robertson, L.; Adam, S.; Ellerton, C.; Ford, S.; Hill, M.; Randles, G.; Woodall, A.; Young, C.; Macdonald, A. Dietetic Management of Adults with Phenylketonuria (PKU) in the UK: A Care Consensus Document. *Nutrients* **2022**, *14*, 576. [CrossRef]
40. Das, A.M.; Goedecke, K.; Meyer, U.; Kanzelmeyer, N.; Koch, S.; Illsinger, S.; Lücke, T.; Hartmann, H.; Lange, K.; Lanfermann, H.; et al. Dietary Habits and Metabolic Control in Adolescents and Young Adults with Phenylketonuria: Self-Imposed Protein Restriction May Be Harmful. *JIMD Rep.* **2014**, *13*, 149–158. [CrossRef]
41. Hochuli, M.; Bollhalder, S.; Thierer, C.; Refardt, J.; Gerber, P.; Baumgartner, M.R. Effects of Inadequate Amino Acid Mixture Intake on Nutrient Supply of Adult Patients with Phenylketonuria. *Ann. Nutr. Metab.* **2017**, *71*, 129–135. [CrossRef]
42. Wild, J.; Shanmuganathan, M.; Hayashi, M.; Potter, M.; Britz-McKibbin, P. Metabolomics for Improved Treatment Monitoring of Phenylketonuria: Urinary Biomarkers for Non-Invasive Assessment of Dietary Adherence and Nutritional Deficiencies. *Analyst* **2019**, *144*, 6595–6608. [CrossRef] [PubMed]
43. Weigel, C.; Kiener, C.; Meier, N.; Schmid, P.; Rauh, M.; Rascher, W.; Knerr, I. Carnitine Status in Early-Treated Children, Adolescents and Young Adults with Phenylketonuria on Low Phenylalanine Diets. *Ann. Nutr. Metab.* **2008**, *53*, 91–95. [CrossRef] [PubMed]
44. Daly, A.; Evans, S.; Pinto, A.; Ashmore, C.; Rocha, J.C.; MacDonald, A. A 3 Year Longitudinal Prospective Review Examining the Dietary Profile and Contribution Made by Special Low Protein Foods to Energy and Macronutrient Intake in Children with Phenylketonuria. *Nutrients* **2020**, *12*, 3153. [CrossRef] [PubMed]
45. Robinson, M.; White, F.J.; Cleary, M.A.; Wraith, E.; Lam, W.K.; Walter, J.H. Increased Risk of Vitamin B12 Deficiency in Patients with Phenylketonuria on an Unrestricted or Relaxed Diet. *J. Pediatr.* **2000**, *136*, 545–547. [CrossRef]
46. Demirdas, S.; van Spronsen, F.J.; Hollak, C.E.M.; van der Lee, J.H.; Bisschop, P.H.; Vaz, F.M.; Ter Horst, N.M.; Rubio-Gozalbo, M.E.; Bosch, A.M. Micronutrients, Essential Fatty Acids and Bone Health in Phenylketonuria. *Ann. Nutr. Metab.* **2017**, *70*, 111–121. [CrossRef]
47. Procházková, D.; Jarkovský, J.; Vinohradská, H.; Konečná, P.; Machačová, L.; Doležel, Z. Controlled Diet in Phenylketonuria and Hyperphenylalaninemia May Cause Serum Selenium Deficiency in Adult Patients: The Czech Experience. *Biol. Trace Elem. Res.* **2013**, *154*, 178–184. [CrossRef]
48. Sumanszki, C.; Kiss, E.; Simon, E.; Galgoczi, E.; Soos, A.; Patocs, A.; Kovacs, B.; Nagy, E.V.; Reismann, P. The Association of Therapy Adherence and Thyroid Function in Adult Patients with Phenylketonuria. *Ann. Nutr. Metab.* **2019**, *75*, 16–23. [CrossRef]
49. Marciniak, A.; Patro-Małysza, J.; Kimber-Trojnar, Ż.; Marciniak, B.; Oleszczuk, J.; Leszczyńska-Gorzelak, B. Fetal Programming of the Metabolic Syndrome. *Taiwan. J. Obstet. Gynecol.* **2017**, *56*, 133–138. [CrossRef]
50. Sedaghat, K.; Zahediasl, S.; Ghasemi, A. Intrauterine Programming. *Iran. J. Basic Med. Sci.* **2015**, *18*, 212–220.
51. Fleming, T.P.; Watkins, A.J.; Velazquez, M.A.; Mathers, J.C.; Prentice, A.M.; Stephenson, J.; Barker, M.; Saffery, R.; Yajnik, C.S.; Eckert, J.J.; et al. Origins of Lifetime Health around the Time of Conception: Causes and Consequences. *Lancet* **2018**, *391*, 1842–1852. [CrossRef]
52. Roseboom, T.; de Rooij, S.; Painter, R. The Dutch Famine and Its Long-Term Consequences for Adult Health. *Early Hum. Dev.* **2006**, *82*, 485–491. [CrossRef]
53. Zhu, Z.; Cao, F.; Li, X. Epigenetic Programming and Fetal Metabolic Programming. *Front. Endocrinol.* **2019**, *10*, 764. [CrossRef] [PubMed]

54. Sookoian, S.; Gianotti, T.F.; Burgueño, A.L.; Pirola, C.J. Fetal Metabolic Programming and Epigenetic Modifications: A Systems Biology Approach. *Pediatr. Res.* **2013**, *73*, 531–542. [CrossRef] [PubMed]
55. Jaenisch, R.; Bird, A. Epigenetic Regulation of Gene Expression: How the Genome Integrates Intrinsic and Environmental Signals. *Nat. Genet.* **2003**, *33*, 245–254. [CrossRef]
56. Reichetzeder, C. Overweight and Obesity in Pregnancy: Their Impact on Epigenetics. *Eur. J. Clin. Nutr.* **2021**, *75*, 1710–1722. [CrossRef] [PubMed]
57. Moreno-Fernandez, J.; Ochoa, J.J.; Lopez-Frias, M.; Diaz-Castro, J. Impact of Early Nutrition, Physical Activity and Sleep on the Fetal Programming of Disease in the Pregnancy: A Narrative Review. *Nutrients* **2020**, *12*, 3900. [CrossRef]
58. Lammi-Keefe, C.J.; Couch, S.C.; Kirwan, J.P. *Handbook on Nutrition and Pregnancy*, 2nd ed.; Humana Press: New York, NY, USA, 2018. [CrossRef]
59. Barker, D.J.P. Fetal Programming of Coronary Heart Disease. *Trends Endocrinol. Metab.* **2002**, *13*, 364–368. [CrossRef]
60. Tiffon, C. The Impact of Nutrition and Environmental Epigenetics on Human Health and Disease. *Int. J. Mol. Sci.* **2018**, *19*, 3425. [CrossRef]
61. Hodgetts, V.A.; Morris, R.K.; Francis, A.; Gardosi, J.; Ismail, K.M. Effectiveness of Folic Acid Supplementation in Pregnancy on Reducing the Risk of Small-for-Gestational Age Neonates: A Population Study, Systematic Review and Meta-Analysis. *BJOG* **2015**, *122*, 478–490. [CrossRef]
62. Rajendram, R.; Preedy, V.R.; Patel, V.B. *Diet, Nutrition, and Fetal Programming;* Humana Press: New York, NY, USA, 2017. [CrossRef]
63. Tian, X.; Anthony, K.; Neuberger, T.; Diaz, F.J. Preconception Zinc Deficiency Disrupts Postimplantation Fetal and Placental Development in Mice. *Biol. Reprod.* **2014**, *90*, 83. [CrossRef]
64. Nafee, T.M.; Farrell, W.E.; Carroll, W.D.; Fryer, A.A.; Ismail, K.M.K. Epigenetic Control of Fetal Gene Expression. *BJOG* **2008**, *115*, 158–168. [CrossRef] [PubMed]
65. Hanson, M.A.; Bardsley, A.; De-Regil, L.M.; Moore, S.E.; Oken, E.; Poston, L.; Ma, R.C.; McAuliffe, F.M.; Maleta, K.; Purandare, C.N.; et al. The International Federation of Gynecology and Obstetrics (FIGO) Recommendations on Adolescent, Preconception, and Maternal Nutrition: "Think Nutrition First". *Int. J. Gynaecol. Obstet. Off. Organ Int. Fed. Gynaecol. Obstet.* **2015**, *131*, S213–S253. [CrossRef]
66. Hvas, A.M.; Nexo, E.; Nielsen, J.B. Vitamin B12 and Vitamin B6 Supplementation Is Needed among Adults with Phenylketonuria (PKU). *J. Inherit. Metab. Dis.* **2006**, *29*, 47–53. [CrossRef] [PubMed]
67. Stølen, L.H.; Lilje, R.; Jørgensen, J.V.; Bliksrud, Y.T.; Almaas, R. High Dietary Folic Acid and High Plasma Folate in Children and Adults with Phenylketonuria. *JIMD Rep.* **2014**, *13*, 83–90. [CrossRef] [PubMed]
68. Vugteveen, I.; Hoeksma, M.; Monsen, A.-L.B.; Fokkema, M.R.; Reijngoud, D.-J.; van Rijn, M.; van Spronsen, F.J. Serum Vitamin B12 Concentrations within Reference Values Do Not Exclude Functional Vitamin B12 Deficiency in PKU Patients of Various Ages. *Mol. Genet. Metab.* **2011**, *102*, 13–17. [CrossRef]
69. Hieronimus, B.; Ensenauer, R. Influence of Maternal and Paternal Pre-Conception Overweight/Obesity on Offspring Outcomes and Strategies for Prevention. *Eur. J. Clin. Nutr.* **2021**, *75*, 1735–1744. [CrossRef]
70. World Health Organization. Anemia in Women and Children. Available online: https://www.who.int/data/gho/data/themes/topics/anaemia_in_women_and_children (accessed on 20 October 2021).
71. Black, R.E.; Victora, C.G.; Walker, S.P.; Bhutta, Z.A.; Christian, P.; de Onis, M.; Ezzati, M.; Grantham-McGregor, S.; Katz, J.; Martorell, R.; et al. Maternal and Child Undernutrition and Overweight in Low-Income and Middle-Income Countries. *Lancet* **2013**, *382*, 427–451. [CrossRef]
72. Young, M.F.; Oaks, B.M.; Tandon, S.; Martorell, R.; Dewey, K.G.; Wendt, A.S. Maternal Hemoglobin Concentrations across Pregnancy and Maternal and Child Health: A Systematic Review and Meta-Analysis. *Ann. N. Y. Acad. Sci.* **2019**, *1450*, 47–68. [CrossRef]
73. Rao, R.; Tkac, I.; Townsend, E.L.; Gruetter, R.; Georgieff, M.K. Perinatal Iron Deficiency Alters the Neurochemical Profile of the Developing Rat Hippocampus. *J. Nutr.* **2003**, *133*, 3215–3221. [CrossRef]
74. Green, B.; Rahman, Y.; Firman, S.; Adam, S.; Jenkinson, F.; Nicol, C.; Adams, S.; Dawson, C.; Robertson, L.; Dunlop, C.; et al. Improved Eating Behaviour and Nutrient Intake in Noncompliant Patients with Phenylketonuria after Reintroducing a Protein Substitute: Observations from a Multicentre Study. *Nutrients* **2019**, *11*, 2035. [CrossRef]
75. Levie, D.; Korevaar, T.I.M.; Bath, S.C.; Murcia, M.; Dineva, M.; Llop, S.; Espada, M.; van Herwaarden, A.E.; de Rijke, Y.B.; Ibarluzea, J.M.; et al. Association of Maternal Iodine Status with Child IQ: A Meta-Analysis of Individual Participant Data. *J. Clin. Endocrinol. Metab.* **2019**, *104*, 5957–5967. [CrossRef] [PubMed]
76. Stroup, B.M.; Ney, D.M.; Murali, S.G.; Rohr, F.; Gleason, S.T.; van Calcar, S.C.; Levy, H.L. Metabolomic Insights into the Nutritional Status of Adults and Adolescents with Phenylketonuria Consuming a Low-Phenylalanine Diet in Combination with Amino Acid and Glycomacropeptide Medical Foods. *J. Nutr. Metab.* **2017**, *2017*, 6859820. [CrossRef] [PubMed]
77. Tian, X.; Diaz, F.J. Acute Dietary Zinc Deficiency before Conception Compromises Oocyte Epigenetic Programming and Disrupts Embryonic Development. *Dev. Biol.* **2013**, *376*, 51–61. [CrossRef] [PubMed]
78. Duttaroy, A.K.; Basak, S. Maternal Dietary Fatty Acids and Their Roles in Human Placental Development. *Prostaglandins. Leukot. Essent. Fatty Acids* **2020**, *155*, 102080. [CrossRef] [PubMed]
79. Yi, S.H.L.; Kable, J.A.; Evatt, M.L.; Singh, R.H. A Cross-Sectional Study of Docosahexaenoic Acid Status and Cognitive Outcomes in Females of Reproductive Age with Phenylketonuria. *J. Inherit. Metab. Dis.* **2011**, *34*, 455–463. [CrossRef]

80. Gramer, G.; Haege, G.; Langhans, C.-D.; Schuhmann, V.; Burgard, P.; Hoffmann, G.F. Long-Chain Polyunsaturated Fatty Acid Status in Children, Adolescents and Adults with Phenylketonuria. *Prostaglandins. Leukot. Essent. Fatty Acids* **2016**, *109*, 52–57. [CrossRef]
81. Stroup, B.M.; Nair, N.; Murali, S.G.; Broniowska, K.; Rohr, F.; Levy, H.L.; Ney, D.M. Metabolomic Markers of Essential Fatty Acids, Carnitine, and Cholesterol Metabolism in Adults and Adolescents with Phenylketonuria. *J. Nutr.* **2018**, *148*, 194–201. [CrossRef]
82. Giovannini, M.; Biasucci, G.; Agostoni, C.; Bellù, R.; Riva, E. Fatty Acid Supplementation in a Case of Maternal Phenylketonuria. *J. Inherit. Metab. Dis.* **1994**, *17*, 630–631. [CrossRef]
83. Koletzko, B.; Boey, C.C.M.; Campoy, C.; Carlson, S.E.; Chang, N.; Guillermo-Tuazon, M.A.; Joshi, S.; Prell, C.; Quak, S.H.; Sjarif, D.R.; et al. Current Information and Asian Perspectives on Long-Chain Polyunsaturated Fatty Acids in Pregnancy, Lactation, and Infancy: Systematic Review and Practice Recommendations from an Early Nutrition Academy Workshop. *Ann. Nutr. Metab.* **2014**, *65*, 49–80. [CrossRef]
84. Aune, D.; Saugstad, O.D.; Henriksen, T.; Tonstad, S. Maternal Body Mass Index and the Risk of Fetal Death, Stillbirth, and Infant Death: A Systematic Review and Meta-Analysis. *JAMA* **2014**, *311*, 1536–1546. [CrossRef]
85. Poston, L.; Caleyachetty, R.; Cnattingius, S.; Corvalán, C.; Uauy, R.; Herring, S.; Gillman, M.W. Preconceptional and Maternal Obesity: Epidemiology and Health Consequences. *Lancet Diabetes Endocrinol.* **2016**, *4*, 1025–1036. [CrossRef]
86. Gesink Law, D.C.; Maclehose, R.F.; Longnecker, M.P. Obesity and Time to Pregnancy. *Hum. Reprod.* **2007**, *22*, 414–420. [CrossRef] [PubMed]
87. Marchi, J.; Berg, M.; Dencker, A.; Olander, E.K.; Begley, C. Risks Associated with Obesity in Pregnancy, for the Mother and Baby: A Systematic Review of Reviews. *Obes. Rev. Off. J. Int. Assoc. Study Obes.* **2015**, *16*, 621–638. [CrossRef] [PubMed]
88. Strauss, A. Obesity in Pregnant Women: Maternal, Fetal, and Transgenerational Consequences. *Eur. J. Clin. Nutr.* **2021**, *75*, 1681–1683. [CrossRef]
89. Edwards, S.M.; Cunningham, S.A.; Dunlop, A.L.; Corwin, E.J. The Maternal Gut Microbiome during Pregnancy. *MCN. Am. J. Matern. Child Nurs.* **2017**, *42*, 310–317. [CrossRef]
90. Raymond, J.L.; Morrow, K. *Krause and Mahan's Food & The Nutrition Care Process*, 15th ed.; Elsevier: Amsterdam, The Netherlands, 2020.
91. Moszak, M.; Szulińska, M.; Bogdański, P. You Are What You Eat-The Relationship between Diet, Microbiota, and Metabolic Disorders-A Review. *Nutrients* **2020**, *12*, 1096. [CrossRef]
92. Verduci, E.; Carbone, M.T.; Borghi, E.; Ottaviano, E.; Burlina, A.; Biasucci, G. Nutrition, Microbiota and Role of Gut-Brain Axis in Subjects with Phenylketonuria (PKU): A Review. *Nutrients* **2020**, *12*, 3319. [CrossRef]
93. Bruinenberg, V.M.; Gordijn, M.C.M.; MacDonald, A.; van Spronsen, F.J.; Van der Zee, E.A. Sleep Disturbances in Phenylketonuria: An Explorative Study in Men and Mice. *Front. Neurol.* **2017**, *8*, 167. [CrossRef]
94. Poggiogalle, E.; Jamshed, H.; Peterson, C.M. Circadian Regulation of Glucose, Lipid, and Energy Metabolism in Humans. *Metabolism* **2018**, *84*, 11–27. [CrossRef]
95. Grandner, M.A. Sleep, Health, and Society. *Sleep Med. Clin.* **2020**, *15*, 319–340. [CrossRef]
96. Dean, S.V.; Lassi, Z.S.; Imam, A.M.; Bhutta, Z.A. Preconception Care: Nutritional Risks and Interventions. *Reprod. Health* **2014**, *11* (Suppl. 3), S3. [CrossRef] [PubMed]
97. World Health Organization. *WHO Guidelines on Physical Activity and Sedentary Behaviour*; World Health Organization: Geneva, Switzerland, 2020.
98. World Health Organization. Increasing Fruit and Vegetable Consumption to Reduce the Risk of Noncommunicable Diseases. Available online: https://www.who.int/elena/titles/fruit_vegetables_ncds/en/ (accessed on 26 November 2021).
99. EFSA. Dietary Reference Values for Nutrients Summary Report. *EFSA Support. Publ.* **2017**, *14*, e15121. [CrossRef]
100. Zmora, N.; Suez, J.; Elinav, E. You Are What You Eat: Diet, Health and the Gut Microbiota. *Nat. Rev. Gastroenterol. Hepatol.* **2019**, *16*, 35–56. [CrossRef] [PubMed]
101. Burda, B.U.; Chambers, A.R.; Johnson, J.C. Appraisal of Guidelines Developed by the World Health Organization. *Public Health* **2014**, *128*, 444–474. [CrossRef]
102. Gao, Y.; Sheng, C.; Xie, R.-H.; Sun, W.; Asztalos, E.; Moddemann, D.; Zwaigenbaum, L.; Walker, M.; Wen, S.W. New Perspective on Impact of Folic Acid Supplementation during Pregnancy on Neurodevelopment/Autism in the Offspring Children—A Systematic Review. *PLoS ONE* **2016**, *11*, e0165626. [CrossRef]
103. Thiele, A.G.; Rohde, C.; Mütze, U.; Arelin, M.; Ceglarek, U.; Thiery, J.; Baerwald, C.; Kiess, W.; Beblo, S. The Challenge of Long-Term Tetrahydrobiopterin (BH4) Therapy in Phenylketonuria: Effects on Metabolic Control, Nutritional Habits and Nutrient Supply. *Mol. Genet. Metab. Rep.* **2015**, *4*, 62–67. [CrossRef]
104. National Institute for Health and Care Excellence. Cardiovascular Disease: Risk Assessment and Reduction, including Lipid Modification. Available online: https://www.nice.org.uk/guidance/cg181 (accessed on 18 January 2022).
105. Watson, N.F.; Badr, M.S.; Belenky, G.; Bliwise, D.L.; Buxton, O.M.; Buysse, D.; Dinges, D.F.; Gangwisch, J.; Grandner, M.A.; Kushida, C.; et al. Recommended Amount of Sleep for a Healthy Adult: A Joint Consensus Statement of the American Academy of Sleep Medicine and Sleep Research Society. *Sleep* **2015**, *38*, 843–844. [CrossRef]

nutrients

MDPI

Article

The Challenges and Dilemmas of Interpreting Protein Labelling of Prepackaged Foods Encountered by the PKU Community

Imogen Hall [1], Alex Pinto [2], Sharon Evans [2], Anne Daly [2], Catherine Ashmore [2], Suzanne Ford [3,4], Sharon Buckley [5] and Anita MacDonald [2,*]

[1] Faculty of Health, Education & Life Sciences, Birmingham City University City South Campus, Westbourne Road, Edgbaston, Birmingham B15 3TN, UK; imogenhall8@outlook.com

[2] Birmingham Women's and Children's NHS Foundation Trust, Steelhouse Lane, Birmingham B4 6NH, UK; alex.pinto@nhs.net (A.P.); sharon.morris6@nhs.net (S.E.); a.daly3@nhs.net (A.D.); catherine.ashmore@nhs.net (C.A.)

[3] National Society for Phenylketonuria, P.O. Box 6046, Sheffield S12 9ET, UK; suzanne.ford@nspku.org

[4] North Bristol NHS Trust, Southmead Road, Bristol BS10 5NB, UK

[5] Department of Psychology, Faculty of Health, Psychology and Social Care, Manchester Campus, Manchester Metropolitan University, 53 Bonsall Street, Manchester M15 6GX, UK; sharon.j.bucley@stu.mmu.ac.uk

* Correspondence: anita.macdonald@nhs.net; Tel.: +44-1213338024

Abstract: Phenylketonuria (PKU) can lead to severe intellectual impairment unless a phenylalanine-restricted diet starts early in life. It requires expert user knowledge about the protein content of foods. The ability of adults or caregivers of children with PKU to calculate protein exchanges from food labels on manufactured foods and any difficulties they encounter in interpreting food labels has not been studied systematically. Individuals with PKU or their caregivers residing in the UK were invited to complete a cross-sectional online survey that collected both qualitative and quantitative data about their experience when calculating protein exchanges from the food labelling on prepackaged foods. Data was available from 246 questionnaire respondents (152 caregivers of patients with PKU aged <18 years, 57 patients with PKU aged ≥18 years or their caregivers (n = 28), and 9 teenagers with PKU). Thirty-one per cent (n = 76/246) found it difficult to interpret food protein exchanges from food labels. The respondents listed that the main issues with protein labelling were the non-specification of whether the protein content was for the cooked or uncooked weight (64%, n = 158/246); labels stating foods contained 0 g protein but then included protein sources in the list of ingredients (56%, n = 137/246); the protein content being given after a product was prepared with regular milk rather than the dry weight of the product (55%, n = 135/246); and the non-clarity of whether the protein content was for the weight of prepared or unprepared food (in addition to non-specification of cooked or uncooked weights on food labelling) (54%, n = 133/246). Over 90% (n = 222/246) of respondents had experienced problems with food labelling in the previous six months. Misleading or confusing protein labelling of manufactured foods was common. The food industry and legislators have a duty to provide accurate and clear protein food labelling to protect populations requiring low protein diets.

Keywords: phenylketonuria; food labelling; protein

Citation: Hall, I.; Pinto, A.; Evans, S.; Daly, A.; Ashmore, C.; Ford, S.; Buckley, S.; MacDonald, A. The Challenges and Dilemmas of Interpreting Protein Labelling of Prepackaged Foods Encountered by the PKU Community. *Nutrients* **2022**, *14*, 1355. https://doi.org/10.3390/nu14071355

Academic Editor: Bahram H. Arjmandi

Received: 23 February 2022
Accepted: 22 March 2022
Published: 24 March 2022

Publisher's Note: MDPI stays neutral with regard to jurisdictional claims in published maps and institutional affiliations.

Copyright: © 2022 by the authors. Licensee MDPI, Basel, Switzerland. This article is an open access article distributed under the terms and conditions of the Creative Commons Attribution (CC BY) license (https://creativecommons.org/licenses/by/4.0/).

1. Introduction

Phenylketonuria (PKU) is a genetic condition in which there is an inability to metabolise the amino acid phenylalanine into tyrosine. The treatment strategy for this condition is a lifelong phenylalanine-restricted diet to prevent adverse neurocognitive and psychological outcomes. This maintains blood phenylalanine levels within a narrow target therapeutic range but still delivers enough phenylalanine to support physiological protein synthesis, growth, and development. Patients with classical phenotypes usually have a natural protein tolerance that limits amounts to only 20% or less of what is expected in a regular diet [1]. High-protein foods such as meat, fish, eggs, cheese, seeds, and nuts are avoided with controlled and measured intakes of cereals, potato, breakfast cereals, and some vegetables allocated in a 1 g protein exchange system (1 exchange is equivalent to ~50 mg phenylalanine) [1]. The amount of natural protein tolerated is individual and influenced by the patient's phenotype, use of adjunct therapy (such as sapropterin), growth rate, and dosage of protein substitute intake.

Since the 1960s, the UK has adopted a straightforward approach to dietary management, allocating foods such as fruit and vegetables containing phenylalanine up to 75 mg/100 g weight without measurement. Although there is a long history of including manufactured foods in a protein-restricted diet, the range of prepackaged foods available has exponentially increased, and food choice is now almost indefinable. Every major British supermarket stocks 30,000 to 40,000 consumable items, including a diverse range of prepackaged foods. The breadth of food additives is continually expanding, and many prepackaged foods contain a multitude of ingredients with some contributing extra protein or phenylalanine, such as artificial sweeteners, spirulina extract as a colour additive; cereal; gelatine thickeners and taste enhancers, e.g., yeast extracts. In particular, aspartame, an artificial sweetener, is a peptide rich in phenylalanine. In the EU and UK, prepackaged foods should list the protein content as one of six mandatory nutrients and state the amount of protein per 100 g or per 100 millilitres [2]. However, it is not mandatory to issue food label warnings if the food product recipe changes and alters the nutritional content. Navigating food labels and understanding the suitability of individual manufactured foods has intensified the complexity of dietary management.

In 2020, the British Inherited Metabolic Disease Dietitians Group (BIMDG-DG) published consensus statements about the suitability of foods in a phenylalanine-restricted diet for PKU to help standardise interpretation, particularly of prepackaged foods [3]. Statements divided food and drink into categories based on defined protein content. It included foods allowed without restriction, which contain protein $\leq$0.5 g/100 g, and foods that should be calculated/weighed as an exchange food if they contain protein exchange ingredients (categorised into foods with a protein content of: >0.1 g/100 g (milk/plant milks only), >0.5 g/100 g (bread/pasta/cereal/flours), >1 g/100 g (cook-in/tabletop sauces/dressings), and >1.5 g/100 g (soya sauces) [3]. The practical statements were endorsed and translated into practical dietary advice for patients and caregivers by the National Society for PKU (NSPKU).

In order for patients/caregivers to fully adhere to dietary management, they are expected to acquire expert knowledge about the protein content of foods. It is the role of dietitians specialising in inherited metabolic disorders to teach parents and patients about the application of the complex set of BIMDG dietary rules. This enables patients/caregivers to understand and interpret food label ingredient lists and explain how to calculate 1 g protein exchanges directly from protein labelling. Patients and caregivers are given a range of dietary resources, including 'pocket' protein exchange calculators, dietary information books, detailed food lists, and a collection of suitable manufactured food picture books.

In practice, reading and interpreting food labels adds an additional task to a dietary regimen already associated with a heavy time burden [4]. The ability of adults with PKU/caregivers to calculate protein exchanges and any difficulties they encounter in interpreting food labels and calculating protein exchanges have not been studied systematically. This project aimed to explore the perception and opinion of patients with PKU and their caregivers about their experiences when calculating protein exchanges from the food labelling of prepackaged foods.

2. Materials and Methods

2.1. Methodology

This was a cross-sectional study using an online survey collecting qualitative and quantitative data from caregivers of children with PKU and adult patients. Respondents were excluded if they did not reside in the UK.

The questionnaire was built in an Online Surveys platform (https://www.onlinesurveys.ac.uk, accessed on 17 July 2020). This was shared on the UK National Society for Phenylketonuria (NSPKU) website, with additional promotion on the NSPKU Twitter, Instagram, and Facebook sites. The questionnaire was open from the 18 July 2020 until the 1 February 2021.

2.2. Questionnaire

The non-validated questionnaire contained 24 questions. There were fourteen multiple-choice, four multiple-responses, and six open-ended questions. Five questions consisted of more than one part (2–7 parts). Four other questions invited additional comments. There were 4 questions about alcohol labelling that were targeted at adults aged $\geq$18 years; these data will be included in a separate publication.

The questionnaire was developed by dietitians with expert practical and scientific knowledge of PKU (AP, SE, CA, AD, AM), a colleague from the NSPKU (SF), and a student dietitian from Birmingham City University (IH). It was reviewed by colleagues and lay people to ensure its readability and then amended according to feedback.

2.3. Data Collected

The questionnaire was divided into 3 sections. Section 1 collected information on patient age, sex, type of supermarket they commonly shopped at, and ease of calculating protein exchanges from food analysis labels for known problems previously identified [5].

These included 4 groups of manufactured foods:

(1) Stock cubes, gravy granules, dried sauce powders, tabletop sauces, cooking sauces, curry paste;
(2) Tinned tomatoes, tomato puree, dried soups, tinned or soup pots;
(3) Dried custard, ready-made custard, instant dessert powders, milkshake powders, milkshake liquids, drinking chocolate powder, ice cream, ice lollies;
(4) Dried rice, cooked rice, microwave rice, dried noodles, pot noodles.

Section 2 contained information about interpreting the protein content of alcohol that was only collected from adults.

Section 3 contained information collected about the problems with food labelling, examples of issues experienced in the previous 6 months, the respondents' approach to dealing with food labelling issues, emotions when identifying misleading labelling, and changes that should be made to food labelling legislation. All data collected were based on the patient's/caregiver's knowledge of their own experiences when interpreting the suitability of foods and calculating protein exchanges from food labelling.

2.4. Statistics

Questions were analysed with descriptive statistics only.

Qualitative data analyses of open-ended responses were carried out in NVIVO v 12 PRO (QSR International Pty Ltd., Australia, New Zealand and Oceania Level 5, Suite 5.11 737 Burwood Road Hawthorn East, Vic 3123). The whole survey dataset was imported into NVIVO so that coding of open-ended responses could be broken down by survey questions. All open-ended question responses were analysed thematically.

2.5. Ethics

Ethical approval was obtained from the Birmingham City University Ethics Committee prior to commencement of the study (Hall/7499/R(B)/2020/Jul/HELS FAEC–MSc Healthcare Project: What are the current issues with protein labelling for PKU patients?). At the beginning of the online questionnaire, respondents gave consent, and it was emphasised that questionnaire completion was voluntary. Potential respondents were advised that data from the survey would be published in an anonymised form. Names or hospitals mentioned in verbatim abstracts were removed from results presented in this manuscript.

3. Results

3.1. Demographics

Two hundred and forty-six respondents from the UK answered the questionnaire. Twenty-three per cent (n = 57/246) were adults with PKU (aged >18 years), 11% (n = 28/246) were parents/caregivers of adults with PKU, 62% (n = 152/246) were parents of children with PKU, and 4% (n = 9/246) were children/teenagers with PKU. Forty-eight per cent (n = 117/246) of the respondents or respondent's children with PKU were male, 50% (n = 124/246) female, and 2% (n = 4/246) non-binary, and one respondent (0.4%) preferred not to answer. The four main regular supermarkets used by respondents were: Tesco (62%, n = 153/246), Asda (54%, n = 132/246), Aldi (39%, n = 97/246), and Sainsbury's (39%, n = 95/246).

3.2. Rating of Food Labelling in General

This received a mixed response from respondents, with 2% (n = 5/246) describing it as very good, 41% (n = 101/246) as fairly good, 30% (n = 74/246) as neither good nor bad; 19% (n = 47/246) as fairly bad, 7% (n = 17/246) as poor, and 1% (n = 2/246) did not know.

3.3. Ease of Calculating Protein Exchanges from Food Labels

There was difficulty in calculating protein exchanges from food labels for food and drinks for at least one-third of the respondents (Table 1). For some individually manufactured foods, increased problems were described, including dried powdered products such as sauces, soups, dessert powders, dried custard powders, drinking chocolates, pot noodles, and noodles. In an open comment question, 398 verbatim comments were received about food labelling. The mixed responses were thematically analysed into the following categories: (1) finding food labelling easy to understand, (2) difficulty with interpreting food content, (3) difficulty with understanding how to calculate protein exchanges, and (4) did not use protein labelling. Examples of responses are given in Table 2.

Many respondents commented that protein labelling was unclear when the protein analysis was given after theoretical preparation, particularly when the manufacturers had assumed a product was prepared with cow's milk or egg. Ice cream was complicated as protein analysis was commonly given by volume as mL rather than weight as g. Some commented that it was difficult when food products such as jelly or yoghurt had to be checked for both protein content and the presence of aspartame. It was also remarked that due to deficits with cognitive functioning, particularly mathematical and reading skills, some respondents were unable to calculate protein exchanges. Some respondents with sight difficulties were unable to read the small font of some food analysis labelling, and some did not calculate protein intake but preferred to use food picture books showing suitable manufactured foods provided to them by their hospitals and NSPKU, as they had confidence that these were likely to be correct. Others did not deviate from the foods they knew were safe and did not try new manufactured foods.

3.4. Main Issues with Protein Labelling

The respondents listed that the main issues with protein content on food labels were (Table 3): not specifying if the protein content is for the cooked or uncooked weight; a manufactured food stating that it contains 0 g protein but the ingredients list contains a source of protein such as milk or gelatine; protein amount given only after a product has been prepared with regular milk; and non-clarity if the protein content was for prepared or unprepared food weight (in addition to cooked or uncooked weight).

3.5. Issues with Protein Food Labelling in the Previous 6 Months

Over 90% (n = 222/246) had experienced problems with food labelling in the previous 6 months. In fact, n = 97/246 (39%) identified having problems at least 10 times in the 6-month period, with n = 68/246 (28%) describing weekly issues with food protein labelling. One hundred and sixteen respondents listed examples of problematic food labelling, and these were thematically analysed into nine categories: (1) inadequate aspartame warning (n = 27); (2) dried products that are made up/served with milk (n = 16); (3) no differentiation of dried, unprepared, or uncooked weight vs. cooked/prepared weight (n = 16), (4) unclear protein labelling in general (n = 13); (5) suspect/doubtful protein content (n = 11); (6) foods purchased in multi-packs with unclear protein labelling (n = 9); (7) recipe change of a food item without warning (n = 9); (8) unclear protein content of imported foods (n = 8); and (9) analysis of protein content by volume rather than weight (n = 7). Examples of verbatim comments by the respondents are given in Table 4.

Table 1. Rating of interpretation of protein exchanges from food labels by respondents (n = 246) for food and drinks.

Food Item	Impossible		Difficult		Neither Easy/Not Easy		Fairly Easy		Very Easy		Do Not Know	
	%	n	%	n	%	n	%	n	%	n	%	n
Any food	7	16	24	60	18	45	35	87	15	37	0.4	1
Any drink	10	25	27	67	19	46	30	75	13	31	0.8	2
Stock cubes	5	13	28	70	13	31	21	51	11	27	22	54
Gravy granules	3	8	31	76	13	32	26	63	10	25	17	42
Gravy Pots	3	8	26	63	15	36	16	40	9	21	32	78
Dried sauce powders (e.g., cheese sauce)	5	12	40	98	12	29	15	37	7	16	22	54
Tabletop sauces, e.g., brown sauce	4	9	21	51	17	42	29	71	22	53	8	20
Ready-to-use cooking sauce	2	6	18	44	19	47	33	81	21	52	7	16
Curry paste	5	11	27	67	15	37	18	44	6	15	29	72
Dried custard powder	5	11	32	78	10	25	22	53	8	19	24	60
Ready-made custard	0.4	1	13	33	13	31	33	82	15	37	25	62
Instant dessert powders	8	20	35	87	13	31	13	32	4	10	27	66
Milkshake powders	4	10	39	95	10	24	19	46	4	10	25	61
Milkshake liquids	4	9	29	72	11	27	20	49	7	18	29	71
Drinking chocolate powder	3	7	39	97	11	26	19	46	7	16	22	54
Ice cream	2	4	33	82	11	28	29	71	15	38	9	23
Ice lollies	1	2	12	29	14	35	38	94	29	72	6	14
Tinned tomatoes	1	3	13	32	15	36	31	76	31	77	9	22
Tomato puree	2	4	15	38	18	44	28	69	29	71	8	20
Dried soups	3	7	33	82	11	26	19	46	9	22	26	63
Tinned or soup pots	0	0	17	41	15	38	31	76	18	44	19	47
Dried rice	3	8	31	75	10	24	24	60	15	36	18	43
Cooked rice	1	3	24	58	15	38	30	74	15	37	15	36
Microwave rice	0.4	1	19	47	12	30	30	74	13	33	25	61
Dried noodles	7	17	36	88	7	16	17	42	5	13	28	70
Pot noodles	6	14	30	74	9	23	14	35	5	12	36	88

Table 2. Verbatim comments about ease of calculating protein exchanges from protein content on manufactured food labels.

General Comments for Food and Drinks (n = 103)			
Find Food Protein Labelling Easy	**Difficulty with Interpreting Protein Content from Food Labelling**	**Difficulty with Understanding How to Calculate Protein Exchanges**	**Do Not Use Food Labels**
23% (n = 24)	49% (n = 50)	12% (n = 12)	17% (n = 17)
• *'Providing the food label is correct, it's okay'* • *'I got used to it now–had problems trying to remember the calculation'* • *'Generally okay–sometimes it's hard'* • *'Its quite easy- I use the NSPKU card calculator for exchanges'* • *'I always use a calculator card and it is taped on my kitchen cupboard'*	• *'It can get confusing, particularly when it tells you per 100 g and per portion'* • *'Print can be very small and easy to misread'* • *'Protein content does not seem to be on alcohol drinks labels'* • *'Sometimes it says no protein and then there is aspartame'* • *'Amounts per 100 g don't readily translate to amount used'*	• *'I have a mild learning disability related to PKU and cannot work out/calculate the exchanges mathematically'* • *'Don't read English well'* • *'Don't know how to interpret protein from food labels'* • *'Find it tricky to read labels. This leads to reducing the variety of foods we can offer.'* • *'I cannot read very well'*	• *'I don't work out exchanges anymore, we tend to stick to the same things and use the hospital picture book'* • *'Tend to use only branded foods my dietitian tells me are safe'* • *'Mostly I don't bother or only buy items where I know what the protein content is'* • *'Eat mainly fresh foods and stick to what we know'*
Comments for Sauces and Gravies (n = 80)			
Find Food Protein Labelling Easy	**Difficulty with Interpreting Protein Content from Food Labelling**	**Difficulty with Understanding How to Calculate Protein Exchanges**	**Do not Use Food Labels/Products**
10% (n = 8)	24% (n = 19)	25% (n = 20)	41% (n = 33)
• *'I find it quite easy looking at these labels to decide if he can eat them or not'* • *'Feel quite happy working out from jars and packets'* • *'As long it tells me per 100 g then it's fine'* • *'Cooking sauces are easier'*	• *'It is a guesstimate with curry paste where you use very little but it is intrinsically high in protein'* • *'The protein amount can change based on what the instructions say to add in.'* • *'It's tricky when they give two different values e.g., they do protein values for both made up and as sold'*	• *'If you don't use the product often sometimes it's hard to remember what to do'* • *'It's confusing when the table lists protein, but all ingredients are exchange free'* • *'I know to aim for a protein cut off of less than 0.5 g per 100 g but I don't know what I'm doing when it comes to sauces'*	• *'If unsure about protein content–don't use the products'* • *'Some of these I don't count due to the small quantity'* • *'Don't take protein from most sauces/gravies into account'* • *'Gravies, ketchups I always buy the same ones as I know they are free'*
Comments for Soups and Tomato Products (n = 73)			
Find Food Protein Labelling Easy	**Difficulty with Interpreting Protein Content from Food Labelling**	**Difficulty with Understanding How to Calculate Protein Exchanges**	**Do not Use Food Labels/Products**
19% (n = 14)	16% (n = 12)	10% (n = 7)	55% (n = 40)
• *'These items are usually well labelled with protein content values shown'*	• *'Tomato based products can be difficult as the actual Phe content doesn't always correspond with the protein content'*	• *'Not always clear when a food is exchange free'* • *'Struggle to recognise some of the ingredients–particularly starches'*	• *'Use homemade soup only'*

Table 2. *Cont.*

Comments for Ice Cream, Custards, Drinking Chocolate (n = 73)			
Find Food Protein Labelling Easy	**Difficulty with Interpreting Protein Content from Food Labelling**	**Difficulty with Understanding How to Calculate Protein Exchanges**	**Do not Use Food Labels/Products**
4% (n = 3)	32% (n = 23)	25% (n = 18)	40% (n = 29)
• *'Ice creams and lollies are easy when packaged'* • *'As a rule of thumb, I know readymade custard is not suitable and we always buy the brand of custard powder that is suitable when made up with protein-free milk'* • *'Brand dependent–some better than others'*	• *'It is very difficult to work out protein exchanges for drinks/desserts where the protein values given have assumed they have been reconstituted with milk–not low protein milk'* • *'Some ice-cream quotes volume in mL which I find unhelpful as we need to know the weight'*	• *'I would not know how to work our exchanges for items which only have the information on them for when made up with milk'* • *'I feel like I need a master's degree in Mathematics to figure out how much milkshake I can drink'* • *'Very difficult to understand labels of milk shake powder'*	• *'Keep to same brand'* • *'I only give her stuff in the picture books that we get from the hospital'* • *'I've always stick with the dessert and milkshake products I was allowed freely in my childhood'*
Comments for Rice and Dried Noodles (n = 69)			
Find Food Protein Labelling Easy	**Difficulty with Interpreting Protein Content from Food Labelling**	**Difficulty with Understanding How to Calculate Protein Exchanges**	**Do not Use Food Labels/Products**
20% (n = 14)	17% (n = 12)	12% (n = 8)	51% (n = 35)
• *'I know the exchange amount for rice'* • *'Can easily find the protein on the label'*	• *'Many protein levels for rice and pasta are given as dried, but serving portions are often given cooked'.* • *'Noodles are confusing. It is not clear if protein content is before or after preparation'*	• *'Just guess protein values'* • *'Dried weight changes when cooked so very difficult'* • *'Stay away from these foods–do not know how to work out protein content'*	• *'Only use low protein rice'* • *'Tend to stick to same things as find protein difficult'*

Table 3. Issues associated with food protein labelling identified by respondents ($n = 246$). NB: respondents could choose more than one answer.

Problem	Number of Responses	% of Responses
Unclear if protein content is for cooked or uncooked weight	158	64
Ingredients contain protein source, e.g., milk, but protein content says 0 g	137	56
Protein content is only given after it has been made with regular milk	135	55
Unclear if protein content is for prepared or or unprepared weight	133	54
No warning on a product that the recipe has changed	118	48
Protein content is only given for a food portion and not per 100 g/food	104	42
Protein content given for mls rather than grams	102	41
Protein content given as <0.5 g/item	98	40
Protein content confusing	93	38
Unable to read the writing on the food labels (too small/too shiny)	93	38
Protein content appears incorrect	80	33
No protein included on the label	77	31
Protein content appears too low	57	23
Lack of knowledge/confidence in interpreting protein content	50	20
No problems experienced	4	2

Table 4. Verbatim comments of respondents in 9 thematically analysed categories explaining their practical problems with protein food labelling in the 6 months prior to completion of the questionnaire.

Inadequate Aspartame Warnings	Dried or Unprepared Weight vs. Cooked/Prepared Weight	Dried Products That Are Made-Up/Served with Milk
n = 27	*n* = 16	*n* = 16
• *'A brand of squash added aspartame but there was no warning, my child had it by mistake'* • *'Iced slush drinks from stall holders or machines have no food labelling on them to identify presence of aspartame. You cannot identify if they contain aspartame'* • *'Unsure if alcoholic drinks contain aspartame–may not be identified on the label'*	• *'Potato products always confusing between cooked and uncooked weights'* • *'Dried noodles gave a protein content/100 g for cooked weight only, so did not know how much to weigh out dry before cooking'* • *'Dried products are difficult e.g., rice. Does not state if cooked or uncooked. It makes me feel anxious'*	• *'Cupcake mixes–difficult to calculate protein content if we make up with egg replacer rather than regular egg. The manufacturer gives protein values after it is assumed that it has been prepared with egg'* • *'Individual sachets of dried porridge. Only gave protein content after they were made up with cow's milk. I did not realise and I calculated them incorrectly in the diet'*
Unclear Protein Labelling	**Suspect Protein Content**	**Foods Purchased in Multi-packs**
n = 13	*n* = 11	*n* = 9
• *'Presentation of protein analysis in very unclear by some food brands–all the nutrient analysis may be given on one or two lines or small print'* • *'Sometimes in Polish shops they put another label over the nutritional info'* • *'Penny sweets/fresh gluten free breads have no protein analysis'*	• *'I purchased an egg replacement on Amazon, where it was listed as being 1 g protein per "yolk". When it arrived the pack said 0 g despite containing nutritional yeast, which is an exchange ingredient'* • *'I bought some sweet potato chips that were covered in rice flour. The protein content per 100 g was lower than the protein content of a portion which was 80 g'*	• *'Some variety packs of mixed breakfast cereals just give an average protein analysis on outer label and no individual protein labelling on the boxes.'* • *'We had popcorn where the protein content on the outer packet was different to the inner packets.'* • *'Multipacks of crisps do not put protein content on individual packs'*
Recipe Change of a Food Item without Warning	**Protein Content of Imported Foods**	**Analysis of Protein Content Is by Volume Rather Than Weight**
n = 9	*n* = 8	*n* = 7
• *'Following the UK sugar tax, the protein content of some breakfast cereals increased, but there was no warning on the labelling'* • *'A child's snack packet had a protein content of 0.5 g/pack. The ingredients changed and they moved to 1.3 g/pack and then have changed again back to 0.5 g/pack with no warning on the label'*	• *'The USA food products are confusing as they state their protein content in portion sizes and not per 100 g'* • *'Some shops e.g., Polish, Chinese only have information in a foreign language so I cannot work out information about the ingredients or protein content'*	• *'Any ice-cream where protein content given in volume rather than weight- it is a bit tedious to work out the protein exchanges'* • *'Any ice cream–I hate this as all the labelling is so confusing'*

If respondents were unsure about the interpretation of food labelling, the majority said they would not use the food products (57%, *n* = 140/246), 47% (*n* = 115/246) would ask their dietitian or other health professional for help, 30% (*n* = 73/246) would ask others on social media, and 14% (*n* = 35/246) would guess the protein content and use it. Eight per cent (*n* = 20/246) said they would either try looking at other sources of information on websites, ask their relatives, or try and calculate it themselves.

3.6. Respondent Emotions Associated with Food Labelling

Respondents reported that misleading or inadequate information on protein food labelling made them feel frustrated (67%, *n* = 165/246), anxious (33%, *n* = 82/246), angry (33%, *n* = 81/246), upset (28%, *n* = 70/246), unhappy (28%, *n* = 68/246), and excluded (27%, *n* = 67/246).

3.7. Suggested Changes to Food Labelling by Adults with PKU/Caregivers

Suggested changes to protein labelling are presented in Table 5.

Table 5. Suggested changes to protein labelling as requested by questionnaire respondents (respondents could choose more than one response), *n* = 246.

Recommendations
• Foods should be labelled with a warning on the packaging if the recipe has changed (60%, *n* = 148/246); • Protein should be given for cooked and uncooked weights (58%, *n* = 142/246); • Protein analysis should be given per 100 g as well as per portion size (55%, *n* = 136/246); • Protein analysis should be given per 100 g rather than per 100 mL (53%, *n* = 130/246); • The ingredients list should be made to be more easily readable (51%, *n* = 125/246); • Protein amount should always be identified, even at 0.1 g/100 g (*n* = 48%, *n* = 119/246); • Protein value should always be given for dried weight (42%, *n* = 103/246); • None (1%, *n* = 2/246).

There were 33 other suggested changes, including that manufacturers should not assume that products are prepared with cow's milk and give the protein analysis only after theoretical preparation; aspartame should always be in bold; all protein analysis should be made available on every supermarket website; and products should state accurate protein analysis and not use protein <0.5 g/100 g, which is unhelpful for low-protein diets. Some suggested that the protein content should always be in a uniform position on the food analysis list. It was also suggested that nutrient analysis should be in a larger font, and the protein content should be included on the front of the packet alongside the energy content.

4. Discussion

This paper highlights the considerable problems faced by both adult PKU patients and caregivers of children with PKU when trying to calculate exchanges from the protein analysis provided on food labels of prepackaged foods. Although there was a consensus that overall food labelling was satisfactory, the findings indicate that many patients/caregivers find protein calculations a complex process and identified several difficulties when interpreting protein labelling.

It was disconcerting that over 90% of respondents described specific issues with food labelling in the previous 6 months. Several respondents were frustrated that some potentially suitable instant dessert mixes and dried cereals had a protein content given on the food analysis after manufacturers had assumed they would be reconstituted/prepared with added cow's milk or egg, rendering the products unsuitable for people with PKU; no data were provided about the protein content of the dry products as purchased. There were many examples of ice creams that gave protein content for volume (in millilitres) rather than weight, and prepackaged foods that only gave a protein content of <0.5 g/100 g.

Some commented that it would make a 'massive difference' if food labelling was clearer as there would be more foods that could be consumed, that the 'confusing protein labelling made it very hard when choosing suitable foods in the supermarket', and 'the problems of interpreting protein labelling will not help my son become independent.' These issues were also identified by Kravela et al. 2020, who examined the accuracy of protein analysis from supermarket websites [5].

It was worrying that some respondents identified that manufacturers changed the recipes of some of their products, affecting the protein content, without any 'front of package' warnings, possibly causing dietary error. This commonly occurred in foods such as breakfast cereals following the Public Health England voluntary sugar-reduction programme (2017), which requested that manufacturers lower the sugar content of foods by 20% [6]. Some manufacturers replaced sugar with other ingredients containing protein. If people with PKU or their caregivers do not detect changes in protein labelling immediately, it may potentially lead to a long-term miscalculation of protein intake. It is well established that some patients with PKU struggle with maintaining satisfactory blood phenylalanine control [7–9]. This is often attributed to poor dietary adherence, but inadequate standards of food protein labelling could contribute to this. Misinterpretation of protein food labelling may cause some of the day to day blood phenylalanine variation that is observed in PKU, although this remains an area not considered by researchers.

Respondents also described an unfortunate trend for average protein labelling on multi-packs of different individually wrapped foods (e.g., small boxes of breakfast cereals, mixed flavoured bags of popcorn and crisps, and sweets and chocolates) with each individual item in the multi-pack having a different protein content per 100 g. For many multi-packs, respondents described how the protein content was given as an average on the outer packs, with no protein content stated on individual packs. For one product, a different protein content was stated on the outer compared with the inner packaging, which suggested careless protein labelling practice by the manufacturer. There appears to be no mandatory law to inform manufacturers that this practice is misleading and unsafe for people with PKU as well as other patient groups following protein-restricted diets. It is extraordinary that the UK Food Standards Agency has allowed this practice to occur.

There was respondent mistrust around the accuracy of protein labelling, with examples given of discrepancies of protein analysis between websites and actual food product labelling. Some food products declared high-protein-containing ingredients in the first two or three items listed on their labels, yet the protein analysis was 0 g/100 g. One product contained less protein per 100 g than was given for an 80 g portion size. There were examples of decimal place typing errors that had clearly not been detected by the proofreaders of the manufacturer's labels; this could have serious consequences for patients with PKU. There were descriptions of protein analyses being hidden/lost in packaging 'folds,' or the protein analysis being written in a linear format with other nutrients listed on the same line, making it difficult to distinguish protein from other nutrients. There were also important concerns about the protein labelling of imported foods. Food labels from the USA state protein content in portion sizes only. Imported foods from the USA only acknowledge the presence of protein on food labels if a prepackaged product contains more than 1 g of protein/portion; otherwise, they inaccurately state that the product contains 0 g of protein/portion. Some imported foods were reported to not include any English-language food analysis on the labels, although all labels need to comply with the UK food labelling laws, and this is mandatory.

Over one-third of respondents found drinks labelling a particular issue. Any alcoholic drink with a volume content above 1.2% does not legally require protein content to be declared, although appropriate allergen information should be given [10]. Importantly, aspartame content is exempted from inclusion in the labelling of alcoholic drinks. [11]. Several examples were given of inconsistent aspartame identification on the labels of fruit squashes or drinks bought from shop vendors. Detailed information about the perceptions of aspartame and food labelling of patients or caregivers of patients with PKU has been reported [11].

Except for the mandatory guidelines that manufacturers should state the product protein analysis per 100 g or 100 mL, there are few legal requirements about protein labelling [12]. The legislation allows manufacturers to use different methods to calculate the protein content of foods. It does not necessarily require laboratory analysis, and it may be possible for a food business operator themselves to perform a calculation from the known, or actual, average values of the ingredients used or to utilize established and accepted data [13]. Food regulations consider that a protein amount of $\leq$0.5 g per 100 g or 100 mL to be negligible, and so neglects the needs of people with PKU. Manufacturers may give the protein content per portion and/or per consumption unit, but this is not mandatory [2]. There is much that is lacking in protein legislation. Legislators must be aware that an inattentive approach to protein food labelling is a source of increased stress and burden for people with PKU and their caregivers. It limits their food choices, may induce unhealthy/repetitive food patterns, reduces variety in the diet, and may contribute to food neophobia [14].

This study has some limitations. Recruitment of participants for the online survey was performed via the NSPKU website and promoted on PKU social media sites, so respondents were limited to individuals with access to the internet using appropriate technology. Hence, it is likely that respondents were people who accessed social media sites frequently, and their views may not fully represent those of the broader population of PKU patients or their caregivers. However, problems deciphering food labels may be just as frequent in non-social media users, and this could be further investigated. Although there was a large response from caregivers (n = 180), there was a low response from adults with PKU (n = 57). It is known that in England alone, there are around 1100 adults on diet therapy with PKU. It is unclear whether this was due to a low interest in this area; unchanging dietary habits; limited reading of food labels; or low usage of websites, or PKU sites in particular, by affected adults [15]. The questionnaire was not validated prior to use, and the respondents' levels of education were unknown. We did not examine the amount of teaching they had received about a phenylalanine-restricted diet, which may have affected their answers, and the data from adult patients were not compared with those of caregivers.

5. Conclusions

Calculating PKU protein exchanges whilst considering portion sizes and checking for ingredients such as aspartame is a complex process with significant health implications. It is crucial that the quantity and presentation of protein and additive information on food labels enable patients with PKU or their caregivers to interpret this correctly. The range and extent of the issues identified around food labelling and interpretation suggest that the food and drinks industry is not currently providing clear and accurate information.

There appears to be no monitoring system examining the reliability of protein analyses on product labelling. Food manufacturers and legislators have a duty to provide a safe environment by ensuring accurate and clear protein labelling for populations requiring therapeutic low-protein diets.

Author Contributions: Conceptualisation, A.M., A.P. and I.H.; methodology, A.M., A.P., S.E., S.F., S.B. and I.H.; formal analysis, A.M. and S.E.; writing original draft preparation, A.M.; writing, review and editing, I.H., A.P., S.E., A.D., C.A., S.F., S.B. and A.M.; supervision, A.M. and A.P. All authors have read and agreed to the published version of the manuscript.

Funding: This research received no external funding.

Institutional Review Board Statement: This study was conducted according to the guidelines of the Declaration of Helsinki and approved by the Birmingham City University Ethics Committee prior to the commencement of the study (Hall/7499/R(B)/2020/Jul/HELS FAEC–MSc Healthcare Project: What are the current issues with protein labelling for PKU patients?).

Informed Consent Statement: Informed consent was given by all subjects when filling in the questionnaire.

Data Availability Statement: The data will be made available from the authors upon reasonable request.

Acknowledgments: We would like to acknowledge and thank all the patients and families that took the time to complete this survey.

Conflicts of Interest: The authors declare no conflict of interest.

References

1. MacDonald, A.; van Wegberg, A.M.J.; Ahring, K.; Beblo, S.; Bélanger-Quintana, A.; Burlina, A.; Campistol, J.; Coşkun, T.; Feillet, F.; Giżewska, M.; et al. PKU dietary handbook to accompany PKU guidelines. *Orphanet J. Rare Dis.* **2020**, *15*, 171. [CrossRef] [PubMed]
2. Department of Health. Technical Guidance on Nutrition Labelling. Available online: https://assets.publishing.service.gov.uk/government/uploads/system/uploads/attachment_data/file/595961/Nutrition_Technical_Guidance.pdf (accessed on 27 December 2021).
3. Evans, S.; Adam, S.; Adams, S.; Allen, H.; Ashmore, C.; Bailey, S.; Banks, J.; Churchill, H.; Cochrane, B.; Cook, J.; et al. Uniformity of Food Protein Interpretation Amongst Dietitians for Patients with Phenylketonuria (PKU): 2020 UK National Consensus Statements. *Nutrients* **2020**, *12*, 2205. [CrossRef] [PubMed]
4. MacDonald, A.; Smith, T.A.; de Silva, S.; Alam, V.; van Loon, J.M. The personal burden for caregivers of children with phenylketonuria: A cross-sectional study investigating time burden and costs in the UK. *Mol. Genet. Metab. Rep.* **2016**, *9*, 1–5. [CrossRef] [PubMed]
5. Kraleva, D.; Evans, S.; Pinto, A.; Daly, A.; Ashmore, C.; Pointon-Bell, K.; Rocha, J.C.; MacDonald, A. Protein Labelling Accuracy for UK Patients with PKU Following a Low Protein Diet. *Nutrients* **2020**, *12*, 3440. [CrossRef] [PubMed]
6. England, P.H. Sugar Reduction and Wider Reformulation. Available online: https://www.gov.uk/government/collections/sugar-reduction (accessed on 27 December 2021).
7. Jurecki, E.R.; Cederbaum, S.; Kopesky, J.; Perry, K.; Rohr, F.; Sanchez-Valle, A.; Viau, K.S.; Sheinin, M.Y.; Cohen-Pfeffer, J.L. Adherence to clinic recommendations among patients with phenylketonuria in the United States. *Mol. Genet. Metab.* **2017**, *120*, 190–197. [CrossRef] [PubMed]
8. Kanufre, V.; Almeida, M.F.; Barbosa, C.S.; Carmona, C.; Bandeira, A.; Martins, E.; Rocha, S.; Guimas, A.; Ribeiro, R.; MacDonald, A.; et al. Metabolic Control of Patients with Phenylketonuria in a Portuguese Metabolic Centre Comparing Three Different Recommendations. *Nutrients* **2021**, *13*, 3118. [CrossRef] [PubMed]
9. Walter, J.H.; White, F.J.; Hall, S.K.; MacDonald, A.; Rylance, G.; Boneh, A.; Francis, D.E.; Shortland, G.J.; Schmidt, M.; Vail, A. How practical are recommendations for dietary control in phenylketonuria? *Lancet* **2002**, *360*, 55–57. [CrossRef] [PubMed]
10. Commission, E. Labelling of Alcoholic Beverages in the EU: Some Facts. Available online: https://ec.europa.eu/food/safety/labelling-and-nutrition/food-information-consumers-legislation/alcohol-labelling_en (accessed on 13 February 2022).
11. Newbould, E.; Pinto, A.; Evans, S.; Ford, S.; O'Driscoll, M.; Ashmore, C.; Daly, A.; MacDonald, A. Accidental Consumption of Aspartame in Phenylketonuria: Patient Experiences. *Nutrients* **2021**, *13*, 707. [CrossRef] [PubMed]
12. Agency, F.S. Labelling Guidance for Prepacked for Direct Sale (PPDS) Food Products. Available online: https://www.food.gov.uk/business-guidance/labelling-guidance-for-prepacked-for-direct-sale-ppds-food-products (accessed on 13 February 2022).
13. Union, E. Regulation (EU) No 1169/2011 of the European Parliament and of the Council of 25 October 2011 on the provision of food information to consumers, amending Regulations (EC) No 1924/2006 and (EC) No 1925/2006 of the European Parliament and of the Council, and repealing Commission Directive 87/250/EEC, Council Directive 90/496/EEC, Commission Directive 1999/10/EC, Directive 2000/13/EC of the European Parliament and of the Council, Commission Directives 2002/67/EC and 2008/5/EC and Commission Regulation (EC) No 608/2004 Text with EEA Relevance. Available online: https://eur-lex.europa.eu/legal-content/EN/ALL/?uri=celex%3A32011R1169 (accessed on 13 February 2022).
14. Ford, S.; O'Driscoll, M.; MacDonald, A. Living with Phenylketonuria: Lessons from the PKU community. *Mol. Genet. Metab. Rep.* **2018**, *17*, 57–63. [CrossRef] [PubMed]
15. Wood, G.; Pinto, A.; Evans, S.; Daly, A.; Adams, S.; Costelloe, S.; Gribben, J.; Ellerton, C.; Emm, A.; Firman, S.; et al. Special Low Protein Foods Prescribed in England for PKU Patients: An Analysis of Prescribing Patterns and Cost. *Nutrients* **2021**, *13*, 3977. [CrossRef] [PubMed]

nutrients

MDPI

Article

Validation of a Low-protein Semi-Quantitative Food Frequency Questionnaire

Sharon Evans [1,*], Catherine Ashmore [1], Anne Daly [1], Richard Jackson [2], Alex Pinto [1] and Anita MacDonald [1]

[1] Birmingham Women's and Children's Hospital NHS Foundation Trust, Birmingham B4 6NH, UK; catherine.ashmore@nhs.net (C.A.); a.daly3@nhs.net (A.D.); alex.pinto@nhs.net (A.P.); anita.macdonald@nhs.net (A.M.)

[2] Liverpool Clinical Trials Centre, University of Liverpool, Brownlow Hill, Liverpool L69 3GL, UK; r.j.jackson@liverpool.ac.uk

* Correspondence: evanss21@me.com; Tel.: +44-121-333-8021

Abstract: Analysis of dietary patterns and their role in long-term health is limited in phenylketonuria (PKU). Food frequency questionnaires (FFQ) are commonly used to assess habitual intake. A semi-quantitative 89-item FFQ with a portion size photographic booklet was developed for children with PKU as a tool for collecting data on habitual intake of foods, food groups, energy and macronutrient intake. Twenty children with PKU aged 11–16 years, 30 parents of children with PKU aged 4–10 years, and 50 age/gender-matched control children were recruited. To test reproducibility, FFQs were completed twice with a mean interval of 5 weeks (range: 4–10). In order to test validity, FFQs were compared with five 24-h dietary recalls with a mean interval of 10 days (range: 6–18). Energy and macronutrient intake and quantity/week of individual food items were calculated and compared. There was good reproducibility for the FFQ with macronutrient correlations $r > 0.6$ and good validity data with most correlations $r > 0.5$. Bland–Altman plots for reproducibility and validity showed mean levels close to 0 and usually within 2 standard deviations. FFQ comparisons of PKU and control groups identified expected differences in % energy from macronutrients (PKU vs. control: carbohydrate 59% vs. 51%, fat 26% vs. 33%, protein 15% vs. 16%). This FFQ for PKU produced comparable data to repeated dietary recalls and is a valid tool for collecting data on habitual food and nutrient intake. It will be useful in assessing changes in dietary phenylalanine tolerance of new pharmacological treatments for PKU.

Keywords: phenylketonuria (PKU); dietary patterns; food frequency questionnaire; validation; reproducibility

Citation: Evans, S.; Ashmore, C.; Daly, A.; Jackson, R.; Pinto, A.; MacDonald, A. Validation of a Low-protein Semi-Quantitative Food Frequency Questionnaire. *Nutrients* **2022**, *14*, 1595. https://doi.org/10.3390/nu14081595

Academic Editor: Shanon L. Casperson

Received: 7 March 2022
Accepted: 9 April 2022
Published: 12 April 2022

Publisher's Note: MDPI stays neutral with regard to jurisdictional claims in published maps and institutional affiliations.

Copyright: © 2022 by the authors. Licensee MDPI, Basel, Switzerland. This article is an open access article distributed under the terms and conditions of the Creative Commons Attribution (CC BY) license (https://creativecommons.org/licenses/by/4.0/).

1. Introduction

Phenylketonuria (PKU) is a rare genetic condition, resulting in the failure to metabolise the amino acid phenylalanine, resulting in severe neurocognitive disability if untreated. It is managed with a low phenylalanine diet supplemented with a protein substitute (either phenylalanine-free L-amino acids or glycomacropeptide (GMP), typically with additional micronutrients), and special low-protein foods (SLPFs). The remaining diet consists of food starches, sugars, fruit and low-protein vegetables.

Dietary pattern analysis is increasingly used to examine food intake and the synergistic effect of food and nutrients [1,2], but this is unreported in PKU. Conceptually, dietary patterns provide a broad picture of food and nutrient consumption and may be more predictive of disease risk than individual foods or nutrients. In dietary pattern analysis, food consumption patterns are characterised by habitual intake [3]. With PKU, although much is known about the dietary prescription, little is known about what is consumed, including food preferences, range of meal choices and food patterns. Whilst it is assumed that patients eat plentiful amounts of low-phenylalanine fruit and vegetables, evidence suggests the converse position [4–6]. Furthermore, food neophobia appears to be more prevalent in children with PKU [5,7,8], and they appear reluctant to eat a wide range of fruits and vegetables [6].

Food frequency questionnaires (FFQ) are common tools used to measure dietary patterns. Respondents are given a list of foods and they describe how often each is eaten, e.g., how many times per day/per week/per month [9]. Compared with traditional dietary assessment methods, such as food diaries or 24-h recalls, FFQs require limited health professional time (both in data collection and analysis), low participant commitment, and may be completed by individuals with lower education or motivation [9,10]. They can be completed on paper or electronically in hospital clinics or the home environment. The results obtained by FFQs represent usual intakes over time and are suitable for ranking subjects into low, medium or high intake groups for individual foods or nutrients.

A FFQ should be tailored to each diet therapy. In PKU, portraying the full range of diverse foods permitted in a phenylalanine-restricted diet is challenging. This includes SLPFs and differing food types and quantities based on natural protein tolerance. Patients with classical phenotypes may tolerate only 3 g/day of natural protein (150 mg/day phenylalanine) but mild phenotypes tolerate $\geq$25 g/day (1250 mg/day phenylalanine), resulting in varying dependencies on SLPFs and protein substitutes. Pharmaceutical treatments, such as sapropterin dihydrochloride (BH4), may increase natural protein intake and the types of foods consumed in a subset of patients with PKU [11].

Ideally, a FFQ should contain no more than 100 commonly eaten foods grouped into sections, as only marginal gain is associated with more detailed questionnaires [12]. All FFQs should be validated to ensure that they measure what is intended and that they yield consistent results from repeated samples over time. This in turn improves the quality of the data collected and enables comparisons between studies using the same tool. There are different types of validity, meaning that a questionnaire is never fully validated but is valid for certain populations under specified conditions [13]. Validation of FFQs can be achieved in various ways and it is suggested that a combination of methods should be used to assess reproducibility and validity [14]. Checking that the questionnaire content is relevant and valid (content validity), that it can differentiate between different subject groups (construct/discriminative validity), that it produces reliable/reproducible results (reproducibility) and compares well with an existing standard (criterion validity), provides more credibility to the resulting data. Similarly, reporting that experts established questionnaire face validity, that the questionnaire was pilot tested on a subset of participants for understanding and relevance, and that appropriate statistical tests were used, also improve the integrity of the data [15].

Any FFQ designed for PKU should be validated by comparing it with a control group population to demonstrate that it is able to distinguish between the variations in macronutrient intake associated with the different food items eaten in a phenylalanine-restricted diet. Food intake will also vary according to the age, ethnic, social, educational, and economic background of the study population. Thus far, only one PKU-specific FFQ has been validated from the USA; 29 adults/adolescents were studied, and they compared the results of a FFQ with a 3-day food diary [10]. Whilst this study found good agreement between the different dietary methods and between repeated measures of the FFQ for protein intake, it was not validated in children, is likely to be specific to the USA population for food types and portion sizes, and it did not report on the validity of energy or other macronutrient intake such as carbohydrates, fat or fibre.

The aim of this study was to develop and validate a semi-quantitative FFQ for use in children with PKU, providing a tool that collects data on habitual intake for foods, food groups, energy and macronutrient intake, which can be utilised for dietary pattern and lifestyle analysis nationally.

2. Materials and Methods

2.1. Construct Validity (Ability to Differentiate between Different Subjects)—Study Subjects

Fifty children with PKU and 50 age and gender-matched healthy control children were recruited to test the FFQ for construct validity, which is the ability to differentiate between the dietary patterns for different groups. For children aged 4–10 years, data was completed by a parent/carer with assistance from an inherited metabolic disorder (IMD) dietitian; and for children aged 11–16 years, data was completed by children with assistance from the parent/carer and IMD dietitian. Inclusion criteria for subjects with PKU comprised the following: diagnosed by newborn screening; dietary treatment only (i.e., not prescribed sapropterin), supplemented with a prescribed free/low phenylalanine protein substitute from diagnosis and SLPFs; and no co-existing medical conditions, other special dietary requirements or intercurrent infection. All subjects with PKU were recruited from Birmingham Children's Hospital over a 30-month period (2018–2020). For control subjects, inclusion criteria comprised the following: age (within 6 months) and gender-matched to subjects with PKU; and on a regular diet (special diets, including vegan, vegetarian, and dairy-free were excluded). Control subjects were recruited from siblings of other children with inherited metabolic disorders, friends, or family of Birmingham Children's Hospital staff during the same time period as PKU subjects. The average nutrient and individual food intake from the 2 FFQs for each group were compared to establish construct validity.

2.2. Content Validity (Checked by Experts in the Field)—Food Frequency Questionnaire Development

Other UK and European PKU centres who were members of the SSIEM (Society for the Study of Inborn Errors of Metabolism) or BIMDG (British Inherited Metabolic Diseases Group) were invited to share their food frequency questionnaires. Five questionnaires were received (2 from England, 2 Scotland and 1 Germany). From these, a draft FFQ was developed for PKU and then adapted for control children (low-protein meal choices were matched with regular foods in the control FFQ, and SPLFs were also added to the PKU FFQ). The FFQs were reviewed by 5 IMD dietitians and piloted on 5 children with PKU and 5 control children to assess content validity. Following minor modifications, an 89-item PKU FFQ (+11 general dietary pattern questions) and a 69-item control FFQ (+5 general dietary pattern questions) were produced, including portion sizes for each item. The difference in the number of food items on the 2 questionnaires was a result of the additional SLPFs on the PKU version. General dietary pattern questions focused on meal frequency, missed meals, addition of salt to food, vitamin and mineral supplements and frequency of eating out (restaurants/cafes). The PKU FFQ also included 6 questions about quantity and frequency of protein substitute dosage and the number and amount of protein exchanges (the weight of food/drink that yields 1 g protein or 50 mg phenylalanine is one exchange).

2.3. Reproducibility—Food Frequency Questionnaire

The FFQ was completed at recruitment and at the end of the study (1–2 months apart) to test for reproducibility (test/retest reliability). This length of time was chosen in order to minimise changes over time but also to minimise recall of previous answers. The questionnaires were administered by one of four trained IMD dietitians using a standard script (see Supplementary Materials Supplementaries S1–S4). The same dietitian completed both questionnaires with each subject. For each food item on the FFQ, both the number of daily and weekly food portions consumed were recorded. Items consumed less than once a week were omitted.

2.4. Portion Size Booklet

Photographic portion size booklets were designed to accompany each FFQ, with pictures of the portion sizes specified in the questionnaire. Of 425 food photographs (captured from a range of distances and angles for each food), 65 were selected for the PKU

FFQ and 58 for the control FFQ (24 foods were the same; 21 were the same but for the PKU FFQ were SLPFs instead of regular foods, e.g., low-protein pasta, bread, burgers; 8 were the same food but differed in portion sizes to show the amount that was equivalent to a 1 g protein/50 mg phenylalanine exchange; the remaining were diet specific foods, e.g., SLPFs with no regular comparative, or meat products with no low-protein comparative). Foods were prepared and presented on a plain white plate or bowl or transparent glass, on a neutral background and photographed immediately to maximise the aesthetic appearance of the food. Foods that were pre-packaged in standard portion sizes or were equivalent in size to another food did not have photographic representation. Portion sizes for all foods (including SLPFs) were described by weight (g), volume (mL) (if applicable), and household measurements, e.g., tablespoons, teaspoons, glasses, slices, packets/sachets, whole items, or a combination of these. Portion sizes for protein/phenylalanine containing exchange foods were usually described as the amount yielding 1 g protein or 50 mg/phenylalanine, often described as a weight or volume range, e.g., 50–70 g to allow for small variations in intake. Average portion sizes were generally used, these could then be multiplied up (e.g., double) or down (e.g., halved) if a larger or smaller portion size was consumed. For dietary analysis purposes, the smaller value was used to calculate nutrient intake.

2.5. FFQ Database

In order to analyse energy and macronutrient intake (protein, fat, carbohydrate and fibre) from the FFQ, food items were assigned a nutrient content based on composition data compiled from *McCance and Widdowson, The Composition of Foods* [16] supermarket nutrient analysis data (Tesco website accessed May 2017) and SLPF nutrient composition data from manufacturers. For each FFQ item, the nutrient analyses were selected from one or more of these sources and the nutrient contents were averaged to obtain a single value for each nutrient. These values were then entered in the *Nutritics* [17] software computer analysis program as 'new foods', including portion sizes, and data from the FFQs were then analysed using the items as entered in *Nutritics*. Data entry was completed by the same dietitian and cross-checked for accuracy by a second dietitian. Nutrient intakes for each of the 2 FFQs for each subject were obtained and converted into average daily intakes for energy (kJ), carbohydrate, protein, fat, dietary fibre, starch and sugars. The 2 FFQs were then compared to establish reproducibility, and the average of the 2 FFQs compared with the average of the 24-h recalls to establish criterion validity.

2.6. Criterion Validity (Comparison with an Existing Standard)—24-h Dietary Recalls

As a means of comparison and to test criterion validity (comparison with an existing standard), five 24-h dietary recalls were completed with subjects by one of 4 trained IMD dietitians experienced in taking diet diaries and using the same standard script for asking questions. All food and drink consumed the day before were recorded including type and quantity, and time of day consumed. All 5 dietary recalls for each subject were completed by the same dietitian during the same time period as the FFQs (over a 1–2-month period) and with a minimum of 5 days between each one including at least 1 weekend day and no more than 2 of the same weekdays, to ensure a representative intake (Figure 1).

Dietary recalls were all analysed by the same dietitian using the *Nutritics* [17] computer analysis program, and results were averaged to obtain an average daily intake. These were then compared with the average of the 2 FFQs to establish criterion validity.

2.7. Anthropometry

Body weight, height and BMI (body mass index) were measured and z-scores calculated at recruitment. Weight was measured to the nearest 10 g using Seca electronic scales; length was recorded to the nearest 1 mm using a Seca 213 portable stadiometer (Seca, Hamburg, Germany).

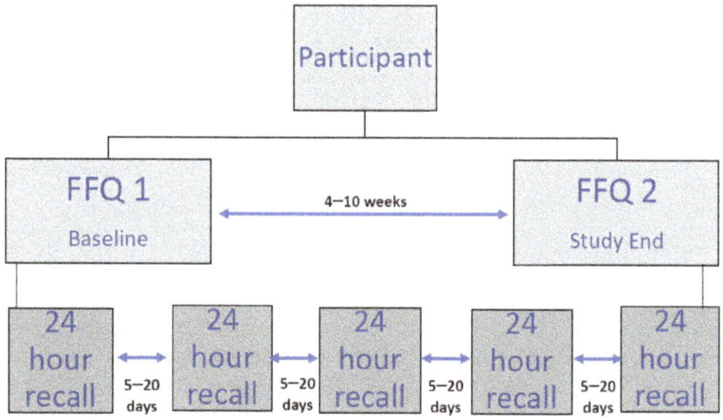

Figure 1. Study design. FFQ: Food frequency questionnaires.

2.8. *Statistics—Sample Size*

Data from a previous study [5] in a subset of children with PKU suggested that for a high fat, high carbohydrate food such as potato fries, power to detect a clinically relevant difference in mean intake of 1.7 days/week (the difference between a PKU group mean, μ_1, of 3.2 days/week and a control group mean, μ_2, of 1.5 days/week). Assuming a common standard deviation (SD) of 2.08 and a two-sided significance level of 0.05, a sample size of 33 in each group will have a power of 90%. Sample size calculations were performed using nQuery Advisor.

2.9. *Data Analysis*

Analyses were performed to evaluate the differences between the FFQs for PKU and control groups, as well as between the FFQ and dietary recalls data using GraphPad Prism version 6.01 for Windows, GraphPad Software, La Jolla California, USA. Continuous data were summarised as median (IQR) and categorical data were summarised as frequencies of counts with associated percentages. The strengths of association between dietary components were estimated using Spearman's correlation coefficients with Wilcoxon signed rank tests used to evaluate any differences between PKU and control groups. Comparisons of continuous data were performed using Wilcoxon sign rank for paired data and rank sum test for unpaired data. Categorical data were compared between groups using a Fisher test. Bland–Altman methods were used to assess the agreement between the FFQ and the 24-h dietary recall data. A *p*-value of 0.05 was used to determine statistical significance throughout.

2.10. *Ethical Approval*

This study was conducted according to the guidelines laid down in the Declaration of Helsinki, and a favourable ethical opinion was obtained from the London—Queens Square National Research Ethics Service (NRES) Committee (REC reference: 15/LO/1463 and IRAS ID: 185896). Written informed consent was obtained from the parent/carer of all subjects, and assent from children was obtained where appropriate, according to their level of understanding.

3. Results

3.1. *Subjects*

3.1.1. PKU Group

Fifty children (24 male) with PKU, (mean age 9.3 years; range: 4–16 years) on a phenylalanine-restricted diet only, were recruited from one specialist PKU centre (Birmingham Children's Hospital, UK). No changes were made to dietary intake during the study period. For 30 of the children aged 4–10 years, their mothers completed the questionnaires with a dietitian, and for the remaining 20 children, (aged 11–16 years) they self-completed the questionnaires with assistance from parents/carers and a dietitian.

3.1.2. Control Group

Fifty control children were age (within 6 months) and gender matched to the subjects with PKU. Questionnaires were either completed by parents/carers or by teenagers, following the same criteria as in the PKU group.

3.2. Demographics and Anthropometry

Most children were white UK/European origin (n = 45 subjects with PKU and n = 47 controls), with the remaining being of either Asian (n = 3 PKU, n = 2 control) or mixed-race (n = 2 PKU, n = 1 control) origin. There was no significant difference between mean z-scores for BMI, weight or height between PKU and control groups (see Supplementary Materials Table S1).

3.3. Meal Patterns—FFQ 1 vs. FFQ 2 (Reproducibility) vs. Dietary Recalls (Criterion Validity-Comparison with an Existing Standard)

The mean time between the two FFQs was 5 weeks (range: 4–10). Recalls were completed at a mean interval of 10 days (range: 6–18).

For meal patterns, there was little difference between the two FFQs, or between the FFQ and dietary recalls (Table 1). For the PKU group, there was a difference in the median number of meals that were consumed for FFQ 2 (4 meals/day) compared with FFQ 1 and the dietary recalls (5 meals/day). The PKU group varied across assessment methods in the percentage consuming mid-morning snacks, and this group was also less likely to consume a mid-morning snack compared with controls (FFQ 2 $p = 0.0005$). However, some children took their protein substitute at this time.

Table 1. Meal patterns—percentage of subjects consuming meals and snacks, or missing meals for FFQ 1, FFQ 2 and dietary recalls (PKU and control groups).

	PKU				CONTROL			
	FFQ 1 n = 50	**FFQ 2 n = 50**	**24-h Dietary Recalls ** n = 50**	***p* Value**	**FFQ 1 n = 50**	**FFQ 2 n = 50**	**24-h Dietary Recalls ** n = 50**	***p* Value**
Median no. meals eaten/day	5	4	5	0.01 #	5	5	5	0.37 #
% eating breakfast (n)	92 (46)	94 (47)	100 (50)	1 *	100 (50)	100 (50)	100 (50)	1 *
% eating midday meal (n)	100 (50)	98 (49)	100 (50)	1 *	100 (50)	100 (50)	100 (50)	1 *
% eating evening meal (n)	100 (50)	100 (50)	100 (50)	1 *	100 (50)	100 (50)	100 (50)	1 *
% eating mid-morning snack (n)	62 (31)	38 (19)	48 (24)	0.027 *	78 (39)	74 (37)	60 (30)	0.815 *
% eating afternoon snack (n)	70 (35)	52 (26)	70 (35)	0.100 *	55 (28)	62 (31)	74 (37)	0.685 *
% eating bedtime snack (n)	40 (20)	46 (23)	52 (26)	0.686 *	46 (23)	54 (27)	58 (29)	0.549 *
% miss meals 1 x/week (n)	10 (5)	6 (3)	8 (4)	0.715 *	10 (5)	8 (4)	8 (4)	1 *
% miss meals > 1 x/week (n)	12 (6)	8 (4)	2 (1)	0.741 *	2 (1)	0 (0)	0 (0)	1 *

** average of 5-day dietary recalls; * Fisher test; # Wilcoxon signed rank; (n) = number of children; FFQ = Food Frequency Questionnaire; PKU = Phenylketonuria

3.4. Protein Exchanges and Protein Substitute Intake

Dietary patterns related to intake of protein substitute and natural protein exchanges were comparable between repeated FFQs and between FFQ and dietary recall data (Table 2). The median daily number of 1 g protein exchanges (50 mg phenylalanine) from the 24-h dietary recalls was 0.5 g protein (25 mg phenylalanine) less than prescribed (5.0 g vs. 5.5 g). Dietary recall data showed that 70% (n = 35) of children with PKU were taking their daily number of 1 g protein exchanges to within 0.5 exchanges of prescribed amounts. Of the remaining 30%, 6% (n = 4) were allocated $\geq$ 10 g/day of protein and 18% (n = 9) were prescribed $\geq$5 g/day, indicating that these patients were less protein restricted.

Table 2. Percentage of subjects consuming natural protein exchanges (1 g protein = 50 mg phenylalanine) and protein substitute at meals and mid meals (PKU group only).

	FFQ 1 n = 50	FFQ 2 n = 50	24-h Dietary Recalls ** n = 50	*p* Value
Median no. 1 g natural protein (50 mg phenylalanine) exchanges/day (range)	5.5 (3–25)	5.5 (3–25)	5.0 (2–23.5)	0.02 #
% eating prescribed protein exchanges at every meal (n)	62 (31)	60 (30)	66 (33)	1 *
% actually eating prescribed protein exchanges at every main meal (n)	74 (37)	60 (30)	46 (23)	0.202 *
Median no. meals/snacks per day that prescribed protein exchanges are consumed (range)	3 (1–6)	3 (1–6)	3 (1–5)	1 *
% eating prescribed protein exchanges at breakfast (n)	78 (39)	72 (36)	68 (34)	0.645 *
% eating prescribed protein exchanges at midday meal (n)	94 (47)	86 (43)	88 (44)	0.318 *
% eating prescribed protein exchanges at evening meal (n)	96 (48)	100 (50)	82 (41)	0.495 *
% eating prescribed protein exchanges at mid-morning snack (n)	4 (2)	2 (1)	4 (2)	1 *
% eating prescribed protein exchanges at mid-afternoon snack (n)	14 (7)	14 (7)	14 (7)	1 *
% eating prescribed protein exchanges at bedtime snack (n)	10 (5)	4 (2)	20 (10)	0.436 *
Median no. times/day protein substitute dose taken (range)	3 (3–5)	3 (3–5)	3 (3–5)	1 *
% taking protein substitute dose at breakfast (n)	100 (50)	100 (50)	100 (50)	1 *
% taking protein substitute dose at midday meal (n)	78 (39)	72 (36)	78 (39)	0.645 *
% taking protein substitute dose at evening meal (n)	74 (37)	68 (34)	62 (31)	0.660 *
% taking protein substitute with morning snack (n)	10 (5)	12 (6)	10 (5)	1 *
% taking protein substitute with afternoon snack (n)	36 (18)	34 (17)	26 (13)	1 *
% taking protein substitute with bedtime snack (n)	48 (24)	56 (28)	66 (33)	0.548 *

** average of 5 days; * Fisher test; (n) = number of children; # Wilcoxon signed rank.

3.5. *Macronutrient Intake*

3.5.1. Reproducibility (A Measure of Whether the FFQ Produces the Same Results at Different Times)—FFQ 1 vs. FFQ 2

There was no significant difference between the two FFQs for PKU for any nutrients or between the two control FFQs except for protein and starch ($p = 0.05$) in control children, with FFQ 1 reporting values slightly higher than FFQ 2 (Table 3). Similarly, correlation r values all exceeded 0.5 for nutrients in both PKU and control groups, showing good correlation between FFQs taken at different intervals. Bland–Altman plots also demonstrated no clinically significant differences with mean levels close to 0 and homogeneous data mostly within the upper and lower levels of agreement (2 standard deviations—SD) (Figure 2).

Table 3. Nutrients—median (IQR) intakes FFQ 1 vs. FFQ 2 and FFQ vs. Dietary Recalls (DR) for PKU and control groups; and FFQ PKU vs. control.

		PKU				CONTROL				Spearman Rank Correlation (r)				Wilcoxon Signed Rank (p value)				
		FFQ 1	FFQ 2	FFQ Average	Diet Recalls	FFQ 1	FFQ 2	FFQ Average	Diet Recalls	PKU FFQ 1 vs. 2	PKU FFQ vs. DR	Ctrl FFQ 1 vs. 2	Ctrl FFQ vs. DR	PKU FFQ 1 vs. 2	PKU FFQ vs. DR	Ctrl FFQ 1 vs. 2	Ctrl FFQ vs. DR	PKU FFQ vs. Ctrl FFQ
KJ/day	median (IQR)	8116.8 (7078–11,288)	8591.1 (6824–9446)	8259.7 (7312–10,064)	7851.8 (6806–8927)	7011.7 (6008–8678)	6779.4 (6274–7804)	7144.2 (6157–8216)	7163.0 (6318–7878)	0.723	0.500	0.633	0.334	0.19	0.005	0.10	0.57	<0.0001
CHO g/day	median (IQR)	277.3 (229–394)	293.7 (237–343)	287.8 (241–356)	267.1 (231–315)	226.9 (191–271)	215.4 (194–247)	221.0 (186–262)	208.0 (183–233)	0.682	0.601	0.651	0.408	0.62	0.01	0.17	0.02	<0.0001
Sugars g/day	median (IQR)	135.7 (105–172)	128.8 (103–173)	131.2 (108–178)	107.0 (92–134)	112.2 (92–138)	111.6 (97–130)	115.7 (94–135)	95.7 (77–108)	0.78	0.464	0.719	0.550	0.78	0.0001	0.38	<0.0001	0.01
Starch g/day	median (IQR)	136.3 (114–183)	146.8 (105–187)	144.2 (114–195)	140.5 (117–167)	102.8 (81–132)	96.7 (79–115)	100.1 (82–124)	116.8 (93–133)	0.731	0.600	0.742	0.493	0.51	0.25	0.05	0.004	<0.0001
Fat g/day	median (IQR)	57.4 (49–76)	55.7 (48–68)	58.2 (49–69)	52.0 (43–61)	61.1 (54–77)	59.4 (54–69)	60.7 (54–73)	65.7 (57–75)	0.691	0.271	0.614	0.381	0.07	0.04	0.11	0.11	0.23
Total protein equivalent g/day	median (IQR)	72.9 (65–84)	73.2 (66–82)	72.6 (65–83)	71.8 (68–80)	66.7 (55–77)	63.4 (55–73)	65.6 (57–74)	64.6 (56–76)	0.913	0.848	0.523	0.344	0.37	0.08	0.04	0.91	0.0009
Fibre g/day	median (IQR)	20.4 (16–27)	22.6 (18–25)	21.5 (19–26)	18.1 (14–22)	18.0 (15–33)	17.4 (15–20)	17.8 (15–21)	17.0 (14–20)	0.673	0.445	0.514	0.427	0.70	0.001	0.26	0.02	0.0005
% Energy CHO	median (IQR)	58.8 (55–62)	58.6 (55–61)	59.3 (56–62)	59.0 (55–64)	51.3 (48–54)	51.1 (48–53)	50.7 (49–53)	49.0 (47–52)	0.768	0.689	0.492	0.414	0.35	0.94	0.85	0.02	<0.0001
% Energy Fat	median (IQR)	26.6 (23–29)	25.4 (23–29)	26.1 (24–29)	25.1 (22–29)	33.1 (31–36)	33.6 (32–35)	33.3 (32–35)	35.7 (32–37)	0.628	0.333	0.422	0.402	0.04	0.44	0.56	0.005	<0.0001
% Energy Protein	median (IQR)	14.5 (12–18)	15.4 (13–17)	14.3 (13–17)	15.3 (14–18)	15.4 (15–17)	15.5 (14–17)	15.5 (15–16)	15.2 (11–15)	0.777	0.729	0.622	0.290	0.34	0.21	0.45	0.89	0.12

DR = Dietary recall, IQR = Interquartile range.

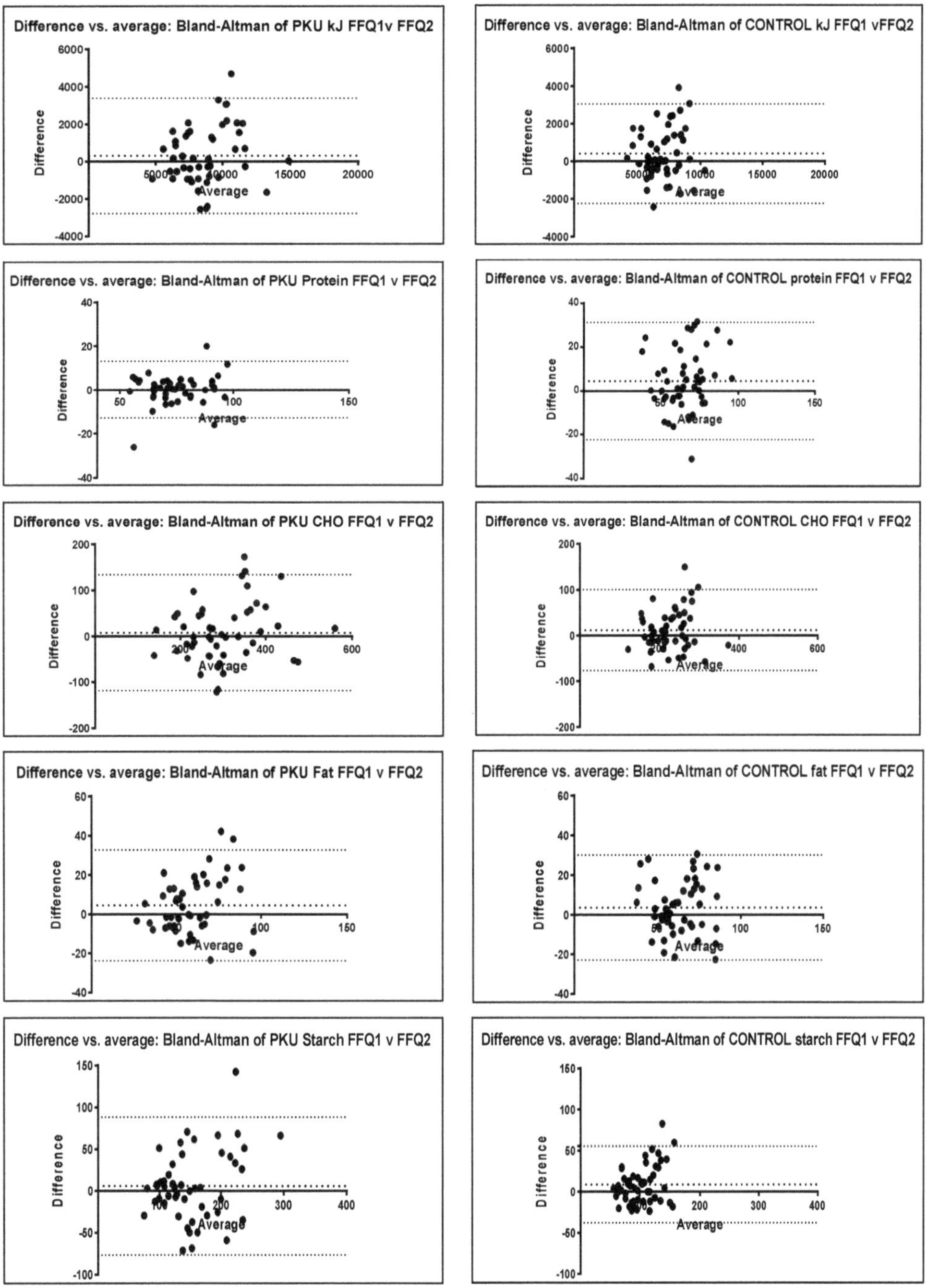

Figure 2. *Cont.*

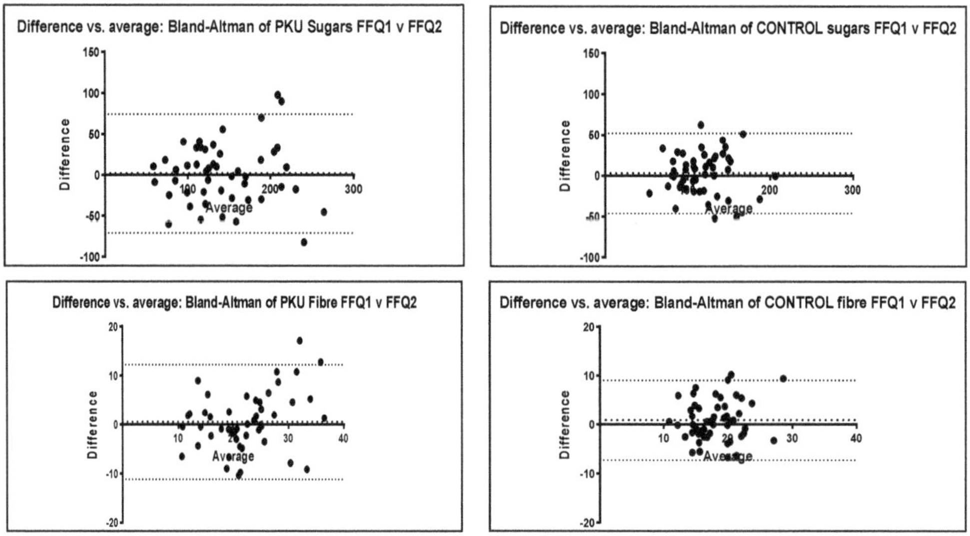

Figure 2. Bland–Altman plots for PKU and control group macronutrient intake FFQ 1 vs. FFQ 2. bias line (mean); upper and lower levels of agreement 95% confidence (2 SD). CHO = carbohydrate.

3.5.2. Criterion Validity (Comparison with an Existing Standard)—FFQ vs. Dietary recalls

For the PKU group, there was a trend for the FFQ to report higher intakes of all nutrients compared to the dietary recalls (Table 3). In the control group, the same was observed except for energy, starch and fat. Nutrient correlations for the PKU group were close to or above 0.5 (r) except for fat. For the control group, correlations were less strong, ranging from 0.33 to 0.55. Conversely, most Wilcoxon p values for the PKU group were significant except for protein and starch, whilst in the control group fewer nutrients showed statistical differences, with only protein, fat and energy not showing a difference. However, from a clinical perspective, differences were not of relevance. For example, the difference in sugar intake between FFQ and dietary recalls for the PKU group was around 25 g or approximately 1 tablespoon per day, whilst the difference in fat intake was around 5 g or 1 teaspoon of fat per day. The Bland–Altman plots show homogeneous data with most values falling within the upper and lower levels of agreement (2 SD) and mean values close to 0 (Figure 3). The exceptions to this were sugar and fibre for the PKU group, and sugar for the control group.

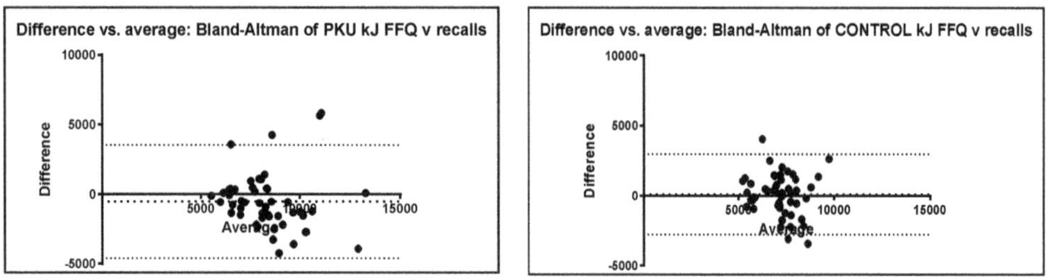

Figure 3. *Cont.*

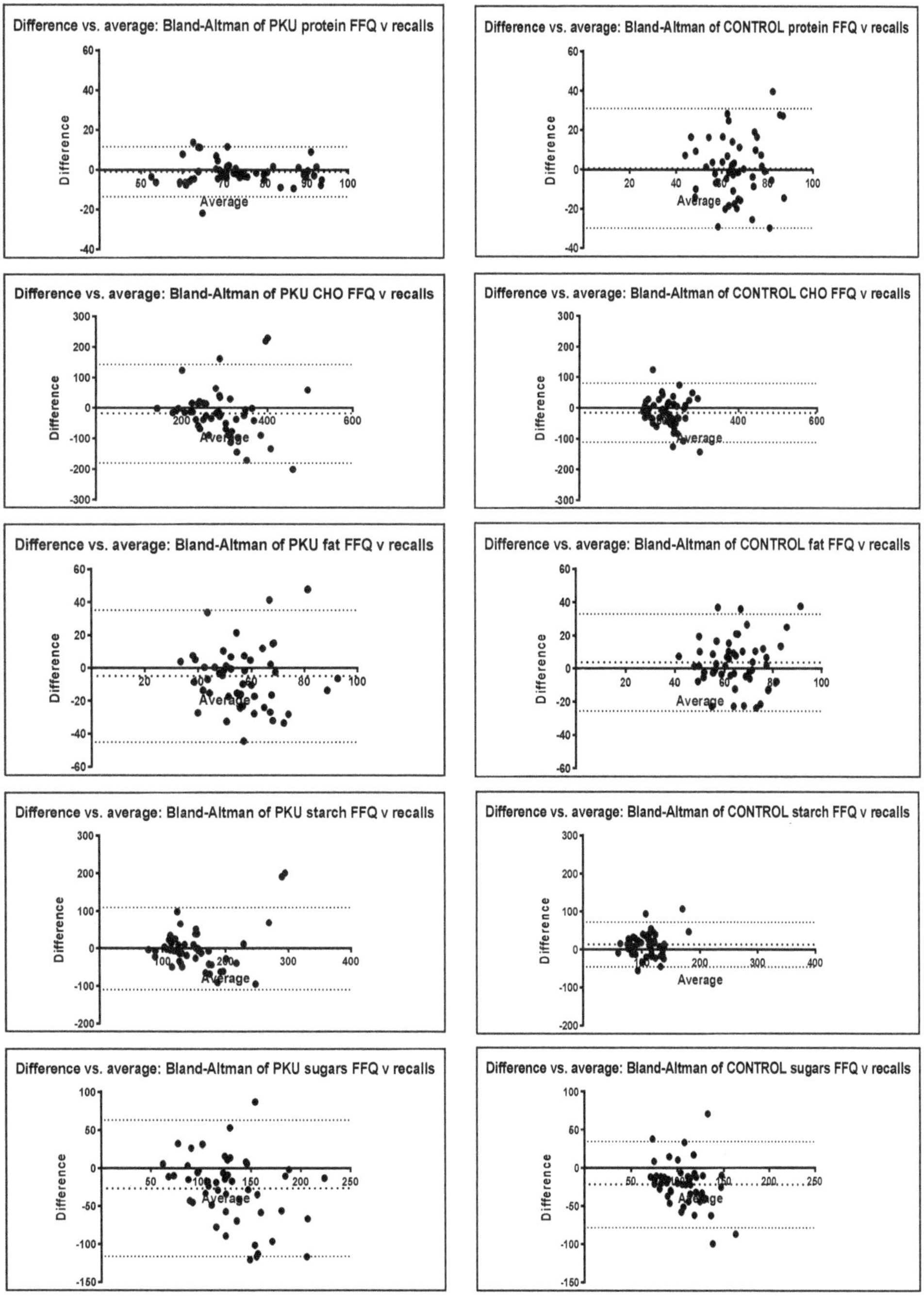

Figure 3. *Cont.*

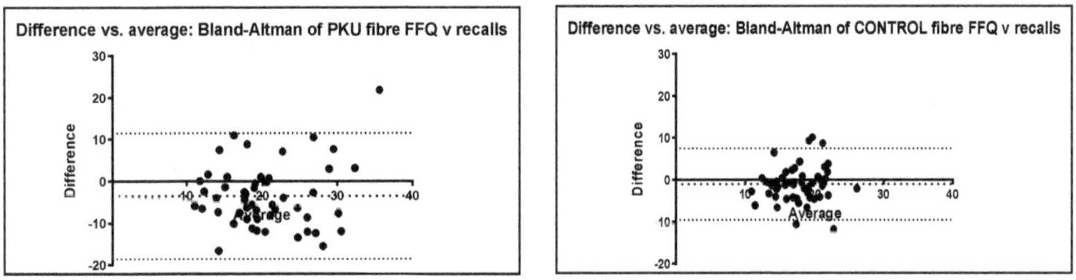

Figure 3. Bland–Altman plots for PKU and control group macronutrient intake FFQ vs. dietary recalls. Bias line (mean); upper and lower levels of agreement 95% confidence (2 SD); CHO = carbohydrate.

3.5.3. Construct Validity (Ability to Distinguish between Different Groups)—PKU FFQ vs. Control FFQ

As expected, and in agreement with previous research, due to the composition of a phenylalanine restricted diet there were significant differences in macronutrient intake between the PKU and control groups when using the FFQ (Table 3). The PKU group had significantly higher carbohydrate and starch intakes, and a higher percentage of energy from carbohydrate and a lower percentage of energy intake from fat compared to controls.

3.6. FFQ Individual Food Items

3.6.1. Reproducibility (A Measure of Whether the FFQ Produces the Same Results at Different Times)—FFQ 1 vs. FFQ 2

Most food items for both PKU and controls showed good correlation between FFQ 1 and 2 ($r > 0.40$), demonstrating good reproducibility (see Supplementary Materials Table S2). Foods with a lower correlation coefficient were usually consumed by fewer than 10 subjects. The exceptions for the PKU group were vegetarian gummy sweets, pasta sauce, dried fruit and regular biscuits. For the control group, exceptions were meat pie, meat curry and butter/margarine.

Similarly, for commonly eaten foods (> 10 subjects) there was no significant difference ($p > 0.05$) between FFQ 1 and 2 for most food items in either group. Exceptions in the PKU group included: corn/rice/oat-based breakfast cereal, sweet drinks and vegetables containing phenylalanine < 75 mg /100 g and those with >100 mg/100 g. Exceptions in the control group included the following: dairy desserts, wheat-based breakfast cereals, mayonnaise/dressings, pizza and crackers. No items with a low r value (<0.40) were significantly different ($p < 0.05$).

3.6.2. Criterion Validity (Comparison with an Existing Standard)—FFQ vs. Dietary Recalls

There was a trend for the FFQ to report higher intakes compared with the dietary recalls for just over half the items (n = 54/89, 61% PKU; n = 35/69, 51% control) (See Supplementary Materials Table S3). Similarly, for most foods, the FFQ reported more people consuming individual foods than the dietary recalls (FFQ 85%, n = 76/89 foods vs. recalls 15%, n = 13/79 foods for PKU; FFQ 83%, n = 57/69 foods vs. recalls 17%, n = 12/69 foods for controls).

Most food items for both PKU and control groups showed good correlation between the FFQ and the dietary recalls ($r > 0.40$), demonstrating satisfactory criterion validity except for items consumed less often (<10 subjects). The exceptions for the PKU group were as follows: vegetarian gummy sweets, chips, vegetarian burgers and low-protein biscuits. For the control group, exceptions were greater in number and similar to those that varied between the 2 FFQ: meat pie, meat curry and butter/margarine, in addition to chips, processed meats, ice cream, cheese, cake, pizza, pasta and chocolate.

Similarly, for commonly eaten foods (>10 subjects) there was no significant difference between the FFQ and dietary recalls for most food items in either group. Those that did tended to be different than the items that had low r values (<0.40). Exceptions in the PKU group included vegetarian gummy sweets, and in the control group cake, gummy sweets and table sauces—which had low r values (<0.40) and were significantly different ($p < 0.05$).

There were some commonly eaten foods (>10 subjects consuming) that showed significant differences in the mean g/week consumed between the FFQ and dietary recalls in both control and PKU groups. These included the following: boiled, mashed and jacket potato, table sauce, crisps and vegetables containing phenylalanine >75 mg /100 g. These foods tended to be considerably higher in the FFQ, except for crisps which were lower compared with the dietary recalls.

3.6.3. Construct Validity (Ability to Distinguish between Different Groups)—PKU FFQ vs. Control FFQ

There were significant differences between the intake of PKU and control groups using the FFQ, particularly in the foods expected to be different (see Supplementary Materials Table S4). This included higher protein foods that were consumed in greater quantities by controls: milk, cheese, soft cheese, dairy desserts, cream, wheat-based breakfast cereal, sandwich spreads, milk sauces, legumes, vegetables containing phenylalanine >75 mg/100 g, eggs, meat pies, meat curries, sugar-free drinks (usually containing aspartame), hot chocolate powder, nuts/seeds, and regular varieties of bread, bread rolls, pasta, pizza, biscuits, cakes, puddings, jelly, chocolate, gummy sweets and crackers. In addition, the higher carbohydrate/fat foods allowed since they are low protein/aspartame-free, were higher in the PKU group; these included: sweet spreads, mayonnaise/dressings, sweetened drinks (aspartame free), sugar, other sweets and butter, in addition to vegetarian varieties of foods such as burgers, pies and curries. Additionally, some foods commonly used as protein exchange foods were higher in the PKU group: tinned pasta, processed potato and potato or corn-based crisps.

4. Discussion

This is the first FFQ validated for children with PKU, with data suggesting that it is an effective, accurate and practical tool for estimating energy and macronutrient intake as an alternative method to dietary recalls. It identified dietary patterns, the quality of natural protein consumed, and adherence with protein prescription.

This FFQ demonstrated excellent reproducibility when administered at a mean time interval of 5 weeks. PKU group meal patterns were similar, and all nutrients showed good correlations ($r > 0.6$). The protein amounts in the PKU group had a correlation of 0.91, demonstrating that the FFQ reliably estimated usual intakes with similar accuracy to repeated 24-h dietary recalls. In addition, individual foods generally showed good correlation ($r > 0.4$) if they were commonly consumed items (eaten by >10 individuals). Discrepancies between FFQ 1 and 2 for individual foods may be explained by differences in interpretation between the two questionnaires or in participant memory of the types of food consumed at the various time points. For example, parents were sometimes ambivalent about the sugar content of drinks their children consumed, particularly if drinks were consumed at school/nursery or outside of the home. However, this could equally vary across the dietary recalls. Furthermore, some foods on the FFQ were rarely consumed (by <3 individuals). In order for a food itemisied on a FFQ to contribute to absolute intake or differentiate between individuals, it should be eaten regularly and by a significant number of the study population [9]. Therefore, some foods were removed from the study FFQ following analysis.

A minimum correlation coefficient of 0.3 to 0.4 has been suggested to detect associations when validating FFQs [9]. In this study, all nutrient correlation coefficients were above 0.5 for the comparison of the 2 FFQs in both groups, and above 0.4 for the comparison of FFQ and dietary recalls, except for fat in both groups and energy and protein for the control group only. Similar correlation results were shown in other validation studies [18–23].

Bland–Altman plots were used to display the stability and direction of the bias across levels of intake [19]. Agreement was considered reliable if the difference between the two measures for reproducibility (FFQ 1 vs. FFQ 2) or validity (FFQ vs. recalls) was within 2 standard deviations (SD) of the mean [10]; the mean was close to 0; and demonstrated homogeneous data. Expert consensus suggests a combination of correlation or regression statistical methods together with Bland–Altman analysis should be used to assess reproducibility and validity of a FFQ, rather than any one single method [14].

To be truly valid, reported dietary intake from any assessment method should not be significantly different to actual intake, however there are practical difficulties with measuring 'absolute validity'; thus alternatively, 'comparative validity' (comparing with an alternative or 'reference method') is reported [24]. There is no gold standard method for recording dietary intake, all have limitations: weighed food records require a high level of subject commitment, adherence and understanding that would have excluded some recruits from this study; 3-day food diaries represent the current diet, rather than typical or usual intake over time; doubly labeled water, a more accurate method for comparison of energy intake, is expensive and requires specialised equipment that may be intimidating to children. Repeated dietary recalls have been previously used for validating FFQs [22,23,25]; whilst single day recalls do not account for day-to-day variability in food intake or episodically consumed foods [19], we chose to complete multiple 24-h recalls over a 4–10-week period to capture a more realistic picture of usual intake over time. This approach is supported by a systematic review of the validity of different dietary assessment methods compared with doubly labeled water, suggesting that multiple 24-h dietary recalls conducted over at least 3 days and using parents as proxy reporters was the most accurate method for children aged 4–11 years [25]. Neither the FFQ nor the multiple 24-h dietary recalls are likely to measure actual macronutrient intake with precision, as both are subject to recall bias; however, both methods produced a similar picture of intake.

Our FFQ designed for PKU demonstrated acceptable criterion validity when compared with the chosen reference standard, repeated 24-h dietary recalls. Total natural protein intake only varied by 0.5 g (25 mg phenylalanine) per day between methods, which is a very good correlation. There was some variability between assessment methods for the percentages of children reporting that they consumed food at mid meals; however, this is likely to be something that varies in individuals from day to day. Some individuals with PKU may also choose to consume their protein substitute in place of a snack between meals so as not to reduce appetite at main meals.

In keeping with other validation studies comparing FFQs with other methods [18–20,22,23] there was a tendency for the FFQ to report higher nutrient and individual food intakes than the dietary recalls. FFQs have been reported to overestimate dietary intake in children resulting from the use of adult portion sizes [10]; however, we overcame this by developing pictorial child-size portions. The main difference between the FFQ and dietary recalls was not so much the quantity of a food consumed, but less variation in the types of foods consumed for the dietary recalls. This reflects one of the limitations of dietary recalls in that they only capture recent intake rather than habitual food intake.

Consistent with previous studies looking at the macronutrient content of the PKU diet [26–28] our results demonstrated that the FFQ can differentiate the differences in macronutrient and individual food intake between children with PKU and children in the general population that would be expected. This substantiates good construct validity.

FFQs rely on recall over a longer assessment period than other methods and hence are associated with less accurate quantification. It is suggested that children under the age of 8 years may have difficulty recalling food intake, estimating portion size and conceptualizing frequency of food consumption [24,25]. The ability to cognitively self-report dietary intake accurately is commonly given as approximately 12 years [24,25]. Previous research has shown that when older children complete a FFQ, they receive less assistance from parents, and this can result in a greater number of inaccuracies [2]. There may also be anomalies in data (from both dietary assessment methods) for adolescents due to inaccurate self-reporting and the highly variable food patterns commonly seen in this age group [19]. In this study, children aged 11–16 years completed the FFQ and recalls themselves which may have led to misreporting, although parents were able to assist. Furthermore, children with PKU have a more repetitive food pattern, receive dietary education and are accustomed to measuring portion sizes and completing dietary assessments. Correlations for the FFQ compared with the dietary recalls were stronger for the PKU subjects than for controls, suggesting that PKU subjects or their parents may have had better dietary recall than the control group [14,29].

Completion of any dietary assessment method may draw participants' attention to their diets [9], and there is also the risk of subjects responding in a way that demonstrates good adherence only in the presence of a dietitian. FFQs were administered by an IMD dietitian trained and experienced in dietary assessment rather than self-completion due to anticipated initial difficulties of comprehension and interpretation. As four different dietitians were involved in administration, there may have been some degree of inter-rater reliability. However, a standard script was used to administer questionnaires and food recalls to minimise this. It is anticipated that with repeated use, parents/carers and adolescents (>12 years) would be able to self-administer the FFQ independently. Recent studies have demonstrated that technology-assisted methods, such as an online FFQs, performed equally as well in estimating intakes as doubly labelled water and other methods [23,24]. As such, further analysis of this tool after regular use and with an online version may be warranted.

5. Conclusions

A FFQ can simplify dietary data collection in PKU, particularly if patients are familiar with the tool and can complete it electronically before clinic appointments. This low-protein FFQ designed for use in patients with PKU yielded comparable data to repeated dietary recalls, and can be validly used to collect data on usual food and nutrient intake in place of other dietary assessment methods. It will also enable assessment of the dietary patterns that may lead to lifestyle diseases, such as obesity in PKU, and in turn will facilitate tailored health messages to the PKU population that will help to reduce the incidence of health-related illness. It could also be particularly important in assessing the impact of dietary changes associated with pharmaceutical treatments in PKU. Further testing of an online version of the FFQ is warranted.

Supplementary Materials: The following supporting information can be downloaded at https://www.mdpi.com/article/10.3390/nu14081595/s1: Table S1: Mean z-scores for BMI, weight and height for PKU and control children; Table S2: FFQ food items mean g/week FFQ 1 vs. FFQ 2 (PKU and Control); Table S3: FFQ items mean g/week FFQ vs. Dietary recalls (PKU and Control); Table S4 Kapp coefficient statistics; Supplementary S1: Script for FFQ (PKU); Supplementary S2: Script for FFQ (Control); Supplementary S3: Script for 24-h dietary recall (PKU); Supplementary S4: Script for 24-h dietary recall (Control).

Author Contributions: Conceptualization, S.E. and A.M.; data curation, S.E., R.J. and A.M.; formal analysis, S.E., R.J. and A.M.; funding acquisition, S.E.; investigation, S.E., C.A., A.D., A.P. and A.M.; methodology, S.E., C.A., A.D., A.P. and A.M.; project administration, S.E. and A.M.; supervision, A.M.; validation, S.E., R.J. and A.M.; visualization, S.E. and A.M.; writing—original draft, S.E. and A.M.; writing—review & editing, S.E., C.A., A.D., R.J., A.P. and A.M. All authors have read and agreed to the published version of the manuscript.

Funding: This study was funded by the Birmingham Children's Hospital Charities Research Foundation (BCHRF407).

Institutional Review Board Statement: The study was conducted in accordance with the Declaration of Helsinki, and approved by the London—Queens Square National Ethics Research Service Committee, UK (protocol version 3, 18 April 2019, approved 16 May 2019; REC reference: 15/LO/1463; IRAS ID: 185896).

Informed Consent Statement: Informed consent was obtained from all subjects involved in the study.

Data Availability Statement: Not applicable.

Acknowledgments: We would like to thank all the children and parents/carers who participated in the study.

Conflicts of Interest: S.E. has received research funding from Nutricia, financial support from Nutricia and Vitaflo to attend study days and conferences; C.A. received honoraria from Nutricia and Vitaflo International to attend study days and conferences; A.D. research funding from Vitaflo, financial support from Nutricia and Vitaflo to attend study days and conferences; A.P. has received an educational grant from Cambrooke Therapeutics and grants from Vitaflo International, Nutricia, Merck Serono, Biomarin and Mevalia to attend scientific meetings; A.M. has received research funding and honoraria from Nutricia, Vitaflo International, and Biomarin. She is a Member of the Advisory Board Element (Danone-Nutricia). The funders had no role in the design of the study; in the collection, analyses, or interpretation of data; in the writing of the manuscript, or in the decision to publish the results.

References

1. Northstone, K.; Emmett, P.M. Are dietary patterns stable throughout early and mid-childhood? A birth cohort study. *Br. J. Nutr.* **2008**, *100*, 1069–1076. [CrossRef] [PubMed]
2. Ambrosini, G.L.; Emmett, P.M.; Northstone, K.; Howe, L.D.; Tilling, K.; Jebb, S.A. Identification of a dietary pattern prospectively associated with increased adiposity during childhood and adolescence. *Int. J. Obes.* **2012**, *36*, 1299–1305. [CrossRef] [PubMed]
3. Hu, F.B.; Rimm, E.; Smith-Warner, S.A.; Feskanich, D.; Stampfer, M.J.; Ascherio, A.; Sampson, L.; Willett, W.C. Reproducibility and validity of dietary patterns assessed with a food-frequency questionnaire. *Am. J. Clin. Nutr.* **1999**, *69*, 243–249. [CrossRef] [PubMed]
4. Rohde, C.; Mütze, U.; Weigel, J.F.W.; Ceglarek, U.; Thiery, J.; Kiess, W.; Beblo, S. Unrestricted consumption of fruits and vegetables in phenylketonuria: No major impact on metabolic control. *Eur. J. Clin. Nutr.* **2012**, *66*, 633–638. [CrossRef]
5. Evans, S.; Daly, A.; Chahal, S.; MacDonald, J.; MacDonald, A. Food acceptance and neophobia in children with phenylketonuria: A prospective controlled study. *J. Hum. Nutr. Diet.* **2016**, *29*, 427–433. [CrossRef]
6. Daly, A.; Evans, S.; Pinto, A.; Ashmore, C.; Rocha, J.C.; MacDonald, A. A 3 Year Longitudinal Prospective Review Examining the Dietary Profile and Contribution Made by Special Low Protein Foods to Energy and Macronutrient Intake in Children with Phenylketonuria. *Nutrients* **2020**, *12*, 3153. [CrossRef]
7. Evans, S.; Daly, A.; Chahal, S.; Ashmore, C.; MacDonald, J.; MacDonald, A. The influence of parental food preference and neophobia on children with phenylketonuria (PKU). *Mol. Genet. Metab. Rep.* **2018**, *14*, 10–14. [CrossRef]
8. MacDonald, A.; Harris, G.; Rylance, G.; Asplin, D.; Booth, I.W. Abnormal feeding behaviours in phenylketonuria. *J. Hum. Nutr. Diet.* **1997**, *10*, 163–170. [CrossRef]
9. Cade, J.; Thompson, R.; Burley, V.; Warm, D. Development, validation and utilisation of food-frequency questionnaires—A review. *Public Health Nutr.* **2002**, *5*, 567–587. [CrossRef]
10. Viau, K.S.; Ernst, S.L. Food Frequency Questionnaire. *Top. Clin. Nutr.* **2014**, *29*, 332–342. [CrossRef]
11. Van Wegberg, A.M.J.; MacDonald, A.; Ahring, K.; Bélanger-Quintana, A.; Blau, N.; Bosch, A.M.; Burlina, A.; Campistol, J.; Feillet, F.; Giżewska, M.; et al. The complete European guidelines on phenylketonuria: Diagnosis and treatment. *Orphanet J. Rare Dis.* **2017**, *12*, 162. [CrossRef] [PubMed]
12. Willett, W.C. *Nutritional Epidemiology*; Oxford University Press: New York, NY, USA, 1998.
13. Howard, K. Validating Questionnaires. 2008. Available online: https://www.researchgate.net/profile/Julio_Torales/post/To_perform_a_questionnaire_study_how_to_validate_the_questionnaire/attachment/59d638ff79197b8077996361/AS%3A399936555372544%401472363713499/download/Validating_Questionnaires.pdf (accessed on 6 March 2022).
14. Cade, J.E.; Burley, V.J.; Warm, D.L.; Thompson, R.L.; Margetts, B.M. Food-frequency questionnaires: A review of their design, validation and utilisation. *Nutr. Res. Rev.* **2004**, *17*, 5–22. [CrossRef]
15. Collingridge, D. Validating a Questionnaire. 2014. Available online: https://www.methodspace.com/blog/validating-a-questionnaire (accessed on 6 March 2022).
16. McCance and Widdowson. The Composition of Foods 7th edn. 2015. Available online: https://www.gov.uk/government/publications/composition-of-foods-integrated-dataset-cofid (accessed on 19 May 2016).
17. Nutritics, Research Edition Professional Premium. 2019. v5.7. Available online: https://www.nutritics.com/p/home (accessed on 7 March 2022).
18. Dumartheray, E.W.; Krieg, M.A.; Cornuz, J.; Whittamore, D.R.; Lovell, D.P.; Burckhardt, P.; Lanham-New, S.A. Validation and reproducibility of a semi-quantitative Food Frequency Questionnaire for use in elderly Swiss women. *J. Hum. Nutr. Diet.* **2006**, *19*, 321–330. [CrossRef] [PubMed]
19. Watson, J.F.; Collins, C.E.; Sibbritt, D.W.; Dibley, M.J.; Garg, M.L. Reproducibility and comparative validity of a food frequency questionnaire for Australian children and adolescents. *Int. J. Behav. Nutr. Phys. Act.* **2009**, *6*, 62. [CrossRef] [PubMed]
20. Craig, L.C.; McNeill, G.; Macdiarmid, J.I.; Masson, L.F.; Holmes, B.A. Dietary patterns of school-age children in Scotland: Association with socio-economic indicators, physical activity and obesity. *Br. J. Nutr.* **2010**, *103*, 319–334. [CrossRef]

21. Vereecken, C.; Covents, M.; Maes, L. Comparison of a food frequency questionnaire with an online dietary assessment tool for assessing preschool children's dietary intake. *J. Hum. Nutr. Diet.* **2010**, *23*, 502–510. [CrossRef]
22. Xia, W.; Sun, C.; Zhang, L.; Zhang, X.; Wang, J.; Wang, H.; Wu, L. Reproducibility and relative validity of a food frequency questionnaire developed for female adolescents in Suihua, North China. *PLoS ONE* **2011**, *6*, e19656. [CrossRef]
23. Azarias, H.G.D.A.; Marques-Rocha, J.L.; Miranda, A.E.D.S.; Dos Santos, L.C.; Gomes Domingos, A.L.; Hermsdorff, H.H.M.; Bressan, J.; Oliveira, F.L.P.D.; Leal, A.C.G.; Pimenta, A.M. Online Food Frequency Questionnaire From the Cohort of Universities of Minas Gerais (CUME Project, Brazil): Construction, Validity, and Reproducibility. *Front Nutr.* **2021**, *8*, 709915. [CrossRef]
24. Burrows, T.; Goldman, S.; Rollo, M. A systematic review of the validity of dietary assessment methods in children when compared with the method of doubly labelled water. *Eur. J. Clin. Nutr.* **2020**, *74*, 669–681. [CrossRef]
25. Burrows, T.L.; Martin, R.J.; Collins, C.E. A Systematic Review of the Validity of Dietary Assessment Methods in Children when Compared with the Method of Doubly Labeled Water. *J. Am. Diet. Assoc.* **2010**, *110*, 1501–1510. [CrossRef]
26. Sailer, M.; Elizondo, G.; Martin, J.; Harding, C.O.; Gillingham, M.B. Nutrient intake, body composition, and blood phenylalanine control in children with phenylketonuria compared to healthy controls. *Mol. Genet. Metab. Rep.* **2020**, *23*, 100599. [CrossRef] [PubMed]
27. Evans, S.; Daly, A.; Wildgoose, J.; Cochrane, B.; Chahal, S.; Ashmore, C.; Loveridge, N.; MacDonald, A. Growth, Protein and Energy Intake in Children with PKU Taking a Weaning Protein Substitute in the First Two Years of Life: A Case-Control Study. *Nutrients* **2019**, *11*, 552. [CrossRef] [PubMed]
28. Rose, H.J.; White, F.; MacDonald, A.; Rutherford, P.J.; Favre, E. Fat intakes of children with PKU on low phenylalanine diets. *J. Hum. Nutr. Diet.* **2005**, *18*, 395–400. [CrossRef]
29. Viau, K.S.; Jones, J.L.; Murtaugh, M.A.; Gren, L.H.; Stanford, J.B.; Bilder, D.A. Phone-based motivational interviewing to increase self-efficacy in individuals with phenylketonuria. *Mol. Genet. Metab. Rep.* **2016**, *6*, 27–33. [CrossRef] [PubMed]

MDPI
St. Alban-Anlage 66
4052 Basel
Switzerland
Tel. +41 61 683 77 34
Fax +41 61 302 89 18
www.mdpi.com

Nutrients Editorial Office
E-mail: nutrients@mdpi.com
www.mdpi.com/journal/nutrients

www.ingramcontent.com/pod-product-compliance
Ingram Content Group UK Ltd.
Pitfield, Milton Keynes, MK11 3LW, UK
UKHW060122010826
14089UKWH00006B/106